Will Wilson
Zoology Department
Duke University
Durham, NC 27708-0325

AN INTRODUCTION TO NUMERICAL ANALYSIS

UNIVERSITY OF IOWA **Kendall E. Atkinson**

AN INTRODUCTION TO NUMERICAL ANALYSIS

JOHN WILEY & SONS, New York • Chichester • Brisbane • Toronto

Library of Congress Cataloging in Publication Data:

Atkinson, Kendall E.
  An introduction to numerical analysis.

  Includes bibliographies and index.
1. Numerical analysis.   I. Title.
QA297.A84   519.4   78-6706
ISBN 0-471-02985-8

Printed in the United States of America

10 9 8 7 6 5 4 3

To my mother, Helen Fleming Hart, and
to the memory of my father, Harold Eugene Atkinson

PREFACE

This introduction to numerical analysis was written for students in mathematics, the physical sciences, and engineering, at the upper undergraduate to beginning graduate level. Prerequisites for using the text are elementary calculus, linear algebra, and an introduction to differential equations. The student's level of mathematical maturity or experience with mathematics should be somewhat higher; I have found that most students do not attain the necessary level until their senior year. Finally, the student should have a knowledge of computer programming. The preferred language for most scientific programming is Fortran.

A truly effective use of numerical analysis in applications requires both a theoretical knowledge of the subject and computational experience with it. The theoretical knowledge should include an understanding of both the original problem being solved and of numerical methods for its solution, including their derivation, error analysis, and an idea of when they will perform well or poorly. This kind of knowledge is necessary even if you are only considering using a package program from your computer center. You must still understand the program's purpose and limitations, to know whether it applies to your particular situation or not. More importantly, a majority of problems cannot be solved by a simple application of some standard program. For such problems you must devise new numerical methods, and this is usually done by adapting standard numerical methods to the new situation. This requires a good theoretical foundation in numerical analysis, both to devise the new methods and to avoid certain numerical pitfalls that occur easily in a number of problem areas.

Computational experience is also very important. It gives a sense of reality to most theoretical discussions; and it brings out the important difference between the exact arithmetic implicit in most theoretical discussions and the finite length arithmetic of actual computation, whether on a computer or a hand calculator. The use of a computer also imposes constraints on the structure of numerical methods, constraints that are not evident and that seem unnecessary from a strictly mathematical viewpoint. For example, iterative procedures are

often preferred over direct procedures because of simpler programming requirements or computer memory size limitations, even though the direct procedure
may seem simpler to explain and use. Many numerical examples are given in
this text to illustrate these points, and there are a number of exercises that will
give the student a variety of computational experience.

The book is organized in a fairly standard manner. Topics which are
simpler, both theoretically and computationally, come first; for example, rootfinding for a single nonlinear equation is covered in Chapter 2. The more
sophisticated topics within numerical linear algebra are left until the last three
chapters. However, if an instructor prefers, Chapters 7 through 9 on numerical
linear algebra can be inserted at any point following Chapter 1. Chapter 1
contains a number of introductory topics, some of which the instructor may
wish to postpone until later in the course. It is important, however, to cover the
mathematical and notational preliminaries of Section 1.1 and the introduction to
computer floating point arithmetic given in Section 1.2 and part of Section 1.3.

The text contains more than enough material for a one year course. In
addition, introductions are given to some topics that instructors may wish to
expand on from their own notes. For example, a brief introduction is given to
stiff differential equations in the last part of Section 6.8 in Chapter 6; and some
theoretical foundation for the least squares data fitting problem is given in
Theorem 7.5 and Problem 15 of Chapter 7. These can easily be expanded using
the references given in the respective chapters.

Each chapter contains a discussion of the research literature and a bibliography of some of the important books and papers on the material of the chapter.
The chapters all conclude with a set of exercises. Some of these exercises are
illustrations or applications of the text material, and others involve the development of new material. As an aid to the student, answers and hints to selected
exercises are given at the end of the book. It is important, however, for students
to solve some problems in which there is no given answer against which they can
check their results. This forces them to develop a variety of other means for
checking their own work; and it will force them to develop some common sense
or judgment as an aid in knowing whether their results are reasonable or not.

I teach a one year course covering much of the material of this book.
Chapters 1 through 5 form the first semester, and Chapters 6 through 9 form the
second semester. In most chapters, a number of topics can be deleted without
any difficulty arising in later chapters. Exceptions to this are Section 2.4 on
linear iteration methods, Sections 3.1–3.3, 3.6 on interpolation theory, Section
4.4 on orthogonal polynomials, and Section 5.1 on the trapezoidal and Simpson
integration rules.

I thank Professor Herb Hethcote of the University of Iowa for his helpful
advice and for having taught from an earlier rough draft of the book. I am also
grateful for the advice of Professors Robert Barnhill, U. of Utah, Herman
Burchard, Oklahoma State U. and Robert J. Flynn, Polytechnic Institute of New

York. I am very grateful to Ada Burns and Lois Friday, who did an excellent job of typing this and earlier versions of the book. I thank the many students who, over the past twelve years, enrolled in my course and used my notes and rough drafts rather than a regular text. They pointed out numerous errors; and their difficulties with certain topics helped me in preparing better presentations of them. The staff of John Wiley have been very helpful, and the text is much better as a result of their efforts. Finally, I thank my wife Alice for her patient and encouraging support, without which the book would probably have not been completed.

Iowa City
August, 1978   **Kendall E. Atkinson**

CONTENTS

AN INTRODUCTION TO NUMERICAL ANALYSIS

ONE

THE
SOURCES AND
PROPAGATION
OF ERROR

The subject of numerical analysis provides computational methods for the study and solution of mathematical problems. In this text we derive numerical methods for the solution of the most common mathematical problems and we analyze the errors present in these methods. Because most computation is now done on digital computers, we also discuss the implementation of these numerical methods as computer programs.

The study of error is a central concern of numerical analysis. Most numerical methods give answers that are only approximations to the true desired solution, and it is important to understand and to be able to estimate or bound the resulting error. This chapter classifies the various kinds of error that may occur in a problem. A brief introduction is given to computer arithmetic. Elementary results are given on the propagation of errors in various kinds of numerical calculations, and the concept of stability is introduced. The first section contains some mathematical preliminaries necessary for the work of later chapters.

1.1 Mathematical Preliminaries

This section contains a review of results from calculus, which will be used in this text. We first give some mean value theorems, and then we present and discuss Taylor's theorem. The section concludes with notation that will be used in later chapters.

The following three theorems are fairly elementary results, but they are important tools used throughout numerical analysis.

Theorem 1.1 (Intermediate Value) Let $f(x)$ be continuous on the finite interval $a \leq x \leq b$, and define

$$m = \underset{a \leq x \leq b}{\text{Infimum}} \, f(x) \qquad M = \underset{a \leq x \leq b}{\text{Supremum}} \, f(x) \qquad (1.1)$$

3

Then for any number γ in the interval $[m, M]$, there is at least one point ζ in $[a, b]$ for which

$$f(\zeta) = \gamma$$

In particular, there are points $\underline{x}$ and $\bar{x}$ in $[a, b]$ for which

$$f(\underline{x}) = m \qquad f(\bar{x}) = M \tag{1.2}$$

Theorem 1.2 (Mean Value) Let $f(x)$ be continuous and differentiable for $a \leqq x \leqq b$. Then there is at least one point ζ in $[a, b]$ for which

$$f(b) - f(a) = f'(\zeta)(b - a) \tag{1.3}$$

Theorem 1.3   (Integral Mean Value) Let $w(x) \geqq 0$ be integrable on $[a, b]$,

$$\int_a^b w(x)\, dx < \infty$$

and let $f(x)$ be continuous on $[a, b]$. Then there is at least one point ζ in $[a, b]$ for which

$$\int_a^b w(x) f(x)\, dx = f(\zeta) \int_a^b w(x)\, dx \tag{1.4}$$

Proof   The first two theorems are discussed in most calculus courses and thus we omit any discussion of them. To prove the integral mean value theorem, introduce the function

$$g(u) = f(u) \int_a^b w(x)\, dx - \int_a^b f(x) w(x)\, dx$$

$$= \int_a^b [f(u) - f(x)] w(x)\, dx \qquad a \leqq u \leqq b$$

This is a continuous function on $[a, b]$ since it is just a constant multiple of $f(u)$ plus another constant. Using the point $\underline{x}$ of (1.2),

$$g(\underline{x}) = \int_a^b [f(\underline{x}) - f(x)] w(x)\, dx \leqq 0.$$

This follows from the fact that $f(\underline{x}) - f(x) \leqq 0$ at all points of $[a, b]$, and consequently the integrand is never positive. Similarly, $g(\bar{x}) \geqq 0$, with $\bar{x}$

taken from (1.2). Combining these two results, the intermediate value theorem implies that $g(u)$ must have a zero ζ between $\underline{x}$ and $\bar{x}$. This proves (1.4). ∎

For a geometric interpretation of the integral mean value theorem, consider the case $w(x) \equiv 1$. Then there is a point ζ in $[a,b]$ for which

$$\int_a^b f(x)\,dx = f(\zeta)(b-a)$$

The left side is the area under the curve $y=f(x)$ between $x=a$ and $x=b$, and the right side is the area of a rectangle with base $[a,b]$ and height $f(\zeta)$. Applications of all three theorems will occur later in the chapter and in the problems.

Taylor's theorem   One of the most important tools of numerical analysis is Taylor's theorem and the associated Taylor series. It will be used in all chapters of this text except those on numerical linear algebra. The theorem gives a relatively simple method for approximating functions $f(x)$ by polynomials, and thereby gives a method for computing $f(x)$.

Theorem 1.4 (Taylor) Let $f(x)$ have $n+1$ continuous derivatives on $[a,b]$ for some $n \geq 0$, and let $x, x_0 \in [a,b]$. Then

$$f(x) = p_n(x) + R_{n+1}(x) \tag{1.5}$$

$$p_n(x) = f(x_0) + \frac{(x-x_0)}{1!} f'(x_0) + \ldots + \frac{(x-x_0)^n}{n!} f^{(n)}(x_0) \tag{1.6}$$

$$R_{n+1}(x) = \frac{1}{n!} \int_{x_0}^x (x-t)^n f^{(n+1)}(t)\,dt = \frac{(x-x_0)^{n+1}}{(n+1)!} f^{(n+1)}(\xi) \tag{1.7}$$

for some ξ between x_0 and x.

Proof The derivation of (1.5) is given in most calculus textbooks. It uses carefully chosen integration by parts in the identity

$$f(x) = f(x_0) + \int_{x_0}^x f'(t)\,dt$$

repeating it n times to obtain (1.5) to (1.7) with the integral form of the remainder formula $R_{n+1}(x)$. The second form of the remainder is obtained by using the integral mean value theorem with $w(t) \equiv (x-t)^n$. ∎

$$f(x) = f(x_0) + f'(x_0)(x-x_0) + \frac{f''(x_0)(x-x_0)^2}{2!} + \ldots$$

Using Taylor's theorem, we obtain the following standard formulas:

$$e^x = 1 + \frac{x}{1!} + \frac{x^2}{2!} + \ldots + \frac{x^n}{n!} + \frac{x^{n+1}}{(n+1)!} e^{\xi_x} \tag{1.8}$$

$$\cos(x) = 1 - \frac{x^2}{2!} + \frac{x^4}{4!} - \ldots + (-1)^n \frac{x^{2n}}{(2n)!} + (-1)^{n+1} \frac{x^{2n+2}}{(2n+2)!} \cos(\xi_x) \tag{1.9}$$

$$\sin(x) = x - \frac{x^3}{3!} + \frac{x^5}{5!} - \ldots + (-1)^{n-1} \frac{x^{2n-1}}{(2n-1)!} + (-1)^n \frac{x^{2n+1}}{(2n+1)!} \cos(\xi_x) \tag{1.10}$$

$$(1+x)^\alpha = 1 + \binom{\alpha}{1} x + \binom{\alpha}{2} x^2 + \ldots + \binom{\alpha}{n} x^n + \binom{\alpha}{n+1} \frac{x^{n+1}}{(1+\xi_x)^{n+1-\alpha}} \tag{1.11}$$

with

$$\binom{\alpha}{k} = \frac{\alpha(\alpha-1)\ldots(\alpha-k+1)}{k!} \qquad k = 1, 2, 3, \ldots$$

for any real number α. For all cases the unknown number ξ_x is located between x and 0.

An important special case of (1.11) is the geometric series

$$\frac{1}{1-x} = 1 + x + x^2 + \ldots + x^n + \frac{x^{n+1}}{1-x}. \tag{1.12}$$

This is the case $\alpha = -1$ with x replaced by $-x$. The remainder has a simpler form than in (1.11); and it is easily proven by multiplying both sides of (1.12) by $1 - x$ and then simplifying. Infinite series representations for the functions on the left side of (1.8) to (1.12) can be obtained by letting $n \to \infty$. The infinite series for (1.8) to (1.10) will converge for all x, and those for (1.11), (1.12) will converge for all $|x| < 1$.

The Taylor series of any function $f(x)$ can be calculated directly from the definition (1.6) with as many terms included as desired. But because of the complexity of the differentiations with many functions, it is often better to obtain the Taylor series indirectly by using one of the above formulas [(1.8) to (1.12)]. We give three examples of this procedure, all of which will also have simpler remainder terms than if (1.7) were used directly.

Example 1.   Let $f(x) = e^{-x^2}$. Replace x by $-x^2$ in (1.8) to obtain

$$e^{-x^2} = 1 - x^2 + \frac{x^4}{2!} - \ldots + (-1)^n \frac{x^{2n}}{n!} + (-1)^{n+1} \cdot \frac{x^{2n+2}}{(n+1)!} e^{\xi_x}$$

with $-x^2 \leq \xi_x \leq 0$.

2. Let $f(x) = \tan^{-1}x$. Begin by setting $x = -u^2$ in (1.12),

$$\frac{1}{1+u^2} = 1 - u^2 + u^4 - \ldots + (-1)^n u^{2n} + (-1)^{n+1} \cdot \frac{u^{2n+2}}{1+u^2}$$

Integrate over $[0, x]$ to get

$$\tan^{-1}x = x - \frac{x^3}{3} + \frac{x^5}{5} - \ldots + (-1)^n \frac{x^{2n+1}}{2n+1} + (-1)^{n+1} \int_0^x \frac{u^{2n+2}\,du}{1+u^2}$$

Applying the integral mean value theorem,

$$\int_0^x \frac{u^{2n+2}\,du}{1+u^2} = \frac{x^{2n+3}}{2n+3} \cdot \frac{1}{1+\xi_x^2}$$

with ξ_x between 0 and x.

3. Let $f(x) = \int_0^1 \sin(xt)\,dt$. Using (1.10),

$$f(x) = \int_0^1 \left[xt - \frac{x^3 t^3}{3!} + \ldots + (-1)^{n-1} \frac{(xt)^{2n-1}}{(2n-1)!} + (-1)^n \frac{(xt)^{2n+1}}{(2n+1)!} \cos(\xi_{xt}) \right] dt$$

$$= \sum_{j=1}^n (-1)^{j-1} \frac{x^{2j-1}}{(2j-1)!(2j)} + (-1)^n \frac{x^{2n+1}}{(2n+1)!} \int_0^1 t^{2n+1} \cos(\xi_{xt})\,dt$$

with ξ_{xt} between 0 and xt. The integral in the remainder is easily bounded by $1/(2n+2)$; but we can also convert it to a simpler form. Although it wasn't proven, it can be shown that $\cos(\xi_{xt})$ is a continuous function of t. Then applying the integral mean value theorem,

$$\int_0^1 \sin(xt)\,dt = \sum_{j=1}^n (-1)^{j-1} \frac{x^{2j-1}}{(2j-1)!(2j)} + (-1)^n \frac{x^{2n+1}}{(2n+1)!(2n+2)} \cos(\zeta_x)$$

for some ζ_x between 0 and x.

Taylor's theorem in two dimensions Let $f(x,y)$ be a given function of the two independent variables x and y. We will show how the earlier Taylor's theorem can be extended to the expansion of $f(x,y)$ about a given point (x_0, y_0). The results will easily extend to functions of more than two variables. As notation, let $L(x_0, y_0; x_1, y_1)$ denote the set of all points (x, y) on the straight line segment joining (x_0, y_0) and (x_1, y_1).

Theorem 1.5 Let (x_0,y_0) and $(x_0+\xi,y_0+\eta)$ be given points, and assume $f(x,y)$ is $n+1$ times continuously differentiable for all (x,y) in some neighborhood of $L(x_0,y_0;\ x_0+\xi,y_0+\eta)$. Then

$$f(x_0+\xi,y_0+\eta)=f(x_0,y_0)+\sum_{j=1}^{n}\frac{1}{j!}\left(\xi\frac{\partial}{\partial x}+\eta\frac{\partial}{\partial y}\right)^{j}f(x,y)\bigg|_{\substack{x=x_0\\y=y_0}}$$

$$+\frac{1}{(n+1)!}\left(\xi\frac{\partial}{\partial x}+\eta\frac{\partial}{\partial y}\right)^{n+1}f(x,y)\bigg|_{\substack{x=x_0+\theta\xi\\y=y_0+\theta\eta}}$$

(1.13)

for some $0\le\theta\le1$. The point $(x_0+\theta\xi,y_0+\theta\eta)$ is an unknown point on the line $L(x_0,y_0;\ x_0+\xi,y_0+\eta)$.

Proof First note the meaning of the derivative notation in (1.13). As an example,

$$\left(\xi\frac{\partial}{\partial x}+\eta\frac{\partial}{\partial y}\right)^{2}f(x,y)=\xi^{2}\frac{\partial^{2}f(x,y)}{\partial x^{2}}+2\xi\eta\frac{\partial^{2}f(x,y)}{\partial x\partial y}+\eta^{2}\frac{\partial^{2}f(x,y)}{\partial y^{2}}$$

The subscript notation, $x=x_0$, $y=y_0$, means the various derivatives are to be evaluated at (x_0,y_0).

The proof of (1.13) is based on applying the earlier Taylor's theorem to

$$F(t)=f(x_0+t\xi,y_0+t\eta)\qquad 0\le t\le1$$

Using (1.5) to (1.7),

$$F(1)=F(0)+\frac{F'(0)}{1!}+\frac{F''(0)}{2!}+\dots+\frac{F^{(n)}(0)}{n!}+\frac{F^{(n+1)}(\theta)}{(n+1)!}$$

for some $0\le\theta\le1$. Clearly, $F(0)=f(x_0,y_0)$, $F(1)=f(x_0+\xi,y_0+\eta)$. For the first derivative,

$$F'(t)=\xi\frac{\partial f(x_0+t\xi,y_0+t\eta)}{\partial x}+\eta\frac{\partial f(x_0+t\xi,y_0+t\eta)}{\partial y}$$

$$=\left(\xi\frac{\partial}{\partial x}+\eta\frac{\partial}{\partial y}\right)f(x,y)\bigg|_{\substack{x=x_0+t\xi\\y=y_0+t\eta}}$$

The higher-order derivatives are calculated similarly. ∎

Example As a simple example, consider expanding $f(x,y) = x/y$ about $(1,1)$. Let $n = 1$. Then

$$\frac{x}{y} = f(1,1) + (x-1)\frac{\partial f(1,1)}{\partial x} + (y-1)\frac{\partial f(1,1)}{\partial y}$$

$$+ \frac{1}{2}\left[(x-1)^2\frac{\partial^2 f(x,y)}{\partial x^2} + 2(x-1)(y-1)\frac{\partial^2 f(x,y)}{\partial x \partial y} + (y-1)^2\frac{\partial^2 f(x,y)}{\partial y^2}\right]_{\substack{x=\delta \\ y=\gamma}}$$

$$= 1 + (x-1) - (y-1) + \frac{1}{2}\left[(x-1)^2 \cdot 0 - 2(x-1)(y-1)\frac{1}{\gamma^2} + (y-1)^2\frac{2\delta}{\gamma^3}\right]$$

with (δ, γ) a point on $L(1,1; x,y)$. For (x,y) close to $(1,1)$,

$$\frac{x}{y} \approx 1 + x - y$$

The graph of $z = 1 + x - y$ is the plane tangent to the graph of $z = x/y$ at $(x,y,z) = (1,1,1)$.

Some mathematical notation There are several concepts that are taken up in this text that are needed in a simpler form in the beginning chapters. These include results on divided differences of functions, vector spaces, and vector and matrix norms. The minimum necessary notation is introduced at this point, and a more complete development is left to other, more natural, places in the text.

For a given function $f(x)$, define

$$f[x_0, x_1] = \frac{f(x_1) - f(x_0)}{x_1 - x_0} \qquad f[x_0, x_1, x_2] = \frac{f[x_1, x_2] - f[x_0, x_1]}{x_2 - x_0} \qquad (1.14)$$

assuming x_0, x_1, x_2 are distinct. These are called the first- and second-order divided differences of $f(x)$, respectively. They are related to derivatives of $f(x)$,

$$f[x_0, x_1] = f'(\xi) \qquad f[x_0, x_1, x_2] = \tfrac{1}{2}f''(\zeta) \qquad (1.15)$$

with ξ between x_0 and x_1, and ζ between the minimum and maximum of x_0, x_1, and x_2. The divided differences are independent of the order of their arguments, contrary to what might be expected from (1.14). More precisely,

$$f[x_0, x_1] = f[x_1, x_0],$$
$$f[x_0, x_1, x_2] = f[x_i, x_j, x_k] \qquad (1.16)$$

for any permutation (i,j,k) of $(0,1,2)$. The proofs of these results are left as an exercise for the reader. A complete development of divided differences is given in Section 3.2 of Chapter 3.

Vector spaces, matrices, and vector and matrix norms are covered in Chapter 7, immediately preceding the chapters on numerical linear algebra. Because some of this material is needed earlier, we introduce it here, while leaving the proofs till Chapter 7.

Roughly speaking, a *vector space* V is a set of objects, called vectors, for which operations of *vector addition* and *scalar multiplication* have been defined. V has a set of scalars associated with it; and in this text the scalars can be either the real numbers or the complex numbers. Two vector spaces are used in a great many applications. They are

$$R^n = \left\{ \begin{bmatrix} x_1 \\ \vdots \\ x_n \end{bmatrix} \middle| x_1, \ldots, x_n \text{ real numbers} \right\} \tag{1.17}$$

$$C[a,b] = \{ f(x) | f(x) \text{ continuous and real valued}, a \leq x \leq b \}. \tag{1.18}$$

For $x, y \in R^n$, and α a real number, define $x + y$ and αx by

$$x + y = \begin{bmatrix} x_1 + y_1 \\ \vdots \\ x_n + y_n \end{bmatrix} \qquad \alpha x = \begin{bmatrix} \alpha x_1 \\ \vdots \\ \alpha x_n \end{bmatrix}$$

For $f, g \in C[a,b]$ and α a real number, define $f + g$ and αf by

$$(f+g)(x) = f(x) + g(x) \qquad (\alpha f)(x) = \alpha f(x) \qquad a \leq x \leq b$$

Vector norms are used to measure the size of a vector. For R^n, define two different norms:

$$\|x\|_\infty = \underset{1 \leq i \leq n}{\text{Maximum}} |x_i| \qquad x \in R^n \tag{1.19}$$

$$\|x\|_2 = \sqrt{x_1^2 + \ldots + x_n^2} \qquad x \in R^n \tag{1.20}$$

For $C[a,b]$, define

$$\|f\|_\infty = \underset{a \leq x \leq b}{\text{Maximum}} |f(x)| \qquad f \in C[a,b] \tag{1.21}$$

These definitions can be shown to satisfy the following three characteristic properties of all norms.

1. $\|v\|=0$ if and only if $v=0$, the zero vector

2. $\|\alpha v\| = |\alpha| \|v\|$, for all vectors v and all real numbers α

3. $\|v+w\| \leq \|v\| + \|w\|$, for all vectors v and w

Property 3 is usually referred to as the triangle inequality. An explanation of this name and a further development of properties and norms for $C[a,b]$ is given in Chapter 4.

Norms can also be introduced for matrices. For an $n \times n$ matrix

$$A = \begin{bmatrix} a_{11} & a_{12}...a_{1n} \\ a_{21} & a_{22}...a_{2n} \\ \cdot \\ \cdot \\ \cdot \\ a_{n1} & a_{n2}...a_{nn} \end{bmatrix}$$

define

$$\|A\|_{\infty} = \underset{1 \leq i \leq n}{\text{Maximum}} \sum_{j=1}^{n} |a_{ij}| \qquad (1.22)$$

With this definition, the properties of a vector norm will be satisfied. And, in addition,

$$\|AB\|_{\infty} \leq \|A\|_{\infty} \|B\|_{\infty}$$

$$\|Ax\|_{\infty} \leq \|A\|_{\infty} \|x\|_{\infty} \qquad \text{all } x \in R^{n} \qquad (1.23)$$

where A and B are arbitrary $n \times n$ matrices. The proofs of these results are given in Section 7.3 of Chapter 7, although most are quite straightforward to show.

Example Consider the vector space R^2 and matrices of order 2×2. In particular, let

$$A = \begin{bmatrix} 1 & -1 \\ 3 & 2 \end{bmatrix} \qquad x = \begin{bmatrix} 1 \\ 2 \end{bmatrix} \qquad y = Ax = \begin{bmatrix} -1 \\ 7 \end{bmatrix}$$

Then

$$\|A\|_{\infty} = 5 \qquad \|x\|_{\infty} = 2 \qquad \|y\|_{\infty} = \|Ax\|_{\infty} = 7$$

and (1.23) is easily satisfied. To show that (1.23) cannot be improved on, take

$$x = \begin{bmatrix} 1 \\ 1 \end{bmatrix} \qquad y = Ax = \begin{bmatrix} 0 \\ 5 \end{bmatrix}$$

Then

$$\|y\|_\infty = 5 = \|A\|_\infty \|x\|_\infty$$

1.2 Computer Representation of Numbers

Digital computers are the principal means of calculation in numerical analysis, and consequently it is useful to understand how they operate. In this section we consider how numbers are represented in computers, and in the next two sections we discuss some of their arithmetic operations.

Most computers have an integer mode and a floating-point mode for representing numbers. The integer mode is used only to represent integers and it will not concern us any further. The floating-point form is used to represent real numbers of greatly varying magnitude, although there is a maximum length to the number of digits that can be stored. It is closely connected to what is called *scientific notation* for writing real numbers in many high school algebra texts.

The number base used in computers is seldom decimal. Most digital computers use the binary number system or some simple variant of it, such as base 8 or base 16. As an example of a binary representation and of its conversion to a decimal base, consider

$$(1101.101)_2 = 2^3 + 2^2 + 2^0 + 2^{-1} + 2^{-3} = (13.625)_{10}$$

The first number is written in binary, and the last number is its decimal equivalent. Let β denote the base of the number system being used. Generally

$$\beta = 2, 8, 10, \text{ or } 16 \tag{1.24}$$

If a number z is represented in floating-point form, then it is stored essentially in the form

$$z = (\operatorname{sign} z) \times (.a_1 a_2 \ldots a_n) \times \beta^e \tag{1.25}$$

The characters a_i are all digits in the base β system, $0 \le a_i \le \beta - 1$, with $a_1 \ne 0$. The number e is called the *exponent* or *characteristic*; and the number $.a_1 a_2 \ldots a_n$ is called the *mantissa*. The characteristic satisfies

$$m \le e \le M$$

with m and M integers, which vary with the computer. Generally, $m = -M$ or

$m = -M \pm 1$. The mantissa contains n digits in the base β system; and all numbers that are longer must be shortened to this length in some way, thus limiting the possible precision of any calculation. We discuss the rounding of numbers in the next section.

Examples (a)   IBM 360 model computers. For this series of machines,

$$\beta = 16 \qquad -64 \leq e \leq 63 \tag{1.26}$$

This exponent range means that numbers z can be represented whose magnitudes satisfy

$$8.6 \times 10^{-78} = 16^{-64} < |z| < 16^{63} = 7.2 \times 10^{75}$$

There are single- and double-precision floating-point numbers, the difference being the length of the mantissa. For single-precision numbers, there are $n = 6$ hexadecimal digits in the mantissa; this is equivalent to 6 to 7 decimal digits of accuracy. For double-precision numbers, $n = 14$; and this is equivalent to over 16 decimal digits.

(b)   CDC 6000 series computers. For these computers with single precision numbers

$$\beta = 2 \qquad n = 48 \qquad -975 \leq e \leq 1071 \tag{1.27}$$

The range in magnitudes is

$$3.1 \times 10^{-294} < |z| < 2.5 \times 10^{322}$$

The mantissa contains just over 14 decimal digits, almost as much as the double precision on the IBM 360 machines.

The finite precision of computer numbers will limit the accuracy of all calculations. The most obvious limitation is that the results of all arithmetic operations must be rounded to the number n of digits that can be stored in the mantissa of a floating-point number. An analysis of this source of error in calculations is included in the next two sections of this chapter. A second and more subtle source of error is in the inherent limitation on the amount of accuracy that can be attained with the number n of digits being used in the floating-point form. This is illustrated with the following simple example.

Example   Consider the calculation of $\pi - (22/7)$ on a decimal computer that has five digits ($n = 5$) in its mantissa. Assuming that π and $22/7$ are evaluated and rounded to the maximum possible precision, we will have $\pi \doteq 3.1416$ and $22/7 \doteq 3.1429$ in the computer. (The symbol "$\doteq$" means "approximately equals.") With these representations for π and 22/7, we will obtain

$$\pi - (22/7) \doteq -.0013 \qquad (1.28)$$

This answer has, at most, two digits of accuracy; and this can be seen clearly from the true answer

$$\pi - (22/7) = -.0012644\ldots$$

Working with five-digit decimal arithmetic, it is not possible to obtain any significantly greater accuracy in $\pi - (22/7)$ than that given in (1.28).

The determination of the inherent limitation on the accuracy with which the solution of a problem can be obtained is clearly important; but it often is a difficult problem. In most practical situations it is linked to a concept called the stability of the problem as well as to the finite length representation of numbers on a computer. The concept of stability is introduced in Section 1.5.

1.3   Sources of Error

We list the major ways in which error is introduced into a problem, including some that fall outside the usual scope of mathematics. We begin with a few simple definitions about error.

The error in a number is its true value minus its approximate value. For the remainder of the chapter, we use the notation x_T and x_A to denote the true and approximate values of a number. Then

$$\text{Error in } x_A = x_T - x_A$$

For many purposes, we prefer to study the percentage or *relative error* in x_A,

$$\text{Relative error in } x_A = \frac{x_T - x_A}{x_T}$$

provided $x_T \neq 0$. This quantity will often be denoted by $\text{Rel}(x_A)$.

Example $x_T = e = 2.7182818\ldots$ $x_A = \dfrac{19}{7} = 2.7142857\ldots$

$\text{Error} = .003996\ldots$ $\text{Relative error} = .00147\ldots$

In place of relative error, we often use the concept of *significant digits*. We say x_A has m significant digits with respect to x_T if the error $x_T - x_A$ has magnitude less than or equal to 5 in the $(m+1)$st digit of x_T, counting to the right from the first nonzero digit in x_T.

Example (a) $x_T = \frac{1}{3}$ $x_A = .333$ $|x_T - x_A| \doteq .00033$

Since the error is less than 5 in the fourth digit to the right of the first nonzero digit in x_T, we say that x_A has three significant digits with respect to x_T.

(b) $x_T = 23.496$ $x_A = 23.494$ $|x_T - x_A| = .002$

x_A has four significant digits with respect to x_T, since the error is less than five in the fifth place to the right of the first nonzero digit in x_T. Note that if x_A is rounded to four places, an additional error is introduced and x_A will no longer have four significant digits.

(c) $x_T = .02138$ $x_A = .02144$ $|x_T - x_A| = .00006$

The number x_A has two significant digits, but not three, with respect to x_T.

We will now give a rough classification of the sources of error, dividing them into five categories.

Mathematical modeling of a physical problem A mathematical model for a physical situation is an attempt to give mathematical relationships between certain quantities of physical interest. Because of the complexity of physical reality, a variety of simplifying assumptions are used to construct more tractable mathematical models. The resulting model has limitations on its accuracy as a consequence of these assumptions; and these limitations may or may not be troublesome, depending on the uses of the model. In the case that the model is not sufficiently accurate, the numerical solution of the model cannot improve upon this basic lack of accuracy.

Example Consider a projectile of mass m to have been fired into the air, its flight path always remaining close to the earth's surface. Let an xyz coordinate system be introduced with origin on the earth's surface and with the positive z-axis perpendicular to the earth and directed upward. Let the position of the projectile at time t be denoted by $\mathbf{r}(t) = x(t)\mathbf{i} + y(t)\mathbf{j} + z(t)\mathbf{k}$, using the standard vector field theory notation. One model for the flight of the projectile is given by Newton's second law as

$$m\frac{d^2\mathbf{r}(t)}{dt^2} = -mg\mathbf{k} - b\frac{d\mathbf{r}(t)}{dt} \tag{1.29}$$

where $b > 0$ is a constant and g is the acceleration due to gravity. This equation says that the only forces acting on the projectile are (1) the gravitational force of the earth, and (2) a frictional force that is directly proportional to the speed $|\mathbf{v}(t)| = |d\mathbf{r}(t)/dt|$ and directed opposite to the path of flight.

In some situations this is an excellent model; and it may not be necessary to include even the frictional term. But the model doesn't include forces of resistance acting perpendicular to the plane of flight, for example, a cross-wind; and it doesn't allow for the Coriolis effect. Also, the frictional force in (1.29) may be proportional to $|v(t)|^\alpha$ with $\alpha \neq 1$.

If a model is adequate for physical purposes, then we wish to use a numerical scheme that preserves this accuracy. But if the model is inadequate, then the numerical analysis cannot improve the accuracy except by chance accident. For books concerned explicitly with mathematical modeling in the sciences, see Lin and Segal [L2] and Maki and Thompson [M1].

Blunders   In precomputer times, chance arithmetic errors were always a serious problem. Check schemes, some quite elaborate, were devised to detect if such errors had occurred and to correct for them before the calculations had proceeded very far past the error. For an example, see Fadeeva [F1] for check schemes used when solving linear systems of equations.

With the introduction of digital computers, the type of *blunder* has changed. Chance arithmetic errors (machine errors) are now relatively rare, and programming errors are now the main difficulty. Often a program error will be repeated many times in the course of executing the program, and its existence becomes obvious because of absurd numerical output (although the source of the error may still be difficult to find). But as computer programs become more complex and lengthy, the existence of a small program error may be hard to detect and correct, even though the error may make a subtle, but crucial difference in the numerical results. This makes good program debugging very important, even though it may not seem very rewarding immediately.

Uncertainty in physical data   Most data from a physical problem contains error or uncertainty within it. This must affect the accuracy of any calculations based on the data, limiting the accuracy of the answers. The techniques for analyzing the effects in other calculations of this error are much the same as those used in analyzing the effects of rounding error. The material of the next section discusses this further.

Rounding error   When working with only a limited number of digits in a number, as on a computer, rounding errors are inevitable with most arithmetic calculations. Using the notation for floating-point numbers from Section 1.2, let

$$z = \sigma \times (.a_1 a_2 \ldots a_n a_{n+1} \ldots)_\beta \times \beta^e, \sigma = \pm 1, a_1 \neq 0$$

And, because of (1.24), we can assume that β is even without any real loss of generality. Then the rounded machine version of z is

$$\mathrm{fl}(z) = \begin{cases} \sigma(.a_1 a_2 \ldots a_n)_\beta \times \beta^e, & 0 \leq a_{n+1} < \dfrac{\beta}{2} \\[2ex] \sigma[(.a_1 \ldots a_n)_\beta + (.00 \cdots 01)_\beta] \times \beta^e, & \dfrac{\beta}{2} \leq a_{n+1} < \beta \end{cases} \tag{1.30}$$

in which $(.0 \cdots 01)_\beta$ means β^{-n}. This is a formal definition of a simple procedure. We round up if the $(n+1)$st digit is greater than or equal to $\frac{1}{2}\beta$, and we round down if it is less than $\frac{1}{2}\beta$.

For the error in $\mathrm{fl}(z)$, first assume $0 \leq a_{n+1} < \beta/2$

$$z - \mathrm{fl}(z) = \sigma(.00 \cdots 0 a_{n+1} a_{n+2} \ldots)_\beta \times \beta^e$$

$$= \sigma(.a_{n+1} a_{n+2} \ldots)_\beta \times \beta^{e-n}$$

$$|z - \mathrm{fl}(z)| \leq \tfrac{1}{2} \beta^{e-n} \tag{1.31}$$

By a similar argument, the same result is true for $\beta/2 \leq a_{n+1} < \beta$.

To calculate the relative error in $\mathrm{fl}(z)$,

$$|z - \mathrm{fl}(z)| \leq \frac{1}{2} \beta^{e-n} \frac{|z|}{|z|} = \frac{1}{2} \cdot \frac{|z| \beta^{-n}}{|z| \beta^{-e}}$$

$$= \frac{1}{2} \cdot \frac{|z| \beta^{-n}}{(.a_1 a_2 \ldots)_\beta} \leq \frac{1}{2} \cdot \frac{|z| \beta^{-n}}{(.100 \ldots)_\beta} = \frac{1}{2} \frac{|z| \beta^{-n}}{\beta^{-1}}$$

$$\frac{|z - \mathrm{fl}(z)|}{|z|} \leq \frac{1}{2} \beta^{-n+1} \tag{1.32}$$

Example   For a binary computer with n digits in the mantissa,

$$\frac{|z - \mathrm{fl}(z)|}{|z|} \leq 2^{-n}$$

and for a decimal computer,

$$\frac{|z - \mathrm{fl}(z)|}{|z|} \leq 5 \times 10^{-n}$$

To obtain another convenient form, let ϵ be defined by

$$\frac{z - \mathrm{fl}(z)}{z} = -\epsilon$$

Then for some ϵ,

$$\mathrm{fl}(z) = (1 + \epsilon)z \qquad |\epsilon| \leq \tfrac{1}{2} \beta^{-n+1} \tag{1.33}$$

This form is much used in the analysis of the propagation of rounding errors, and its introduction and application are due principally to J. H. Wilkinson; for example, see [W1]. An example is given in the next section.

Many digital computers do not use rounding. Instead they truncate or *chop*, using the first part of definition (1.30), regardless of the size of the remainder $.0\cdots 0a_{n+1}a_{n+2}\cdots$. For this case, the result (1.31) must be changed to

$$|z - \text{fl}(z)| \leq \beta^{e-n}, \tag{1.34}$$

and similarly the bounds must be doubled for (1.32) and (1.33).

Example (a)   The CDC 6000 series computers use a chopped binary arithmetic, and thus the error in fl(z)satisfies

$$\frac{|z - \text{fl}(z)|}{|z|} \leq 2^{-n+1} = 2^{-47}$$

with $n = 48$.

(b)   The IBM 360 series computers also using chopping, but the base is $\beta = 16$, and

$$\frac{|z - \text{fl}(z)|}{|z|} \leq 16^{-n+1} = 16^{-5}$$

with $n = 6$. The much larger base makes the chopping much more harmful than with a binary computer. A *guard digit* has been added to the arithmetic section to help mitigate some of the difficulties of a short word length, the base 16 arithmetic, and the lack of rounding; but the performance is still not ideal.

Mathematical truncation error   This name refers to the error of approximation in numerically solving a mathematical problem, and it is the error generally associated with the subject of numerical analysis. It involves the approximation of infinite processes by finite ones, replacing noncomputable problems with computable ones. Following are some examples to make the idea more precise.

Example (a)   Using the first two terms of the Taylor series from (1.11),

$$\sqrt{1+x} \approx 1 + \tfrac{1}{2}x \tag{1.35}$$

which is a good approximation when x is small. See Chapter 4 for the general area of approximation of functions.

(b)   For evaluating an integral on $[0, 1]$, use

$$\int_0^1 f(x)\,dx \approx \frac{1}{n} \sum_{j=1}^{n} f\left(\frac{2j-1}{2n}\right) \qquad n = 1, 2, 3, \ldots \tag{1.36}$$

This is called the *midpoint numerical integration rule*; see the last part of Section

5.2 for more detail. The topic of numerical integration is examined in Chapter 5.

(c)   For the differential equation problem

$$Y'(t) = f(t, Y(t)) \qquad Y(t_0) = Y_0 \qquad (1.37)$$

use the approximation of the derivative

$$Y'(t) \approx \frac{Y(t+h) - Y(t)}{h}$$

for some small h. Let $t_j = t_0 + jh$ for $j \geq 0$, and define an approximate solution function $y(t_j)$ by

$$\frac{y(t_{j+1}) - y(t_j)}{h} = f(t_j, y(t_j))$$

so we have

$$y(t_{j+1}) = y(t_j) + h f(t_j, y(t_j)) \qquad j \geq 0$$

This is Euler's method for solving a differential equation initial value problem; a complete discussion of it is given in Section 6.2. Chapter 6 gives a complete development of numerical methods for solving the initial value problem (1.37).

Most numerical analysis problems in the following chapters will involve principally mathematical truncation errors. The major exception is numerical linear algebra in which rounding errors and their propagation are the major source of error.

1.4   Propagation of Errors

In this section we consider the effect of calculations with numbers that are in error. We begin by considering the basic arithmetic operations. Let ω denote an arithmetic operation, such as $+$, $-$, $\times$, or $/$; and let ω^* be the computer version of the same operation, which will usually include rounding or truncation. Let x_A and y_A be the numbers being used for calculations, and suppose they are in error, with true values

$$x_T = x_A + \epsilon \qquad y_T = y_A + \eta$$

Then $x_A \omega^* y_A$ is the number actually computed; and for its error,

$$x_T \omega y_T - x_A \omega^* y_A = (x_T \omega y_T - x_A \omega y_A) + (x_A \omega y_A - x_A \omega^* y_A) \qquad (1.38)$$

The first term in parentheses is called the *propagated error*, and the second term is usually rounding error. For the second term we often have

$$x_A \omega^* y_A = \mathrm{fl}(x_A \omega y_A), \tag{1.39}$$

which means that $x_A \omega y_A$ is computed exactly and then rounded (or truncated). With (1.32) and (1.39),

$$|x_A \omega y_A - x_A \omega^* y_A| \leqq \frac{\beta}{2} |x_A \omega y_A| \beta^{-n} \tag{1.40}$$

provided rounding is used.

For the propagated error, we examine particular cases.

Case (a)   Multiplication. For the error in $x_A y_A$,

$$x_T y_T - x_A y_A = x_T y_T - (x_T - \epsilon)(y_T - \eta)$$

$$= x_T \eta + y_T \epsilon - \epsilon \eta$$

$$\mathrm{Rel}(x_A y_A) \equiv \frac{x_T y_T - x_A y_A}{x_T y_T} = \frac{\eta}{y_T} + \frac{\epsilon}{x_T} - \frac{\epsilon}{x_T} \cdot \frac{\eta}{y_T}$$

$$= \mathrm{Rel}(x_A) + \mathrm{Rel}(y_A) - \mathrm{Rel}(x_A)\mathrm{Rel}(y_A) \tag{1.41}$$

For $|\mathrm{Rel}(x_A)|$, $|\mathrm{Rel}(y_A)| \ll 1$

$$\mathrm{Rel}(x_A y_A) \approx \mathrm{Rel}(x_A) + \mathrm{Rel}(y_A) \tag{1.42}$$

The symbol "$\ll$" means *much less than*.

Case (b)   Division. By a similar argument,

$$\mathrm{Rel}(x_A / y_A) = \frac{\mathrm{Rel}(x_A) - \mathrm{Rel}(y_A)}{1 - \mathrm{Rel}(y_A)} \tag{1.43}$$

For $|\mathrm{Rel}(y_A)| \ll 1$,

$$\mathrm{Rel}(x_A / y_A) \approx \mathrm{Rel}(x_A) - \mathrm{Rel}(y_A) \tag{1.44}$$

For both multiplication and division, relative errors do not propagate rapidly.

Case (c)   Addition and subtraction, and loss of significance errors.

$$(x_T \pm y_T) - (x_A \pm y_A) = (x_T - x_A) \pm (y_T - y_A) = \epsilon \pm \eta$$

$$\mathrm{Error}(x_A \pm y_A) = \mathrm{Error}(x_A) \pm \mathrm{Error}(y_A) \tag{1.45}$$

This is misleading since the relative error in $x_A \pm y_A$ may be quite poor when compared with $\mathrm{Rel}(x_A)$ and $\mathrm{Rel}(y_A)$. To illustrate this, we introduce the notion of a loss of significance error using an example.

Example   Solve $x^2 - 26x + 1 = 0$ using the quadratic formula. Define

$$x_T^{(1)} = 13 + \sqrt{168} \qquad x_T^{(2)} = 13 - \sqrt{168}$$

the two solutions of the equation. From a five-place table of square roots, $\sqrt{168} \doteq 12.961$; and thus

$$|\sqrt{168} - 12.961| \leqq .0005$$

Define

$$x_A^{(1)} = 25.961 \qquad x_A^{(2)} = .039$$

the result of substituting 12.961 for $\sqrt{168}$ in the true formulas. Then

$$|\text{Error}(x_A^{(1)})| = |\text{Error}(x_A^{(2)})| \leqq .0005$$

$$|\text{Rel}(x_A^{(1)})| \leqq \frac{.0005}{25.9605} \doteq 1.9 \times 10^{-5}$$

$$|\text{Rel}(x_A^{(2)})| \leqq \frac{.0005}{x_T^{(2)}} \leqq \frac{.0005}{.0385} \doteq 1.3 \times 10^{-2}$$

$x_A^{(2)}$ has a large relative error in spite of good accuracy in the component parts. The significant digits of 12.961 were lost in computing $x_A^{(2)}$.

In this example there is a simple solution. Define a new $x_A^{(2)}$ based on the following.

$$x_T^{(2)} = 13 - \sqrt{168} = \frac{1}{13 + \sqrt{168}}$$

$$\frac{1}{13 + \sqrt{168}} \doteq \frac{1}{25.961} \doteq .03851932 \equiv x_A^{(2)}$$

a better approximation to $x_T^{(2)}$. For the error with this new $x_A^{(2)}$,

$$|x_T^{(2)} - x_A^{(2)}| \leqq \left| x_T^{(2)} - \frac{1}{25.961} \right| + \left| \frac{1}{25.961} - .03851932 \right|$$

$$\leqq \left| \text{Error}\left(\frac{1}{25.961} \right) \right| + 5 \times 10^{-9}$$

$$|\text{Rel}(x_A^{(2)})| \leqq \left| \text{Rel}\left(\frac{1}{25.961} \right) \right| + \frac{5 \times 10^{-9}}{.0385 \cdots}$$

$$\leqq 1.9 \times 10^{-5} + 1.3 \times 10^{-7},$$

$$|\text{Rel}(x_A^{(2)})| \leqq 1.91 \times 10^{-5},$$

essentially the same as $x_A^{(1)}$.

As other examples of loss of significance errors, consider computing

$$\frac{e^x - 1}{x}, 1 - \cos(x)$$

for x near 0. Such functions should be computed by producing a Taylor series approximation, analytically avoiding the loss of significance error.

In a lengthy computer program, it is necessary to be quite careful in avoiding loss of significance errors that can occur in subtle ways, thus destroying the accuracy in a calculation. For example, if many numbers of large magnitude and varying sign are being added, and if the result is to be a number that is much smaller, then loss of significance errors are occurring.

Propagation of errors in function evaluations   Let $f(x_T)$ be approximated by $f(x_A)$. Then the mean-value theorem gives

$$f(x_T) - f(x_A) \approx f'(x_T)(x_T - x_A). \tag{1.46}$$

This assumes that x_A and x_T are relatively close and that $f'(x)$ does not vary greatly for x between x_A and x_T.

Example $\sin(\pi/5) - \sin(.628) \approx \cos(\pi/5)[(\pi/5) - .628] = .00026$, which is an excellent estimate of the error.

Using Taylor's theorem (1.13) for functions of two variables,

$$f(x_T, y_T) - f(x_A, y_A) \approx f_x(x_T, y_T)(x_T - x_A) + f_y(x_T, y_T)(y_T - y_A) \tag{1.47}$$

with $f_x \equiv \partial f / \partial x$. We are assuming that $f_x(x,y)$ and $f_y(x,y)$ do not vary greatly for (x,y) between (x_T, y_T) and (x_A, y_A).

Example   For $f(x,y) = x^y$, $f_x = yx^{y-1}$, $f_y = x^y \log x$, and

$$x_T^{y_T} - x_A^{y_A} \approx \epsilon_T x_T^{y_T - 1} + \eta(\log x_T) x_T^{y_T}$$

$$\text{Rel}(x_A^{y_A}) \approx y_T \left[\text{Rel}(x_A) + \text{Rel}(y_A) \log x_T \right] \tag{1.48}$$

The relative error in $x_A^{y_A}$ may be large, even though $\text{Rel}(x_A)$ and $\text{Rel}(y_A)$ are small. As a further illustration, take $y_T = y_A = 500$, $x_T = 1.2$, $x_A = 1.2001$. Then $x_T^{y_T} = 3.89604 \times 10^{39}$, $x_A^{y_A} = 4.06179 \times 10^{39}$, $\text{Rel}(x_A^{y_A}) = -.0425$. Compare this with $\text{Rel}(x_A) = 8.3 \times 10^{-5}$.

These arguments give the errors in the function f due to errors in the variables, but they do not take into account possible new errors resulting from the evaluation of $f(x_A)$ or $f(x_A, y_A)$. Refer back to (1.38) through (1.40) for the comparable statements with the simple arithmetic operations.

Propagation of error in a sum   We analyze the computation of the sum

$$S = \sum_{1}^{m} x_i \qquad (1.49)$$

in a computer, with $x_1, \ldots, x_m$ in floating-point form, that is,

$$x_i = \text{fl}(x_i) \qquad i = 1, \ldots, m$$

Define

$$S_2 = \text{fl}(x_1 + x_2) = (x_1 + x_2)(1 + \epsilon_2)$$

with

$$|\epsilon_2| \leq \frac{\beta}{2}\beta^{-n}$$

if numbers are rounded. The definition of S_2 and of the following sums is consistent with the earlier formula (1.40) for the approximation of the simple arithmetic operations.

Define recursively

$$S_{r+1} = \text{fl}(S_r + x_{r+1}) \qquad r = 1, 2, \ldots, m-1$$

Then

$$S_{r+1} = (S_r + x_{r+1})(1 + \epsilon_{r+1}), \qquad |\epsilon_{r+1}| \leq \frac{\beta}{2}\beta^{-n}$$

Expanding the first few sums, we obtain the following:

$$S_2 - (x_1 + x_2) = \epsilon_2(x_1 + x_2)$$

$$S_3 - (x_1 + x_2 + x_3) = (x_1 + x_2)\epsilon_2 + (x_1 + x_2)(1 + \epsilon_2)\epsilon_3 + x_3\epsilon_3$$

$$\approx (x_1 + x_2)\epsilon_2 + (x_1 + x_2 + x_3)\epsilon_3$$

$$S_4 - (x_1 + x_2 + x_3 + x_4) \approx (x_1 + x_2)\epsilon_2 + (x_1 + x_2 + x_3)\epsilon_3 + (x_1 + x_2 + x_3 + x_4)\epsilon_4$$

We have neglected cross-product terms $\epsilon_i\epsilon_j$ since they will be of much smaller magnitude. By induction, we obtain

$$S_m - \sum_{1}^{m} x_i \approx (x_1 + x_2)\epsilon_2 + \ldots + (x_1 + x_2 + \ldots + x_m)\epsilon_m$$

$$= x_1(\epsilon_2 + \epsilon_3 + \ldots + \epsilon_m) + x_2(\epsilon_2 + \epsilon_3 + \ldots + \epsilon_m)$$

$$+ x_3(\epsilon_3 + \ldots + \epsilon_m) + \ldots + x_m\epsilon_m \qquad (1.50)$$

From this formula we deduce that the best strategy for addition is to add from the smallest to the largest. Of course, counterexamples can be produced; but over a large number of examples the above rule should be best, especially if the numbers x_i are all of one sign so that no cancellation occurs in the calculation of the intermediate sums $x_1 + \ldots + x_m$, $m = 1, \ldots, n$. If chopping is used rather than rounding, and if all $x_i > 0$, then there is no cancellation in the sums of ϵ_i. With the strategy of adding from smallest to largest, we minimize the effect of these chopping errors.

The effect of inaccurate data   Data from a physical experiment will usually have a limited precision, say r digits of accuracy. It does not follow from this that only r digits should be carried in the computation. Instead, by carrying n digits, $n > r$, any new errors due to rounding will be introduced in the nth place, thus not making worse the limited precision in the data. Of course the considerations of the first part of this section on propagation of errors will apply to the errors in the data; and thus these errors will propagate and will cause the eventual answers in the computation to have fewer than r digits of accuracy.

As an illustration of the above, consider the following somewhat contrived example. Consider evaluating the sum

$$\sum_{1}^{m} x_i y_i$$

with all numbers x_i and y_i in the interval $[.1, 1]$ and accurate to r significant digits. Using decimal arithmetic, we compute the sum in two ways and analyze the error. In both cases X_i and Y_i denote the true values,

$$X_i = x_i + \epsilon_i, \quad Y_i = y_i + \eta_i$$

$$|\epsilon_i|, |\eta_i| \leq 5 \times 10^{-(r+1)}$$

Case (a)   Form $x_i y_i$, round to r digits, and then sum. For the error in the sum,

$$E_1 = \sum_{1}^{m} X_i Y_i - \text{Computed value of } \sum_{1}^{m} x_i y_i$$

$$= \sum_{1}^{m} X_i Y_i - \sum_{1}^{m} \left[(X_i - \epsilon_i)(Y_i - \eta_i) + \gamma_i \right]$$

with $|\gamma_i| \leq 5 \times 10^{-(r+1)}$, γ_i the rounding error in the ith product. Discarding the terms $\epsilon_i \eta_i$ because of their small size, we have

$$E_1 \approx \sum_{1}^{m} (\epsilon_i Y_i + \eta_i X_i - \gamma_i)$$

$$|E_1| \leq 3m(5 \times 10^{-(r+1)}) \tag{1.51}$$

Case (b)   Form $x_i y_i$, $i = 1, \ldots, m$, and add these products of length $2r$, keeping all the digits. Then round the final sum to r digits. For the error,

$$E_2 = \sum_1^m X_i Y_i - \text{Computed value of } \sum_1^m x_i y_i$$

$$= \sum_1^m X_i Y_i - \left\{ \left[\sum_1^m (X_i - \epsilon_i)(Y_i - \eta_i) \right] + \gamma \right\}$$

$$\approx \left(\sum_1^m \epsilon_i Y_i + \eta_i X_i \right) - \gamma$$

with $|\gamma| \leq 5 \times 10^{-(r+1)}$. Then

$$|E_2| \leq (2m + 1)(5 \times 10^{-(r+1)}) \tag{1.52}$$

a smaller bound than for E_1 in (1.51). The error E_2 is the result of the propagation of the errors in the original data; and E_1 also contains the errors of rounding all intermediate results to r digits. Clearly, it is sensible to calculate with a larger number of digits than r to avoid making the propagated error much worse.

A statistical treatment of the error   We consider a statistical treatment of propagated error in calculating a sum of numbers. To simplify the work we use exact arithmetic on data with error in it, thus avoiding the introduction of any new errors. Let $\{x_i\}$ be the data, and let the true values be

$$X_i = x_i + \epsilon_i \qquad i = 1, \ldots, m \tag{1.53}$$

with $|\epsilon_i| \leq e, i = 1, \ldots, m$.

Define the error E by

$$E = \sum_1^m X_i - \sum_1^m x_i = \sum_1^m \epsilon_i \tag{1.54}$$

Using bounds, as in the calculation of (1.51) and (1.52), we have

$$|E| \leq me. \tag{1.55}$$

This bound is for the worst possible case in which all the errors ϵ_i are as large as possible and are of the same sign.

A better model is to assume that the errors ϵ_i are uniformly distributed random variables in the interval $[-e, e]$ and that they are independent. Then

$$E = m \left(\frac{1}{m} \sum_1^m \epsilon_i \right) = m \bar{\epsilon}$$

The sample mean $\bar{\epsilon}$ is a new random variable, having a probability distribution with mean 0 and variance $e^2/3m$. To calculate probabilities for statements involving $\bar{\epsilon}$, it is important to note that the probability distribution for it is well approximated by the normal distribution with the same mean and variance, even for small values such as $m \geq 10$. This follows from the central limit theorem of probability theory; see Hogg and Craig (H2, Chapter 7) Using the approximating normal distribution, the probability is $1/2$ that

$$|\bar{\epsilon}| \leq .39e/\sqrt{m} \qquad |E| \leq .39e\sqrt{m}$$

and probability is .99 that

$$|\bar{\epsilon}| \leq 1.49e/\sqrt{m} \qquad |E| \leq 1.49e\sqrt{m} \qquad (1.56)$$

The result (1.56) is a considerable improvement on (1.55) if m is at all large.

Although statistical analyses give more realistic bounds, they are usually much more difficult to compute. As a more sophisticated example, see Henrici [H1, pp. 41-59] for a statistical analysis of the error in the numerical solution of differential equations.

1.5   Stability in Numerical Analysis

A number of mathematical problems have solutions that are quite sensitive to small computational errors, for example, rounding errors. To deal with this phenomenon, we introduce the concepts of stability and condition number. The condition number of a problem is closely related to the maximum accuracy that can be attained in the solution when using finite length numbers and computer arithmetic. These concepts are then extended to the numerical methods that are used to calculate the solution. Generally we want to use numerical methods that have no greater sensitivity to small errors than was true of the original mathematical problem.

To simplify the presentation, the discussion is limited to problems that have the form of an equation

$$F(x,y)=0 \qquad (1.57)$$

The variable x is the unknown being sought, and the variable y is data on which the solution depends. This equation may represent many different kinds of problems. For example, (1) F may be a real-valued function of the real variable x, and y may be a vector of coefficients present in the definition of F; or (2) the equation may be an integral or differential equation, with x an unknown function and y a given function or given boundary values.

We will say that the problem (1.57) is *stable* if the solution x depends in a continuous way on the variable y. If $\{y_n\}$ is a sequence of values approaching y in some sense, then the associated solution values $\{x_n\}$ must also approach x in some way. Or equivalently, if we make increasingly smaller changes in y, these must lead to increasingly smaller changes in x. The sense in which the changes are small will depend on the norm being used to measure the sizes of the vectors x and y; and there are many possible choices, varying with the problem. Stable problems are also called *well-posed problems*, and we use the two terms interchangeably. If a problem is not stable, it is called *unstable* or *ill-posed*.

Example (a) Consider the solution of

$$ax^2 + bx + c = 0 \qquad a \neq 0$$

The solution x is a real number. For the data in this case, we use $y = (a, b, c)$, the vector of coefficients. It should be clear from the quadratic formula

$$x = \frac{-b \pm \sqrt{b^2 - 4ac}}{2a}$$

that the two solutions for x will vary in a continuous way with the data $y = (a, b, c)$.

(b) Consider the integral equation problem

$$\int_0^1 \frac{.75x(t)\,dt}{1.25 - \cos(2\pi(s + t))} = y(s) \qquad 0 \leq s \leq 1 \tag{1.58}$$

This is an unstable problem. There are perturbations $\delta_n(s) = y_n(s) - y(s)$ for which

$$\max_{0 \leq s \leq 1} |\delta_n(s)| \to 0 \qquad \text{as} \quad n \to \infty \tag{1.59}$$

and the corresponding solutions $x_n(s)$ satisfy

$$\max_{0 \leq s \leq 1} |x_n(s) - x(s)| = 1 \qquad \text{all} \quad n \geq 1 \tag{1.60}$$

Specifically, define $y_n(s) = y(s) + \delta_n(s)$,

$$\delta_n(s) = \frac{1}{2^n} \cos(2n\pi s) \qquad 0 \leq s \leq 1, \quad n \geq 1$$

Then it can be shown that

$$x_n(s) - x(s) = \cos(2n\pi s)$$

thus proving (1.60).

If a problem (1.57) is unstable, then there are serious difficulties in attempting to solve it. It is usually not possible to solve such problems without first attempting to understand more about the properties of the solution, usually by returning to the context in which the mathematical problem was formulated. This is currently a very active area of research in applied mathematics and numerical analysis, for example see Lavrentiev [L1], Nashed [N1], and Tikhonov and Arsenin [T1].

For practical purposes there are many problems that are stable in the above sense, but that are still very troublesome as far as numerical computations are concerned. To deal with this difficulty, we introduce a measure of stability called a *condition number*. It shows that practical stability represents a continuum of problems, some better behaved than others.

The condition number attempts to measure the worst possible effect on the solution x of (1.57) when the variable y is perturbed by a small amount. Let δy be a perturbation of y, and let $x + \delta x$ be the solution of the perturbed equation

$$F(x + \delta x, y + \delta y) = 0 \tag{1.61}$$

Define

$$K(x) = \underset{\delta y}{\text{Supremum}} \frac{\|\delta x\|/\|x\|}{\|\delta y\|/\|y\|} \tag{1.62}$$

We have used the notation $\|\cdot\|$ to denote a measure of size. Recall the definition (1.19), (1.20), (1.21) for vectors from R^n and $C[a,b]$. The example (1.58) uses the norm (1.21) for measuring the perturbations in both x and y. Commonly x and y may be different kinds of variables, and then different norms are appropriate. The supremum in (1.62) is taken over all small perturbations δy for which the perturbed problem (1.61) will still make sense. Problems which are unstable lead to $K(x) = \infty$.

The number $K(x)$ is called the condition number for (1.57). It is a measure of the sensitivity of the solution x to small changes in the data y. If $K(x)$ is quite large, then there exists small relative changes δy in y that lead to large relative changes δx in x. But if $K(x)$ is small, say $K(x) \leq 10$, then small relative changes in y always lead to correspondingly small relative changes in x. Since numerical calculations almost always involve a variety of small computational errors, we do not want problems with a large condition number. Such problems are called ill-conditioned, and they are generally very hard to solve accurately.

Example Consider solving

$$x - a^y = 0 \qquad a > 0 \tag{1.63}$$

Perturbing y by δy, we have

$$\frac{\delta x}{x} = \frac{a^{y+\delta y} - a^y}{a^y} = a^{\delta y} - 1$$

For the condition number for (1.63),

$$K(x) = \underset{\delta y}{\text{Supremum}} \frac{|\delta x/x|}{|\delta y/y|} = \underset{\delta y}{\text{Supremum}} \left| y\left(\frac{a^{\delta y} - 1}{\delta y}\right)\right|$$

Restricting δy to be small, we have

$$K(x) \doteq |y \ln(a)| \tag{1.64}$$

Regardless of how we compute x in (1.63), if $K(x)$ is large, then small relative changes in y leads to much larger relative changes in x. If $K(x) = 10^4$ and if the value of y being used has relative error 10^{-7} due to using finite length computer arithmetic and rounding, then it is likely that the resulting value of x will have relative error of about 10^{-3}. This is a large drop in accuracy, and there is little way to avoid it except perhaps by doing all computations in longer precision computer arithmetic, provided y can then be obtained with greater accuracy.

Example Consider the $n \times n$ nonsingular matrix

$$Y = \begin{bmatrix} 1 & \dfrac{1}{2} & \dfrac{1}{3} & \cdots & \dfrac{1}{n} \\[2ex] \dfrac{1}{2} & \dfrac{1}{3} & \dfrac{1}{4} & \cdots & \dfrac{1}{n+1} \\[1ex] \vdots & & & & \vdots \\[1ex] \dfrac{1}{n} & \dfrac{1}{n+1} & \cdots & & \dfrac{1}{2n-1} \end{bmatrix} \tag{1.65}$$

which is called the Hilbert matrix. The problem of calculating the inverse of Y, or equivalently of solving $YX = I$ with I the identity matrix, is a well-posed problem. The solution X can be obtained in a finite number of steps using only simple arithmetic operations. But the problem of calculating X is increasingly ill conditioned as n increases.

The ill-conditioning of the numerical inversion of Y is shown in a practical setting. Let $\hat{Y}$ denote the result of entering the matrix Y into an IBM 360 computer and storing the matrix entries using single-precision floating-point format. The fractional elements of Y will be expanded in the hexadecimal (base 16) number system and then chopped after six hexadecimal digits (about seven decimal digits). Since most of the entries in Y do not have finite hexadecimal expansions, there will be a relative error of about 10^{-6} in each such element of $\hat{Y}$.

Using higher precision arithmetic, we can calculate the exact value of $\hat{Y}^{-1}$. The inverse Y^{-1} is known analytically, and we can compare it with $\hat{Y}^{-1}$. For $n=6$, some of the elements of $\hat{Y}^{-1}$ differ from the corresponding elements in Y in the first nonzero digit. For example, the entries in row 6, column 2 are

$$(Y^{-1})_{6,2}=83160.00 \qquad (\hat{Y}^{-1})_{6,2}=73866.34$$

This makes the calculation of Y^{-1} an ill-conditioned problem, and it becomes increasingly so as n increases. The condition number in (1.62) will be at least 10^6 as a reflection of the poor accuracy in $\hat{Y}^{-1}$ compared with Y^{-1}. And lest this be thought of as an odd pathological example that couldn't occur in practice, this particular example occurs very naturally when doing least squares approximation theory; for example, see Section 4.3. The general area of ill-conditioned problems for linear systems and matrix inverses is considered in greater detail in Chapter 8.

Stability of numerical algorithms A numerical method for solving a mathematical problem is considered stable if the sensitivity of the numerical answer to the data is no greater than in the original mathematical problem. We make this more precise, again using (1.57) as a model for the problem. A numerical method for solving (1.57) generally results in a sequence of approximate problems

$$F_n(x_n,y_n)=0 \qquad\qquad (1.66)$$

depending on some parameter, say n. The data y_n are to approach y as $n\to\infty$, the function values $F_n(z,w)$ are to approach $F(z,w)$ as $n\to\infty$, for all z near x and w near y; and hopefully the resulting approximate solutions x_n will then approach x as $n\to\infty$. For example, (1.57) may represent a differential equation initial value problem, and (1.66) may present a sequence of finite difference approximations depending on $h=1/n$, as in and following (1.37). Or n may represent the number of digits being carried in the calculations, and we may be solving $F(x,y)=0$ as exactly as possible within this finite precision arithmetic.

For each of the problems (1.66) we can define a condition number $K_n(x_n)$, just as in (1.62). Using these condition numbers, define

$$\tilde{K}(x)= \underset{\substack{n\to\infty \\ k\geq n}}{\text{Limit Supremum }} K_k(x_k) \qquad\qquad (1.67)$$

We say the numerical method is stable if $\tilde{K}(x)$ is of about the same magnitude as $K(x)$ from (1.62), for example if

$$\tilde{K}(x)\leq 2K(x)$$

If this is true, then the sensitivity of (1.66) to changes in the data is about the same as that of the original problem (1.57).

Some problems and numerical methods may not fit easily within the framework of (1.57), (1.62), (1.66), and (1.67), but there is a general idea of stable problems and condition numbers that can be introduced and given similar meaning. The main use of these concepts in this text will be (1) rootfinding for polynomial equations, (2) solving differential equations, and (3) problems in numerical linear algebra. Generally there is little problem with unstable numerical methods in this text. The main difficulty will be the solution of ill-conditioned problems.

Example Consider the evaluation of a Bessel function,

$$x = J_m(y) = \left(\tfrac{1}{2}y\right)^m \sum_{k=0}^{\infty} \frac{\left(-\tfrac{1}{4}y^2\right)^k}{k!(m+k)!} \qquad m \geq 0 \qquad (1.68)$$

This series converges very rapidly, and the evaluation of x is easily shown to be a well-conditioned problem in its dependence on y.

Now consider the evaluation of $J_m(y)$ using the triple recursion relation

$$J_{m+1}(y) = \frac{2m}{y} J_m(y) - J_{m-1}(y) \qquad m \geq 1 \qquad (1.69)$$

assuming $J_0(y)$ and $J_1(y)$ are known. We demonstrate numerically that this is an unstable numerical method for evaluating $J_m(y)$, for even moderately large m. We take $y = 1$, so that (1.69) becomes

$$J_{m+1}(1) = 2m J_m(1) - J_{m-1}(1) \qquad m \geq 1 \qquad (1.70)$$

We use values for $J_0(1)$ and $J_1(1)$, which are accurate to 10 significant digits. The subsequent values $J_m(1)$ are calculated from (1.70) using exact arithmetic, and the results are given in Table 1.1. The true values are given for comparison, and they show the rapid divergence of the approximate values from the true values.

Table 1.1 Computed values of $J_m(1)$

m	Computed $J_m(1)$	True $J_m(1)$
0	.7651976866	.7651976866
1	.4400505857	.4400505857
2	.1149034848	.1149034849
3	.195633535E−1	.1956335398E−1
4	.24766362E−2	.2476638964E−2
5	.2497361E−3	.2497577302E−3
6	.207248E−4	.2093833800E−4
7	−.10385E−5	.1502325817E−5

The only errors introduced were the rounding errors in $J_0(1)$ and $J_1(1)$, and they cause an increasingly large perturbation in $J_m(1)$ as m increases.

The use of three term recursion relations

$$f_{m+1}(x) = a_m(x)f_m(x) - b_m(x)f_{m-1}(x), \qquad m \geq 1$$

is a common tool in applied mathematics and numerical analysis. But as shown above, they can lead to unstable numerical methods. For a general analysis of triple recursion relations, see Gautschi [G1]. In the case of (1.69) and (1.70), large loss of significance errors are occurring.

Discussion of the Literature

A knowledge of the details of computer arithmetic is important for programmers who are concerned with accuracy and who are writing programs that will be widely used. Also, in the conversion of programs from one type of computer to another (e.g., from an IBM 360 model machine to a CDC 6000 series machine), it is usually necessary to take the differing precision of the two machines into account. For a complete introduction to the arithmetic of several current computers, including ways to compensate for their disadvantages, see Sterbenz [S1]. To increase the portability of mathematical programs between different computers, much current mathematical software uses a reference to the machine unit roundoff error to determine machine precision. This was made popular in Shampine and Gordon [S1, p. 157], and it is introduced in the following Problem 7.

The topic of error propagation, especially that due to rounding error, has been difficult to treat in a precise, but useful manner. There are some important early papers, but the current approach to the subject and its main results are due principally to J. H. Wilkinson. Most of his work is in the area of numerical linear algebra, but he has made important contributions to most areas of numerical analysis. For a general introduction to his techniques for analyzing the propagation of errors, with applications to several important problems, see his book [W1].

Another very different approach to the control of error is called interval analysis. With it, we carry along an interval $[x_l, x_u]$ in our calculations, rather than a single number x_A; and the numbers x_l and x_u are guaranteed to bound the true desired value x_T. The difficulty with the approach is that the size of $x_u - x_l$ is generally a magnitude or more larger than the error $x_T - x_A$, mainly because possible cancellation of errors of opposite sign is not considered when computing x_l and x_u. For a complete introduction, see Moore [M2], which includes techniques for improving the usual poor error bounds that would be produced by interval analysis.

The topic of ill-posed problems has been of increasing interest in recent years. There are many problems of indirect physical measurement that lead to such problems, and in this form they are often called *inverse problems*; for example, geophysics is an important area of application. The book by Lavrentiev [L1] gives a general introduction, although it discusses mainly (1) analytic continuation of analytic functions of a complex variable, and (2) inverse problems for differential equations. As an example of the current literature on numerical methods for ill-posed problems, see Nashed [N1] including the bibliography and the book Tikhonov and Arsenin [T1].

Bibliography

[F1] Fadeeva, V., *Computational Methods of Linear Algebra*, Dover Press, New York, 1959.

[G1] Gautschi, W., Computational aspects of three term recurrence relations, *SIAM Rev.*, **9** (1967), pp. 24–82.

[H1] Henrici, P., *Discrete Variable Methods in Ordinary Differential Equations*, John Wiley, New York, 1962.

[H2] Hogg, R. and A. Craig, *Introduction to Mathematical Statistics*, 2nd ed., Macmillan, New York, 1965.

[L1] Lavrentiev, M., *Some Improperly Posed Problems of Mathematical Physics*, Springer Verlag, New York, 1967.

[L2] Lin, C. and L. Segal, *Mathematics Applied to Deterministic Problems in the Natural Sciences*, Macmillan, New York, 1974.

[M1] Maki, D. and M. Thompson, *Mathematical Models and Applications*, Prentice-Hall, Englewood Cliffs, N. J., 1973.

[M2] Moore, R., *Interval Analysis*, Prentice-Hall, Englewood Cliffs, N. J., 1966.

[N1] Nashed, Z., Approximate regularized solutions to improperly posed linear integral and operator equations, in *Constructive and Computational Methods for Differential and Integral Equations*, D. Colton and R. Gilbert, eds., Springer Lec. Notes #430, 1974, pp. 289–332.

[S1] Shampine, L. and M. Gordon, *Computer Solution of Ordinary Differential Equations*, Freeman Publications, San Francisco, 1975.

[S2] Sterbenz, P., *Floating-Point Computation*, Prentice-Hall, Englewood Cliffs, New Jersey, 1974.

[T1] Tikhonov, A. and Arsenin, V., *Solutions of Ill-posed Problems*, John Wiley and Sons, New York, 1977.

[W1] Wilkinson, J. H., *Rounding Errors in Algebraic Processes*, Prentice-Hall, Englewood Cliffs, N. J., 1963.

Problems

1. Assume $f(x)$ is continuous for $a \leq x \leq b$, and consider the sum

$$S = \sum_{j=1}^{n} f(x_j)$$

with all points x_j in the interval $[a, b]$. Show that

$$S = nf(\zeta)$$

for some ζ in $[a, b]$. *Hint*: use the intermediate value theorem in Section 1.1, and consider the range of values for $(1/n)S$.

2. Construct a Taylor series for the following functions, and bound the error when truncating after n terms:

(a) $\dfrac{1}{x} \displaystyle\int_0^x e^{-t^2} dt$ 　　　　 (b) $\sin^{-1}(x)$, $|x| < 1$

(c) $\dfrac{1}{x} \displaystyle\int_0^x \dfrac{\tan^{-1} t \, dt}{t}$ 　　　 (d) $\cos(x) + \sin(x)$

3. Use Taylor's theorem for two variables to find a simple approximation to $\sqrt{1 + x - y}$ for small values of x and y.

4. Consider the second-order divided difference $f[x_0, x_1, x_2]$ defined in (1.14).

(a) Prove the property (1.16), that the order of the arguments x_0, x_1, x_2 does not affect the value of the divided difference.

(b) Prove formula (1.15),

$$f[x_0, x_1, x_2] = \tfrac{1}{2} f''(\zeta)$$

for some ζ between the minimum and maximum of x_0, x_1, and x_2. *Hint*: from part (a), there is no loss of generality in assuming $x_0 < x_1 < x_2$. Use Taylor's theorem to reduce $f[x_0, x_1, x_2]$, expanding about x_1; and then use the intermediate value theorem to simplify the error term.

5. (a) Show that the vector norms of (1.19) and (1.21) satisfy the three general properties of norms that are listed following (1.21).

(b) Show that the matrix norm (1.22) satisfies (1.23). For simplicity, consider only matrices of order 2×2.

6. For the following numbers x_A and x_T, how many significant digits are there in x_A with respect to x_T?

   (a) $x_A = 451.023$, $\quad x_T = 451.01$

   (b) $x_A = -.045113$, $\quad x_T = -.04518$

   (c) $x_A = 23.4213$, $\quad x_T = 23.4604$

7. What is the smallest positive number u for which $1 + u > 1$ on a CDC 6000 series computer? Recall that truncation rather than rounding is used. Repeat the problem with reference to an IBM 360 model computer, using single-precision arithmetic. These numbers give a fairly precise idea of the maximum accuracy of a number stored in floating-point form. The number u is called the unit round-off error, and it can be used to increase the portability of computer programs between computers of different numerical characteristics. For example, see Shampine and Gordon [S1, p. 157].

8. Let all of the numbers below be correctly rounded to the number of digits shown: (a) $1.1062 + .947$, (b) $23.46 - 12.753$, (c) $(2.747)(6.83)$, (d) $8.473/.064$. For each calculation, determine the smallest interval in which the result, using true instead of rounded values, must lie.

9. Prove the formula (1.43) for $\text{Rel}(x_A/y_A)$.

10. Given the equation $x^2 - 40x + 1 = 0$, find its roots to five significant digits. Use $\sqrt{399} \doteq 19.975$, correctly rounded to five digits.

11. Let $f(x) = [\ln(1 - x) + xe^{x/2}]/x^3$. Calculate $f(x)$ for $x = 10^{-m}$, $m = 1, 2, 3, 4, 5, 6, 7$; and explain the resulting behaviour. According to theory, what should be the value of $f(0)$? For x near zero, what is a better way to calculate $f(x)$? *Hint*: use the Taylor series.

12. Suppose you use the series $e^x = \sum_0^\infty (x^n/n!)$ to evaluate e^{-5}. How many terms are necessary to evaluate e^{-5} with a relative error of less than 10^{-3}? If you were restricted to a decimal computer that carried only four decimal places, could you accomplish this? What would be a better way to compute e^{-5} with such a computer, again with a desired relative error less than 10^{-3}? *Hint*: consider the magnitudes of the terms in the series for e^{-5}, along with the size of e^{-5} itself.

13. Suppose you wish the values of (a) $\cos(1.473)$, (b) $\tan^{-1}(2.621)$, (c) $\ln(1.471)$, (d) $e^{2.653}$. But in each case, you have only a table of values of the function with the argument x given in increments of .01. Obtain estimates

of the error made by using the nearest values in the tables to the given arguments.

14. Assume that $x_A = .937$ has three significant digits with respect to x_T. Bound the relative error in x_A. For $f(x) = \sqrt{1-x}$, bound the error and relative error in $f(x_A)$ with respect to $f(x_T)$.

15. The numbers given below are correctly rounded to the number of digits shown. Estimate the errors in the function values in terms of the errors in the arguments. Bound the relative errors.

    (a) $\sin[(3.14)(2.685)]$ (b) $\ln(1.712)$

    (c) $(1.56)^{3.414}$

16. Consider the product $a_0 a_1 \ldots a_m$ where $a_0, a_1, \ldots, a_m$ are $m+1$ numbers stored in a computer with n digit base β arithmetic. Assume the computer uses rounding rather than truncation. Define $p_1 = \text{fl}(a_0 a_1), p_2 = \text{fl}(a_2 p_1), p_3 = \text{fl}(a_3 p_2), \ldots, p_m = \text{fl}(a_m p_{m-1})$. If we write $p_m = a_0 a_1 \ldots a_m (1 + W)$, determine an estimate for W. Assume that $a_i = \text{fl}(a_i), i = 0, 1, \ldots, m$. What is a statistical estimate for the size of W?

TWO

ROOTFINDING
FOR NONLINEAR
EQUATIONS

Finding one or more roots of an equation

$$f(x) = 0 \tag{2.1}$$

is one of the most commonly occurring problems of applied mathematics. In most cases explicit solutions are not available and we must be satisfied with being able to find a root to any specified degree of accuracy. The numerical methods for finding the roots are called iterative methods, and they are the main subject of this chapter.

We begin with iterative methods for solving (2.1) when $f(x)$ is any continuously differentiable real-valued function of a real variable x. The iterative methods for this quite general class of equations requires knowledge of one or more initial guesses x_0 for the desired root α of $f(x)$. An initial guess x_0 can usually be found by using the context in which the problem first arose; otherwise, a simple graph of $y = f(x)$ will often suffice for estimating x_0.

The second major problem is that of finding one or more roots of a polynomial equation,

$$p(x) \equiv a_0 + a_1 x + \ldots + a_n x^n = 0 \qquad a_n \neq 0 \tag{2.2}$$

The methods of the first problem are often specialized to deal with (2.2), and that is our approach. But there is a large literature of methods that have been developed especially for polynomial equations, using their special properties in an essential way. These are the most important methods used in creating automatic computer programs for solving (2.2), and we reference some of them in the discussion of the literature at the end of the chapter.

The third problem to be discussed is the solution of nonlinear systems of equations. These systems are very diverse in form, and the associated numerical analysis is both extensive and sophisticated. We just touch on this subject, indicating some successful methods that are fairly general in applicability. An

39

adequate development of the subject requires a good knowledge of both theoretical and numerical linear algebra, and these topics are not taken up until Chapters 7 through 9.

To illustrate the concept of an iterative method for finding a root of (2.1), we begin with an example. Consider solving

$$f(x) \equiv a - 1/x = 0 \tag{2.3}$$

for a given $a > 0$. This problem has a practical application to computers without a machine divide operation, as was true with some of the early computers.

Let $x_0 \approx 1/a$ be an approximate solution of the equation. At the point $(x_0, f(x_0))$, draw the tangent line to the graph of $y = f(x)$; see Figure 2.1. Let x_1 be the point at which the tangent line intersects the x-axis. It should be an improved approximation of the root α.

To obtain an equation for x_1, match the slopes obtained from the tangent line and the derivative of $f(x)$ at x_0;

$$f'(x_0) = \frac{f(x_0) - 0}{x_0 - x_1}$$

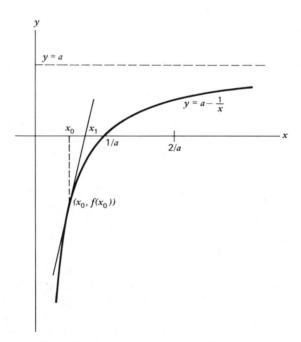

Figure 2.1   Iterative solution of $a - (1/x) = 0$.

Substituting from (2.3) and manipulating, we obtain

$$x_1 = x_0(2 - ax_0).$$

The general iteration formula is then obtained by repeating the process, with x_1 replacing x_0, ad infinitum, to get

$$x_{n+1} = x_n(2 - ax_n) \qquad n \geq 0$$

A more convenient form for the theory is obtained by introducing the *residual*

$$r_n = 1 - ax_n \tag{2.4}$$

Using it,

$$x_{n+1} = x_n(1 + r_n) \qquad n \geq 0 \tag{2.5}$$

For the error,

$$e_n = \frac{1}{a} - x_n = \frac{r_n}{a} \tag{2.6}$$

We now analyze the convergence of this method, its speed and its dependence on x_0. First,

$$r_{n+1} = 1 - ax_{n+1} = 1 - ax_n(1 + r_n) = 1 - (1 - r_n)(1 + r_n)$$

$$r_{n+1} = r_n^2 \tag{2.7}$$

Inductively,

$$r_n = r_0^{2^n} \qquad n \geq 0 \tag{2.8}$$

From (2.6), the error e_n converges to zero as $n \to \infty$ if and only if r_n converges to zero. From (2.8), r_n converges to zero if and only if $|r_0| < 1$, or equivalently,

$$-1 < 1 - ax_0 < 1$$

$$0 < x_0 < 2/a \tag{2.9}$$

In order that x_n converge to $1/a$, it is necessary and sufficient that x_0 be chosen to satisfy (2.9).

To examine the speed of convergence when (2.9) is satisfied, we obtain formulas for the error and relative error. For the speed of convergence when

(2.9) is satisfied,

$$e_{n+1} = r_{n+1}/a = r_n^2/a = e_n^2 a^2/a$$

$$e_{n+1} = ae_n^2$$

$$\frac{e_{n+1}}{1/a} = e_n^2 a^2 = \left[\frac{e_n}{1/a}\right]^2 \tag{2.10}$$

$$\text{Rel}(x_{n+1}) = [\text{Rel}(x_n)]^2 \qquad n \geq 0 \tag{2.11}$$

The notation $\text{Rel}(x_n)$ denotes the relative error in x_n. The equation (2.10) shows that e_n *converges to zero quadratically*. To illustrate how rapidly the error will decrease, suppose that $\text{Rel}(x_0) = 0.1$. Then $\text{Rel}(x_4) = 10^{-16}$. Each iteration doubles the number of significant digits.

This example illustrates the construction of an iterative method for solving an equation, and a complete convergence analysis is given. This analysis included a proof of convergence, a determination of the *interval of convergence* for the choice of x_0, and a determination of the speed of convergence. These ideas are examined in detail in the following sections using a more general approach to (2.1).

Definition   A sequence of iterates $\{x_n | n \geq 0\}$ is said to converge with order $p \geq 1$ to a point α if

$$|\alpha - x_{n+1}| \leq c|\alpha - x_n|^p \qquad n \geq 0 \tag{2.12}$$

for some $c > 0$. If $p = 1$, the sequence is said to converge linearly to α. In that case, it is necessary that $c \leq 1$ (generally $c < 1$ in practice); and the constant c is called the rate of linear convergence of x_n to α.

Using this definition, the earlier example (2.4), (2.5) has order of convergence $p = 2$. This definition of order is not always a convenient one for some linear iterative methods. Using induction on (2.12) with $p = 1$ and $c < 1$, we obtain

$$|\alpha - x_n| \leq c^n|\alpha - x_0| \qquad n \geq 0. \tag{2.13}$$

This shows directly the convergence of x_n to α. For some iterative methods we can show (2.13) directly, whereas (2.12) may not be true for any $c < 1$. In such a case, the method will still be said to converge linearly with a rate of c.

2.1   Simple Enclosure Methods

Assume an interval $[a, b]$ is given, and that on it $f(x)$ is continuous and satisfies

$$f(a)f(b) < 0 \tag{2.14}$$

Using the intermediate value theorem 1.1 from Chapter 1, the function $f(x)$ must have at least one root in $[a, b]$. Usually $[a, b]$ is chosen to contain only one root α; but the following algorithm for the bisection method will always converge to some root α in $[a, b]$, because of (2.14).

Algorithm Bisect $(f, a, b, \text{root}, \epsilon)$

    1. Define $c := (a + b)/2$.

    2. If $b - c \leqq \epsilon$, then accept root $:= c$, and exit.

    3. If sign $(f(b))\,\text{sign}\,(f(c)) \leqq 0$, then $a := c$; otherwise $b := c$.

    4. Return to step 1.

The interval $[a, b]$ is halved in size for every pass through the algorithm; and because of step 3 $[a, b]$ will always contain a root of $f(x)$. Since a root α is in $[a, b]$, it must lie within either $[a, c]$ or $[c, b]$; and consequently

$$|c - \alpha| \leqq b - c = c - a$$

This is justification for the test in step 2. On completion of the algorithm, c will be an approximation to the root with

$$|c - \alpha| \leqq \epsilon$$

Example Find the largest real root α of

$$f(x) \equiv x^6 - x - 1 = 0 \tag{2.15}$$

It is straightforward to show that $1 < \alpha < 2$, and we use this as our initial interval $[a, b]$. The algorithm Bisect was used with $\epsilon = .00005$; the results are shown in Table 2.1. The first two iterates give the initial interval enclosing α; and the remaining values $c_n, n \geqq 2$, denote the successive midpoints found using Bisect. The final value $c_{16} = 1.13474$ was accepted as an approximation to α with $|\alpha - c_{16}| \leqq .00004$.

Table 2.1 Example for bisection method

n	c_n	$f(c_n)$	n	c_n	$f(c_n)$
0	$2.0 = b$	61.0	9	1.13672	.02062
1	$1.0 = a$	-1.0	10	1.13477	.00043
2	1.5	8.89063	11	1.13379	$-.00960$
3	1.25	1.56470	12	1.13428	$-.00459$
4	1.125	$-.09771$	13	1.13452	$-.00208$
5	1.1875	.61665	14	1.13464	$-.00083$
6	1.15625	.23327	15	1.13470	$-.00020$
7	1.14063	.06158	16	1.13474	.00111
8	1.13281	$-.01958$			

The true solution is

$$\alpha \doteq 1.134724138 \tag{2.16}$$

The true error in c_{16} is

$$\alpha - c_{16} = -.000016$$

It is much smaller than the predicted error bound. It might seem as though we could have saved some computation by stopping with an earlier iterate. But there is no way to predict the possibly greater accuracy in an earlier iterate, and thus there is no way we can know the iterate is sufficiently accurate. For example, c_{10} is sufficiently accurate, but there was no way of telling that fact during the computation.

To examine the speed of convergence, let c_n denote the nth value of c in the algorithm. Then it is easy to see that

$$\alpha = \operatorname*{limit}_{n \to \infty} c_n$$

$$|\alpha - c_n| \leq \left(\tfrac{1}{2}\right)^n (b-a) \tag{2.17}$$

where $(b-a)$ denote the length of the original interval input into Bisect. Using the variant (2.13) for defining linear convergence, we say that the bisection method converges linearly with a rate of $1/2$. The actual error may not decrease by a factor of $1/2$ at each step, but the average rate of decrease is $1/2$, based on (2.17). The preceding example illustrates the result (2.17).

The Regula Falsi method   Because the bisection method converges at a fairly slow speed, people have attempted to produce more rapidly converging methods. An advantage of the bisection method is that it is guaranteed to converge. Our next method is an attempt to produce a method that is more rapid and is still guaranteed to converge.

As before we assume (2.14). The graph of $f(x)$ is approximated by a straight line on $[a,b]$, connecting $(a,f(a))$ and $(b,f(b))$; and its root c is used as an approximation to a root α of $f(x)$. See Figure 2.2. We discard a or b as before, to obtain a new interval containing the root. To obtain a formula for c, use the slope equality

$$\frac{f(b)-f(a)}{b-a} = \frac{f(b)-0}{b-c}$$

$$c := b - f(b) \left[\frac{b-a}{f(b)-f(a)} \right] \tag{2.18}$$

The following algorithm is called the *Regula Falsi method.*

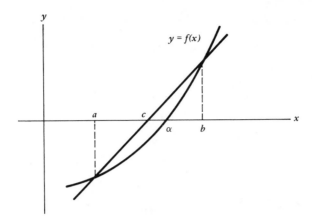

Figure 2.2 The Regula Falsi method for calculating α.

Algorithm Regfalsi $(f, a, b, \text{root}, \epsilon)$

1. pastc: $=2b-a$, to make the first pass through error test fail in step 4.

2. Set

$$c := b - f(b)\left[\frac{b-a}{f(b)-f(a)}\right]$$

3. If sign $(f(b))$ sign $(f(c)) \leq 0$, then $a := c$; otherwise $b := c$.

4. If $|c - \text{pastc}| \leq \epsilon$, then accept root: $=c$ as an approximate root of $f(x)$.

5. Otherwise pastc: $=c$, and return to step 2.

There are several serious problems with this algorithm, and we analyze them below. But the use of the straight line approximation to $y = f(x)$ and the enclosure of the root are basically good ideas. A very successful modification is presented in Section 2.8 later in the chapter.

Example The equation

$$f(x) \equiv x^6 - x - 1 = 0$$

was solved using the algorithm Regfalsi. The initial interval was [1.0, 1.2], and the desired error tolerance was $\epsilon = .00005$. The results are shown in Table 2.2, along with the errors. The column labeled a_n gives the successive values of a in the algorithm, since the value of b never changed. The column labeled "Ratio"

Table 2.2   Example of Regula Falsi method

n	a_n	$f(a_n)$	$\alpha - a_n$	Ratio
0	$1.2 = b$	.78598		
1	1.0	-1.0	1.35E-1	
				.169
2	1.11198	$-.22143$	2.27E-2	
				.149
3	1.13133	$-.03464$	3.40E-3	
				.146
4	1.13423	$-.00510$	4.96E-4	
				.146
5	1.13465	$-.00074$	7.23E-5	
				.146
6	1.13471	$-.00011$	1.05E-5	
				.146
7	1.13472	$-.00002$	1.53E-6	

gives the ratio of successive errors

$$\frac{\alpha - a_{n+1}}{\alpha - a_n} \tag{2.19}$$

with $\alpha \doteq 1.134724138$.

The ratios (2.19) approach the constant .146. Recalling the definition (2.12), this shows the method is converging linearly with a rate of .146. This is faster than the average rate of $1/2$ for the bisection method. But the Regula Falsi method is not always superior to the bisection method. If the initial interval had been chosen as $[a,b]=[1.0, 2.0]$, then the iterates would have converged linearly with the much poorer rate of .855. This is much slower than with bisection.

In order to understand the convergence behavior of the Regula Falsi method, an error formula is needed. For the formula (2.18), we can prove

$$\alpha - c = -(\alpha - b)(\alpha - a)\frac{f''(\zeta)}{2f'(\xi)} \tag{2.20}$$

for some ξ between a and b and some ζ between the maximum and minimum of a, b, and α. This assumes that $f(x)$ is twice continuously differentiable on some interval containing a, b, and α.

To obtain (2.20), begin by multiplying (2.18) by -1 and adding α to each side of the equation,

$$\alpha - c = \alpha - b + f(b)\frac{b-a}{f(b)-f(a)}$$

Factoring $(\alpha - b)(\alpha - a)$ from the right side and doing some algebraic manipulation (left as a problem), we obtain

$$\alpha - c = -(\alpha - b)(\alpha - a)\cdot\frac{f[a,b,\alpha]}{f[a,b]} \tag{2.21}$$

The quantities $f[a,b]$ and $f[a,b,\alpha]$ are divided differences, and they were introduced in (1.14) of Section 1.1. Using the result (1.15) for these differences leads directly to (2.20).

The most common case of the Regula Falsi method is when $f''(\alpha)\neq0$ and α is the only root in $[a,b]$. To simplify the analysis, assume $f''(x)\geq0$ for $a\leq x\leq b$; replace f by $-f$ if $f''(\alpha)<0$. With the assumption $f''(x)\geq0$, the straight line segment connecting $(x_1,f(x_1))$ and $(x_2,f(x_2))$ lies above the graph of $y=f(x)$ for all choices $a\leq x_1<x_2\leq b$.

Case 1 $f'(\alpha)>0$. Using (2.20) or the geometry of the last sentence, the iterate c must always lie between a and α, as in Figure 2.2. Thus in step 3 of the algorithm Regfalsi, c always replaces a rather than b. The error $\alpha-b$ remains constant. Let a_n denote the nth value of a in the algorithm. Then

$$\alpha-a_{n+1}=-(\alpha-a_n)(\alpha-b)\frac{f''(\zeta_n)}{2f'(\xi_n)}\qquad n\geq0$$

with ξ_n and ζ_n between a_n and b. Thus the method is linear. It is not difficult to produce examples for which the multiplying constant,

$$\left|(\alpha-b)\frac{f''(\zeta_n)}{2f'(\xi_n)}\right|\qquad n\geq1$$

is always greater than $1/2$. This makes the bisection method faster for such examples, as in the following.

Example Consider

$$f(x)=(x-1)\left(x-\tfrac{1}{2}\right)$$

with

$$\alpha=1\qquad\tfrac{1}{2}<a_0<1\qquad b=2$$

By direct calculation from (2.18),

$$1-a_{n+1}=\left(\frac{2-a_n}{1+\tfrac{3}{2}a_n-a_n^2}\right)(1-a_n).$$

The multiplying constant is bounded below by $2/3$; and the convergence is linear with a limiting rate of $2/3$.

Case 2 $f'(\alpha)<0$. The iterates c will always satisfy $\alpha<c<b$. The point a will remain unchanged, with c replacing b at each iteration. All other points of case (1) apply here.

There are other problems with Regfalsi. The error test can be poor, predicting too large or too small an error. The general problem of estimating the error in iterative methods is examined in Section 2.6.

2.2 The Secant Method

As with the last method, the graph of $y=f(x)$ is approximated by a secant line in the vicinity of the root α. But now we do not force the iterates to bracket the root; see Figure 2.3. The method is

$$x_{n+1}=x_n-f(x_n)\frac{x_n-x_{n-1}}{f(x_n)-f(x_{n-1})} \qquad n\geq 1, \tag{2.22}$$

with x_0 and x_1 given. The method is not guaranteed to converge; but it does so with a greater speed than the previous methods when it does converge.

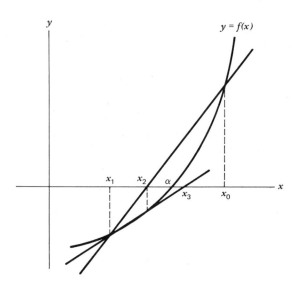

Figure 2.3 The Secant method.

Example Consider again finding the largest root of

$$f(x)\equiv x^6-x-1=0.$$

The secant method (2.22) was used, and the computation continued until the successive differences x_n-x_{n-1} were considered sufficiently small. The numerical results are given in Table 2.3.

The convergence is increasingly rapid as n increases. One way of measuring this is to calculate the ratios

$$\frac{\alpha-x_{n+1}}{\alpha-x_n} \qquad n\geq 0$$

Table 2.3 Example of Secant method

n	x_n	$f(x_n)$	$\alpha - x_n$
0	2.0	61.0	8.65E-1
1	1.0	-1.0	1.35E-1
2	1.0161290	-9.154E-1	1.16E-1
3	1.1905778	6.575E-1	-5.59E-2
4	1.1176558	-1.685E-1	1.71E-2
5	1.1325316	-2.244E-2	2.19E-3
6	1.1348168	9.536E-4	-9.27E-5
7	1.1347236	-5.066E-6	4.92E-7
8	1.134724138	-4.0E-9	0.0

For a linear method these are generally constant as x_n converges to α. But in this example, these ratios become smaller with increasing n. One intuitive explanation is that the straight line connecting $(x_{n-1}, f(x_{n-1}))$ and $(x_n, f(x_n))$ becomes an increasingly accurate approximation to the graph of $y = f(x)$ in the vicinity of $x = \alpha$, and consequently the root x_{n+1} of the straight line is a much improved estimate of α. Also note that the iterates x_n move above and below the root α in an apparently random fashion as n increases. An explanation of this will come from the error formula (2.23) given below.

The derivation of (2.20) did not depend on the order of the elements a, b, c, and α, and it can be applied equally well to the secant method. From (2.20), with $e_n = \alpha - x_n$ for $n \geq 1$, we obtain the error formula

$$e_{n+1} = -e_{n-1}e_n \frac{f''(\zeta_n)}{2f'(\xi_n)} \tag{2.23}$$

with ξ_n between x_{n-1} and x_n, and ζ_n between x_{n-1}, x_n, and α. This will be used to prove convergence of x_n to α, under suitable assumptions.

Theorem 2.1 Assume that $f(x)$, $f'(x)$, and $f''(x)$ are continuous for all values of x in some interval containing α, and assume $f'(\alpha) \neq 0$. Then if the initial guesses x_0 and x_1 are chosen sufficiently close to α, the iterates x_n of (2.22) will converge to α. The order of convergence will be $p = (1 + \sqrt{5})/2 \doteq 1.62$.

Proof For some neighborhood $I = [\alpha - \epsilon, \alpha + \epsilon]$ with some $\epsilon > 0$, let

$$M = \frac{\underset{x \in I}{\text{Max}} |f''(x)|}{2 \underset{x \in I}{\text{Min}} |f'(x)|}.$$

Then for all $x_0, x_1 \in [\alpha - \epsilon, \alpha + \epsilon]$, using (2.23),

$$|e_2| \leq |e_1||e_0|M$$

$$M|e_2| \leq M|e_1|M|e_0|$$

Further assume that x_1 and x_0 are so chosen that

$$\delta \equiv \text{Max}\{M|e_0|, M|e_1|\} < 1 \tag{2.24}$$

Then $M|e_2| < 1$ since

$$M|e_2| \leq \delta^2$$

Also $M|e_2| \leq \delta^2 < \delta$ implies

$$e_2 < \frac{\delta}{M} = \text{Max}\{|e_1|, |e_0|\} \leq \epsilon$$

and thus $x_2 \in [\alpha - \epsilon, \alpha + \epsilon]$. We apply this argument inductively to show that $x_n \in [\alpha - \epsilon, \alpha + \epsilon]$ for every $n \geq 2$.

To prove convergence and obtain the order of convergence, continue applying (2.23) to get

$$M|e_3| \leq M|e_2|M|e_1| \leq \delta^2 \cdot \delta = \delta^3$$

$$M|e_4| \leq M|e_3|M|e_2| \leq \delta^5$$

For

$$M|e_n| \leq \delta^{q_n}$$

$$M|e_{n+1}| \leq M|e_n|M|e_{n-1}| \leq \delta^{q_n + q_{n-1}} = \delta^{q_{n+1}} \tag{2.25}$$

Thus

$$q_{n+1} = q_n + q_{n-1} \qquad n \geq 1 \tag{2.26}$$

with $q_0 = q_1 = 1$. This is a Fibonacci sequence of numbers, and an explicit formula can be given:

$$q_n = \frac{1}{\sqrt{5}}\left[r_0^{n+1} - r_1^{n+1}\right] \qquad n \geq 0$$

$$r_0 = \frac{1 + \sqrt{5}}{2} \approx 1.618 \qquad r_1 = \frac{1 - \sqrt{5}}{2} \approx -.618 \tag{2.27}$$

Thus

$$q_n \approx \frac{1}{\sqrt{5}}(1.618)^{n+1} \text{ for large } n$$

For example $q_5 = 8$, and the last formula gives 8.025. Returning to (2.25), we obtain the error bound

$$|e_n| \leq \frac{1}{M}\delta^{q_n} \qquad n \geq 0 \tag{2.28}$$

with q_n given by (2.27). Since $q_n \to \infty$ as $n \to \infty$, we have $x_n \to \alpha$.

By doing a more careful derivation, we can actually show the order of convergence is $p = (1 + \sqrt{5})/2$. To simplify the presentation, we will instead show that this is the rate at which the bound in (2.28) decreases. Let B_n denote the upper bound in (2.28). Then

$$\frac{B_{n+1}}{B_n^{r_0}} = \frac{\frac{1}{M}\delta^{q_{n+1}}}{\left(\frac{1}{M}\right)^{r_0}\delta^{r_0 q_n}} = M^{r_0 - 1}\delta^{q_{n+1} - r_0 q_n}$$

$$\leq \delta^{-1/2}M^{r_0 - 1} \equiv c$$

since $q_{n+1} - r_0 q_n = r_1^{n+1} < 1$. Thus

$$B_{n+1} \leq cB_n^{r_0}$$

which implies an order of convergence $p = r_0 = (1 + \sqrt{5})/2$. A similar result holds for the actual errors e_n; and moreover,

$$\underset{n \to \infty}{\text{Limit}} \frac{|e_{n+1}|}{|e_n|^{r_0}} = \left|\frac{f''(\alpha)}{2f'(\alpha)}\right|^{\frac{\sqrt{5}-1}{2}} \tag{2.29}$$

∎

The error formula (2.23) can be used to explain the oscillating behavior of the iterates x_n about the root α in the last example. For x_n, x_{n-1} near to α, (2.23) implies

$$\alpha - x_{n+1} \doteq -(\alpha - x_n)(\alpha - x_{n-1})\frac{f''(\alpha)}{2f'(\alpha)}$$

The sign of $\alpha - x_{n+1}$ is determined from that of the previous two errors, together with the sign of $f''(\alpha)/f'(\alpha)$.

The condition (2.24) gives some information on how close the initial values x_0 and x_1 should be to α in order to have convergence. If the quantity M is large, or more specifically, if

$$\left|\frac{f''(\alpha)}{2f'(\alpha)}\right|$$

is very large, then $\alpha - x_0$ and $\alpha - x_1$ must be correspondingly smaller. Convergence can occur without (2.24), but it is likely to be initially quite haphazard in such a case.

There are a variety of difficulties that can occur with the Secant method. The method may diverge if the initial guesses are not well chosen; and the fraction

$$\frac{x_n - x_{n-1}}{f(x_n) - f(x_{n-1})}$$

will be increasingly inaccurate as $x_n \to \alpha$ because of loss of significance errors in the evaluation of the numerator and denominator. Since the iterates do not necessarily enclose the root, there is also difficulty in knowing when to stop the iteration. This last point will be returned to later in a more general setting with other methods; see Section 2.6.

In spite of these difficulties, the Secant method is well recommended as an efficient procedure for a large variety of problems. And it forms an essential part of the algorithm in Section 2.8, which is guaranteed to converge.

2.3   Newton's Method

As with the two previous methods, we approximate the graph of $y = f(x)$ in a vicinity of the root α by a straight line, but now we use a tangent line rather than a secant line. Use the root of the straight line as an approximation to the root α. Given an initial guess $x_0 \approx \alpha$, form the line tangent to the graph of $y = f(x)$ at the point $(x_0, f(x_0))$. As with the example beginning this chapter, we obtain the formula

$$x_{n+1} = x_n - \frac{f(x_n)}{f'(x_n)} \qquad n \geq 0 \qquad (2.30)$$

Note that the secant method is related to (2.30) by using

$$f'(x_n) \approx \frac{f(x_n) - f(x_{n-1})}{x_n - x_{n-1}} \qquad n \geq 1$$

Newton's method (Figure 2.4) is the best known procedure for finding the roots of an equation. It has been generalized in many ways for the solution of other more difficult nonlinear problems, for example nonlinear systems and nonlinear integral and differential equations. It is not always the best method for a given problem, but its formal simplicity and its great speed often lead it to be the first method that people use in attempting to solve a nonlinear problem.

An another approach to (2.30), we will use a Taylor series development. Expanding $f(x)$ about x_n,

$$f(x) = f(x_n) + (x - x_n)f'(x_n) + \frac{(x - x_n)^2}{2} f''(\xi)$$

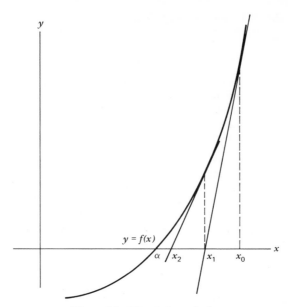

Figure 2.4 Newton's method.

with ξ between x and x_n. Letting $x = \alpha$ and then solving for α,

$$\alpha = x_n - \frac{f(x_n)}{f'(x_n)} - \frac{(\alpha - x_n)^2}{2} \cdot \frac{f''(\xi_n)}{f'(x_n)}$$

with ξ_n between x_n and α. We can drop the error term to obtain a better approximation to α than x_n, and we recognize this approximation as x_{n+1} from (2.30). Then

$$\alpha - x_{n+1} = -(\alpha - x_n)^2 \frac{f''(\xi_n)}{2f'(x_n)} \qquad n \geqq 0 \qquad (2.31)$$

Note the similarity to the secant error formula (2.23). The formula (2.31) will be used to show that Newton's method has a quadratic order of convergence, $p = 2$.

Example We again solve for the largest root of

$$f(x) \equiv x^6 - x - 1 = 0$$

Newton's method (2.30) is used, and the results are shown in Table 2.4. After examining these results, the reader should also compare them with the earlier results for this equation obtained using the Bisection method, the Regula Falsi method, and the secant method.

Table 2.4 Example of Newton's method

n	x_n	$f(x_n)$	$\alpha - x_n$
0	2.0	61.0	$-8.652E\text{-}1$
1	1.680628273	19.85	$-5.459E\text{-}1$
2	1.430738989	6.147	$-2.960E\text{-}1$
3	1.254970957	1.652	$-1.202E\text{-}1$
4	1.161538433	$2.943E\text{-}1$	$-2.681E\text{-}2$
5	1.136353274	$1.683E\text{-}2$	$-1.629E\text{-}3$
6	1.134730528	$6.574E\text{-}5$	$-6.390E\text{-}6$
7	1.134724138	$-4.0E\text{-}9$	0.0

The Newton method converges very rapidly once an iterate is fairly close to the root. This is illustrated in iterates x_4, x_5, x_6, x_7. The iterates x_0, x_1, x_2, x_3 show the slow initial convergence that is possible with a poor initial guess x_0. If the initial guess $x_0 = 1$ had been chosen, then x_4 would have been accurate to 7 significant digits and x_5 to 10 digits.

A convergence result will be given, showing the speed of convergence and also an interval from which the initial guesses can be chosen. The theorem is similar to the earlier Theorem 2.1 for the secant method.

Theorem 2.2 Assume $f(x), f'(x)$, and $f''(x)$ are continuous for all x in some neighborhood of α, and assume $f(\alpha) = 0$, $f'(\alpha) \neq 0$. Then if x_0 is chosen sufficiently close to α, the iterates $x_n, n \geq 0$, of (2.30) will converge to α. Moreover,

$$\underset{n \to \infty}{\text{Limit}} \frac{\alpha - x_{n+1}}{(\alpha - x_n)^2} = -\frac{f''(\alpha)}{2f'(\alpha)} \tag{2.32}$$

proving that the iterates have an order of convergence $p = 2$.

Proof Pick an interval $I = [\alpha - \epsilon, \alpha + \epsilon]$, and let

$$M = \frac{\underset{x \in I}{\text{Max}} |f''(x)|}{2 \underset{x \in I}{\text{Min}} |f'(x)|}.$$

Using (2.31),

$$|\alpha - x_1| \leq M |\alpha - x_0|^2$$

$$M|\alpha - x_1| \leq (M|\alpha - x_0|)^2$$

Pick $|\alpha - x_0| \leq \epsilon$ and $M|\alpha - x_0| < 1$. Then $M|\alpha - x_1| < 1$ and $M|\alpha - x_1| \leq$

$M|\alpha - x_0|$, which says $|\alpha - x_1| \leq \epsilon$. We can apply the same argument to $x_1, x_2, \ldots,$ inductively, showing that $|\alpha - x_n| \leq \epsilon$ and $M|\alpha - x_n| < 1$ for all $n \geq 1$.

To show convergence, use (2.31) to give

$$|\alpha - x_{n+1}| \leq M|\alpha - x_n|^2$$

$$M|\alpha - x_{n+1}| \leq (M|\alpha - x_n|)^2 \qquad (2.33)$$

and inductively,

$$M|\alpha - x_n| \leq (M|\alpha - x_0|)^{2^n}$$

$$|\alpha - x_n| \leq \frac{1}{M}(M|\alpha - x_0|)^{2^n}$$

Since $M|\alpha - x_0| < 1$, this shows that $x_n \to \alpha$ as $n \to \infty$.

In formula (2.31), the unknown point ξ_n is between x_n and α, and thus $\xi_n \to \alpha$ as $n \to \infty$. Thus

$$\operatorname*{Limit}_{n \to \infty} \frac{\alpha - x_{n+1}}{(\alpha - x_n)^2} = -\operatorname*{Limit}_{n \to \infty} \frac{f''(\xi_n)}{2f'(x_n)} = \frac{-f''(\alpha)}{2f'(\alpha)} \qquad \blacksquare$$

The error column in Table 2.4 can be used to illustrate (2.32). In particular for that example,

$$-\frac{f''(\alpha)}{2f'(\alpha)} = -2.417 \qquad \frac{\alpha - x_6}{(\alpha - x_5)^2} = -2.41$$

Let M denote the limit on the right side of (2.32). Then if x_n is near α, (2.32) implies

$$M(\alpha - x_{n+1}) \approx [M(\alpha - x_n)]^2$$

In order to have convergence of x_n to α, this statement says that we should probably have

$$|\alpha - x_0| < \frac{1}{M}$$

Thus M is a measure of how close x_0 must be chosen to α to insure convergence to α. Some examples with large values of M are given in the problems at the end of the chapter.

The following is a simple but general algorithm for Newton's method.

Algorithm   Newton $(f, df, x_0, \epsilon, \text{root}, \text{itmax}, \text{ier})$

1. **Remark:** *df* is the derivative function $f'(x)$, itmax is the maximum number of iterates to be computed, and ier is an error flag to the user.

2. itnum: $= 1$

3. denom: $= df(x_0)$.

4. If denom $= 0$, then ier: $= 2$ and exit.

5. $x_1 := x_0 - f(x_0)/\text{denom}$

6. If $|x_1 - x_0| \leq \epsilon$, then set ier: $= 0$, root: $= x_1$, and exit.

7. If itnum $=$ itmax, set ier: $= 1$, root: $= x_1$, and exit.

8. Otherwise, itnum: $=$ itnum $+ 1$, $x_0: = x_1$, and go to step 3.

The error test in step 6 is discussed in Section 2.6, and it is shown that $x_1 - x_0 \approx \alpha - x_0$. Thus the root returned by Newton will have greater accuracy than requested. A relative error test could be included by using

$$\frac{x_1 - x_0}{x_1} \approx \frac{\alpha - x_0}{\alpha}$$

as an approximation to the relative error in x_0.

Comparison of Newton's method and the Secant method   We will compare the computational efficiency of Newton's method and the secant method. At a first glance, it might seem obvious that Newton's method is faster. But the secant method needs only one function evaluation per step, provided the previous value of the function is not recomputed unnecessarily; and the Newton method requires two evaluations per step. Thus Newton's method will generally require fewer iterates to obtain a desired accuracy, but it will require more computation per iterate.

We consider the expenditure of time necessary to reach a desired root α within a desired tolerance of ϵ. To simplify the analysis, we assume the initial guesses are quite close to the desired root. Define

$$x_{n+1} = x_n - \frac{f(x_n)}{f'(x_n)} \qquad n \geq 0$$

$$\bar{x}_{n+1} = \bar{x}_n - f(\bar{x}_n) \frac{\bar{x}_n - \bar{x}_{n-1}}{f(\bar{x}_n) - f(\bar{x}_{n-1})} \qquad n \geq 1$$

and let $x_0 = \bar{x}_0$; we will define $\bar{x}_1$ based on the following convergence formula. From (2.32) and (2.29), respectively,

$$|\alpha - x_{n+1}| \approx c|\alpha - x_n|^2 \qquad n \geq 0 \qquad c = \left| \frac{f''(\alpha)}{2f'(\alpha)} \right|$$

$$|\alpha - \bar{x}_{n+1}| \approx d|\alpha - \bar{x}_n|^r \qquad r = \frac{1 + \sqrt{5}}{2} \qquad d = c^{r-1}$$

Inductively for the error in the Newton iterates,

$$c|\alpha - x_{n+1}| \approx (c|\alpha - x_n|)^2$$

$$c|\alpha - x_n| \approx (c|\alpha - x_0|)^{2^n}$$

$$|\alpha - x_n| \approx \frac{1}{c}(c|\alpha - x_0|)^{2^n} \qquad n \geq 0$$

Similarly for the secant method iterates,

$$|\alpha - \bar{x}_n| \approx d|\alpha - \bar{x}_{n-1}|^r$$

$$\approx d^{1 + r + \dots + r^{n-1}}|\alpha - \bar{x}_0|^{r^n}$$

Using the finite geometric series

$$1 + r + \dots + r^{n-1} = \frac{r^n - 1}{r - 1}$$

we obtain

$$d^{1 + r + \dots + r^{n-1}} = d^{\frac{r^n - 1}{r - 1}} = c^{r^n - 1}$$

and thus

$$|\alpha - \bar{x}_n| \approx c^{r^n - 1}|\alpha - \bar{x}_0|^{r^n} = \frac{1}{c}[c|\alpha - x_0|]^{r^n}$$

To satisfy $|\alpha - x_n| \leq \epsilon$, for the Newton iterates, we must have

$$(c|\alpha - x_0|)^{2^n} \leq c\epsilon$$

$$n \geq \frac{K}{\log 2}, \qquad K = \log\left(\frac{\log \epsilon c}{\log c|\alpha - x_0|} \right)$$

Let m be the time to evaluate $f(x)$, and let $s \cdot m$ be the time to evaluate $f'(x)$.

Then the minimum time to obtain the desired accuracy with Newton's method is

$$T_N = (m + ms)n = \frac{(1+s)mK}{\log 2} \tag{2.34}$$

For the secant method, a similar calculation shows that $|\alpha - \bar{x}_n| \leq \epsilon$ if

$$n \geq \frac{K}{\log r}$$

Thus the minimum time necessary to obtain the desired accuracy is

$$T_S = mn = \frac{mK}{\log r} \tag{2.35}$$

To compare the times for the secant method and Newton's method, we have

$$\frac{T_S}{T_N} = \frac{\log 2}{(1+s)\log r}$$

The secant method is faster than the Newton method if the ratio is less than one,

$$\frac{T_S}{T_N} < 1$$

$$s > \frac{\log 2}{\log r} - 1 \doteq .44 \tag{2.36}$$

If the time to evaluate $f'(x)$ is more than 44 percent that which is necessary to evaluate $f(x)$, then the secant method is more efficient.

The above argument is useful in illustrating that the mathematical speed of convergence is not the complete picture. Total computing time, ease of use of an algorithm, stability, and other factors also bear on the relative desirability of one algorithm over another one.

2.4   A General Theory for One-Point Iteration Methods

We consider solving an equation $x = g(x)$ for a root α by defining

$$x_{n+1} = g(x_n) \qquad n \geq 0 \tag{2.37}$$

with x_0 an initial guess to α. The Newton method fits in this pattern with

$$g(x) \equiv x - \frac{f(x)}{f'(x)}$$

and the Regula Falsi method generally fits the pattern since one of the endpoints becomes fixed, as was pointed out previously in Section 2.1.

The solutions of $x = g(x)$ are called fixed points of g. We are usually interested in solutions of an equation $f(x) = 0$, and there are many ways this can be reformulated as a fixed point problem. At this point, we will just illustrate this with an example.

Example   Consider $x^2 - a = 0$ for $a > 0$.

1. $x = x^2 + x - a$, or more generally, $x = x + c(x^2 - a)$ for some $c \neq 0$

2. $x = a/x$

3. $x = \frac{1}{2}(x + \frac{a}{x})$

We give a numerical example with $a = 3$, $x_0 = 2$, and $\alpha = \sqrt{3} = 1.732051$. With $x_0 = 2$, the numerical results for (2.37) in these cases are given in Table 2.5.

Table 2.5   Iteration examples for $x^2 - 3 = 0$

	Case(1)	Case(2)	Case(3)
n	x_n	x_n	x_n
0	2.0	2.0	2.0
1	3.0	1.5	1.75
2	9.0	2.0	1.732143
3	87.0	1.5	1.732051

It is natural to ask what makes the various iterative schemes behave in the way they do in this example. We will develop a general theory to explain this behavior and to aid in analyzing new iterative methods.

Lemma 1   Let $g(x)$ be continuous on the interval $a \leq x \leq b$, and assume that $a \leq g(x) \leq b$ for every $a \leq x \leq b$. (We say g sends $[a,b]$ into $[a,b]$, and denote it by $g([a,b]) \subset [a,b]$.) Then $x = g(x)$ has at least one solution in $[a,b]$.

Proof   Consider the continuous function $g(x) - x$. At $x = a$ it is positive and at $x = b$ it is negative. Thus by the intermediate value theorem it must have a root in the interval $[a,b]$. In Figure 2.5, the roots are the intersection points of $y = x$ and $y = g(x)$. ∎

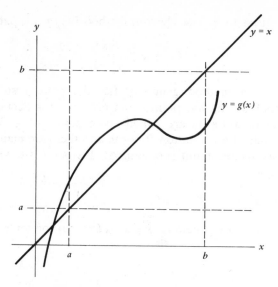

Figure 2.5 Example of Lemma 1.

Lemma 2 Let $g(x)$ be continuous on $[a,b]$, and assume $g([a,b])\subset[a,b]$. Furthermore, assume there is a constant $0<\lambda<1$ with

$$|g(x)-g(y)|\leq\lambda|x-y| \qquad \text{for all } x,y\in[a,b] \qquad (2.38)$$

Then $x=g(x)$ has a unique solution α in $[a,b]$. Also, the iterates

$$x_n=g(x_{n-1}) \qquad n\geq0$$

will converge to α for any choice of x_0 in $[a,b]$.

Proof Suppose $x-g(x)$ has two solutions α and β in $[a,b]$. Then

$$|\alpha-\beta|=|g(\alpha)-g(\beta)|\leq\lambda|\alpha-\beta|$$

$$(1-\lambda)|\alpha-\beta|\leq0$$

Since $0<\lambda<1$, this implies $\alpha=\beta$. Also we know by the earlier lemma that there is at least one root α in $[a,b]$.

To examine the convergence of the iterates, first note that they all remain in $[a,b]$. The result

$$x_n\in[a,b] \text{ implies } x_{n+1}=g(x_n)\in[a,b]$$

can be used with mathematical induction to prove $x_n\in[a,b]$ for all n. For the convergence,

$$|\alpha - x_{n+1}| = |g(\alpha) - g(x_n)| \leq \lambda|\alpha - x_n|$$

and by induction

$$|\alpha - x_n| \leq \lambda^n|\alpha - x_0| \qquad n \geq 0 \tag{2.39}$$

As $n \to \infty$, $\lambda^n \to 0$ and thus $x_n \to \alpha$. ∎

Note that if $g(x)$ is differentiable on $[a,b]$, then

$$g(x) - g(y) = g'(\xi)(x - y), \qquad \xi \text{ between } x \text{ and } y$$

for all $x, y \in [a,b]$. Define

$$\lambda = \underset{a \leq x \leq b}{\text{Maximum}} |g'(x)|$$

Then

$$|g(x) - g(y)| \leq \lambda|x - y| \qquad \text{all } x, y \in [a,b]$$

Theorem **2.3** Assume $g(x)$ is continuously differentiable on $[a,b]$, that $g([a,b]) \subset [a,b]$, and that

$$\lambda = \underset{a \leq x \leq b}{\text{Maximum}} |g'(x)| < 1. \tag{2.40}$$

Then (1) $x = g(x)$ has a unique solution α in $[a,b]$;
(2) for any choice of x_0 in $[a,b]$, with $x_{n+1} = g(x_n)$, $n \geq 0$,

$$\underset{n \to \infty}{\text{Limit}} \, x_n = \alpha$$

(3)
$$|\alpha - x_n| \leq \lambda^n|\alpha - x_0|$$

$$\underset{n \to \infty}{\text{Limit}} \, \frac{\alpha - x_{n+1}}{\alpha - x_n} = g'(\alpha) \tag{2.41}$$

Proof Every result comes from the preceding lemmas, except for the rate of convergence (2.41). For it

$$\alpha - x_{n+1} = g(\alpha) - g(x_n) = g'(\xi_n)(\alpha - x_n) \qquad n \geq 0$$

with ξ_n an unknown point between α and x_n. Since $x_n \to \alpha$, we must have $\xi_n \to \alpha$, and thus

$$\underset{n \to \infty}{\text{Limit}} \, \frac{\alpha - x_{n+1}}{\alpha - x_n} = \text{Limit} \, g'(\xi_n) = g'(\alpha) \qquad ∎$$

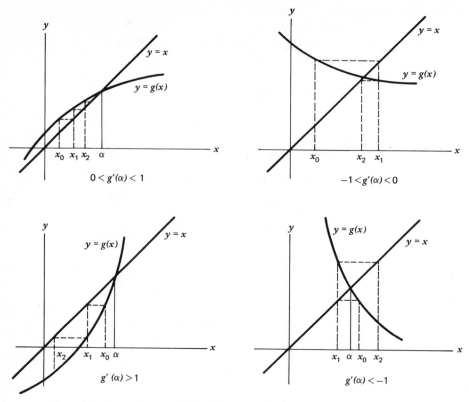

Figure 2.6 Examples of convergent and nonconvergent sequences $x_{n+1} = g(x_n)$.

This theorem generalizes to systems of m nonlinear equations in m unknowns. Just regard x as an element of R^m, $g(x)$ as a function from R^m to R^m, replace the absolute values by vector norms, and replace $g'(x)$ by the Jacobian matrix for $g(x)$. The assumption $g([a,b]) \subset [a,b]$ must be replaced with a stronger assumption, and care must be exercised in the choice of a region generalizing $[a,b]$. The lemmas generalize, but they are nontrivial to prove.

To see the importance of the assumption (2.40) on the size of $g'(x)$, suppose $|g'(\alpha)| > 1$. Then if we had a sequence of iterates $x_{n+1} = g(x_n)$ and a root $\alpha = g(\alpha)$,

$$\alpha - x_{n+1} = g(\alpha) - g(x_n) = g'(\xi_n)(\alpha - x_n).$$

If x_n becomes sufficiently close to α, then $|g'(\xi_n)| > 1$ and the error $|\alpha - x_{n+1}|$ will be greater than $|\alpha - x_n|$. Thus convergence is not possible if $|g'(\alpha)| > 1$. We graphically portray the computation of the iterates in four cases; see Figure 2.6.

To simplify the application of the previous theorem, we give the following result.

Theorem 2.4 Assume α is a solution of $x=g(x)$, and suppose that $g(x)$ is continuously differentiable in some neighboring interval about α with $|g'(\alpha)|<1$. Then the results of Theorem 2.3 are still true, provided x_0 is chosen sufficiently close to α.

Proof Pick an interval $I=[\alpha-\epsilon, \alpha+\epsilon]$ with

$$\underset{x \in I}{\text{Max}} |g'(x)| \leq \lambda < 1$$

Then $g(I) \subset I$, since $|\alpha-x| \leq \epsilon$ implies

$$|\alpha-g(x)|=|g(\alpha)-g(x)|=|g'(\xi)||\alpha-x| \leq \lambda|\alpha-x| \leq \epsilon$$

Now apply the preceding theorem using $[a,b]=[\alpha-\epsilon, \alpha+\epsilon]$. ∎

Example Referring back to the earlier example in this section, calculate $g'(\alpha)$.

1. $g(x)=x^2+x-3$, $g'(\alpha)=g'(\sqrt{3})=2\sqrt{3}+1>1$
2. $g(x)=3/x$, $g'(\sqrt{3})=-3/(\sqrt{3})^2=-1$
3. $g(x)=\frac{1}{2}(x+\frac{3}{x})$, $g'(x)=\frac{1}{2}(1-\frac{3}{x^2})$, $g'(\sqrt{3})=0$

Example For $x=x+c(x^2-3)$, pick c to insure convergence. Since the solution is $\alpha=\sqrt{3}$, and since $g'(x)=1+2cx$, pick c so that

$$-1<1+2c\sqrt{3}<1$$

and for a good rate of convergence, pick c so that

$$1+2c\sqrt{3} \approx 0$$

This gives

$$c \approx \frac{-1}{2\sqrt{3}}$$

Use $c=-\frac{1}{4}$. Then $g'(\sqrt{3})=1-(\sqrt{3}/2) \doteq .134$. This gives the iteration scheme

$$x_{n+1}=x_n-\tfrac{1}{4}(x_n^2-3) \qquad n \geq 0 \tag{2.42}$$

The numerical results are given in Table 2.6.
  The ratio column gives the values of

$$\frac{\alpha-x_n}{\alpha-x_{n-1}} \qquad n \geq 1$$

The results agree closely with the theoretical value of $g'(\sqrt{3})$.

Table 2.6 Numerical example of iteration (2.42)

n	x_n	$\alpha - x_n$	Ratio
0	2.0	$-2.68\text{E}-1$	
1	1.75	$-1.79\text{E}-2$	.0668
2	1.7343750	$-2.32\text{E}-3$	.130
3	1.7323608	$-3.10\text{E}-4$	.134
4	1.7320923	$-4.15\text{E}-5$	.134
5	1.7320564	$-5.56\text{E}-6$	.134
6	1.7320516	$-7.45\text{E}-7$	.134
7	1.7320509	$-1.00\text{E}-7$	.134

Higher-order one-point methods We complete the development of the theory for one-point iteration methods by considering methods with an order of convergence greater than one, for example Newton's method.

Theorem 2.5 Assume that α is a root of $x = g(x)$, and that $g(x)$ is p times continuously differentiable for all x near to α, for some $p \geq 2$. Furthermore, assume

$$g'(\alpha) = \ldots = g^{(p-1)}(\alpha) = 0 \qquad (2.43)$$

Then if the initial guess x_0 is chosen sufficiently close to α, the iteration

$$x_{n+1} = g(x_n) \qquad n \geq 0$$

will have order of convergence p, and

$$\operatorname*{Limit}_{n \to \infty} \frac{\alpha - x_{n+1}}{(\alpha - x_n)^p} = (-1)^{p-1} \frac{g^{(p)}(\alpha)}{p!} \qquad (2.44)$$

Proof Expand $g(x_n)$ about α

$$x_{n+1} = g(x_n) = g(\alpha) + (x_n - \alpha) g'(\alpha) +$$

$$\ldots + \frac{(x_n - \alpha)^{p-1}}{(p-1)!} g^{(p-1)}(\alpha) + \frac{(x_n - \alpha)^p}{p!} g^{(p)}(\xi_n)$$

for some ξ_n between x_n and α. Using (2.43) and $\alpha = g(\alpha)$,

$$\alpha - x_{n+1} = -\frac{(x_n - \alpha)^p}{p!} g^{(p)}(\xi_n)$$

Use Theorem 2.4 and $x_n \to \alpha$ to complete the proof. ∎

The Newton method can be analyzed by this result.

$$g(x) = x - \frac{f(x)}{f'(x)} \qquad g'(x) = \frac{f(x)f''(x)}{[f'(x)]^2}$$

$$g'(\alpha) = 0 \qquad g''(\alpha) = \frac{f''(\alpha)}{f'(\alpha)}$$

This and (2.44) give the previously obtained convergence result (2.32) for Newton's method. For another one-point method of order two, see Problem 17 describing Steffenson's method. It has the advantage of not needing the derivative $f'(x)$, although there are still two function evaluations per iterate.

2.5 Aitken Extrapolation for Linearly Convergent Sequences

From (2.41)

$$\frac{\alpha - x_{n+1}}{\alpha - x_n} \approx \lambda = \text{constant} \qquad n \geq 0 \tag{2.45}$$

with the approximation improving as n increases. Using (2.45),

$$\frac{\alpha - x_{n+1}}{\alpha - x_n} \approx \frac{\alpha - x_{n+2}}{\alpha - x_{n+1}} \qquad n \geq 0$$

$$(\alpha - x_{n+1})^2 \approx (\alpha - x_n)(\alpha - x_{n+2})$$

$$\alpha \approx a_{n+2} \equiv x_{n+2} - \frac{(x_{n+2} - x_{n+1})^2}{(x_n - x_{n+1}) - (x_{n+1} - x_{n+2})}. \tag{2.46}$$

This is Aitken's extrapolation formula. If (2.45) is well satisfied, then the extrapolate a_{n+2} is a much improved estimate over x_{n+2} of the root α.

The ratio λ in (2.45) can be computed by combining (2.46) and (2.45). But more directly,

$$\frac{x_{n+2} - x_{n+1}}{x_{n+1} - x_n} = \frac{(\alpha - x_{n+1}) - (\alpha - x_{n+2})}{(\alpha - x_n) - (\alpha - x_{n+1})} = \frac{1 - \dfrac{\alpha - x_{n+2}}{\alpha - x_{n+1}}}{\dfrac{\alpha - x_n}{\alpha - x_{n+1}} - 1} \approx \frac{1 - \lambda}{\dfrac{1}{\lambda} - 1} = \lambda$$

$$\frac{x_{n+2} - x_{n+1}}{x_{n+1} - x_n} \approx \lambda \tag{2.47}$$

The accuracy of (2.46) and (2.47) is not dependent on the size of λ, but instead upon the accuracy of the statement (2.45) for the values of x_n being used.

Example   Recall the fixed point iteration (2.42),

$$x_{n+1} = x_n - \tfrac{1}{4}(x_n^2 - 3) \qquad n \geq 0$$

and the resulting iterates given in Table 2.6. In Table 2.7 we give the same iterates, but now we include the successive differences $x_n - x_{n-1}$ and their successive ratios. The true root is $\alpha = \sqrt{3}$.

Table 2.7   Example of (2.47)

n	x_n	$x_n - x_{n-1}$	Ratio
0	2.0		
1	1.75	$-2.50E-1$	
2	1.7343750	$-1.56E-2$	.0625
3	1.7323608	$-2.01E-3$	.129
4	1.7320923	$-2.68E-4$	.133
5	1.7320564	$-3.60E-5$	.134
6	1.7320516	$-4.82E-6$	.134
7	1.7320509	$-6.45E-7$	.134

The values in the ratio column illustrate the result (2.47), since $\lambda = 1 - (\sqrt{3}/2) \doteq .134$ for this iteration. To illustrate the value of the extrapolation formula (2.46), use the iterates x_3, x_4, and x_5 for extrapolation. Then we obtain

$$a_5 = x_5 - \frac{(x_5 - x_4)^2}{(x_3 - x_4) - (x_4 - x_5)} = 1.73205086$$

The error in a_5 is

$$\alpha - a_5 = -5.10E - 8$$

Compare this with the error in x_5,

$$\alpha - x_5 = -5.56E - 6$$

There has been an improvement in the accuracy by a factor of about 100, and only a few simple arithmetic operations were used.

We now combine linear iteration and Aitken extrapolation in a simple algorithm.

Algorithm   Aitken $(g, x_0, \epsilon, \text{root})$

1.   Remark: It is assumed that $|g'(\alpha)| < 1$ and that ordinary linear iteration using x_0 will converge to α.

2. $x_1 := g(x_0)$, $x_2 := g(x_1)$

3. $a := x_2 - \dfrac{(x_2 - x_1)^2}{(x_2 - x_1) - (x_1 - x_0)}$

4. If $|a - x_2| \le \epsilon$, then root: $= a$ and exit.

5. Set $x_0 := a$ and go to step 2.

This routine will usually converge provided the assumptions of step 1 are satisfied. A somewhat more efficient algorithm can be produced using the error estimates of Section 2.6 to estimate the error in x_1 and x_2 in the algorithm, thus possibly stopping with an earlier iterate.

Example To illustrate algorithm Aitken, we repeat the previous example (2.42),

$$x_{n+1} = x_n - \tfrac{1}{4}(x_n^2 - 3) \qquad n \ge 0, \, x_0 = 2$$

The numerical results are given in Table 2.8.

Table 2.8 Example of algorithm Aitken

n	x_n	$\alpha - x_n$	**Ratio**
0	2.0	$-2.68\text{E}-1$	
1	1.75	$-1.79\text{E}-2$	.067
2	1.7343750	$-2.32\text{E}-3$	.129
3	1.7333333	$-1.28\text{E}-3$	.552
4	1.7322222	$-1.71\text{E}-4$	.134
5	1.7320738	$-2.30\text{E}-5$	.134
6	1.732050924	$-1.16\text{E}-7$	.0050

The Aitken iterates are x_3 and x_6. Note that x_6 is more of an improvement on x_5 than is x_3 compared to x_2. This is because the assumption (2.45) is satisfied better when computing x_6 than with x_3. For x_3, there is some decrease in computation due to use of extrapolation, but it is not a significant savings.

If the value of λ is near to ± 1, then the speed of convergence will be quite slow. But the estimate (2.45) may still be valid, and then (2.46) can result in a great improvement in the speed of convergence. To illustrate this point, consider the following.

Example Solve $x = 1.6 + .99 \cos(x)$ using the linear iteration

$$x_{n+1} = 1.6 + .99 \cos(x_n) \qquad n = 0, 1, 2, \ldots,$$

and $x_0 = \pi/2$, without and with Aitken extrapolation. The correct solution is

Table 2.9   Example of linear iteration

	Without extrapolation			With extrapolation	
n	x_n	$x_n - x_{n-1}$	n	x_n	$x_n - x_{n-1}$
0	1.57080		0	1.57080000	
1	1.60000	.0292	1	1.59999636	.0292
2	1.57109	−.0289	2	1.57109607	−.0289
3	1.59971	.0286	3	1.58547286	.0144
4	1.57138	−.0283	4	1.58547075	−.00000211
5	1.59942	.0280	5	1.58547284	.00000209
6	1.57167	−.0278	6	1.58547180	−.00000104

$\alpha = 1.585471802$. The results are given in Table 2.9, and the results on the left side show that the linear iteration is converging very slowly. The use of Aitken extrapolation greatly accelerates the convergence.

Extrapolation is often used with slowly convergent linear iteration methods for solving large systems of simultaneous linear equations. The actual methods are different; but they also are based on the general idea of finding the qualitative behavior of the error, as in (2.45), and of then using that to produce an improved estimate of the answer. This idea is also pursued in developing numerical methods for integration, differential equations, and other mathematical problems.

2.6   Error Tests

In using an iterative method for solving $f(x) = 0$, a decision must be made as to when to stop the iteration. This depends on two quantities: (1) the accuracy desired by the user, generally specified by giving an error tolerance ϵ to the algorithm and (2) the best accuracy that can be expected in the root based on the accuracy with which $f(x)$ can be computed and the multiplicity of the root. This second quantity is investigated in the next section on multiple roots, as it is a significant consideration for such problems.

A commonly used error test is to check whether

$$|f(x_n)| \leq \epsilon \tag{2.48}$$

But used alone, this is generally not a good test for checking the accuracy of x_n compared with α. To understand why, suppose this test is satisfied. Using the mean value theorem 1.2,

$$f(x_n) = f(x_n) - f(\alpha) = f'(\xi_n)(x_n - \alpha)$$

$$\alpha - x_n = -\frac{f(x_n)}{f'(\xi_n)} \tag{2.49}$$

for some ξ_n between x_n and α. Thus $|\alpha - x_n| \approx |f(x_n)|$ if $|f'(\alpha)| \approx 1$, and the approximate root x_n will be sufficiently accurate. But if $|f'(\alpha)| \ll 1$, then $|\alpha - x_n|$ is likely to be much larger than the desired accuracy ϵ; and if $|f'(\alpha)| \gg 1$, then $|\alpha - x_n|$ is much smaller than desired, causing unnecessary computation. The two latter cases are illustrated in Figure 2.7.

For Newton's method, and using (2.49),

$$x_{n+1} - x_n = -\frac{f(x_n)}{f'(x_n)} \approx -\frac{f(x_n)}{f'(\xi_n)} = \alpha - x_n$$

Thus

$$|x_{n+1} - x_n| \leqq \epsilon \qquad \text{implies} \qquad |\alpha - x_n| \leqq \epsilon \qquad (2.50)$$

provided x_n is close enough to α to make valid the approximation $f'(x_n) \approx f'(\xi_n)$.

A second natural test to use is to terminate the iteration when

$$|x_{n+1} - x_n| \leqq \epsilon$$

This is a valid test for Newton's method, as shown in (2.50). But it may be inaccurate for many linearly convergent methods. For example, consider a linearly convergent one-point method $x_{n+1} = g(x_n)$, and suppose $g'(\alpha) \approx 1$. For iterates x_n that are sufficiently close to α,

$$\alpha - x_{n+1} = g(\alpha) - g(x_n) = g'(\xi_n)(\alpha - x_n) \approx g'(\alpha)(\alpha - x_n)$$

$$x_{n+1} - x_n = (\alpha - x_n) - (\alpha - x_{n+1}) \approx [1 - g'(\alpha)](\alpha - x_n)$$

$$\alpha - x_n \approx \frac{1}{1 - g'(\alpha)}(x_{n+1} - x_n) \qquad (2.51)$$

Thus $|\alpha - x_n| \gg |x_{n+1} - x_n|$ if $g'(\alpha) \approx 1$.

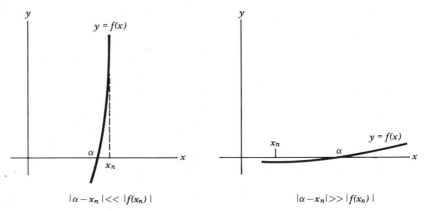

Figure 2.7 Comparisons of $\alpha - x_n$ and $f(x_n)$.

If a value of $\lambda \approx g'(\alpha)$ is known, say from (2.47), then (2.51) can be used to estimate the error in x_n and also the error in the more accurate iterate x_{n+1}:

$$\alpha - x_{n+1} \approx \frac{g'(\alpha)}{1 - g'(\alpha)}(x_{n+1} - x_n) \tag{2.52}$$

Example   Use the last example of Section 2.5,

$$x_{n+1} = 1.6 + .99\cos(x_n)$$

without Aitken extrapolation. Using (2.47) and the results in the left half of Table 2.9, we obtain

$$g'(\alpha) \doteq -.98966$$

Formula (2.52) gives

$$\alpha - x_6 \doteq .0138$$

which is the true error. This is just an error estimate based implicitly on using the Aitken extrapolation a_6 to approximate α, and then using $\alpha - x_6 \approx a_6 - x_6$.

If we assume only the weaker result

$$|\alpha - x_{n+1}| \leq \lambda |\alpha - x_n| \tag{2.53}$$

for all x_n of interest, we can get a result similar to (2.52).

$$|x_{n+1} - x_n| = |(\alpha - x_n) - (\alpha - x_{n+1})| \geq |\alpha - x_n| - |\alpha - x_{n+1}|$$

$$\geq (1 - \lambda)|\alpha - x_n|$$

$$|\alpha - x_n| \leq \frac{1}{1 - \lambda}|x_{n+1} - x_n|$$

$$|\alpha - x_{n+1}| \leq \frac{\lambda}{1 - \lambda}|x_{n+1} - x_n| \tag{2.54}$$

In order to have $|\alpha - x_{n+1}| \leq \epsilon$, choose x_{n+1} to satisfy

$$|x_{n+1} - x_n| \leq \left(\frac{1 - \lambda}{\lambda}\right)\epsilon \tag{2.55}$$

This is a useful error test for linearly convergent methods, and it generalizes to iteration methods for systems of equations. The value of λ is determined empirically, using (2.47) or a related formula.

In writing a general purpose algorithm, it is best to include several stopping criteria. First, stop if $|f(x_n)|$ becomes sufficiently small. Second, stop if $|\alpha - x_n|$ is sufficiently small; the algorithm will relate this to some test involving $|x_{n+1} - x_n|$. And third, have the user specify a maximum number of iterates to be computed, to avoid an infinite loop if that is possible. The algorithm should contain a mechanism for comparing the desired error against the smallest error that is possible based on the computer arithmetic being used. A number of such automatic algorithms have now been written for various methods, and they are becoming widely distributed. For example, see R. Brent's method in Section 2.8.

2.7   The Numerical Evaluation of Multiple Roots

We say that the function $f(x)$ has a root α of multiplicity $p > 1$ if

$$f(x) = (x - \alpha)^p h(x) \tag{2.56}$$

with $h(\alpha) \neq 0$ and $h(x)$ continuous at $x = \alpha$. We will restrict p to be an integer, although some of the following is equally valid for nonintegral values. If $h(x)$ is sufficiently differentiable at $x = \alpha$, then (2.56) is equivalent to

$$f(\alpha) = f'(\alpha) = \ldots = f^{(p-1)}(\alpha) = 0 \qquad f^{(p)}(\alpha) \neq 0 \tag{2.57}$$

When finding a root of any function on a computer, there is always an interval of uncertainty about the root; and this is made worse when the root is multiple. To see this more clearly, consider evaluating the two functions $f_1(x) = x^2 - 3$ and $f_2(x) = 9 + x^2(x^2 - 6)$. $\alpha = \sqrt{3}$ has multiplicity one as a root of f_1 and multiplicity two as a root of f_2. Using four-digit arithmetic, $f_1(x) < 0$ for $x \leqq 1.731$, $f_1(1.732) = 0$, and $f_1(x) > 0$ for $x \geqq 1.733$. But $f_2(x) = 0$ for $1.726 \leqq x \leqq 1.738$, thus limiting the amount of accuracy that can be attained in finding a root of $f_2(x)$. Also, consider the following more realistic example.

Example.   Evaluate

$$p(x) = (x - 1.1)^3(x - 2.1)$$

$$= x^4 - 5.4x^3 + 10.56x^2 - 8.954x + 2.7951 \tag{2.58}$$

on an IBM 360 model computer using single-precision arithmetic. The coefficients will not enter exactly because they do not have finite binary expansions (except for the x^4 term). The polynomial $p(x)$ was evaluated in its above expanded form (2.58) and also using the nested multiplication scheme

$$p(x) = 2.7951 + x\left(-8.954 + x\left(10.56 + x\left(-5.4 + x\right)\right)\right) \tag{2.59}$$

Table 2.10   Evaluation of a polynomial

x	Form (2.58)	Form (2.59)
1.090	2.86E-6	9.54E-7
1.092	−3.81E-7	0.0
1.094	9.54E-7	0.0
1.096	−9.54E-7	0.0
1.098	9.54E-7	9.54E-7
1.100	0.0	0.0
1.101	−1.91E-6	9.54E-7
1.103	0.0	0.0
1.105	0.0	0.0
1.107	2.86E-6	9.54E-7
1.109	−2.86E-6	−9.54E-7

The numerical results are given in Table 2.10. There is considerable variation in the value of $p(x)$, and it changes as the form of evaluation is changed. There is uncertainty in evaluating any function, due to using finite precision arithmetic with its resultant rounding or truncation error. For multiple roots, this leads to considerable uncertainty as to the location of the root.

In Figure 2.8, the solid line indicates the graph of $y = f(x)$; and the dotted lines give the region of uncertainty in the evaluation of $f(x)$, which is due to rounding errors and finite digit arithmetic. The interval of uncertainty in finding the root of $f(x)$ is given by the intersection of the band about the graph of $y = f(x)$ and the x-axis. It is clearly greater with the double root than with the simple root, even though the widths of the bands about $y = f(x)$ are the same. Below we show that the earlier iterative methods will not perform well for finding multiple roots, and we derive modified versions that will perform well.

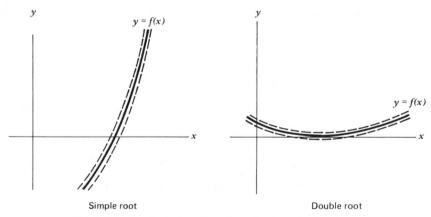

Figure 2.8   Band of uncertainty in evaluation of a function.

But the above inherent limitation on the possible accuracy of the root will remain. The only way to increase the possible accuracy is to analytically remove the multiple root, changing it into a simple root.

To see the effect of (2.56) on Newton's method, consider Newton's method as a fixed point method,

$$x_{n+1} = g(x_n), \quad g(x) = x - \frac{f(x)}{f'(x)} \qquad x \neq \alpha$$

Before calculating $g'(\alpha)$, we first simplify $g(x)$ using (2.46);

$$f'(x) = (x-\alpha)^p h'(x) + p(x-\alpha)^{p-1} h(x)$$

$$g(x) = x - \frac{(x-\alpha)h(x)}{ph(x) + (x-\alpha)h'(x)}$$

Differentiating,

$$g'(x) = 1 - \frac{h(x)}{ph(x) + (x-\alpha)h'(x)} - (x-\alpha)\frac{d}{dx}\left[\frac{h(x)}{ph(x) + (x-\alpha)h'(x)}\right]$$

and

$$g'(\alpha) = 1 - \frac{1}{p} \neq 0 \text{ for } p > 1 \tag{2.60}$$

Thus Newton's method is a linear method with rate of convergence $(p-1)/p$.

Example. Find the smallest root of $f(x) = x^3 - 5.56x^2 + 9.1389x - 4.68999$ using Newton's method. The results are shown in Table 2.11. The method is clearly converging with a linear rate of $\frac{1}{2}$, as would be predicted from (2.60) since $\alpha = 1.23$ is a double root.

Table 2.11 Example of Newton's method for a multiple root

n	x_n	$f(x_n)$	$x_n - x_{n-1}$	R_n
0	1.0	$-.111$		
1	1.10903	$-.0291$	.109	.539
2	1.16773	$-.00749$	.0587	.539
3	1.19837	$-.00190$	.0306	.513
4	1.21406	$-.000479$	.0157	.506
5	1.22199	$-.000120$	.00794	.506
6	1.22599	$-.0000302$	.00399	.501
7	1.22799	$-.00000755$	.00200	.500
8	1.22900	$-.00000189$	.00100	.500

To improve Newton's method, we would like a function $g(x)$ for which $g'(\alpha)=0$. Based on the derivation of (2.60), define

$$g(x)=x-p\frac{f(x)}{f'(x)}$$

Then easily, $g'(\alpha)=0$; and thus

$$\alpha-x_{n+1}=g(\alpha)-g(x_n)$$

$$=-g'(\alpha)(x_n-\alpha)-\tfrac{1}{2}(x_n-\alpha)^2 g''(\xi_n)$$

with ξ_n between x_n and α. Thus

$$\alpha-x_{n+1}=-\tfrac{1}{2}(\alpha-x_n)^2 g''(\xi_n)$$

showing that the method

$$x_{n+1}=x_n-p\frac{f(x_n)}{f'(x_n)}\qquad n=0,1,2,\ldots \tag{2.61}$$

has order of convergence two, the same as the original Newton method for simple roots.

Example. Apply (2.61) to the preceding example. The results are shown in Table 2.12. This example was done on a calculator using 13 digits and floating-point decimal arithmetic. The example shows the rapid convergence of (2.61). But it also shows the inherent limitation on the accuracy with which the root can be found, since the final iterate is accurate to only seven places.

Table 2.12   Example of modified Newton method (2.61)

n	x_n	$f(x_n)$	x_n-x_{n-1}
0	1.0	-1.111E-1	
1	1.219629012	-2.022E-4	2.196E-1
2	1.229971477	-1.51E-9	1.034E-2
3	1.229999787	-3.0E-11	2.83E-5
4	1.230075194	-1.056E-8	7.54E-5
5	1.230000091	0.0	-7.51E-5

If the multiplicity is unknown, define

$$K(x)=\frac{f(x)}{f'(x)} \tag{2.62}$$

Using (2.56),

$$K(x) = \frac{(x-\alpha)h(x)}{ph(x) + (x-\alpha)h'(x)} \qquad K'(\alpha) = \frac{1}{p} \neq 0$$

Thus α is a simple root of $K(x)$, and a regular iterative method can be applied to solving $K(x) = 0$. But one higher order of derivative is necessary as compared with the same method applied to $f(x) = 0$. The secant method requires $f(x)$ and $f'(x)$; and Newton's method requires $f(x)$, $f'(x)$, and $f''(x)$. The speed of convergence for solving $K(x) = 0$ will be better than with the same method applied to $f(x) = 0$. But the interval of uncertainty for the root will remain virtually unchanged, because the formula for $K(x)$ still contains the multiplicity within it computationally.

Once it is recognized that a multiple root is present, analytic differentiation can be applied to $f(x)$ to remove it. If α is a root of $f(x)$ of multiplicity p, then α is a simple root of $f^{(p-1)}(x)$.

Example. The previous example had a root of multiplicity $p = 2$. Then it is a simple root of

$$f'(x) = 3x^2 - 11.12x + 9.1389$$

Use the previously obtained root as an initial guess, $x_0 = 1.230000091$. Using Newton's method and the same 13-digit decimal calculator,

$$x_1 = 1.230000000000$$

accurate to 13 digits.

2.8 Brent's Rootfinding Algorithm

We now describe an algorithm that combines the advantages of the bisection method and the secant method, while avoiding the disadvantages of each. The algorithm is due to R. Brent and is given in Chapter 4 of his book [B1]. It is a further development of an earlier algorithm due to T. J. Dekker [D3]. The algorithm results in a small interval that will contain the root. And if the function is sufficiently smooth around the desired root ζ, then the order of convergence will be greater than linear, as with the secant method.

In describing the algorithm we use the notation of R. Brent from [B1, p. 48]. The program is entered with two values a and b for which (1) there is at least one root ζ of $f(x)$ between a and b, and (2) $f(a)f(b) \leq 0$. The program is also entered with a desired tolerance t, from which a stopping tolerance δ is produced:

$$\delta = t + 2\epsilon |b|$$

$$\epsilon = \frac{\beta}{2} \beta^{-n} \quad \text{or} \quad \beta^{1-n}$$

with β the base of the computer arithmetic, n the number of digits in the mantissa, and the choice of ϵ dependent on whether rounding or truncation is used. In single precision arithmetic on an IBM 360 machine,

$$\epsilon = 16^{-5} \doteq 9.54 \times 10^{-7}$$

For additional information on computer arithmetic, refer back to Sections 1.2 and 1.3 and Problem 7 of Chapter 1.

In the algorithm, b is the best current estimate of the root ζ, a is the previous value of b, and c is a past iterate so chosen that the root ζ lies between b and c (initially $c = a$). Define $m = \frac{1}{2}(c - b)$.

Stop the algorithm if (1) $f(b) = 0$, or (2) $|m| \leq \delta$. In either case set the approximate root $\hat{\zeta} = b$. For case (2), because b will generally have been obtained by the secant method, the root ζ will generally be closer to b than to c. Thus usually,

$$|\zeta - \hat{\zeta}| \leq \delta$$

although all that can be guaranteed is that

$$|\zeta - \hat{\zeta}| \leq 2\delta$$

If the error test is not satisfied, set

$$i = b - f(b) \frac{b - a}{f(b) - f(a)}$$

Then set

$$b'' = \begin{cases} i, & \text{if } i \text{ lies between } b \text{ and } b + m = \dfrac{b + c}{2} \\ b + m, & \text{otherwise (which is the bisection method)} \end{cases}$$

$$b' = \begin{cases} b'', & \text{if } |b - b''| > \delta \\ b + \delta \cdot \text{sign}(m), & |b - b''| \leq \delta \end{cases}$$

If you are some distance from the root, then $b' = b''$; and the method is linear interpolation or the bisection method, usually the former for a smooth function $f(x)$. This generally results in a value of m that does not become small, because of a phenomena like that with the Regula Falsi method. To obtain a small interval containing the root ζ, once we are close to it, we use $b' = b + \delta \, \text{sign}(m)$, a step of δ in the direction of c. Because of the way in which a new c is chosen, this will usually result in a new small interval about ζ. Brent makes an additional important, but technical step before choosing a new b, usually the b' of above.

Having obtained the new b', we set $b=b'$, $a=$ the old value of b. If the sign of $f(b)$, using the new b, is the same as with the old b, the c is unchanged; and otherwise c is set to the old value of b, resulting in a smaller interval about ζ. The accuracy of the value of b is now tested, as described earlier.

Brent has taken great care to avoid underflow and overflow difficulties with his method; but the program is somewhat complicated to read as a consequence.

Example.   Each of the following cases was computed on an IBM 360 computer in single-precision arithmetic. The tolerance was $t=10^{-5}$, and thus

$$\delta = 10^{-5} + 2|b| \times 9.54 \times 10^{-7}$$

$$\approx 1.19 \times 10^{-5}$$

since the root is $\zeta = 1$ in all cases. The functions were evaluated in the form in which they are given here.

Case 1 $f(x)=(x-1)[1+(x-1)^2]$. The numerical results are given in Table 2.13.

Table 2.13   Example (1) of Brent's method

b	$f(b)$	c
0.0	-2.0	3.0
0.5	$-.625$	3.0
.7139037	$-.310$	3.0
.9154506	-8.51E-2	3.0
.9901778	-9.82E-3	3.0
.9998566	-1.43E-4	3.0
.9999998	-1.79E-7	3.0
.9999998	-1.79E-7	1.000010

This illustrates the necessity of using $b'=b+\delta\cdot\text{sign}(m)$ in order to obtain a small interval enclosing the root ζ.

Case 2 $f(x)=x^2-1$. The numerical results are given in Table 2.14.

Table 2.14   Example (2) of Brent's method

b	$f(b)$	c
.90	$-.19$	2.0
.9655172	-6.78E-2	2.0
1.001045	2.09E-3	.9655172
.9999818	-3.65E-5	1.001045
.9999999	-1.19E-7	1.001045
.9999999	-1.19E-7	1.000010

Case 3 $f(x) = -1 + x(3 + x(-3 + x))$

The root is $\zeta = 1$, of multiplicity three, and it took 23 iterations to converge. The initial values were $a = 3.0, b = 0$. The bisection method would use only 19 iterations for the same accuracy. If Brent's method is compared with the bisection method over the class of all continuous functions, then the number of necessary iterates for an error tolerance of δ is approximately

$$\log_2\left(\frac{b-a}{\delta}\right) \quad \text{for bisection method}$$

$$\left[\log_2\left(\frac{b-a}{\delta}\right)\right]^2 \quad \text{for Brent's algorithm}$$

Thus there are cases for which bisection is better, as our example shows. But for sufficiently smooth functions with $f'(\alpha) \neq 0$, Brent's algorithm is far faster.

Case 4 $f(x) = (x-1)\exp\left[-\dfrac{1}{(x-1)^2}\right]$. The root $x = 1$ has *infinite multiplicity* since $f^{(r)}(1) = 0$ for all $r \geq 0$. The numerical results are given in Table 2.15. Note that the routine has found an "exact" root, due to the inherent imprecision in the evaluation of the function as described in the preceding section on multiple roots. It is, of course, very inaccurate, but this is nothing that the program can treat.

Table 2.15   Example (4) of Brent's method

b	f(b)	c
0.0	−.368	3.0
.5731754	−1.76E-3	3.0
.5759330	−1.63E-3	3.0
.6098446	−5.47E-4	3.0
.6098446	−5.47E-4	1.804921
.6136357	−4.76E-4	1.804921
.6389259	−1.68E-4	1.804921
1.221923	3.37E-10	.6389259
1.221911	3.37E-10	.6389259
1.216574	1.19E-10	.6389259
.9277498	0.0	1.216574

2.9   Roots of Polynomials

We now consider solving the polynomial equation

$$p(x) \equiv a_0 + a_1 x + \ldots + a_n x^n = 0 \qquad a_n \neq 0 \tag{2.63}$$

This problem arises in many ways, and a large literature has been created to deal with it. Sometimes a particular root is wanted and an initial guess is known. In that case, the best approach is to modify one of the earlier iterative methods to take advantage of the special form of polynomials. In other cases little may be known about the location of the roots; and other methods must be used, of which there is a large variety. We merely give a brief excursion into the area here without any pretense at completeness. Modifications of the methods of earlier sections are emphasized, and numerical stability questions are considered.

We begin with an efficient way to evaluate a polynomial. It is called *nested multiplication* or Horner's method, and it is written most simply as

$$p(x) = a_0 + x(a_1 + x(a_2 + \ldots + x(a_{n-1} + a_n x)\ldots) \tag{2.64}$$

With form (2.63), there are n additions and $2n - 1$ multiplications; and with (2.64) there are n additions and n multiplications, a considerable savings. For later work, it is convenient to introduce the following auxiliary coefficients. Let $b_n = a_n$,

$$b_k = a_k + z b_{k+1} \qquad k = n-1, n-2, \ldots, 0$$

By considering (2.64), it is easy to see that

$$p(z) = b_0. \tag{2.65}$$

Introduce the polynomial

$$q(x) = b_1 + b_2 x + \ldots + b_n x^{n-1} \tag{2.66}$$

Then

$$b_0 + (x - z)q(x) = b_0 + (x - z)\left[b_1 + b_2 x + \ldots + b_n x^{n-1} \right]$$

$$= (b_0 - b_1 z) + (b_1 - b_2 z)x + \ldots + (b_{n-1} - b_n z)x^{n-1} + b_n x^n$$

$$= a_0 + a_1 x + \ldots + a_n x^n = p(x)$$

$$p(x) = b_0 + (x - z)q(x). \tag{2.67}$$

$q(x)$ is the quotient and b_0 the remainder when $p(x)$ is divided by $x - z$. If z is a root of $p(x)$, then $b_0 = 0$ and $p(x) = (x - z)q(x)$. To find additional roots of $p(x)$, we can restrict our search to the roots of $q(x)$. This reduction process is called *deflation*; and it must be used with caution, a point we will return to later.

If we want to apply Newton's method to find a root of $p(x)$, we must be able to evaluate both $p(x)$ and $p'(x)$ at any point z. From (2.67),

$$p'(x) = (x - z)q'(x) + q(x),$$

$$p'(z) = q(z). \tag{2.68}$$

We will use (2.65) and (2.68) in the following adaption of Newton's method to polynomial rootfinding.

Algorithm Polynew $(a, n, x_0, \epsilon, \text{itmax}, \text{root}, b, \text{ier})$

1. Remark: a is the vector of coefficients, itmax the maximum number of iterates to be computed, b the vector of coefficients for the deflated polynomial, and ier an error indicator.

2. itnum: $= 1$

3. $z := x_0$, $b_n := c := a_n$

4. For $k = n-1, \ldots, 1$, set
$$b_k := a_k + zb_{k+1}, \qquad c = b_k + zc$$

5. $b_0 := a_0 + zb_1$

6. $x_1 := x_0 - b_0/c$

7. If $|x_1 - x_0| \leq \epsilon$, then set ier: $= 0$, root: $= x_1$, and exit.

8. If itnum $=$ itmax, set ier: $= 1$, root: $= x_1$, and exit.

9. Otherwise, itnum: $=$ itnum $+ 1$, $x_0 := x_1$, and go to step 3.

The initial guess may have to be quite close to the desired root to attain convergence. This is especially true of the important problem of finding the roots of higher degree orthogonal polynomials, a subject taken up in Chapter 4. The algorithm Polynew was used in computing the example given in Table 2.18 later in this section.

Stability problems There are many polynomials in which the roots are quite sensitive to small changes in the coefficients. Some of these are problems with multiple roots, and it is not surprising that these roots are quite sensitive to small changes in the coefficients. But there are many polynomials with only simple roots that appear to be well separated, and for which the roots are still quite sensitive to small perturbations. Formulas will be derived below that explain this sensitivity, and numerical examples are also given.

For the theory, introduce
$$p(x) = a_0 + a_1 x + \ldots + a_n x^n \qquad a_n \neq 0$$
$$q(x) = b_0 + b_1 x + \ldots + b_n x^n$$

and define a perturbation of $p(x)$ by
$$p(x; \epsilon) = p(x) + \epsilon q(x) \tag{2.69}$$

Denote the zeroes of $p(x;\epsilon)$ by $z_1(\epsilon),\ldots,z_n(\epsilon)$, repeated according to their multiplicity; and let $z_j = z_j(0), i = 1,\ldots,n$ denote the corresponding n zeroes of $p(x) = p(x;0)$. It is well known that the zeroes of a polynomial are continuous functions of the coefficients of the polynomial, for example see Henrici [H1, p. 281]. Consequently $z_j(\epsilon)$ is a continuous function of ϵ. What we want to determine is how rapidly the root $z_j(\epsilon)$ varies with ϵ, for ϵ near 0.

Example. Define

$$p(x;\epsilon) = (x-1)^3 - \epsilon \qquad p(x) = (x-1)^3 \qquad \epsilon > 0$$

Then the roots of $p(x)$ are $z_1 = z_2 = z_3 = 1$. The roots of $p(x;\epsilon)$ are

$$z_1(\epsilon) = 1 + \sqrt[3]{\epsilon} \qquad z_2(\epsilon) = 1 + \omega\sqrt[3]{\epsilon} \qquad z_3 = 1 + \omega^2\sqrt[3]{\epsilon}$$

with $\omega = \frac{1}{2}(-1 + i\sqrt{3})$. For all three roots of $p(x;\epsilon)$,

$$|z_j(\epsilon) - 1| = \sqrt[3]{\epsilon}$$

To illustrate this, let $\epsilon = .001$. Then

$$p(x;\epsilon) = x^3 - 3x^2 + 3x - 1.001$$

which is a relatively small change in $p(x)$. But for the roots,

$$|z_j(\epsilon) - 1| = .1$$

a relatively large change in the roots $z_j = 1$.

We now give some general estimates for $z_j(\epsilon) - z_j$.

Case 1 z_j is a simple root of $p(x)$, and thus $p'(z_j) \neq 0$. Using the theory of functions of a complex variable, it is known that $z_j(\epsilon)$ can be written as a power series,

$$z_j(\epsilon) = z_j + \sum_{l=1}^{\infty} \gamma_l \epsilon^l \tag{2.70}$$

To estimate $z_j(\epsilon) - z_j$, we obtain a formula for the first term $\gamma_1\epsilon$ in the series. To begin, it is easy to see that

$$\gamma_1 = z_j'(0)$$

To calculate $z_j'(\epsilon)$, differentiate the identity

$$p(z_j(\epsilon)) + \epsilon q(z_j(\epsilon)) = 0, \text{ all } \epsilon$$

We obtain

$$p'(z_j(\epsilon))z_j'(\epsilon) + q(z_j(\epsilon)) + \epsilon q'(z_j(\epsilon))z_j'(\epsilon) = 0$$

$$z_j'(\epsilon) = \frac{-q(z_j(\epsilon))}{p'(z_j(\epsilon)) + \epsilon q'(z_j(\epsilon))}. \tag{2.71}$$

Substituting $\epsilon = 0$, we obtain

$$\gamma_1 = z_j'(0) = \frac{-q(z_j)}{p'(z_j)}$$

Returning to (2.70),

$$z_j(\epsilon) = z_j - \frac{q(z_j)}{p'(z_j)}\epsilon + \sum_{l=2}^{\infty}\gamma_l\epsilon^l$$

$$\left|z_j(\epsilon) - \left[z_j - \frac{q(z_j)}{p'(z_j)}\epsilon\right]\right| \le K\epsilon^2 \qquad |\epsilon| \le \epsilon_0 \tag{2.72}$$

for some constants $\epsilon_0 > 0$, $K > 0$. To estimate $z_j(\epsilon)$ for small ϵ, we use

$$z_j(\epsilon) \approx z_j - \frac{q(z_j)}{p'(z_j)}\epsilon. \tag{2.73}$$

The coefficient of ϵ determines how rapidly $z_j(\epsilon)$ changes relative to ϵ; and if it is large, the root z_j is called ill conditioned.

Case 2 z_j has multiplicity $m > 1$. By using techniques related to those used in Case 1, we can obtain

$$\left|z_j(\epsilon) - \left(z_j + \gamma_1\epsilon^{1/m}\right)\right| \le K|\epsilon|^{2/m} \qquad |\epsilon| \le \epsilon_0 \tag{2.74}$$

for some $\epsilon_0 > 0, K > 0$. There are m possible values to γ_1, given as the m complex roots of

$$\gamma_1^m = \frac{-m!\, q(z_j)}{p^{(m)}(z_j)}$$

Example. Consider the simple polynomial

$$p(x) = (x-1)(x-2)\ldots(x-7)$$

$$= x^7 - 28x^6 + 322x^5 - 1960x^4 + 6769x^3 - 13132x^2$$

$$+ 13068x - 5040 \tag{2.75}$$

For the perturbation, take

$$q(x) = x^6 \qquad \epsilon = -.002$$

Then for the root $z_j = j$,

$$p'(z_j) = \prod_{l \neq j} (j - l) \qquad q(z_j) = j^6$$

From (2.73) we have the estimate

$$z_j(\epsilon) \approx j + \frac{(-1)^{j-1}.002j^6}{(j-1)!(7-j)!} \equiv j + \delta(j). \qquad (2.76)$$

The numerical values of $\delta(j)$ are given in Table 2.16. The relative error in the coefficient of x^6 is $.002/28 = 7.1\text{E-}5$; but the relative errors in the roots is much larger. In fact, the size of some of the perturbations $\delta(j)$ casts doubt on the validity of using the linear estimate (2.76). The actual roots of $p(x) + \epsilon q(x)$ are given in Table 2.17, and they correspond closely to the predicted perturbations.

Table 2.16 Values of $\delta(j)$ from (2.76)

j	$\delta(j)$
1	2.78E-6
2	−1.07E-3
3	3.04E-2
4	−2.28E-1
5	6.51E-1
6	−7.77E-1
7	3.27E-1

Table 2.17 Roots of $p(x; \epsilon)$ for (2.75)

j	$z_j(\epsilon)$
1	1.0000028
2	1.9989382
3	3.0331253
4	3.8195692
5	$5.4586758 + .54012578i$
6	$5.4586758 - .54012578i$
7	7.2330128

We say that a polynomial whose roots are unstable with respect to small relative changes in the coefficients is ill-conditioned. Many such polynomials occur naturally in applications. The above example should illustrate the difficulty in determining whether or not a polynomial is ill-conditioned with a cursory examination.

Polynomial deflation   Another problem occurs with deflation of a polynomial to a lower-degree polynomial. Since the zeros will not be found exactly, the lower-degree polynomial (2.66) found by extracting the latest root will generally be in error in all of its coefficients. Clearly, from the past example, this can cause a significant perturbation in the roots for some classes of polynomials. J. H. Wilkinson has analyzed the effects of deflation and has recommended the following general strategy. (1) Solve for the roots of smallest magnitude first, ending with those of largest size. (2) After obtaining approximations to all roots, iterate again using the original polynomial and using the previously calculated values as initial guesses. A complete discussion can be found in Wilkinson [W2, pp. 55–65].

Example.   Consider finding the roots of the degree 6 Laguerre polynomial

$$p(x) = 720 - 4320x + 5400x^2 - 2400x^3 + 450x^4 - 36x^5 + x^6$$

The Newton algorithm of the last section was used to solve for the roots, with deflation following the acceptance of each new root. The roots were calculated in two ways: (1) from largest to smallest and (2) from smallest to largest. The calculations were in single precision arithmetic on an IBM 360, and the numerical results are given in Table 2.18. A comparison of the columns headed Method (1) and Method (2) shows clearly the superiority of calculating the roots in the order of increasing magnitude. If the results of method (1) are used as initial guesses for further iteration with the original polynomial, then approximate roots are obtained with an accuracy better than that of method (2); see method (3) in the table. This shows the importance of iterating again with the original polynomial to remove the effects of the deflation process.

Table 2.18   Examples involving polynomial deflation

True	Method (1)	Method (2)	Method (3)
15.98287	15.98280	15.98279	15.98287
9.837467	9.837471	9.837469	9.837467
5.775144	5.775764	5.775207	5.775144
2.992736	2.991080	2.992710	2.992736
1.188932	1.190937	1.188932	1.188932
.2228466	.2219429	.2228465	.2228466

There are other ways to deflate a polynomial, one of which favors finding roots of largest magnitude first. For a complete discussion see Peters and Wilkinson [P1, Section 5]. An algorithm is given for *composite deflation*, which removes the need to find the roots in any particular order.

2.10   Muller's Method

This is a useful method for obtaining both real and complex roots of a polynomial, and it is reasonably easy to program for a computer. It can also be applied to functions other than polynomials. When the initial guesses are chosen appropriately, and when the roots are removed by deflation, the roots can be found roughly in order of increasing magnitude.

Muller's method is a generalization of the approach that led to the secant method. Given three points x_0, x_1, x_2, a quadratic polynomial is constructed which passes through the three points $(x_i, f(x_i))$, $i = 0, 1, 2$; and one of the roots of this polynomial is used as an improved estimate for a root α of $f(x)$.

The quadratic polynomial is given by

$$p(x) = f(x_2) + (x - x_2) f[x_2, x_1] + (x - x_2)(x - x_1) f[x_2, x_1, x_0] \qquad (2.77)$$

The divided differences $f[x_2, x_1]$ and $f[x_2, x_1, x_0]$ were defined in (1.14) of Chapter 1. To check that

$$p(x_i) = f(x_i) \qquad i = 0, 1, 2$$

just substitute into (2.77) and then reduce the resulting expression using (1.14). There are other formulas for $p(x)$, given in Chapter 3, but the form (2.77) is the most convenient for defining Muller's method. The formula (2.77) is called Newton's divided difference form of the interpolating polynomial, and it is developed in general in Section 3.2 of Chapter 3.

To find the zeros of (2.77), we first rewrite it in the more convenient form

$$y = f(x_2) + w(x - x_2) + f[x_2, x_1, x_0](x - x_2)^2$$

$$w = f[x_2, x_1] + (x_2 - x_1) f[x_2, x_1, x_0]$$

$$= f[x_2, x_1] + f[x_2, x_0] - f[x_0, x_1]$$

We want to find the smallest value of $x - x_2$ that satisfies this equation. The solution is

$$x - x_2 = \frac{-w \pm [w^2 - 4f(x_2) f[x_2, x_1, x_0]]^{1/2}}{2f[x_2, x_1, x_0]}$$

with the sign chosen to make the numerator as small as possible. But because of the loss of significance errors implicit in this formula, we *rationalize the numerator* to obtain the new iteration formula

$$x_3 = x_2 - \frac{2f(x_2)}{w \pm [w^2 - 4f(x_2) f[x_2, x_1, x_0]]^{1/2}} \qquad (2.78)$$

with the sign chosen to maximize the magnitude of the denominator.

Repeat (2.78) recursively to define a sequence of iterates $\{x_n\}, n \geq 0$. If they converge to a point α, and if $f'(\alpha) \neq 0$, then α is a root of $f(x)$. To see this, use (1.15) of Chapter 1 and (2.78) to give

$$w \to f'(\alpha) \qquad \text{as } n \to \infty$$

$$\alpha = \alpha - \frac{2f(\alpha)}{f'(\alpha) \pm \left\{ \left[f'(\alpha) \right]^2 - 2f(\alpha)f''(\alpha) \right\}^{1/2}}$$

The numerator must be zero, that is, $f(\alpha) = 0$; and the denominator will be nonzero if $f'(\alpha) \neq 0$, that is, α is a simple root of $f(x)$.

By an argument similar to that used for the secant method, it can be shown that

$$\lim_{n \to \infty} \frac{|\alpha - x_{n+1}|}{|\alpha - x_n|^p} = \left| \frac{f'''(\alpha)}{6f'(\alpha)} \right|^{(p-1)/2} \qquad p \doteq 1.84 \qquad (2.79)$$

provided $f(x)$ is three times continuously differentiable in a neighborhood of α and $f'(\alpha) \neq 0$. The order p is the positive root of

$$x^3 - x^2 - x - 1 = 0$$

All of the above arguments assume $f'(\alpha) \neq 0$, and α is a simple root of $f(x)$. If α is not a simple root, Muller's method will still converge, but at a slower rate.

With the secant method, real choices of x_0 and x_1 will lead to a real value of x_2. But with Muller's method, real choices of x_0, x_1, and x_2 can and do lead to complex roots of $f(x)$. This is an important aspect of Muller's method, one reason it is widely used for polynomial rootfinding.

The following examples were computed using an automatic package program implementation of Muller's method. With no initial guesses given, it found the roots of $f(x)$ in roughly increasing order. The deflation used after $z_1, \ldots, z_r$ had been found as roots was to evaluate

$$g(x) = \frac{f(x)}{(x - z_1) \ldots (x - z_r)}. \qquad (2.80)$$

See Peters and Wilkinson [P1] for a discussion of this implicit deflation. The error test for an approximate root z was to have either (1) $|f(z)| \leq 10^{-10}$, or (2) z should have eight significant digits of accuracy. In Tables 2.19 and 2.20, the roots are given in the order in which they were found. The column headed IT gives the number of iterates that were calculated. The calculations use double-precision arithmetic on an IBM 360 computer.

Example 1. $f(x) = x^{20} - 1$. All of the 20 roots were found with an accuracy of 10 or more significant digits; and in all cases of a root z, $|f(z)| < 10^{-10}$, generally much less. IT ranged from 1 to 18, with an average of 8.5 iterates per root.

2. $f(x) =$ Laguerre polynomial of degree 12. The roots as shown are correct, rounded to the number of places shown; but the imaginary parts should all be zero. Note that $f(x)$ is quite large for many of the roots. The numerical results are given in Table 2.19.

Table 2.19 Muller's method, example 2

IT	Root	f(root)
9	1.1572211736E-1+0i	5.96E-8
10	6.1175748452E-1+9.10E-20i	−2.98E-7+9.06E-11i
14	2.8337513377E0−5.05E-17i	2.55E-5−4.78E-8i
13	4.5992276394E0−5.95E-15i	7.16E-5+9.37E-6i
8	1.5126102698E0+2.98E-16i	3.34E-6−2.35E-7i
19	1.3006054993E1+9.04E-18i	2.32E-1+4.15E-7i
16	9.6213168425E0−4.97E-17i	−3.66E-2+5.38E-7i
14	1.7116855187E1−8.48E-17i	−1.68E 0+2.40E-5i
13	2.2151090379E1+9.35E-18i	8.61E-1+2.60E-5i
7	6.8445254531E0−3.43E-28i	−4.49E3−1.22E-18i
4	2.8487967251E1+5.77E-25i	−6.34E 1−2.96E-11i
4	3.7099121044E1+2.80E-24i	2.12E 3+7.72E-9i

3. $f(x) = x^6 - 12x^5 + 63x^4 - 216x^3 + 567x^2 - 972x + 729$
$$= (x^2 + 9)(x - 3)^4.$$
The numerical results are given in Table 2.20. Note the inaccuracy in the first four roots, which is inherent due to the multiplicity of the root $\alpha = 3$.

Table 2.20 Muller's method, example 3

IT	Root	f(root)
41	2.9987526−6.95E-4i	−3.33E-11+6.70E-11i
17	2.9997591−2.68E-4i	5.68E-14−6.48E-14i
31	3.0003095−3.17E-4i	−3.41E-13+3.22E-14i
10	3.0003046+3.14E-4i	3.98E-13−3.83E-14i
6	5.97E-15+3.000000000i	4.38E-11−1.19E-11i
3	5.97E-15−3.000000000i	4.38E-11+1.19E-11i

The last two cases demonstrate why two error tests are necessary, and they indicate why the routine requests a maximum on the number of iterations to be allowed per root. The form (2.78) of Muller's method is due to Traub [T1, pp. 210–213]. For a computational discussion, see Whitley [W1].

2.11   Nonlinear Systems of Equations

This section is concerned with the solution of n simultaneous nonlinear equations in n unknowns. Such problems arise in a large variety of ways, and a variety of methods are necessary to treat them. We just introduce the subject, giving two methods that are fairly general and can be easily programmed. To do a complete development of the numerical theory for solving nonlinear systems, we need a number of results from numerical linear algebra, which is not taken up until Chapters 7 to 9.

For simplicity of presentation and ease of understanding, the theory will be presented for only two equations

$$f_1(x, y) = 0 \qquad f_2(x, y) = 0 \tag{2.81}$$

in the two unknowns, x and y. The generalization to n equations in n unknowns should be straightforward once the principal ideas have been grasped. As an additional aid, we simultaneously consider the solution of (2.81) in vector notation:

$$f(w) = 0 \qquad w = \begin{bmatrix} x \\ y \end{bmatrix} \qquad f(w) \equiv \begin{bmatrix} f_1(x, y) \\ f_2(x, y) \end{bmatrix} \tag{2.82}$$

The solution of (2.81) can be looked on as a two step process: (1) find the zero curves in the xy-plane of the surfaces $z = f_1(x, y)$ and $z = f_2(x, y)$, and (2) find the points of intersection of these zero curves in the xy-plane. This perspective will be used in the next section to generalize Newton's method to solve (2.81).

We begin by generalizing some of the fixed point iteration theory of Section 2.4. Assume that (2.81) has been reformulated in an equivalent form as

$$x = g_1(x, y) \qquad y = g_2(x, y) \tag{2.83}$$

having the same solutions (ξ, η) as (2.81). We will study the fixed-point iteration method

$$x_{n+1} = g_1(x_n, y_n) \qquad y_{n+1} = g_2(x_n, y_n), \quad n = 0, 1, 2\ldots. \tag{2.84}$$

Using vector notation, we write this as

$$w_{n+1} = g(w_n) \qquad w_n = \begin{bmatrix} x_n \\ y_n \end{bmatrix} \tag{2.85}$$

with

$$g(w) \equiv \begin{bmatrix} g_1(x, y) \\ g_2(x, y) \end{bmatrix} \qquad \alpha = \begin{bmatrix} \xi \\ \eta \end{bmatrix}$$

Example. Consider solving

$$3x^2 + 4y^2 = 1 \qquad y^3 - 8x^3 = 1 \tag{2.86}$$

for the solution near $(-.5, .25)$. This system is solved iteratively with

$$\begin{bmatrix} x_{n+1} \\ y_{n+1} \end{bmatrix} = \begin{bmatrix} x_n \\ y_n \end{bmatrix} - \begin{bmatrix} .016 & -.17 \\ .52 & -.26 \end{bmatrix} \begin{bmatrix} 3x_n^2 + 4y_n^2 - 1 \\ y_n^3 - 8x_n^3 - 1 \end{bmatrix} \tag{2.87}$$

The origin of this reformulation of (2.86) is given later. The numerical results for (2.87) are given in Table 2.21. Clearly, the iterates are converging rapidly.

Table 2.21   Example (2.87) of fixed-point iteration

n	x_n	y_n	$f_2(x_n, y_n)$	$f_2(x_n, y_n)$
0	$-.5$	.25	0.0	1.56E-2
1	$-.497343750$	.254062500	2.43E-4	5.46E-4
2	$-.497254794$	.254077922	9.35E-6	2.12E-5
3	$-.497251343$	.254078566	3.64E-7	8.26E-7
4	$-.497251208$	.254078592	1.50E-8	3.30E-8

To analyze the convergence of (2.84), begin by subtracting $x_{n+1} = g_1(x_n, y_n)$ from $\xi = g_1(\xi, \eta)$. Use the mean-value theorem for functions of two variables (Taylors Theorem 1.5, formula (1.13) of Chapter 1, with $n = 1$) to obtain

$$\xi - x_{n+1} = g_1(\xi, \eta) - g_1(x_n, y_n)$$

$$= \frac{\partial g_1(\bar{x}_n, \bar{y}_n)}{\partial x}(\xi - x_n) + \frac{\partial g_1(\bar{x}_n, \bar{y}_n)}{\partial y}(\eta - y_n)$$

with $(\bar{x}_n, \bar{y}_n)$ on the line segment joining (x_n, y_n) and (ξ, η). Similarly,

$$\eta - y_{n+1} = \frac{\partial g_2(\hat{x}_n, \hat{y}_n)}{\partial x}(\xi - x_n) + \frac{\partial g_2(\hat{x}_n, \hat{y}_n)}{\partial y}(\eta - y_n)$$

with $(\hat{x}_n, \hat{y}_n)$ also on the line joining (x_n, y_n) and (ξ, η). In matrix form, these error equations become

$$\begin{bmatrix} \xi - x_{n+1} \\ \eta - y_{n+1} \end{bmatrix} = \begin{bmatrix} \partial g_1(\bar{x}_n, \bar{y}_n)/\partial x & \partial g_1(\bar{x}_n, \bar{y}_n)/\partial y \\ \partial g_2(\hat{x}_n, \hat{y}_n)/\partial x & \partial g_2(\hat{x}_n, \hat{y}_n)/\partial y \end{bmatrix} \begin{bmatrix} \xi - x_n \\ \eta - y_n \end{bmatrix} \tag{2.88}$$

Let G_n denote the matrix in (2.88). Then we can write

$$\alpha - w_{n+1} = G_n(\alpha - w_n) \qquad n \geq 0 \tag{2.89}$$

It is convenient to introduce the Jacobian matrix for the functions g_1 and g_2,

$$G(w) = \begin{bmatrix} \partial g_1(w)/\partial x & \partial g_1(w)/\partial y \\ \partial g_2(w)/\partial x & \partial g_2(w)/\partial y \end{bmatrix} \tag{2.90}$$

In (2.88) or (2.90), if w_n is close to α, then G_n will be close to $G(\alpha)$. This will make the size of $G(\alpha)$ crucial in analyzing the convergence in (2.89). It plays the role of $g'(\alpha)$ in the theory of Section 2.4. To measure the size of the errors $\alpha - w_n$ and of the matrices G_n and G, we will use the vector and matrix norms $\|\cdot\|_\infty$ of (1.19) and (1.22) in Chapter 1.

Returning to (2.89), we have

$$\|\alpha - w_{n+1}\|_\infty \leq \|G_n\|_\infty \|\alpha - w_n\|_\infty \tag{2.91}$$

We now prove the two-dimensional analog of Theorem 2.4.

Theorem 2.6 Let $g_1(x, y)$ and $g_2(x, y)$ be continuously differentiable for all (x, y) near $\alpha = (\xi, \eta)$, a fixed point of the two functions. Moreover, suppose that

$$\|G(\alpha)\|_\infty < 1 \tag{2.92}$$

Then the iterative method (2.84) will converge to α, provided the initial guess (x_0, y_0) is chosen sufficiently close to α. In addition, for (x_n, y_n) near to α,

$$\alpha - w_{n+1} \approx G(\alpha)(\alpha - w_n)$$

with the Jacobian matrix $G(\alpha)$ of (2.90).

Proof The function $\|G(w)\|_\infty$ is a continuous function of $w = (x, y)$. Consequently the assumption (2.92) implies

$$\|G(w)\| \leq c < 1$$

for all w sufficiently close to α, for some $c > 0$. From the same argument, it will follow that

$$\|G_n\| \leq c < 1 \qquad n \geq 0$$

if w_0 is chosen sufficiently close to α. Use this in (2.91) to complete the proof. ∎

Based on results in Chapter 7 the theorem will still be true if all eigenvalues of $G(\alpha)$ are less than one in magnitude, which can be shown to be a weaker assumption that (2.92).

Example.   Continue the earlier example (2.87). It is straightforward to compute

$$G(\alpha) \doteq \begin{bmatrix} .0389 & .0004 \\ .0085 & -.0066 \end{bmatrix}$$

and therefore

$$\|G(\alpha)\|_\infty \doteq .0393$$

Thus the condition (2.92) of the theorem is satisfied. From (2.91), it will be true that

$$\frac{\|\alpha - w_{n+1}\|_\infty}{\|\alpha - w_n\|_\infty} \leqq \|G_n\|_\infty \doteq .0393$$

for all sufficiently large n.

Suppose that A is a constant nonsingular matrix of order 2×2. We can then reformulate (2.81) or (2.82) as

$$w = w + Af(w) \equiv g(w) \tag{2.93}$$

The example (2.87) illustrates this procedure. To see the requirements on A, we produce the Jacobian matrix. Easily,

$$G(w) = I + AF(w) \tag{2.94}$$

where $F(w)$ is the Jacobian matrix of f_1 and f_2,

$$F(w) = \begin{bmatrix} \partial f_1(w)/\partial x & \partial f_1(w)/\partial y \\ \partial f_2(w)/\partial x & \partial f_2(w)/\partial y \end{bmatrix} \tag{2.95}$$

We want to choose A so that (2.92) is satisfied. And for rapid convergence, we want $\|G(\alpha)\|_\infty \approx 0$, or

$$A \approx -F(\alpha)^{-1}$$

The matrix in (2.87) was chosen in this way using

$$A \approx -F(w_0)^{-1}$$

This suggests using a continual updating of A, say $A = -F(w_n)^{-1}$. The resulting method is

$$w_{n+1} = w_n - F(w_n)^{-1} f(w_n) \qquad n \geqq 0 \tag{2.96}$$

We consider this method in the next section.

2.12   Newton's Method for Nonlinear Systems

As with Newton's method for a single equation, there is more than one way of viewing and deriving the method. Using Taylor's theorem for functions of two variables (see Theorem 1.5 in Chapter 1),

$$0 = f_i(\xi, \eta) = f_i(x_0, y_0) + (\xi - x_0)\frac{\partial f_i(x_0, y_0)}{\partial x} + (\eta - y_0)\frac{\partial f_i(x_0, y_0)}{\partial y}$$

$$+ \frac{1}{2}\left[(\xi - x_0)\frac{\partial}{\partial x} + (\eta - y_0)\frac{\partial}{\partial y}\right]^2 f_i(a_i, b_i) \qquad i = 1, 2 \qquad (2.97)$$

where (a_i, b_i) is some point on the line joining (x_0, y_0) and (ξ, η). If we drop the second-order terms in (2.97), we obtain the approximation

$$0 \approx f_1(x_0, y_0) + (\xi - x_0)\frac{\partial f_1(x_0, y_0)}{\partial x} + (\eta - y_0)\frac{\partial f_1(x_0, y_0)}{\partial y}$$

$$0 \approx f_2(x_0, y_0) + (\xi - x_0)\frac{\partial f_2(x_0, y_0)}{\partial x} + (\eta - y_0)\frac{\partial f_2(x_0, y_0)}{\partial y} \qquad (2.98)$$

In vector and matrix notation,

$$0 \approx f(w_0) + F(w_0)(\alpha - w_0) \qquad w_0 = \begin{bmatrix} x_0 \\ y_0 \end{bmatrix}$$

where F is the Jacobian matrix in (2.95). Solving for α,

$$\alpha \approx w_0 - F(w_0)^{-1}f(w_0) \equiv w_1$$

The approximation w_1 should be an improvement on w_0. This leads to the iteration method first obtained at the end of the last section,

$$w_{n+1} = w_n - F(w_n)^{-1}f(w_n) \qquad n \geq 0 \qquad (2.99)$$

This is Newton's method for solving the nonlinear system (2.81). In actual practice, we do not invert $F(w_n)$. Instead we solve a linear system for a correction term to w_n,

$$F(w_n)\delta_{n+1} = -f(w_n) \qquad w_{n+1} = w_n + \delta_{n+1} \qquad (2.100)$$

This is faster in computation time, requiring only about one-third as many operations as inverting $F(w_n)$. See Sections 8.1 and 8.2 of Chapter 8 for a discussion of the numerical solution of linear systems of equations.

There is a geometrical derivation for the method (2.99), in analogy to the tangent line approximation used with the single nonlinear equation in Section

2.3. The graph in xyz-space of the equation

$$z = f_i(x_0, y_0) + (x - x_0)\frac{\partial f_i(x_0, y_0)}{\partial x} + (y - y_0)\frac{\partial f_i(x_0, y_0)}{\partial y} \equiv p_i(x, y)$$

is a plane which is tangent to the graph of $z = f_i(x, y)$ at the point (x_0, y_0), $i = 1, 2$. If (x_0, y_0) is near (ξ, η), these tangent planes should be good approximations to the associated surfaces of $z = f_i(x, y)$ for all (x, y) near (ξ, η). And the intersection of the zero curves of the tangent planes $z = p_i(x, y)$ should be a good approximation to the corresponding intersection (ξ, η) of the zero curves of the original surfaces $z = f_i(x, y)$. This results in the statement (2.98). The intersection of the zero curves of $z = p_i(x, y)$, $i = 1, 2$, is the point w_1.

Example. Consider the system

$$f_1 \equiv 4x^2 + y^2 - 4 = 0 \qquad f_2 \equiv x + y - \sin(x - y) = 0$$

There are only two roots, one near $(1, 0)$ and its reflection through the origin near $(-1, 0)$. Using (2.99) with $(x_0, y_0) = (1, 0)$, we obtain the results given in Table 2.22.

Table 2.22 Example of Newton's method

n	x_n	y_n	$f_1(x_n, y_n)$	$f_2(x_n, y_n)$
0	1.0	0.0	0.0	1.59E-1
1	1.0	$-.1029207154$	1.06E-2	4.55E-3
2	.9986087598	$-.1055307239$	1.46E-5	6.63E-7
3	.9986069441	$-.1055304923$	2.20E-11	-1.03E-11

For the convergence analysis of (2.99), regard it as a fixed point iteration method with

$$g(w) = w - F(w)^{-1} f(w) \qquad (2.101)$$

Also assume

$$\text{determinant}(F(\alpha)) \neq 0$$

which is the analog of assuming α is a simple root when dealing with a single equation, as in Section 2.3. It can then be shown that the Jacobian $G(w)$ of (2.101) is zero at $w = \alpha$ (see Problem 28); and consequently the condition (2.92) is easily satisfied.

Theorem 2.6 then implies that w_n converges to α, provided w_0 is chosen sufficiently close to α. In addition, it can be shown that the iteration is quadratic. For example, the formulas (2.97) and (2.99) can be combined to obtain

$$\|\alpha - w_{n+1}\|_\infty \leq B \|\alpha - w_n\|_\infty^2 \qquad n \geq 0 \qquad (2.102)$$

for some constant $B > 0$.

Variations of Newton's method Newton's method has a number of both advantages and disadvantages when compared with other methods for solving nonlinear systems of equations. Among its advantages, it is very simple in form and there is great flexibility in using it on a large variety of problems. If we don't want to bother supplying partial derivatives to be evaluated by a computer program, we can use a divided difference approximation. For example, we commonly use

$$\frac{\partial f_i(x, y)}{\partial x} \approx \frac{f_i(x + \epsilon, y) - f_i(x, y)}{\epsilon}$$

with some very small number ϵ.

The principal disadvantage of Newton's method is that there are other methods that (1) are less expensive to use, and/or (2) are easier to use for some special classes of problems. For a system of n nonlinear equations in n unknowns, each iterate for Newton's method requires $n^2 + n$ function evaluations, in general. There are other methods that are equally rapid in their speed of convergence, but which require fewer function evaluations per iterate. These methods are usually related to Newton's method, and are often called quasi-Newton or Newtonlike methods. A general survey of the theory for many of these methods is given in Dennis and More [D4].

For the problem of writing an algorithm that will solve general systems of n nonlinear equations, a well-known method was given by K. Brown in [B3] with a Fortran program given in Brown [B4]. Another method was given by R. Brent in [B2], and it has since been implemented in More and Cosnard [M1]. Both Brent's method and Brown's method require $(n^2 + 3n)/2$ function evaluations per iterate, and they both converge quadratically like Newton's method. Brown's method requires only about one-fourth the computer storage of Brent's method and Newton's method, which is about n^2 locations.

More and Cosnard's paper [M1] contains a discussion of Brown's method, Brent's method, and Newton's method, including a comparative discussion of costs and accuracy. A number of numerical examples are included, and they illustrate the general conclusion that Brent's method seems somewhat superior in efficiency.

Brent's method and Brown's method are quite complicated to implement, and they are not as flexible as Newton's method for solving many nonlinear equations that arise in the solution of boundary value problems for ordinary partial differential equations. Generally these problems require the use of Newton's method or of some other Newtonlike method. Because of the large variety of such problems, along with those arising in many other areas, the subject of solving systems of nonlinear equations has become quite large. For a general introduction to it and its mathematical foundation, see the important book of J. Ortega and W. Rheinboldt [O1].

Discussion of the literature

There is a large literature on methods for calculating the roots of a single equation. See the books by Householder [H2], Ostrowski [O2], and Traub [T1] for a more extensive development than we have given, and also see their bibliographies. The automatic algorithm of Brent [B1], described in Section 2.8, has been widely adopted; it is well recommended as a general purpose rootfinder. To see the difficulties of implementing an automatic algorithm on a computer, even for the supposedly easy task of solving a quadratic equation, see Forsythe's article [F1].

The development of methods for solving polynomial equations is an extremely old area, going back to at least the ancient Babylonians. There are many methods and a large literature for them; and many new methods having been developed in the past decade. As an introduction to the area, see Dejon and Henrici [D1], Henrici [H1, Chapter 6], Householder [H2], and Traub [T1] and their bibliographies. Accurate, efficient, and reliable automatic computer programs are now being produced. One set of programs has been produced by Jenkins and Traub; see [J1] for a development of the method, and see [J2] and [J3] for actual programs for polynomials with complex and real coefficients, respectively. Another automatic program is given in Dejon and Nickel [D2]; and the basis of an automatic program is described in Dunaway [D5]. These automatic programs are much too sophisticated, both mathematically and in their computer implementations, to discuss in an introductory text such as this. But they are well worth using; most people could not write a program that would be competitive in either speed or accuracy with the currently available automatic programs. The attainable accuracy will still be subject to the problems of instability and uncertainty described in the text in Sections 2.7 and 2.9.

The study of numerical methods for solving nonlinear systems of equations is currently a popular area, along with the associated subject of optimization problems. For a general introduction to the area, see Rheinboldt [R1]; and for a more detailed development of the theory see Ortega and Rheinboldt [O1]. For general purpose algorithms written in Fortran, see Brown [B4] and More and Cosnard [M1].

Bibliography

[B1] Brent, R., *Algorithms for Minimization Without Derivatives*, Prentice-Hall, Englewood Cliffs, N.J., 1973.

[B2] Brent, R., *Some efficient algorithms for solving systems of nonlinear equations*, SIAM J. Num. Anal., **10**, 1973, *pp.* 327–344.

[B3] Brown, K. M., A quadratically convergent Newton-like method based on Gaussian elimination, *SIAM J. Num. Anal.*, **6**, 1969, pp. 560–569.

[B4] Brown, K. M., Computer oriented algorithms for solving systems of simultaneous nonlinear algebraic equations, in [B5], pp. 281–348.

[B5] Byrne, G. and C. Hall, eds., *Numerical Solution of Systems of Nonlinear Algebraic Equations*, Academic Press, New York, 1973.

[D1] Dejon, B. and P. Henrici, eds., *Constructive Aspects of the Fundamental Theorem of Algebra*, Wiley, New York, 1969.

[D2] Dejon, B. and K. Nickel, A never failing, fast convergent rootfinding algorithm, in [D1], pp. 1–36.

[D3] Dekker, T., Finding a zero by means of successive linear interpolation, in [D1], pp. 37–51.

[D4] Dennis, J. and J. More, Quasi-Newton methods, Motivation and Theory, *SIAM Rev.*, **19**, 1977, pp. 46–89.

[D5] Dunaway, D. K., Calculation of zeros of a real polynomial through factorization using Euclid's algorithm, *SIAM J. Num. Anal.*, **11**, 1974, pp. 1087–1104.

[F1] Forsythe, G., What is a satisfactory quadratic equation solver?, in [D1], pp. 53–61.

[H1] Henrici, P., *Applied and Computational Complex Analysis*, Wiley-Interscience, New York, 1974.

[H2] Householder, A. S., *The Numerical Treatment of a Single Nonlinear Equation*, McGraw-Hill, New York, 1970.

[J1] Jenkins, M. and J. Traub, A three stage algorithm for real polynomials using quadratic iteration, *SIAM J. Numer. Anal.*, **7**, 1970, pp. 545–566.

[J2] Jenkins, M. and J. Traub, Algorithm 419—Zeros of a complex polynomial, *Comm. ACM*, **15**, 1972, pp. 97–99.

[J3] Jenkins, M. and J. Traub, Program ZRPOLY, International Mathematics and Statistics Library, Houston, Texas.

[M1] More, J. and M. Cosnard, Numerical comparison of three nonlinear equation solvers, Argonne National Lab., Appl. Math. Div., Tech. Rep. TM-286, 1976.

[O1] Ortega, J. and W. Rheinboldt, *Iterative Solution of Nonlinear Equations in Several Variables*, Academic Press, New York, 1970.

[O2] Ostrowski, A., *Solution of Equations in Euclidean and Banach Spaces* 3rd edition, Academic Press, New York, 1973.

[P1] Peters, G. and J. Wilkinson, Practical problems arising in the solution of polynomial equations, *J. Inst. Math. Applic.*, **8**, 1971, pp. 16–35.

[R1] Rheinboldt, W., *Methods for Solving Systems of Nonlinear Equations*, SIAM, Philadelphia, 1974.

[T1] Traub, J. F., *Iterative Methods for the Solution of Equations*, Prentice-Hall, Englewood Cliffs, N.J., 1964.

[W1] Whitley, V., Certification of algorithm 196: Muller's method for finding roots of an arbitrary function, *Comm. ACM*, **11**, 1968, pp. 12–14.

[W2] Wilkinson, J., *Rounding Errors in Algebraic Processes*, Prentice-Hall, Englewood Cliffs, N.J., 1963.

Problems

1. The introductory example for $f(x) = a - (1/x)$ is related to the infinite product

$$\prod_{j=0}^{\infty} (1 + r^{2^j}) \equiv \underset{n \to \infty}{\text{Limit}} \left[(1+r)(1+r^2)(1+r^4)\ldots(1+r^{2^n}) \right]$$

By using formulas (2.5) and (2.8), we can calculate the value of the infinite product. What is this value, and what condition on r is required for the infinite product to converge?

2. Write a program implementing the algorithm Bisect given in Section 2.1. Use this program to calculate the smallest positive root of $x - \tan(x) = 0$.

3. Derive the error formula (2.21) for the secant method and Regula Falsi method,

$$\alpha - c = -(\alpha - b)(\alpha - a)\frac{f[a,b,\alpha]}{f[a,b]}$$

4. Using either the secant method or Newton's method, find the indicated roots of the following equations.
    (a) The positive root of $x^3 - x^2 - x - 1 = 0$.
    (b) All roots of $x = 1 + .3\cos(x)$.
    (c) The smallest positive root of $\cos(x) = \frac{1}{2} + \sin(x)$.

5. Use Newton's method to calculate the unique root of

    $$x + e^{-Bx^2}\cos(x) = 0$$

    with $B > 0$ a parameter to be set. Use a variety of increasing values of B, for example, $B = 1, 5, 10, 25, 50$. Among the choices of x_0 used, choose $x_0 = 0$ and explain any anomalous behavior. Theoretically, the Newton method will converge for any value of x_0 and B. Compare this with actual computations for larger values of B.

6. An interesting polynomial rootfinding problem occurs in the computation of annuities. An amount of P_1 dollars is put into an account at the beginning of years $1, 2, \ldots, N_1$. It is compounded annually at a rate of r (e.g., $r = .05$ means a 5 percent rate of interest). At the beginning of years $N_1 + 1, \ldots, N_1 + N_2$, a payment of P_2 dollars is removed from the account. After the last payment, the account is exactly zero. The relationship of the variables is

    $$P_1\left[(1+r)^{N_1} - 1\right] = P_2\left[1 - (1+r)^{-N_2}\right]$$

    If $N_1 = 30$, $N_2 = 20$, $P_1 = 2000$, and $P_2 = 8000$, then what is r? Use either the secant method or Newton's method.

7. (a) Use Newton's method to calculate the smallest positive root of $x - \tan x = 0$.
    (b) Calculate the root of $x - \tan x = 0$ closest to $x = 100$. Explain the difference in the behavior of Newton's method as compared with that in part (a).

8. Consider Newton's method for finding the positive square root of $a > 0$. Derive the following results, assuming $x_0 > 0$, $x_0 \neq \sqrt{a}$.
    (a) $x_{n+1} = \frac{1}{2}\left(x_n + \frac{a}{x_n}\right)$.
    (b) $x_{n+1}^2 - a = \left[\frac{x_n^2 - a}{2x_n}\right]^2$, $n \geq 0$, and thus $x_n > \sqrt{a}$ for all $n > 0$
    (c) The iterates $\{x_n\}$ are a strictly decreasing sequence for $n \geq 1$.

(d) $e_{n+1} = -e_n^2/(2x_n)$, with $e_n = \sqrt{a} - x_n$

$$\text{Rel}(x_{n+1}) = \frac{-\sqrt{a}}{2x_n} \left[\text{Rel}(x_n) \right]^2, \quad n \geq 0$$

with $\text{Rel}(x_n)$ the relative error in x_n

(e) If $x_0 > \sqrt{a}$ and $|\text{Rel}(x_0)| \leq 0.1$, bound $\text{Rel}(x_4)$.

9. Newton's method is the commonly used method for calculating square roots on a computer. To use Newton's method to calculate $\sqrt{a}$, an initial guess x_0 must be chosen; and it would be most convenient to use a fixed number of iterates rather than having to test for convergence. For definiteness, suppose that the computer arithmetic is binary and that the mantissa contains 48 binary bits. Write

$$a = \hat{a} \cdot 2^e \qquad \tfrac{1}{2} \leq \hat{a} < 1$$

This can be easily modified to the form

$$a = b \cdot 2^f \qquad \tfrac{1}{4} \leq b < 1$$

with f an even integer. Then

$$\sqrt{a} = \sqrt{b} \cdot 2^{f/2} \qquad \tfrac{1}{2} \leq \sqrt{b} < 1$$

and the number $\sqrt{a}$ will be in standard floating point form, once $\sqrt{b}$ is known.

This reduces the problem to that of calculating $\sqrt{b}$ for $\tfrac{1}{4} \leq b < 1$. Use the linear interpolating formula

$$x_0 = \tfrac{1}{3}(2b+1) \qquad \tfrac{1}{4} \leq b \leq 1$$

as an initial guess for the Newton iteration for calculating $\sqrt{b}$. Bound the error $x_0 - \sqrt{b}$. Decide how many iterates are necessary in order that

$$0 < x_n - \sqrt{b} \leq 2^{-48}$$

which is the limit of machine precision for b. Note that the effect of rounding errors is being ignored. How might the choice of x_0 be improved?

10. **(a)** Apply Newton's method to the function

$$f(x) = \begin{cases} \sqrt{x} & x \geq 0 \\ -\sqrt{-x} & x \leq 0 \end{cases}$$

with the root $\alpha = 0$. What is the behavior of the iterates? Do they converge, and if so, at what rate?

(b) Do the same as in (a), but with

$$
f(x) = \begin{cases} \sqrt[3]{x^2} & x \geq 0 \\ -\sqrt[3]{x^2} & x \leq 0 \end{cases}
$$

11. (a) Newton's method can be used with complex-valued functions $f(z)$ of a complex variable $z = x + iy$, x and y real,

$$
z_{n+1} = z_n - \frac{f(z_n)}{f'(z_n)} \qquad n \geq 0
$$

If it is desired to avoid complex arithmetic, show that

$$
x_{n+1} = x_n - \frac{A_n C_n + B_n D_n}{C_n^2 + D_n^2} \qquad y_{n+1} = y_n + \frac{A_n D_n - B_n C_n}{C_n^2 + D_n^2}
$$

with

$$
f(z_n) = A_n + iB_n \qquad f'(z_n) = C_n + iD_n
$$

the decomposition into real and imaginary parts.

(b) Use Newton's method to calculate a zero of

$$
p(z) = z^4 - 3z^3 + 20z^2 + 44z + 54,
$$

located near $z_0 = 2.5 + 4.5i$.

12. Show that for Newton's method, the ratio

$$
R_n = (x_n - x_{n-1})/(x_{n-1} - x_{n-2})^2
$$

converges to the value $-f''(\alpha)/2f'(\alpha)$, the same as in (2.32). Hint: Use

$$
x_n - x_{n-1} = (\alpha - x_{n-1}) - (\alpha - x_n)
$$

and use the error formula (2.32) for Newton's method.

13. Show that $x = \frac{1}{2}\cos(x)$ has a solution α. Find an interval $[a,b]$ containing α such that for every $x_0 \in [a,b]$, the iteration

$$
x_{n+1} = \tfrac{1}{2}\cos(x_n) \qquad n \geq 0
$$

will converge to α. Calculate the first few iterates and estimate the rate of convergence.

14. Do the same as in Problem 13, but with the iteration

$$x_{n+1} = 1 + \tan^{-1} x_n \qquad n \geq 0$$

15. To find a root for $f(x) = 0$ by iteration, rewrite the equation as

$$x = x + cf(x) \equiv g(x)$$

for some constant $c \neq 0$. If α is a root of $f(x)$ and if $f'(\alpha) \neq 0$, how should c be chosen in order that the sequence $x_{n+1} = g(x_n)$ converge to α?

16. Show that

$$x_{n+1} = \frac{x_n(x_n^2 + 3a)}{3x_n^2 + a} \qquad n \geq 0$$

is a third-order method for computing $\sqrt{a}$. Calculate

$$\operatorname*{Limit}_{n \to \infty} \frac{\sqrt{a} - x_{n+1}}{\left(\sqrt{a} - x_n\right)^3}$$

assuming x_0 has been chosen sufficiently close to α.

17. There is another modification of Newton's method, similar to the secant method, but using a different approximation to the derivative $f'(x_n)$. Define

$$x_{n+1} = x_n - \frac{f(x_n)}{D_n} \qquad D_n = \frac{f(x_n + f(x_n)) - f(x_n)}{f(x_n)} \qquad n \geq 0$$

This one-point method is called *Steffenson's method*. Assuming $f'(\alpha) \neq 0$, show that this is a second-order method. Hint: Write the iteration as $x_{n+1} = g(x_n)$. Use $f(x) = (x - \alpha)h(x)$ with $h(\alpha) \neq 0$, and then compute the formula for $g(x)$ in terms of $h(x)$. Having done so, apply Theorem 2.5.

18. The algorithm Aitken, given in Section 2.5, can be shown to be second order in its speed of convergence. Let the original iteration be $x_{n+1} = g(x_n)$, $n \geq 0$. The formula (2.46) can be rewritten in the equivalent form

$$\alpha \approx a_{n+2} = x_n + \frac{(x_{n+1} - x_n)^2}{(x_{n+1} - x_n) - (x_{n+2} - x_{n+1})} \qquad n \geq 0$$

To examine the speed of convergence of the iterates in Aitken, we consider the sequence

$$z_{n+1} = z_n + \frac{[g(z_n) - z_n]^2}{[g(z_n) - z_n] - [g(g(z_n)) - g(z_n)]} \qquad n \geq 0$$

The values z_n are the successive values of a produced in the algorithm.

For $g'(\alpha)\neq0$ or 1, show that z_n converges to α quadratically. This is true even if $|g'(\alpha)|>1$ and the original iteration $x_{n+1}=g(x_n)$ is divergent. Hint: do not attempt to use Theorem 2.5, for it will be too complicated. Instead, write

$$g(x)=\alpha+(x-\alpha)h(x) \qquad h(\alpha)=g'(\alpha)\neq0$$

Use this to show that

$$\alpha-z_{n+1}=H(z_n)(\alpha-z_n)^2 \qquad n\geq0$$

for some function $H(x)$ bounded about $x=\alpha$.

19. Use the program from Problem 2 to solve the following equations for the root $\alpha=1$. Use the given interval $[a,b]$, and in all cases use $\epsilon=10^{-5}$ as the stopping tolerance. Compare the results with those obtained in Section 2.8 using Brent's algorithm.
    (a) $(x-1)[1+(x-1)^2]=0$, $a=0$, $b=3$
    (b) $x^2-1=0$, $a=.9$, $b=2$
    (c) $-1+x(3+x(-3+x))=0$, $a=0$, $b=3$
    (d) $(x-1)\exp(-1/(x-1)^2)=0$, $a=0$, $b=3$

20. For the polynomial

$$p(x)=a_0+a_1x+\ldots+a_nx^n, \qquad a_n\neq0$$

define

$$R=(|a_0|+|a_1|+\ldots+|a_{n-1}|)/|a_n|$$

Show that every root x of $p(x)=0$ satisfies

$$|x|\leq \text{Max}\{R, \sqrt[n]{R}\}$$

21. Write a computer program to evaluate the following polynomials $p(x)$ for the given values of x and to evaluate the noise in the values $p(x)$. For each x, evaluate $p(x)$ in both single- and double-precision arithmetic; and use their difference as the noise in the single-precision value, due to rounding errors in the evaluation of $p(x)$. Use both the ordinary formula (2.63) and Horner's rule (2.64) to evaluate each polynomial; this should show that the noise is different in the two cases.
    (a) $p(x)=x^4-5.7x^3-.47x^2+29.865x-26.1602$, $-3\leq x\leq5$, with steps of $h=0.1$ for x.
    (b) $p(x)=x^4-5.4x^3+10.56x^2-8.954x+2.7951$, $1\leq x\leq1.2$, in steps of $h=.001$ for x.

22. Write a program to find the roots of the following polynomials as accurately as possible.
    (a) $676039x^{12} - 1939938x^{10} + 2078505x^8 - 1021020x^6 + 225225x^4$
    $- 18018x^2 + 231$
    (b) $x^4 - 2.054929659x^3 + 1.354386175x^2 - .3666114404x + .03547668788$

23. For the example $f(x) = (x-1)(x-2)\ldots(x-7)$ of Section 2.9, consider perturbing the coefficient of x^i by $\epsilon_i x^i$, in which ϵ_i is chosen so that the relative perturbation in the coefficient of x^i is the same as that of the example in the text in the coefficient of x^6. What does the linearized theory (2.73) predict for the perturbations in the roots? Which coefficient is most sensitive to change?

24. Use a package program for polynomials to find the roots of the polynomials in Problem 22.

25. Using Newton's method, solve the nonlinear system

$$x^2 + y^2 = 4 \qquad x^2 - y^2 = 1$$

The true solutions are easily determined to be $x = \pm\sqrt{2.5}$, $y = \pm\sqrt{1.5}$. As an initial guess, use $(x_0, y_0) = (1.6, 1.2)$.

26. Solve the system

$$x^2 + xy^3 = 9 \qquad 3x^2y - y^3 = 4$$

using Newton's method for nonlinear systems. Use each of the initial guesses

$$(x_0, y_0) = (1.2, 2.5), (-2, 2.5), (-1.2, -2.5), (2, -2.5)$$

Observe which root the method converges to, the number of iterates required, and the speed of convergence.

27. Using Newton's method, find all roots of the system

$$x^2 + y^2 - 2x - 2y + 1 = 0 \qquad x + y - 2xy = 0$$

28. Prove that the Jacobian of (2.101) is zero at the root α.

THREE

INTERPOLATION THEORY

The concept of interpolation is to select a function $p(x)$ from a given class of functions in such a way that the graph of $y = p(x)$ passes through the given data points $(x_i, y_i), i = 1, 2, \ldots, n$. In most of this chapter we limit the interpolating function $p(x)$ to be a polynomial.

Polynomial interpolation theory has several important uses. In this text, its primary use is to furnish some mathematical tools that are used in developing methods in the areas of approximation theory, numerical integration, and the numerical solution of differential equations. The second important use is in developing means for working with functions that are stored in tabular form. For example, almost everyone is familiar from high school algebra with linear interpolation in a logarithm table. We derive computationally convenient forms for polynomial interpolation with tabular data and analyze the resulting error. The chapter concludes with an introduction to piecewise polynomial interpolating functions, including interpolation by spline functions.

3.1 Polynomial Interpolation Theory

Let $x_0, x_1, \ldots, x_n$ be distinct real or complex numbers, and let $y_0, y_1, \ldots, y_n$ be associated function values. We will study the problem of finding a polynomial $p(x)$ that interpolates to the given data,

$$p(x_i) = y_i \qquad i = 0, 1, \ldots, n \tag{3.1}$$

Does such a polynomial exist, and if so, what is its degree? Is it unique? What is a formula for producing $p(x)$ from the given data?

By writing

$$p(x) = a_0 + a_1 x + \ldots + a_m x^m$$

for a general polynomial of degree m, we see there are $m + 1$ independent parameters $a_0, a_1, \ldots, a_m$. Since (3.1) imposes $n + 1$ conditions on $p(x)$ it is

107

reasonable to first consider the case when $m=n$. Then we want to find $a_0, a_1, \ldots, a_n$ such that

$$a_0 + a_1 x_0 + a_2 x_0^2 + \ldots + a_n x_0^n = y_0$$

$$\vdots \tag{3.2}$$

$$a_0 + a_1 x_n + a_2 x_n^2 + \ldots + a_n x_n^n = y_n$$

This is a system of $n+1$ linear equations in $n+1$ unknowns, and solving it is completely equivalent to solving the polynomial interpolation problem. In vector and matrix notation, the system is

$$Xa = y$$

with

$$X = \begin{bmatrix} x_i^j \end{bmatrix} \qquad i, j = 0, 1, \ldots, n$$

$$a = \begin{bmatrix} a_0, a_1, \ldots, a_n \end{bmatrix}^T, \qquad y = \begin{bmatrix} y_0, \ldots, y_n \end{bmatrix}^T \tag{3.3}$$

The matrix X is called a Vandermonde matrix.

Theorem 3.1 Given $n+1$ distinct points $x_0, \ldots, x_n$ and $n+1$ ordinates $y_0, \ldots, y_n$, there is a polynomial $p(x)$ of degree $\leq n$ that interpolates to y_i at $x_i, i = 0, 1, \ldots, n$. This polynomial $p(x)$ is unique among the set of all polynomials of degree at most n.

Proof Three proofs will be given of this important result. Each one will furnish some useful information and has important uses in other interpolation problems.

1. It can be shown that for the matrix X in (3.3),

$$\det(X) = \prod_{0 \leq j < i \leq n} (x_i - x_j) \tag{3.4}$$

See Problem 1. This shows that $\det(X) \neq 0$ since the points x_i are distinct. Thus X is nonsingular and the system $Xa = y$ has a unique solution a. This proves the existence and uniqueness of an interpolating polynomial of degree $\leq n$.

2. By a standard theorem of linear algebra (see Theorem 7.2 of Chapter 7), the system $Xa = y$ has a unique solution if and only if the homogeneous system $Xb = 0$ has only the trivial solution $b = 0$. Therefore, assume $Xb = 0$ for some b. Using that b, define

$$q(x) = b_0 + b_1 x + \ldots + b_n x^n$$

From the system $Xb=0$, we have

$$q(x_i)=0, \qquad i=0,1,\ldots,n$$

The polynomial $q(x)$ has $n+1$ zeroes and degree $q(x)\leq n$. This is not possible unless $q(x)\equiv 0$. But then all coefficients $b_i=0, i=0,1,\ldots,n$, completing the proof.

3. We will exhibit explicitly the interpolating polynomial. To begin, we consider the special interpolation problem in which

$$y_i=1, \qquad y_j=0 \text{ for } j\neq i$$

for some $i, 0\leq i\leq n$. We want a polynomial, $l_i(x)$, of degree $\leq n$ with the n zeroes $x_j, j\neq i$. Then

$$l_i(x)=c(x-x_0)\ldots(x-x_{i-1})(x-x_{i+1})\ldots(x-x_n)$$

for some constant c. The condition $l_i(x_i)=1$ implies

$$c=\left[(x_i-x_0)\ldots(x_i-x_{i-1})(x_i-x_{i+1})\ldots(x_i-x_n)\right]^{-1}$$

This special polynomial is written as

$$l_i(x)= \prod_{j\neq i}\left(\frac{x-x_j}{x_i-x_j}\right) \qquad i=0,1,\ldots,n \tag{3.5}$$

To solve the general interpolation problem (3.1), write

$$p(x)=y_0 l_0(x)+y_1 l_1(x)+ \ldots +y_n l_n(x)$$

With the special properties of the polynomials $l_i(x)$, $p(x)$ easily satisfies (3.1). Also, degree $p(x)\leq n$ since all $l_i(x)$ have degree n.

To prove uniqueness, suppose $q(x)$ is another polynomial of degree $\leq n$ which satisfies (3.1). Define

$$r(x)=p(x)-q(x)$$

Then degree $r(x)\leq n$, and

$$r(x_i)=p(x_i)-q(x_i)=y_i-y_i=0 \qquad i=0,1,\ldots,n$$

Since $r(x)$ has $n+1$ zeroes, we must have $r(x)\equiv 0$. This proves $p(x)\equiv q(x)$, completing the proof. ∎

Uniqueness is a property that is of practical use in much that follows. We will derive other formulas for the interpolation problem (3.1), and uniqueness will say they are the same polynomial. Also, without uniqueness the linear system (3.2) would not be uniquely solvable; and from results in linear algebra, this would imply the existence of data vectors y for which there was no interpolating polynomial of degree $\leq n$.

The formula

$$p_n(x) = \sum_{i=0}^{n} y_i l_i(x) \tag{3.6}$$

is called Lagrange's form of the interpolating polynomial.

Examples

$$p_1(x) = \frac{x - x_1}{x_0 - x_1} y_0 + \frac{x - x_0}{x_1 - x_0} y_1 = \frac{(x_1 - x) y_0 + (x - x_0) y_1}{x_1 - x_0}$$

$$p_2(x) = \frac{(x - x_1)(x - x_2)}{(x_0 - x_1)(x_0 - x_2)} y_0 + \frac{(x - x_0)(x - x_2)}{(x_1 - x_0)(x_1 - x_2)} y_1 + \frac{(x - x_0)(x - x_1)}{(x_2 - x_0)(x_2 - x_1)} y_2$$

The polynomial of degree ≤ 2, which passes through the three points $(0,1)$, $(-1,2)$, and $(1,3)$, is

$$p_2(x) = \frac{(x+1)(x-1)}{(0+1)(0-1)} \cdot 1 + \frac{(x-0)(x-1)}{(-1-0)(-1-1)} \cdot 2 + \frac{(x-0)(x+1)}{(1-0)(1+1)} \cdot 3$$

$$= 1 + \tfrac{1}{2}x + \tfrac{3}{2}x^2$$

If a function $f(x)$ is given, then we can form an approximation to it using the interpolating polynomial

$$p_n(x) = \sum_{i=0}^{n} f(x_i) l_i(x) \tag{3.7}$$

For example, later we consider $f(x) = \log_{10} x$ and linear interpolation to it. The basic result used in analyzing the error of interpolation is the following theorem.

As notation, $\mathcal{H}\{a, b, c, \dots\}$ denotes the smallest interval containing all of the real numbers $a, b, c, \dots$.

Theorem 3.2 Let $x_0, x_1, \dots, x_n$ be distinct real numbers, and let f be a given real-valued function with $n+1$ continuous derivatives on the interval $I_t = \mathcal{H}\{t, x_0, x_1, \dots, x_n\}$, with t some given real number. Then there exists $\xi \in I_t$ with

$$f(t) - \sum_{j=0}^{n} f(x_j) l_j(t) = \frac{(t - x_0) \dots (t - x_n)}{(n+1)!} f^{(n+1)}(\xi) \tag{3.8}$$

Proof   Note that the result is trivially true if t is any node point, since then both sides of (3.8) are zero. Assume t does not equal any node point. Then define

$$E(x)=f(x)-p_n(x) \qquad p_n(x)=\sum_0^n f(x_j)l_j(x)$$

$$G(x)=E(x)-\frac{\Psi(x)}{\Psi(t)}E(t) \qquad \text{for all } x \in I_t$$

with

$$\Psi(x)=(x-x_0)\dots(x-x_n)$$

The function $G(x)$ is $n+1$ times continuously differentiable on the interval I_t, as are $E(x)$ and $\Psi(x)$. Also,

$$G(x_i)=E(x_i)-\frac{\Psi(x_i)}{\Psi(t)}E(t)=0 \qquad i=0,1,\dots,n$$

$$G(t)=E(t)-E(t)=0$$

Thus G has $n+2$ distinct zeroes in I_t. Using the mean value theorem, G' has $n+1$ distinct zeroes. Inductively, $G^{(j)}(x)$ has $n+2-j$ zeroes in I_t, for $j=0,1,\dots,n+1$. Let ξ be the zero of $G^{(n+1)}(x)$,

$$G^{(n+1)}(\xi)=0$$

Since

$$E^{(n+1)}(x)=f^{(n+1)}(x)$$

$$\Psi^{(n+1)}(x)=(n+1)!$$

we obtain

$$G^{(n+1)}(x)=f^{(n+1)}(x)-\frac{(n+1)!}{\Psi(t)}E(t)$$

Substituting $x=\xi$ and solving for $E(t)$,

$$E(t)=\frac{\Psi(t)}{(n+1)!}f^{(n+1)}(\xi)$$

the desired result.

This may seem a tricky derivation, but it is a commonly used technique for obtaining some error formulas. ∎

Example   For $n=1$, using x in place of t,

$$f(x) - \frac{(x_1 - x)f(x_0) + (x - x_0)f(x_1)}{x_1 - x_0} = \frac{(x - x_0)(x - x_1)}{2} f''(\xi_x) \qquad (3.9)$$

for some $\xi_x \in \mathcal{K}\{x_0, x_1, x\}$. The subscript x on ξ_x shows explicitly that ξ depends on x; usually we delete the subscript for convenience.

We will apply the $n=1$ case to the common high-school technique of linear interpolation in a logarithm table. Let

$$f(x) = \log_{10} x$$

Then $f''(x) = -\log_{10} e / x^2$, $\log_{10} e \doteq 0.434$. In a table, we generally would have $x_0 < x < x_1$. Then

$$E(x) = \frac{(x - x_0)(x_1 - x)}{2} \cdot \frac{\log_{10} e}{\xi^2} \qquad x_0 \leq \xi \leq x_1$$

This gives the upper and lower bounds

$$\frac{\log_{10} e}{x_1^2} \cdot \frac{(x - x_0)(x_1 - x)}{2} \leq E(x) \leq \frac{\log_{10} e}{x_0^2} \cdot \frac{(x - x_0)(x_1 - x)}{2}$$

This shows that the error function $E(x)$ looks very much like a quadratic polynomial, especially if the distance $h = x_1 - x_0$ is reasonably small. For a uniform bound on $[x_0, x_1]$,

$$\max_{x_0 \leq x \leq x_1} (x_1 - x)(x - x_0) = \frac{h^2}{4}$$

$$|\log_{10} x - p_1(x)| \leq \frac{h^2}{8} \cdot \frac{.434}{x_0^2} = \frac{.0542 h^2}{x_0^2} \leq .0542 h^2 \qquad (3.10)$$

for $x_0 \geq 1$, as is usual in a logarithm table. Note that the interpolation error in a standard table is much less for x near 10 than near 1. Also, the maximum error is near the midpoint of $[x_0, x_1]$.

For a four place table, $h = .01$,

$$|\log_{10} x - p_1(x)| \leq 5.42 \times 10^{-6} \qquad 1 \leq x_0 < x_1 \leq 10$$

Since the entries in the table are given to four digits, (e.g., $\log_{10} 2 = .3010$), this is sufficiently accurate. Why do we need a more accurate 5-place table if the above is so accurate? Because we have neglected to include the effects of the rounding errors present in the table entries. For example, with $\log_{10} 2 \doteq .3010$,

$$|\log_{10} 2 - .3010| \leq .00005$$

and this will dominate the interpolation error if x_0 or $x_1 = 2$.

Rounding error analysis for linear interpolation   Let

$$f(x_0) = f_0 + \epsilon_0, \qquad f(x_1) = f_1 + \epsilon_1$$

with f_0 and f_1 the table entries and ϵ_0, ϵ_1 the rounding errors. We will assume

$$|\epsilon_0|, |\epsilon_1| \leq \epsilon$$

for a known ϵ. In the case of the four place logarithm table, $\epsilon = .00005$.
We want to bound

$$\mathcal{E}(x) = f(x) - \frac{(x_1 - x)f_0 + (x - x_0)f_1}{x_1 - x_0}, \qquad x_0 \leq x \leq x_1$$

Using $f_i = f(x_i) - \epsilon_i$,

$$\mathcal{E}(x) = f(x) - \frac{(x_1 - x)f(x_0) + (x - x_0)f(x_1)}{x_1 - x_0} + \frac{(x_1 - x)\epsilon_0 + (x - x_0)\epsilon_1}{x_1 - x_0}$$

$$\equiv E(x) + R(x)$$

$$E(x) = \frac{(x - x_0)(x - x_1)}{2} f''(\xi) \qquad \xi \in [x_0, x_1] \tag{3.11}$$

The error $\mathcal{E}(x)$ is the sum of the theoretical interpolation error and $R(x)$, which depends on ϵ_0, ϵ_1. Since $R(x)$ is a straight line, its maximum on $[x_0, x_1]$ is attained at an endpoint,

$$\text{Max}|R(x)| = \text{Max}\{|\epsilon_0|, |\epsilon_1|\} \leq \epsilon \tag{3.12}$$

With $x_1 = x_0 + h, x_0 \leq x \leq x_1$,

$$|\mathcal{E}(x)| \leq \frac{h^2}{8} \underset{x_0 \leq t \leq x_1}{\text{Max}} |f''(t)| + \text{Max}\{|\epsilon_0|, |\epsilon_1|\} \tag{3.13}$$

Example   For the logarithm example using a four-place table,

$$|\mathcal{E}(x)| \leq 5.42 \times 10^{-6} + 5 \times 10^{-5} \doteq 5.5 \times 10^{-5}$$

For a five-place table, $h = .001$, $\epsilon = .000005$, and

$$|\mathcal{E}(x)| \leq 5.42 \times 10^{-8} + 5 \times 10^{-6} \doteq 5.05 \times 10^{-6} \qquad x_0 \leq x_1 \leq x_1$$

The rounding error is the only significant error in using linear interpolation in a five-place logarithm table. In fact, it seems worthwhile to increase the five-place table to a six-place table, without changing the mesh size h. Then we would have a maximum error for $\mathcal{E}(x)$ of 5.5×10^{-7}, without any significant increase in computation.

These arguments on rounding error generalize to higher degree polynomial interpolation, although the result on $\text{Max}|R(x)|$ is slightly more complicated; see Problem 8. None of the results of this section take into account new rounding errors which occur in the evaluation of $p_n(x)$. These are minimized by results given in the next section.

3.2 Newton Divided Differences

The Lagrange form of the interpolation polynomial can be used for interpolation to a function given in tabular form; and tables in Abramowitz and Stegun [Al, Chapter 25] can be used to evaluate the functions $l_i(x)$ more easily. But there are other forms that are much more convenient, and they are developed in this and the following section. With the Lagrange form, it is inconvenient to pass from one interpolation polynomial to another of degree one greater. But such a comparison of different degree interpolation polynomials is a useful technique in deciding what degree polynomial to use. The formulas developed in this section are for nonevenly spaced grid points $\{x_i\}$. As such they are convenient for inverse interpolation in a table, a point we will illustrate later. These formulas are specialized in Section 3.3 to the case of evenly spaced grid points.

We would like to write

$$p_n(x) = p_{n-1}(x) + C(x) \qquad C(x) = \text{correction term}$$

Then in general $C(x)$ is a polynomial of degree n, since usually degree $(p_{n-1}) = n-1$ and degree $(p_n) = n$. Also we have

$$C(x_i) = p_n(x_i) - p_{n-1}(x_i) = f(x_i) - f(x_i) = 0 \qquad i = 0, 1, \ldots, n-1$$

Thus

$$C(x) = a_n(x - x_0)\ldots(x - x_{n-1})$$

Since $p_n(x_n) = f(x_n)$, we have

$$a_n = \frac{f(x_n) - p_{n-1}(x_n)}{(x_n - x_0)\ldots(x_n - x_{n-1})}$$

For reasons to be derived below, this coefficient a_n is called the nth order Newton divided difference of f, and it is denoted by

$$a_n \equiv f[x_0, x_1, \ldots, x_n]$$

Thus our interpolation formula becomes

$$p_n(x) = p_{n-1}(x) + (x - x_0) \ldots (x - x_{n-1}) f[x_0, \ldots, x_n] \tag{3.14}$$

To obtain more information on a_n, we return to the Lagrange formula for $p_n(x)$. Write

$$\Psi_n(x) = (x - x_0) \ldots (x - x_n) \tag{3.15}$$

Then

$$\Psi_n'(x_i) = (x_i - x_0) \ldots (x_i - x_{i-1})(x_i - x_{i+1}) \ldots (x_i - x_n)$$

and if x is not a node point

$$p_n(x) = \sum_{j=0}^{n} \frac{\Psi_n(x)}{(x - x_j)\Psi_n'(x_j)} f(x_j) \tag{3.16}$$

Since a_n is the coefficient of x^n in $p_n(x)$, we use the Lagrange formula to obtain the coefficient of x^n. By looking at each nth degree term in the formula (3.16), we obtain

$$f[x_0, x_1, \ldots, x_n] = \sum_{j=0}^{n} \frac{f(x_j)}{\Psi_n'(x_j)} \tag{3.17}$$

From this formula, we obtain an important property of the divided difference. Let $(i_0, i_1, \ldots, i_n)$ be some permutation of $(0, 1, \ldots, n)$. Then easily

$$\sum_{j=0}^{n} \frac{f(x_j)}{\Psi_n'(x_j)} = \sum_{j=0}^{n} \frac{f(x_{i_j})}{\Psi_n'(x_{i_j})}$$

since the second sum is merely a rearrangement of the first one. But then

$$f[x_0, x_1, \ldots, x_n] = f[x_{i_0}, x_{i_1}, \ldots, x_{i_n}] \tag{3.18}$$

for any permutation $(i_0, \ldots, i_n)$ of $(0, 1, \ldots, n)$.

A second useful formula can also be derived from (3.17), mostly by judicious manipulation (see Problem 11):

$$f[x_0, x_1, \ldots, x_n] = \frac{f[x_1, \ldots, x_n] - f[x_0, \ldots, x_{n-1}]}{x_n - x_0} \tag{3.19}$$

the justification for the name "divided difference." This agrees with the definitions (1.14) used earlier in Chapter 1.

Returning to the formula (3.14), we have the formulas

$$p_0(x) = f(x_0)$$

$$p_1(x) = f(x_0) + (x - x_0)f[x_0, x_1]$$

$$\vdots$$

$$p_n(x) = f(x_0) + (x - x_0)f[x_0, x_1] + (x - x_0)(x - x_1)f[x_0, x_1, x_2] + \cdots$$

$$+ (x - x_0)\ldots(x - x_{n-1})f[x_0, x_1, \ldots, x_n] \tag{3.20}$$

This is called Newton's divided difference formula for the interpolating polynomial. It is much better for computation than the Lagrange formula.

To construct the divided differences, use the format shown in Table 3.1. Each numerator of a difference is obtained by differencing the two adjacent entries in the column to the left of the column you are constructing.

Table 3.1   Format for constructing divided differences of $f(x)$

x_i	$f(x_i)$	$f[x_i, x_{i+1}]$	$f[x_i, x_{i+1}, x_{i+2}]\ldots$
x_0	f_0		
		$f[x_0, x_1]$	
x_1	f_1		$f[x_0, x_1, x_2]\cdots$
		$f[x_1, x_2]$	
x_2	f_2		$f[x_1, x_2, x_3]$
		$f[x_2, x_3]$	
x_3	f_3		$f[x_2, x_3, x_4]$
		$f[x_3, x_4]$	
x_4	f_4		$f[x_3, x_4, x_5]$
		$f[x_4, x_5]$	
x_5	f_5		
$\vdots$	$\vdots$	$\vdots$	$\vdots$

Example   We construct a divided difference table for $f(x) = \sqrt{x}$, shown in Table 3.2. We have used the notation $D^r f(x_i) = f[x_i, x_{i+1}, \ldots, x_{i+r}]$

A simple algorithm can be given for constructing the differences

$$f(x_0), f[x_0, x_1], f[x_0, x_1, x_2], \ldots, f[x_0, x_1, \ldots, x_n]$$

which are necessary for evaluating the Newton form (3.20).

Table 3.2 Example of constructing divided differences

x_i	$f(x_i)$	$f[x_i, x_{i+1}]$	$D^2 f[x_i]$	$D^3 f[x_i]$	$D^4 f(x_i)$
2.0	1.414214		0.34102 − 0.34924		
		.34924	0, 2		
2.1	1.449138		−.0411		
		.34102		.009167	
2.2	1.483240		−.03835		−.002084
		.33335		.008333	
2.3	1.516575		−.03585		
		.32618			
2.4	1.549193				

(handwritten annotations: $\frac{f(2.1) - f(2)}{.1}$ over the .34924 cell)

Algorithm Divdif (d, x, n)

1. Remark. d and x are vectors with entries $f(x_i)$ and $x_i, i = 0, 1, \ldots, n$, respectively. On exit, d_i will contain $f[x_0, \ldots, x_i]$.

2. Do thru step 4 for $i = 1, 2, \ldots, n$.

3. Do thru step 4 for $j = n, n - 1, \ldots, i$.

4. $d_j := (d_j - d_{j-1})/(x_j - x_{j-i})$.

5. Exit from the algorithm.

To evaluate the Newton form of the interpolating polynomial (3.20), we give a simple variant of the nested polynomial multiplication (2.64) of Chapter 2.

Algorithm Interp(d, x, n, t, p)

1. Remark. On entrance, d and x are vectors containing $f[x_0, \ldots, x_i]$ and $x_i, i = 0, 1, \ldots, n$, respectively. On exit, p will contain the value $p_n(t)$ of the nth degree polynomial interpolating to f on x.

2. $p := d_n$

3. Do through step 4 for $i = n - 1, n - 2, \ldots, 0$.

4. $p := d_i + (t - x_i)p$

5. Exit the algorithm.

Example For $f(x) = \sqrt{x}$, we give in Table 3.3 the values of $p_n(x)$ for various values of x and n. The highest degree polynomial $p_4(x)$ uses function values at the grid points $x_0 = 2.0$ through $x_4 = 2.4$. The necessary divided differences are given in the last example, Table 3.2.

Table 3.3 Example of use of Newton's formula (3.20)

	$p_1(x)$	$p_2(x)$	$p_3(x)$	$p_4(x)$	$\sqrt{x}$
2.05	1.431676	1.431779	1.431782	1.431782	1.431782
2.15	1.466600	1.466292	1.466288	1.466288	1.466288
2.45	1.571372	1.564899	1.565260	1.565247	1.565248

When a value of x falls outside $\mathcal{H}\{x_0, x_1, \ldots, x_n\}$, we often say that $p_n(x)$ extrapolates to $f(x)$. In the last example, note the greater inaccuracy of the extrapolated value $p_3(2.45)$ as compared with $p_3(2.05)$ and $p_3(2.15)$. In this text, the word interpolation will always include the possibility that x falls outside the interval $\mathcal{H}\{x_0, \ldots, x_n\}$.

Often we know the value of the function $f(x)$, and we want to compute the corresponding value of x. This is called "inverse interpolation." It is commonly known to users of logarithm tables as computing the antilog of a number. To compute x, we treat it as the dependent variable and $y = f(x)$ as the independent variable. Given table values $(x_i, y_i), i = 0, \ldots, n$, we produce a polynomial $p_n(y)$ that interpolates to x_i at $y_i, i = 0, \ldots, n$. In effect, we are interpolating to the inverse function $g(y) \equiv f^{-1}(y)$; and in the error formula of Theorem 3.2, with $x = f^{-1}(y)$,

$$x - p_n(y) = \frac{(y - y_0) \ldots (y - y_n)}{(n+1)!} g^{(n+1)}(\zeta) \tag{3.21}$$

for some $\zeta \in \mathcal{H}\{y, y_0, y_1, \ldots, y_n\}$. If they are needed, the derivatives of $g(y)$ can be computed by differentiating the composite formula

$$g(f(x)) = x$$

for example,

$$g'(y) = 1/f'(x) \qquad \text{for } y = f(x)$$

Example Consider the Table 3.4 of values of the Bessel function $J_0(x)$, taken from Abramowitz and Stegun [A1, Chapter 9]. We will calculate the value of x for which $J_0(x) = 0.1$. Table 3.5 gives values of $p_n(y)$ for $n = 0, 1, \ldots, 6$, with $x_0 = 2.0$. The polynomial $p_n(y)$ is interpolating to the inverse function for $J_0(x)$, call it $g(y)$. The answer $x = 2.2186838$ is correct to eight significant digits.

An interpolation error formula using divided differences Let t be a real number, distinct from the node points $x_0, x_1, \ldots, x_n$. Construct the polynomial interpolat-

Table 3.4 Values of Bessel function $J_0(x)$

x	$J_0(x)$
2.0	.2238907791
2.1	.1666069803
2.2	.1103622669
2.3	.0555397844
2.4	.0025076832
2.5	−.0483837764
2.6	−.0968049544
2.7	−.1424493700
2.8	−.1850360334
2.9	−.2243115458

Table 3.5 Example of inverse interpolation

n	$p_n(y)$	$p_n(y) - p_{n-1}(y)$	$g[y_0, \ldots, y_n]$
0	2.0		2.0
1	2.216275425	2.16E−1	−1.745694282
2	2.218619608	2.34E−3	.2840748405
3	2.218686252	6.66E−5	−.7793711812
4	2.218683344	−2.91E−6	.7648986704
5	2.218683964	6.20E−7	−1.672357264
6	2.218683773	−1.91E−7	3.477333126

ing to $f(x)$ at $x_0, \ldots, x_n$, and t:

$$p_{n+1}(x) = f(x_0) + (x - x_0)f[x_0, x_1] + \ldots + (x - x_0)\ldots(x - x_{n-1})f[x_0, \ldots, x_n]$$

$$+ (x - x_0)\ldots(x - x_n)f[x_0, x_1, \ldots, x_n, t]$$

$$= p_n(x) + (x - x_0)\ldots(x - x_n)f[x_0, \ldots, x_n, t]$$

Since $p_{n+1}(t) = f(t)$, we have

$$f(t) = p_n(t) + (t - x_0)\ldots(t - x_n)f[x_0, \ldots, x_n, t] \qquad (3.22)$$

and this gives us another formula for the error $f(t) - p_n(t)$. Comparing this with the earlier error formula (3.8), and canceling the multiplying polynomial $\Psi_n(t)$, we have

$$f[x_0, x_1, \ldots, x_n, t] = \frac{f^{(n+1)}(\xi)}{(n+1)!}$$

for some $\xi \in \mathcal{K}\{x_0, x_1, \ldots, x_n, t\}$. To make this result symmetric in the arguments, we generally let $t = x_{n+1}, n = m-1$, and obtain

$$f[x_0, x_1, \ldots, x_m] = \frac{f^{(m)}(\xi)}{m!} \qquad \text{some } \xi \in \mathcal{K}\{x_0, \ldots, x_m\} \qquad (3.23)$$

With this result, the Newton formula (3.20) looks like a truncated Taylor series for $f(x)$, expanded about x_0, provided the size $x_n - x_0$ is not too large.

Example   From the Table 3.2 of divided differences for $f(x) = \sqrt{x}$,

$$f[2.0, 2.1, \ldots, 2.4] = -.002084$$

Since $f^{(4)}(x) = -15/(16x^3\sqrt{x})$, it is easy to show that

$$\frac{f^{(4)}(2.31)}{4!} \doteq -.002084$$

so $\xi \doteq 2.31$ in (3.23) for this case.

Example   If the formula (3.23) is used to estimate the derivatives of the inverse function $g(y)$ of $J_0(x)$, given in the previous example, then the derivatives $g^{(n)}(x)$ are growing rapidly with n. For example,

$$g[y_0, \ldots, y_6] \doteq 3.48$$

$$g^{(6)}(\xi) \doteq (6!)(3.48) \doteq 2500$$

for some ξ in $[-.0968, .2239]$. Similar estimates can be computed for the other derivatives.

To extend the definition of Newton divided difference to the case in which some or all of the nodes coincide, we introduce another formula for the divided difference.

Theorem 3.3   (Hermite-Gennochi).   Let $x_0, x_1, \ldots, x_n$ be distinct, and let $f(x)$ be n times continuously differentiable on the interval $\mathcal{K}\{x_0, x_1, \ldots, x_n\}$. Then

$$f[x_0, x_1, \ldots, x_n] = \int \cdots \int_{\tau_n} f^{(n)}(t_0 x_0 + \ldots + t_n x_n) dt_1 \ldots dt_n \qquad (3.24)$$

in which

$$T_n = \left\{ (t_1, t_2, \ldots, t_n) | t_1 \geq 0, \ldots, t_n \geq 0, \sum_1^n t_i \leq 1 \right\}$$

(3.25)

$$t_0 = 1 - \sum_1^n t_i$$

Note that $t_0 \geq 0$ and $\sum_0^n t_i = 1$.

Proof We will show that (3.24) is true for $n=1$ and 2, and these two cases should suggest the general induction proof.

1. $n=1$. Then $\tau_1 = [0, 1]$.

$$\int_0^1 f'(t_0 x_0 + t_1 x_1)\, dt_1 = \int_0^1 f'(x_0 + t_1(x_1 - x_0))\, dt_1$$

$$= \frac{1}{x_1 - x_0} f(x_0 + t_1(x_1 - x_0)) \Big]_{t_1=0}^{t_1=1}$$

$$= \frac{f(x_1) - f(x_0)}{x_1 - x_0} = f[x_0, x_1]$$

2. $n=2$. Then τ_2 is the triangle with vertices $(0,0)$, $(0,1)$, and $(1,0)$, shown in Figure 3.1

$$\iint_{\tau_2} f''(t_0 x_0 + t_1 x_1 + t_2 x_2)\, dt_1\, dt_2$$

$$= \int_0^1 \int_0^{1-t_1} f''(x_0 + t_1(x_1 - x_0) + t_2(x_2 - x_0))\, dt_2\, dt_1$$

$$= \int_0^1 \frac{1}{x_2 - x_0} \{ f'(x_0 + t_1(x_1 - x_0) + t_2(x_2 - x_0)) \}_{t_2=0}^{t_2=1-t_1}\, dt_1$$

$$= \frac{1}{x_2 - x_0} \left\{ \int_0^1 f'(x_2 + t_1(x_1 - x_2))\, dt_1 - \int_0^1 f'(x_0 + t_1(x_1 - x_0))\, dt_1 \right\}$$

$$= \frac{1}{x_2 - x_0} \{ f[x_1, x_2] - f[x_0, x_1] \} = f[x_0, x_1, x_2].$$

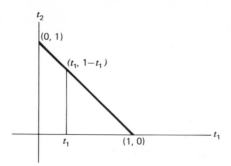

Figure 3.1 Region τ_2.

Do the general case by a similar procedure. Integrate once and reduce to one lower dimension. Then invoke the induction hypothesis and use (3.19) to complete the proof. ∎

We can now look at $f[x_0, x_1, \ldots, x_n]$ using (3.24); and doing so we see that it is a continuous function of the n variables $x_0, x_1, \ldots, x_n$, regardless of whether they are distinct or not. For example, if we let all points coalesce to x_0, then for the nth-order divided difference,

$$f[x_0, \ldots, x_0] = \int \cdots \int_{\tau_n} f^{(n)}(x_0) \, dt_1 \ldots dt_n$$

$$= f^{(n)}(x_0) \cdot \mathrm{Vol}(\tau_n)$$

From Problem 14, $\mathrm{Vol}(\tau_n) = 1/n!$, and thus

$$f[x_0, \ldots, x_0] = \frac{f^{(n)}(x_0)}{n!} \tag{3.26}$$

This could have been predicted directly from (3.23). But if only some of the nodes coalesce, we must use (3.24).

In applications to numerical integration, we need to know whether

$$\frac{d}{dx} f[x_0, \ldots, x_n, x] \tag{3.27}$$

exists. If f is $n+2$ times continuously differentiable, then we can apply (3.24). By applying theorems on differentiating an integral with respect to a parameter in the integrand, we can conclude the existence of (3.27). More directly,

$$\frac{d}{dx} f[x_0, x_1, \ldots, x_n, x] = \underset{h \to 0}{\text{Limit}} \frac{f[x_0, x_1, \ldots, x_n, x+h] - f[x_0, x_1, \ldots, x_n, x]}{h}$$

$$= \underset{h \to 0}{\text{Limit}} \frac{f[x_0, \ldots, x_n, x+h] - f[x, x_0, x_1, \ldots, x_n]}{h}$$

$$= \underset{h \to 0}{\text{Limit}} f[x, x_0, x_1, \ldots, x_n, x+h]$$

$$= f[x, x_0, x_1, \ldots, x_n, x],$$

$$\frac{d}{dx} f[x_0, x_1, \ldots, x_n, x] = f[x_0, x_1, \ldots, x_n, x, x] \tag{3.28}$$

The existence and continuity of the right-hand side is guaranteed using (3.24).

There is a rich theory involving polynomial interpolation and divided differences, but we conclude at this point with one final easy result. If $f(x)$ is a polynomial of degree m, then

$$f[x_0, x_1, \ldots, x_n, x] = \begin{cases} \text{polynomial of degree} & m-n-1, n \leq m-1 \\ a_m & , n = m-1 \\ 0 & , n > m-1 \end{cases} \tag{3.29}$$

where $f(x) = a_m x^m + \text{lower-degree terms}$. For the proof, see Problem 15.

3.3 Finite Differences and Table-Oriented Interpolation Formulas

We introduce forward, backward, and central differences in this section. These are used to produce interpolation formulas for tables in which the abscissas $\{x_i\}$ are evenly spaced. In addition, the differences can be used to determine the maximum degree of interpolation polynomial that can be used safely, based on the accuracy of the table entries. And finite differences can be used to detect *noise* in data, when the noise is large with respect to the rounding errors or uncertainty errors of physical measurement. This is developed in Section 3.4.

For a given $h > 0$, define

$$\Delta_h f(z) = f(z+h) - f(z)$$

Generally the h is understood from the context, and we write

$$\Delta f(z) = f(z+h) - f(z) \tag{3.30}$$

This is called the *forward difference* of f at z, and Δ is called the *forward difference operator*. We will always be working with evenly spaced node points $x_i = x_0 + ih$, $i = 0, 1, 2, 3, \ldots$. Then we write

$$\Delta f(x_i) = f(x_{i+1}) - f(x_i)$$

or more concisely,

$$\Delta f_i = f_{i+1} - f_i \qquad f(x_i) = f_i$$

For $r \geq 0$, define

$$\Delta^{r+1} f(z) = \Delta^r f(z+h) - \Delta^r f(z) \tag{3.31}$$

with $\Delta^0 f(z) = f(z)$. $\Delta^r f(z)$ is the rth order forward difference of f at z. Forward differences are quite easy to compute, and examples will be given later in connection with an interpolation formula.

We first derive results for the forward difference operator by applying the results of the Newton divided difference.

Lemma 1 For $k \geq 0$,

$$f[x_0, x_1, \ldots, x_k] = \frac{1}{k! h^k} \Delta^k f_0 \tag{3.32}$$

Proof For $k = 0$, the result is trivially true. For $k = 1$,

$$f[x_0, x_1] = \frac{f_1 - f_0}{x_1 - x_0} = \frac{1}{h} \Delta f_0$$

which shows (3.32). Assume the result (3.32) is true for all forward differences of order $k \leq r$. Then for $k = r+1$, using (3.19),

$$f[x_0, x_1, \ldots, x_{r+1}] = \frac{f[x_1, \ldots, x_{r+1}] - f[x_0, \ldots, x_r]}{x_{r+1} - x_0}$$

Applying the induction hypothesis, this equals

$$\frac{1}{(r+1)h} \left(\frac{1}{r! h^r} \Delta^r f_1 - \frac{1}{r! h^r} \Delta^r f_0 \right) = \frac{1}{(r+1)! h^{r+1}} \Delta^{r+1} f_0$$

We now modify the Newton interpolation formula (3.20) to a formula involving forward differences in place of divided differences. For a given value of x at which we will evaluate the interpolating polynomial, define

$$\mu = \frac{x - x_0}{h}$$

to indicate the position of x relative to x_0. For example, $\mu = 1.6$ means x is $\frac{6}{10}$ of the distance from x_1 to x_2. We need a formula for

$$(x - x_0)\ldots(x - x_k)$$

with respect to the variable μ. Since

$$x - x_j = x_0 + \mu h - (x_0 + jh) = (\mu - j)h$$

$$(x - x_0)\ldots(x - x_k) = \mu(\mu - 1)\ldots(\mu - k)h^{k+1} \tag{3.33}$$

Combining (3.33) and (3.32) with the divided difference interpolation formula (3.20), we obtain

$$p_n(x) = f_0 + \mu h \frac{\Delta f_0}{h} + \mu(\mu - 1)h^2 \frac{\Delta^2 f_0}{2!h^2} + \cdots$$

$$+ \mu(\mu - 1)\ldots(\mu - n + 1)h^n \frac{\Delta^n f_0}{n!h^n}$$

Define the binomial coefficients,

$$\binom{\mu}{k} = \frac{\mu(\mu - 1)\ldots(\mu - k + 1)}{k!} \qquad k > 0 \tag{3.34}$$

and $\binom{\mu}{0} = 1$. Then

$$p_n(x) = \sum_{j=0}^{n} \binom{\mu}{j} \Delta^j f_0 \qquad \mu = \frac{x - x_0}{h} \tag{3.35}$$

This is the Newton forward difference form of the interpolating polynomial.

Example For $n = 1$,

$$p_1(x) = f_0 + \mu \Delta f_0$$

This is the formula that most people use when doing linear interpolation in a

Table 3.6 Format for constructing forward differences

x_i	f_i	Δf_i	$\Delta^2 f_i$	$\Delta^3 f_i$	$\cdots$
x_0	f_0				
		Δf_0			
x_1	f_1		$\Delta^2 f_0$		
		Δf_1		$\Delta^3 f_0$	
x_2	f_2		$\Delta^2 f_1$		
		Δf_2		$\Delta^3 f_1$	
x_3	f_3		$\Delta^2 f_2$		
		Δf_3		$\Delta^3 f_2$	
x_4	f_4		$\Delta^2 f_3$		
		Δf_4			
x_5	f_5				

table. For $n=2$,

$$p_2(x) = f_0 + \mu \Delta f_0 + \frac{\mu(\mu-1)}{2} \Delta^2 f_0$$

which is an easily computable form of the quadratic interpolating polynomial.

The forward differences are constructed in a pattern like that for divided differences, but now there are no divisions; see Table 3.6.

Example The forward differences for $f(x) = \sqrt{x}$ are given in Table 3.7. The values of the interpolating polynomial will be the same as those obtained using the Newton divided difference formula, but the forward difference formula (3.35) is much easier to compute.

Table 3.7 Forward difference table for $f(x) = \sqrt{x}$

x_i	f_i	Δf_i	$\Delta^2 f_i$	$\Delta^3 f_i$	$\Delta^4 f_i$
2.0	1.414214				
		.034924			
2.1	1.449138		−.000822		
		.034102		.000055	
2.2	1.483240		−.000767		−.000005
		.033335		.00005	
2.3	1.516575		−.000717		
		.032618			
2.4	1.549193				

Example Evaluate $p_n(x)$ using Table 3.7 for $n = 1, 2, 3, 4$, with $x = 2.15$. Note that $\sqrt{2.15} = 1.4662878$ and $\mu = 1.5$.

$$p_1(x) = 1.414214 + 1.5(.034924)$$

$$= 1.414214 + .052386 = 1.4666$$

$$p_2(x) = p_1(x) + \frac{(1.5)(.5)}{2}(-.000822)$$

$$= 1.4666 - .00030825 = 1.466292$$

$$p_3(x) = p_2(x) + \frac{(1.5)(.5)(-.5)}{6}(.000055)$$

$$= 1.466292 - .0000034 = 1.466288$$

$$p_4(x) = p_3(x) + \frac{(1.5)(.5)(-.5)(-1.5)}{24}(.000005)$$

$$= 1.466288 + .00000012 = 1.466288$$

The correction terms are easily computed; and by observing their size, you obtain a generally accurate idea of when the degree n is sufficiently large. Note that the seven place accuracy in the table values of $\sqrt{x}$ has lead to at most one place of accuracy in the forward difference $\Delta^4 f_0$. The forward differences of order greater than three are almost entirely the result of differencing the rounding errors in the table entries; and consequently interpolation in this table should be limited to polynomials of degree less than four. This is given further theoretical justification in the next section.

There are other forms of differences and associated interpolation formulas. Define the *backward difference* by

$$\nabla f(z) = f(z) - f(z - h)$$

$$\nabla^{r+1} f(z) = \nabla^r f(z) - \nabla^r f(z - h) \qquad r \geq 1 \qquad (3.36)$$

Completely analogous results to those for forward differences can be derived. And we obtain the Newton backward difference interpolation formula,

$$p_n(x) = f_0 + \binom{-\nu}{1}\nabla f_0 + \binom{-\nu+1}{2}\nabla^2 f_0 + \dots + \binom{-\nu+n-1}{n}\nabla^n f_0 \qquad (3.37)$$

In this formula, the interpolation nodes are $x_0, x_{-1}, x_{-2}, \ldots, x_{-n}$, with $x_{-j} = x_0 - jh$, as before. The value v is given by

$$v = \frac{x_0 - x}{h}$$

reflecting the fact that x will generally be less than x_0 when using this formula. A backward difference diagram can be constructed in an analogous way to that for forward differences. The backward difference formula will be used in Chapter 6 to develop the Adams family of formulas for the numerical solution of differential equations.

Because we generally prefer to have interpolation node points distributed symmetrically about the point x at which we are interpolating, we introduce *central differences*. Define

$$\delta f(z) = f\left(z + \frac{h}{2}\right) - f\left(z - \frac{h}{2}\right) \tag{3.38}$$

We will be interested in only the even-powered terms of δ,

$$\delta^2 f_j = f(x_{j+1}) - 2f(x_j) + f(x_{j-1})$$

$$\delta^{2k} f_j = \delta^{2k-2} f_{j+1} - 2\delta^{2k-2} f_j + \delta^{2k-2} f_{j-1} \qquad k \geq 2 \tag{3.39}$$

Simple diagrams can be constructed to calculate these differences, just as with the forward differences. The following formula is called Everett's formula, and it uses the interpolation nodes

$$x_{-m}, x_{-m+1}, \ldots, x_0, x_1, \ldots, x_{m+1}$$

This is an even number of node points, and we use $n = 2m + 1$. Generally we would have $x_0 < x < x_1$; and then define

$$\mu = \frac{x - x_0}{h} \qquad v = 1 - \mu$$

The interpolation formula is

$$P_n(x) = vf_0 + \frac{v(v^2 - 1)}{3!} \delta^2 f_0 + \ldots + \frac{v(v^2 - 1)(v^2 - 4) \ldots (v^2 - m^2)}{(2m+1)!} \delta^{2m} f_0$$

$$+ \mu f_1 + \frac{\mu(\mu^2 - 1)}{3!} \delta^2 f_1 + \ldots + \frac{\mu(\mu^2 - 1) \ldots (\mu^2 - m^2)}{(2m+1)!} \delta^{2m} f_1 \tag{3.40}$$

For a derivation, see Hildebrand [H1, p. 103].

3.4   Errors in Data and Forward Differences

We can use a forward difference table to detect *noise* in physical data, as long as the noise is large relative to the usual limits of experimental error. We must begin with some preliminary lemmas.

Lemma 2 $\Delta^r f(x_i) = h^r f^{(r)}(\xi_i)$, for some $x_i \leqq \xi_i \leqq x_{i+r}$.

Proof $\Delta^r f_i = h^r r! f[x_i, \ldots, x_{i+r}] = h^r r! \dfrac{f^{(r)}(\xi_i)}{r!} = h^r f^{(r)}(\xi_i)$,   using   Lemma   (1)   and   (3.23).   ∎

Lemma 3   For any two functions f and g and any two constants α and β,

$$\Delta^r \big[\alpha f(x) + \beta g(x) \big] = \alpha \Delta^r f(x) + \beta \Delta^r g(x) \qquad r \geqq 0$$

Proof   The result is trivial if $r=0$ or $r=1$. Assume the result is true for all $r \leqq n$, and prove it for $r = n+1$.

$$\Delta^{n+1}\big[\alpha f(x) + \beta g(x) \big] = \Delta^n \big[\alpha f(x+h) + \beta g(x+h) \big] - \Delta^n \big[\alpha f(x) + \beta g(x) \big]$$

$$= \alpha \Delta^n f(x+h) + \beta \Delta^n g(x+h) - \alpha \Delta^n f(x) - \beta \Delta^n g(x)$$

using the definition (3.31) of Δ^{n+1} and the induction hypothesis. Then by recombining, we obtain

$$\alpha \big[\Delta^n f(x+h) - \Delta^n f(x) \big] + \beta \big[\Delta^n g(x+h) - \Delta^n g(x) \big] = \alpha \Delta^{n+1} f(x) + \beta \Delta^{n+1} g(x)$$

∎

Lemma (2) says that if the derivatives of $f(x)$ are bounded, or if they do not increase rapidly compared with h^{-n}, then the forward differences $\Delta^n f(x)$ should become smaller as n increases. We will look at the effect of rounding errors and of other errors of a larger magnitude than rounding errors. Let

$$f(x_i) = \tilde{f}_i + e(x_i) \qquad i = 0, 1, 2, \ldots \tag{3.41}$$

with $\tilde{f}_i$ a table value that we use in constructing the forward difference table. Then

$$\Delta^r \tilde{f}_i = \Delta^r f(x_i) - \Delta^r e(x_i)$$

$$= h^r f^{(r)}(\xi_i) - \Delta^r e(x_i) \tag{3.42}$$

The first term becomes smaller as r increases, as illustrated in the earlier forward difference table for $f(x) = \sqrt{x}$.

To better understand the behaviour of $\Delta^r e(x_i)$, consider the simple case in which

$$e(x_i) = \begin{cases} 0, & i \neq k \\ \epsilon, & i = k \end{cases} \tag{3.43}$$

The forward differences of this function are given in Table 3.8. It can be proven that the column for $\Delta^r e(x_i)$ will look like

$$0,\ldots,0,\epsilon, -\binom{r}{1}\epsilon, \binom{r}{2}\epsilon, -\binom{r}{3}\epsilon, \ldots, (-1)^{r+1}\epsilon, 0, \ldots \tag{3.44}$$

Thus the effect of a single rounding error will propagate and increase in value as larger-order differences are formed.

With rounding errors defining a general error function as in (3.41), their effect can be looked on as a sum of functions of the form (3.43). Since the values of $e(x_i)$ in general will vary in sign and magnitude, their effects will overlap in a seemingly random manner. But their differences will still grow in size, and the higher-order differences of table values $\tilde{f}_i$ will eventually become useless. When differences $\Delta^r \tilde{f}_i$ begin to increase in size with increasing r, then these differences are most likely dominated by rounding errors and should not be used. An interpolation formula of degree less than r should be used.

Table 3.8   Forward differences of error function $e(x)$

x_i	$e(x_i)$	$\Delta e(x_i)$	$\Delta^2 e(x_i)$	$\Delta^3 e(x_i)$
$\vdots$	$\vdots$		$\vdots$	
		0		0
			$\vdots$	
		$\vdots$		$\vdots$
x_{k-2}	0		0	
		0		ϵ
x_{k-1}	0		ϵ	
		ϵ		-3ϵ
x_k	ϵ		-2ϵ	
		$-\epsilon$		3ϵ
x_{k+1}	0		ϵ	
		0		$-\epsilon$
x_{k+2}	0	$\vdots$	0	0
$\vdots$	$\vdots$	$\vdots$	$\vdots$	$\vdots$

Table 3.9 Example of detecting an isolated error in data

f_i	$\Delta \tilde{f}_i$	$\Delta^2 \tilde{f}_i$	$\Delta^3 \tilde{f}_i$	Error Guess	Guess $\Delta^3 f(x_i)$
.10396					
	.01700				
.12096		−.00014			
	.01686		−.00003	0	−.00003
.13782		−.00017			
	.01669		−.00002	0	−.00002
.15451		−.00019			
	.01650		.00006	ϵ	−.00002
.17101		−.00013			
	.01637		−.00025	-3ϵ	−.00002
.18738		−.00038			
	.01599		.00021	3ϵ	−.00002
.20337		−.00017			
	.01582		−.00010	$-\epsilon$	−.00002
.21919		−.00027			
	.01555				
.23474					

Detecting noise in data This same analysis can be used to detect and correct isolated errors that are large relative to rounding error. Since (3.42) says that the effect of the errors will eventually dominate, we look for a pattern like (3.44). The general technique is illustrated in Table 3.9. From (3.42),

$$\Delta^r e(x_i) = \Delta^r f(x_i) - \Delta^r \tilde{f}_i$$

Using $r = 3$ and one of the error entries chosen arbitrarily, say the first,

$$\epsilon = -.00002 - (.00006) = -.00008$$

Try this to see how it will alter the column of $\Delta^r \tilde{f}_i$; see Table 3.10. This will not be improved on by another choice of ϵ, say $\epsilon = -.00007$, although the results

Table 3.10 Correcting a data error

$\Delta^3 \tilde{f}_i$	$\Delta^3 e(x_i)$	$\Delta^3 f(x_i)$
−.00002	0	−.00002
.00006	−.00008	−.00002
−.000025	.00024	−.00001
.00021	−.00024	−.00003
−.00010	.00008	−.00002

may be equally good. Tracing backward, the entry $\tilde{f}_i = .18738$ should be

$$f(x_i) = \tilde{f}_i + e(x_i) = .18708 + (-.00008) = .18730$$

In a table in which there are two or three isolated errors, their higher-order differences may overlap, making it more difficult to discover the errors; see Problem 18.

3.5 Further Results on Interpolation Error

Consider again the error formula

$$f(x) - p_n(x) = \frac{(x - x_0) \dots (x - x_n)}{(n+1)!} f^{(n+1)}(\xi_x), \quad \xi_x \in \mathcal{H}\{x_0, \dots, x_n, x\} \quad (3.45)$$

We assume that $f(x)$ is $n+1$ times continuously differentiable on an interval I that contains $\mathcal{H}\{x_0, \dots, x_n, x\}$ for all values of x of interest. Since ξ_x is unknown, we must replace $f^{(n+1)}(\xi_x)$ by

$$c_{n+1} = \underset{t \in I}{\text{Max}} |f^{(n+1)}(t)| \quad (3.46)$$

in order to evaluate (3.45). We concentrate our attention on the polynomial

$$\Psi_n(x) = (x - x_0) \dots (x - x_n)$$

From (3.45),

$$\underset{x \in I}{\text{Max}} |f(x) - p_n(x)| \leq \frac{c_{n+1}}{(n+1)!} \underset{x \in I}{\text{Max}} |\Psi_n(x)| \quad (3.47)$$

We consider only the case with the nodes evenly spaced, $x_j = x_0 + jh$ for $j = 0, 1, \dots, n$.

Case 1 $n = 1$. $\Psi_1(x) = (x - x_0)(x - x_1)$

$$\underset{x_0 \leq x \leq x_1}{\text{Max}} |\Psi_1(x)| = \frac{h^2}{4}$$

See Figure 3.2 for an illustration.

Case 2 $n = 2$. Shift the polynomial $\Psi_2(x) = (x - x_0)(x - x_1)(x - x_2)$ along the x-axis to obtain the polynomial

$$\Psi_2(x) = (x + h)x(x - h)$$

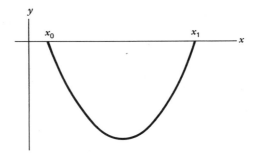

Figure 3.2 $y = \Psi_1(x)$

whose shape and size is exactly the same as with the original polynomial. Using this modification,

$$\underset{x_1 - \frac{h}{2} \leq x \leq x_1 + \frac{h}{2}}{\text{Max}} |\Psi_2(x)| = .375h^3$$

$$\underset{x_0 \leq x \leq x_2}{\text{Max}} |\Psi_2(x)| = \frac{2\sqrt{3}}{9} h^3 \doteq .385h^3$$

The graph of $\Psi_2(x)$ is shown in Figure 3.3. Thus it doesn't matter if x is located near x_1 in the interval $[x_0, x_2]$. But it will make a difference for higher-degree interpolation. Combining the above with (3.47),

$$\underset{x_0 \leq x \leq x_2}{\text{Max}} |f(x) - p_2(x)| \leq \frac{\sqrt{3}}{27} h^3 \cdot \underset{x_0 \leq x \leq x_2}{\text{Max}} |f'''(t)| \qquad (3.48)$$

and $\sqrt{3}/27 \doteq .064$.

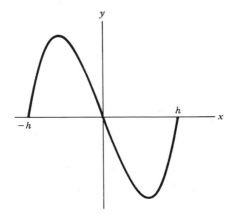

Figure 3.3 $y = \Psi_2(x)$

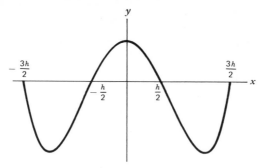

Figure 3.4 $y = \Psi_3(x)$

Case 3 $n=3$. As above, shift the polynomial to make the nodes symmetric about the origin, to obtain

$$\Psi_3(x) = \left(x^2 - \tfrac{9}{4}h^2\right)\left(x^2 - \tfrac{1}{4}h^2\right)$$

The graph of $\Psi_3(x)$ is shown in Figure 3.4. Using this modification,

$$\underset{x_1 \leq x \leq x_2}{\text{Max}} |\Psi_3(x)| = \frac{9}{16}h^4 \doteq 0.56h^4$$

$$\underset{x_0 \leq x \leq x_3}{\text{Max}} |\Psi_3(x)| = h^4$$

Thus in interpolation to $f(x)$ at x, the nodes should be so chosen that $x_1 < x < x_2$. Then from (3.45),

$$\underset{x_1 \leq x \leq x_2}{\text{Max}} |f(x) - p_3(x)| \leq \frac{3h^4}{128} \underset{x_0 \leq t \leq x_3}{\text{Max}} |f^{(4)}(t)| \tag{3.49}$$

Case 4 For a general $n>3$, the behavior exhibited above with $n=3$ is accentuated. For example, consider the graph of $\Psi_6(x)$ in Figure 3.5. As earlier, we can show

$$\underset{x_2 \leq x \leq x_4}{\text{Max}} |\Psi_6(x)| \doteq 12.36h^7$$

$$\underset{x_0 \leq x \leq x_6}{\text{Max}} |\Psi_6(x)| \doteq 95.8h^7$$

To minimize the interpolation error, the interpolation nodes should be so chosen that the interpolation point x is as near as possible to the midpoint of $[x_0, x_n]$. It is inadvisable to use high-degree polynomial interpolation, say $n \geq 8$, on evenly spaced nodes unless the point x of interpolation is near the midpoint of the nodes being used.

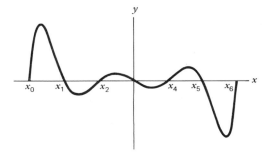

Figure 3.5 $y = \Psi_6(x)$

This result points out the disadvantage of the Newton forward difference interpolation formula (3.35), especially if you do not initially know the desired n. If you begin with a linear formula and $x_0 < x < x_1$ and then you add additional points to obtain smaller correction terms, the point x moves further away from the middle of $[x_0, x_n]$ with each increase in n. It would be desirable to have a formula in which new node points are added symmetrically about the points already being used. Everett's formula (3.40) is such a formula, and there are others, for example, Hildebrand [H1, pp. 97–105].

Example Consider interpolation to $J_0(x)$ at $x = 2.45$, using the table of values of $J_0(x)$ given in an example in Section 3.2. Using the Newton forward difference formula with $x_0 = 2.4$, we obtain the results in Table 3.11. The interpolation nodes are 2.4, 2.5,..., 2.9, and x lies near the left-hand node point. The correct answer is $J_0(2.45) = -.023226743305$, and $p_5(2.45)$ is in error by $-4.9\text{E}-9$.

We now use Everett's formula. Let $x_0 = 2.4$, as before; but the interpolation interval is now $[2.2, 2.7]$. We obtain the results in Table 3.12. The answer p_5 (2.45) is now in error by $1.0\text{E}-9$, about one fifth that obtained with Newton's forward difference formula.

If we know in advance what degree of interpolation polynomial to use, then we can pick $[x_0, x_n]$ so that x is near its midpoint; and we can then use Newton's

Table 3.11 Example of Newton Forward difference interpolation formula (3.35)

n	$p_n(x)$	$\Delta^n f_0$
0	.0025076832	
1	−.0229380466	−.0508914596
2	−.0232468318	.0024702816
3	−.0232276767	.0003064808
4	−.023226681	−.000025491
5	−.0232267384	−.0000021

Table 3.12   Example of Everett's formula (3.40)

n	$p_n(x)$
1	$-.0229380466$
3	$-.0232262293$
5	$-.0232267423$

forward difference formula, which is more convenient. Everett's formula becomes important when we don't know what n to use, especially if it is large, say greater than 5. The tables in Abramowitz and Stegun [A1] are given to many significant digits, and the grid spacing h is not correspondingly small. Consequently, high-degree interpolation must be used. Although this results in more work for the user of the tables, it allows the table to be compacted into a much smaller space; and more tables of more functions can then be included in the volume.

The approximation problem   In using a computer, we generally prefer to store an analytic approximation to a function rather than a table of values from which we interpolate. Consider approximating a given function $f(x)$ on a given interval $[a,b]$ by using interpolating polynomials. In particular, consider the polynomial $p_n(x)$ produced by interpolating to $f(x)$ at an evenly spaced grid in $[a,b]$.

For each $n \geq 1$, define $h = (b-a)/n$, $x_j = a + jh, j = 0, 1, \ldots, n$. Let $p_n(x)$ be the polynomial interpolating to $f(x)$ at $x_0, \ldots, x_n$. Then does

$$\underset{a \leq x \leq b}{\text{Max}} |f(x) - p_n(x)| \tag{3.50}$$

tend to zero as $n \to \infty$? Not necessarily. For many functions, for example, e^x on $[0,1]$, the error in (3.50) does converge to zero as $n \to \infty$; see Problem 21. But there are other functions, which are quite well behaved, for which convergence does not occur. The most famous example, due to Runge, is that of

$$f(x) = \frac{1}{1+x^2} \qquad -5 \leq x \leq 5 \tag{3.51}$$

In Isaacson and Keller [I1, pp. 275–279], it is shown that for any $|x| > 3.64$,

$$\underset{n \geq k}{\text{Sup}} |f(x) - p_n(x)| = \infty, \text{ any } k \geq 0 \tag{3.52}$$

Thus $p_n(x)$ does not converge to $f(x)$ as $n \to \infty$, for any $|x| > 3.64$, even though the set of all node points $\{x_{j,n} | 0 \leq j \leq n, n \geq 1\}$ is dense in $[-5,5]$.

Although interpolation on an even grid may not produce a convergent sequence of interpolation polynomials, there are suitable sets of grid points $\{x_j\}$

that do result in good approximations for all continuously differentiable functions. This will be developed in Section 4.8 of the next chapter on approximation theory.

3.6 Hermite Interpolation

For a variety of applications, it is convenient to consider polynomials $p(x)$ that interpolate to a function $f(x)$ and in addition to have the derivative polynomial $p'(x)$ interpolate to the derivative function $f'(x)$. In this text, the primary application is as a mathematical tool in the development of Gaussian numerical integration in Chapter 5. But it is also a convenient approximation method for developing some numerical methods for solving differential equation problems.

We begin by considering an existence theorem for the basic interpolation problem

$$p(x_i) = y_i \qquad p'(x_i) = y'_i \qquad i = 1, \ldots, n \tag{3.53}$$

in which $x_1, \ldots, x_n$ are distinct nodes (real or complex) and $y_1, \ldots, y_n, y'_1, \ldots, y'_n$ are given data. (The notation has been changed from $n+1$ nodes $\{x_0, \ldots, x_n\}$ to n nodes $\{x_1, \ldots, x_n\}$ in line with the eventual application in Chapter 5.) There are $2n$ conditions imposed in (3.53), and thus we look for a polynomial $p(x)$ of degree at most $2n-1$.

To deal with the existence and uniqueness for $p(x)$, we generalize the third proof of Theorem 3.1. In line with previous notation in Section 3.1, let

$$\Psi_n(x) = (x - x_1) \ldots (x - x_n)$$

$$l_i(x) = \frac{(x-x_1)\ldots(x-x_{i-1})(x-x_{i+1})\ldots(x-x_n)}{(x_i-x_1)\ldots(x_i-x_{i-1})(x_i-x_{i+1})\ldots(x_i-x_n)} = \frac{\Psi_n(x)}{(x-x_i)\Psi'_n(x_i)} \tag{3.54}$$

$$\tilde{h}_i(x) = (x - x_i)[l_i(x)]^2$$

$$h_i(x) = [1 - 2l'_i(x_i)(x - x_i)][l_i(x)]^2 \tag{3.55}$$

Then for $i, j = 1, \ldots, n$,

$$h'_i(x_j) = \tilde{h}_i(x_j) = 0$$

$$h_i(x_j) = \tilde{h}'_i(x_j) = \begin{cases} 0, & i \neq j \\ 1, & i = j \end{cases}$$

The interpolating polynomial for (3.53) is given by

$$H_n(x) = \sum_{1}^{n} y_i h_i(x) + \sum_{1}^{n} y'_i \tilde{h}_i(x) \tag{3.56}$$

To show the uniqueness of $H_n(x)$, suppose there is a second polynomial $G(x)$ that satisfies (3.53) with degree $\leq 2n-1$. Define $R = H_n - G$. Then from (3.53),

$$R(x_i) = R'(x_i) = 0 \qquad i = 1, 2, \ldots, n$$

R is a polynomial of degree $\leq 2n-1$, and it has n double roots, $x_1, x_2, \ldots, x_n$. This can be true only if

$$R(x) = q(x)(x-x_1)^2 \ldots (x-x_n)^2$$

for some polynomial $q(x)$. If $q(x) \not\equiv 0$, then degree $R(x) \geq 2n$, a contradiction. Therefore we must have $R(x) \equiv 0$.

To obtain a more computable form than (3.56) and an error term, first consider the polynomial interpolating to $f(x)$ at $z_1, z_2, \ldots, z_{2n}$:

$$p_{2n-1}(x) = f(z_1) + (x-z_1)f[z_1, z_2] + \ldots$$

$$+ (x-z_1) \ldots (x-z_{2n-1}) f[z_1, \ldots, z_{2n}] \qquad (3.57)$$

For the error,

$$f(x) - p_{2n-1}(x) = (x-z_1) \ldots (x-z_{2n}) f[z_1, \ldots, z_{2n}, x]. \qquad (3.58)$$

In the formula (3.57), we can let nodes coincide and the formula will still exist. In particular, let

$$z_1 = z_2 = x_1, z_3 = z_4 = x_2, \ldots, z_{2n-1} = z_{2n} = x_n$$

to obtain

$$p_{2n-1}(x) = f(x_1) + (x-x_1)f[x_1, x_1] + (x-x_1)^2 f[x_1, x_1, x_2] + \ldots$$

$$+ (x-x_1)^2 \ldots (x-x_{n-1})^2 (x-x_n) f[x_1, x_1, \ldots, x_n, x_n] \qquad (3.59)$$

This is a polynomial of degree $\leq 2n-1$. For its error, take limits in (3.58) as $z_1, z_2 \to x_1, \ldots, z_{2n-1}, z_{2n} \to x_n$. By the continuity of the divided difference, assuming f is sufficiently differentiable,

$$f(x) - p_{2n-1}(x) = (x-x_1)^2 \ldots (x-x_n)^2 f[x_1, x_1, \ldots, x_n, x_n, x] \qquad (3.60)$$

Claim: $p_{2n-1}(x) = H_n(x)$. To prove this, assume $f(x)$ is $2n+1$ times continuously differentiable. Then note that

$$f(x_i) - p_{2n-1}(x_i) = 0 \qquad i = 1, 2, \ldots, n$$

Also,

$$f'(x) - p'_{2n-1}(x) = (x-x_1)^2 \dots (x-x_n)^2 \frac{d}{dx} f[x_1, x_1, \dots, x_n, x_n, x]$$

$$+ 2f[x_1, x_1, \dots, x_n, x_n, x] \sum_{i=1}^{n} \left[(x-x_i) \prod_{\substack{j=1 \\ j \neq i}}^{n} (x-x_j)^2 \right]$$

and

$$f'(x_i) - p'_{2n-1}(x_i) = 0 \qquad i = 1, \dots, n$$

Thus degree $p_{2n-1} \leq 2n-1$ and it satisfies (3.53), relative to the data $y_i = f(x_i), y'_i = f'(x_i)$. By the uniqueness of the Hermite interpolating polynomial, $p_{2n-1} = H_n$. Thus (3.59) gives a divided difference formula for calculating $H_n(x)$, and (3.60) gives an error formula,

$$f(x) - H_n(x) = [\Psi_n(x)]^2 f[x_1, x_1, \dots, x_n, x_n, x] \tag{3.61}$$

Using (3.24) to generalize (3.23), we obtain

$$f(x) - H_n(x) = [\Psi_n(x)]^2 \frac{f^{(2n)}(\xi_x)}{(2n)!} \qquad \xi_x \in \mathcal{K}\{x_1, \dots, x_n, x\} \tag{3.62}$$

Example The most widely used form of Hermite interpolation is probably the cubic Hermite polynomial, which solves

$$p(a) = f(a) \qquad p'(a) = f'(a)$$

$$p(b) = f(b) \qquad p'(b) = f'(b) \tag{3.63}$$

The formula (3.56) becomes

$$H_2(x) = \left(1 + 2\frac{x-a}{b-a}\right)\left(\frac{b-x}{b-a}\right)^2 f(a) + \left(1 + 2\frac{b-x}{b-a}\right)\left(\frac{x-a}{b-a}\right)^2 f(b)$$

$$+ \frac{(x-a)(b-x)^2}{(b-a)^2} f'(a) - \frac{(x-a)^2(b-x)}{(b-a)^2} f'(b) \tag{3.64}$$

The divided difference formula (3.59) is

$$H_2(x) = f(a) + (x-a)f'(a) + (x-a)^2 f[a,a,b]$$

$$+ (x-a)^2(x-b)f[a,a,b,b] \tag{3.65}$$

in which

$$f[a,a,b] = \frac{f[a,b] - f'(a)}{b-a}$$

$$f[a,a,b,b] = \frac{f'(b) - 2f[a,b] + f'(a)}{(b-a)^2}$$

The formula (3.65) can be evaluated by a nested multiplication algorithm analogous to Interp in Section 3.2.

The error formula for (3.64) or (3.65) is

$$f(x) - H_2(x) = (x-a)^2(x-b)^2 f[a,a,b,b,x]$$

$$= \frac{(x-a)^2(x-b)^2}{24} f^{(4)}(\xi_x), \xi_x \in \mathcal{K}\{a,b,x\} \tag{3.66}$$

$$\underset{a \le x \le b}{\text{Max}} |f(x) - H_2(x)| \le \frac{(b-a)^4}{384} \underset{a \le t \le b}{\text{Max}} |f^{(4)}(t)| \tag{3.67}$$

Further use will be made of the cubic Hermite polynomial in the next section, and a numerical example is given at the end of that section.

The general Hermite interpolation problem We generalize the simple Hermite problem (3.53) to the following: find a polynomial $p(x)$ to satisfy

$$p^{(i)}(x_1) = y_1^{(i)} \quad i = 0, 1, \ldots, \alpha_1 - 1$$

$$\vdots \tag{3.68}$$

$$p^{(i)}(x_n) = y_n^{(i)} \quad i = 0, 1, \ldots, \alpha_n - 1$$

The numbers $y_j^{(i)}$ are given data; and the number of conditions on p at x_j is α_j, $j = 1, \ldots, n$.

Define

$$N = \alpha_1 + \ldots + \alpha_n$$

Then there is a polynomial $p(x)$, unique among those of degree $\le N-1$, which satisfies (3.68). The proof is left as problem 23; and all of the earlier results for (3.53) can also be generalized. As an interesting special case, consider $\alpha_1 = N$, $n = 1$. This means that $p(x)$ is to satisfy

$$p^{(i)}(x_1) = f^{(i)}(x_1) \quad i = 0, 1, \ldots, N-1$$

We have replaced $y_1^{(i)}$ by $f^{(i)}(x_1)$ for notational convenience. Then

$$p(x) = f(x_1) + (x - x_1)f'(x_1) + \frac{(x - x_1)^2}{2!} f''(x_1) + \ldots + \frac{(x - x_1)^{N-1}}{(N-1)!} f^{(N-1)}(x_1)$$

$$= f(x_1) + (x - x_1)f[x_1, x_1] + (x - x_1)^2 f[x_1, x_1, x_1] + \ldots$$

$$+ (x - x_1)^{N-1} f[x_1, \ldots, x_1]$$

the Taylor polynomial of f about x_1.

3.7 Piecewise Polynomial Interpolation

Since the early 1960s, the subject of piecewise polynomial functions has become increasingly popular, especially spline functions. They have been used in a large variety of ways in approximation theory, data fitting, numerical integration and differentiation, and the numerical solution of integral, differential, and partial differential equations. We look at piecewise polynomial functions only from the viewpoint of interpolation theory; but much of their useful application occurs in some other area, with interpolation occurring only in a peripheral way.

For a piecewise polynomial function $p(x)$, there is an associated grid (usually finite),

$$-\infty < x_0 < x_1 < \ldots < x_n < \infty$$

The function $p(x)$ is a polynomial on each of the subintervals

$$(-\infty, x_0], [x_0, x_1], \ldots, [x_n, \infty)$$

and usually the polynomial degree is the same on each subinterval. No restrictions of continuity need be placed on $p(x)$ or its derivatives, although usually $p(x)$ is continuous.

There are essentially two types of interpolation problems, local and global. For the local type, the polynomial $p(x)$ on each subinterval $[x_{i-1}, x_i]$ is completely determined by the interpolation data at node points neighboring $[x_{i-1}, x_i]$. But for a global problem, the choice of $p(x)$ on each $[x_{i-1}, x_i]$ is dependent on all of the interpolation data. The global problems are more complicated to study; and the best known examples are the spline functions, to be defined later.

Local interpolation problems To clarify the presentation, we look at just piecewise cubic polynomial functions; this is also the case of most interest in applications. There will often be an interval $[a, b]$ on which we will approximate a function

$f(x)$, and we choose

$$a = x_0 < x_1 < \ldots < x_n = b \tag{3.69}$$

We use even spacing to further simplify the formulas. Let $h = (b-a)/n$, $x_j = a + jh$, $j = 0, 1, \ldots, n$. We first consider ordinary polynomial interpolation; and second we give a formula involving Hermite interpolation.

Assume that n is divisible by 3. Use Lagrange's formula (3.7) to define $f_n(x)$ on each subinterval $[x_{3j-3}, x_{3j}]$:

$$f_n(x) = -\frac{1}{6h^3}(x - z_1)(x - z_2)(x - z_3)f(z_0)$$

$$+ \frac{1}{2h^3}(x - z_0)(x - z_2)(x - z_3)f(z_1)$$

$$- \frac{1}{2h^3}(x - z_0)(x - z_1)(x - z_3)f(z_2)$$

$$+ \frac{1}{6h^3}(x - z_0)(x - z_1)(x - z_2)f(z_3) \tag{3.70}$$

for $x_{3j-3} \leq x \leq x_{3j}$, $j = 1, 2, \ldots, n/3$, where

$$z_0 = x_{3j-3}, \quad z_1 = x_{3j-2}, \quad z_2 = x_{3j-1}, \quad z_3 = x_{3j}.$$

The function $f_n(x)$ is continuous on $[a, b]$, and it interpolates to $f(x)$ at all of the nodes x_j, $j = 0, 1, \ldots, n$; we call it a Lagrange function. Generally, $f_n'(x)$ is discontinuous at the nodes x_{3j}, $j = 1, 2, \ldots, n/3 - 1$, although it still approximates $f'(x)$; see Problem 26(a).

For the error $f(x) - f_n(x)$, use interpolation error formulas for each interval $[x_{3j-3}, x_{3j}]$. This leads to the bound

$$\underset{a \leq x \leq b}{\text{Max}} |f(x) - f_n(x)| \leq \frac{h^4}{24} \underset{a \leq t \leq b}{\text{Max}} |f^{(4)}(t)| \tag{3.71}$$

To approximate $f(x)$ with a better accuracy, we merely decrease h, keeping the same degree of polynomial on each subinterval. In many situations, this is much superior to increasing the degree of the interpolating polynomial, which often may be a poor approximation due to large oscillations between the interpolation nodes, as with the example (3.51) of Runge.

Example   For $f(x) = e^x$ on $[0, 1]$, suppose we want even spacing and a maximum error less than 10^{-6}. Using (3.71),

$$\frac{h^4}{24}e \leq 10^{-6}$$

$$h \leq .055 \qquad n \geq 18.3$$

It will probably be sufficient to use $n = 18$ because of the conservative nature of the bound (3.71). There will be six subintervals of cubic interpolation.

For the storage requirements for piecewise Lagrange interpolation, we need to save four pieces of information for each subinterval $[x_{3j-3}, x_{3j}]$. This means a total storage requirements of $4n/3$ locations in computer memory locations. The choice of coefficients depends on how $f_n(x)$ is to be evaluated. If derivatives of $f_n(x)$ are desired, then it is most convenient to store $f_n(x)$ in the Taylor form on each subinterval $[x_{3j-3}, x_{3j}]$,

$$f_n(x) = a_j + b_j(x - x_{3j-3}) + c_j(x - x_{3j-3})^2 + d_j(x - x_{3j-3})^3 \qquad (3.72)$$

This should be evaluated in nested form. The coefficients are easily produced from the interpolation polynomial.

A second widely used piecewise polynomial interpolation function is the piecewise cubic Hermite function. On each interval $[x_{j-1}, x_j]$, define $h_n(x)$ as the Hermite cubic polynomial interpolating to $f(x)$ and $f'(x)$ at x_{j-1} and x_j, $j = 1, \ldots, n$. The formulas (3.64) or (3.65) can be used; and a Taylor approximation (3.72) can be prepared for each interval $[x_{j-1}, x_j]$. The function $h_n(x)$ and its derivative $h_n'(x)$ are continuous on $[a, b]$, and they interpolate to $f(x)$ and $f'(x)$, respectively, at the nodes $x_0, \ldots, x_n$.

For the error in $h_n(x)$, use (3.66) and (3.67). Then

$$\underset{a \le x \le b}{\text{Max}} |f(x) - h_n(x)| \le \frac{h^4}{384} \underset{a \le t \le b}{\text{Max}} |f^{(4)}(t)| \qquad (3.73)$$

When this is compared with (3.71), it might seem that Hermite interpolation is superior. But this is deceptive. In obtaining the Hermite function $h_n(x)$, we used $2n + 2$ pieces of data on $f(x)$; and in constructing the Lagrange approximation, we used only $n + 1$ pieces of data. If the functions $f_{2n}(x)$ and $h_n(x)$ are compared, the error bounds are the same, that of (3.73).

Since there is no difference in error, the form of piecewise polynomial function used will depend on the application for which it is to be used. In numerical integration applications, the piecewise Lagrange function is most suitable; it is also used in solving some singular integral equations, by means of the product integration methods of Chapter 5, Section 5.7. The piecewise Hermite function is useful for solving some differential equation problems. For example, it is a popular function used with the finite element method for solving boundary value problems for second order differential equations; see Strang and Fix [S3, Chapter 1]. Numerical examples comparing $f_{2n}(x)$ and $h_n(x)$ are given following the material on spline functions.

Spline functions   As before, consider a grid

$$a = x_0 < x_1 < \ldots < x_n = b$$

but do not assume it is evenly spaced. For $m \geq 1$, $s(x)$ is a spline function of order m on $(-\infty, \infty)$ if

1. $s(x)$ is a polynomial of degree $\leq m$ on each of the subintervals $(-\infty, x_0], [x_0, x_1], \ldots, [x_n, \infty)$

2. $s^{(r)}(x)$ is a continuous function on $(-\infty, \infty)$ for $0 \leq r \leq m - 1$.

The derivative of a spline of order m is another spline of order $m - 1$, and similarly for antiderivatives. Often we work on just $[a, b]$ rather than $(-\infty, \infty)$.

Cubic spline functions are the most popular spline functions, for a variety of reasons. They are a smooth function with which to fit data; and when used for interpolation, they do not have the oscillatory behavior that characterizes high-degree polynomial interpolation. Further motivation for their use is given below. Cubic splines are also fairly simple to calculate and use.

For our interpolation problem, we wish to find a cubic spline $s(x)$ for which

$$s(x_i) = y_i \qquad i = 0, 1, \ldots, n \tag{3.74}$$

We begin by investigating how many degrees of freedom are left in the choice of $s(x)$, once it satisfies (3.74). The technique used will not lead directly to a practical means for calculating $s(x)$ but will furnish additional insight. Write

$$s(x) = a_i + b_i x + c_i x^2 + d_i x^3 \qquad x_{i-1} \leq x \leq x_i \qquad i = 1, \ldots, n \tag{3.75}$$

There are $4n$ unknown coefficients $\{a_i, b_i, c_i, d_i\}$. The constraints are (3.74) and the continuity restrictions

$$s^{(j)}(x_i + 0) = s^{(j)}(x_i - 0) \qquad i = 1, \ldots, n - 1 \qquad j = 0, 1, 2 \tag{3.76}$$

Together this gives

$$n + 1 + 3(n - 1) = 4n - 2$$

constraints, as compared with $4n$ unknowns. Thus there are at least two degrees of freedom in choosing the coefficients of (3.75). Note that nothing was said about the value of $s(x)$ on $(-\infty, x_0]$ and $[x_n, \infty)$. Generally this is of no concern; and given $s(x)$ on $[x_0, x_n]$, it is easy to construct an extension to $(-\infty, \infty)$, although it will not be unique.

We will now give a method for constructing $s(x)$. Introduce the notation

$$M_i = s''(x_i) \qquad i = 0, 1, \ldots, n$$

Since $s(x)$ is cubic on $[x_i, x_{i+1}]$, $s''(x)$ is linear and thus

$$s''(x) = \frac{(x_{i+1} - x) M_i + (x - x_i) M_{i+1}}{h_i} \qquad i = 0, 1, \ldots, n - 1 \tag{3.77}$$

where $h_i = x_{i+1} - x_i$. With this formula, $s''(x)$ is continuous on $[x_0, x_n]$. Integrate twice to get

$$s(x) = \frac{(x_{i+1} - x)^3 M_i + (x - x_i)^3 M_{i+1}}{6h_i} + C(x_{i+1} - x) + D(x - x_i)$$

with C and D arbitrary. The interpolating condition (3.74) implies

$$C = \frac{y_i}{h_i} - \frac{h_i M_i}{6} \qquad D = \frac{y_{i+1}}{h_i} - \frac{h_i M_{i+1}}{6}$$

$$s(x) = \frac{(x_{i+1} - x)^3 M_i + (x - x_i)^3 M_{i+1}}{6h_i} + \frac{(x_{i+1} - x)y_i + (x - x_i)y_{i+1}}{h_i}$$

$$- \frac{h_i}{6}\left[(x_{i+1} - x)M_i + (x - x_i)M_{i+1}\right], \quad x_i \le x \le x_{i+1}, \quad 0 \le i \le n-1 \quad (3.78)$$

The equation (3.78) implies the continuity on $[a,b]$ of $s(x)$, as well as the interpolating condition (3.74). To determine the constants $M_0, \ldots, M_n$, we require $s'(x)$ to be continuous at $x_1, \ldots, x_{n-1}$:

$$\lim_{x \nearrow x_i} s(x) = \lim_{x \searrow x_i} s(x) \qquad i = 1, \ldots, n-1 \tag{3.79}$$

On $[x_i, x_{i+1}]$,

$$s'(x) = \frac{-(x_{i+1} - x)^2 M_i + (x - x_i)^2 M_{i+1}}{2h_i} + \frac{y_{i+1} - y_i}{h_i} - \frac{(M_{i+1} - M_i)h_i}{6}$$

and on $[x_{i-1}, x_i]$

$$s'(x) = \frac{-(x_i - x)^2 M_{i-1} + (x - x_{i-1})^2 M_i}{2h_{i-1}} + \frac{y_i - y_{i-1}}{h_{i-1}} - \frac{(M_i - M_{i-1})h_{i-1}}{6}$$

Using (3.79) and some manipulation, we obtain

$$\frac{h_{i-1}}{6} M_{i-1} + \frac{h_i + h_{i-1}}{3} M_i + \frac{h_i}{6} M_{i+1} = \frac{y_{i+1} - y_i}{h_i} - \frac{y_i - y_{i-1}}{h_{i-1}} \tag{3.80}$$

for $i = 1, \ldots, n-1$. This gives $n-1$ equations for the $n+1$ unknowns $M_0, \ldots, M_n$. We generally specify an endpoint condition, at x_0 and x_n, to remove the two degrees of freedom present in (3.80).

Case 1 The cubic natural interpolating spline. One particular choice of endpoint conditions is to use

$$M_0 = M_n = 0 \tag{3.81}$$

This may appear strange, but it is based on the solution of the following important problem. Among all functions $g(x)$, which are twice continuously differentiable on $[a,b]$ and which satisfy

$$g(x_i) = y_i \qquad i = 0, 1, \ldots, n \tag{3.82}$$

choose g to minimize the integral

$$\int_a^b [g''(x)]^2 dx$$

The solution to this problem will be an interpolating function to the data in (3.82); and it should contain a minimum of oscillatory behavior because $g''(x)$ is small. The solution is the cubic spline $s(x)$, which satisfies (3.80) and (3.81). The condition $s''(x_0) = s''(x_n) = 0$ has the interpretation that $s(x)$ is linear on $(-\infty, x_0]$ and $[x_n, \infty)$. This can be given additional physical meaning by using the theory of small deformations of a thin beam of an elastic material; for example, see Ahlberg, Nilson, and Walsh [A2, pp. 1–3].

With the choice of (3.81), we have the system of equations

$$AM = D \tag{3.83}$$

with

$$D^T = \left[\frac{y_2 - y_1}{h_1} - \frac{y_1 - y_0}{h_0}, \ldots, \frac{y_n - y_{n-1}}{h_{n-1}} - \frac{y_{n-1} - y_{n-2}}{h_{n-2}} \right]$$

$$M^T = [M_1, \ldots, M_{n-1}],$$

$$A = \begin{bmatrix} \dfrac{h_0 + h_1}{3} & \dfrac{h_1}{6} & 0 & \cdots & & 0 \\[2mm] \dfrac{h_1}{6} & \dfrac{h_1 + h_2}{3} & \dfrac{h_2}{6} & & & \\[2mm] 0 & & \cdot & & & \\ \vdots & & & \cdot & & \\ & & & & \cdot & \\ 0 & & \cdots & & \dfrac{h_{n-2}}{6} & \dfrac{h_{n-2} + h_{n-1}}{3} \end{bmatrix}$$

This matrix is symmetric, positive definite, and diagonally dominant; and the associated linear system (3.83) can be easily and rapidly solved. See the material on tridigonal systems in Theorem 8.2, Section 8.3 of Chapter 8. The definitions of the matrix terms used are given in Chapters 7 and 8.

An error analysis requires a knowledge of matrix and vector norms, and in addition, some perturbation theorems from Chapter 7. For that reason and for brevity, we merely quote some results from Kershaw [K1]. Let

$$h = \text{Max}\, h_i$$

and assume that $f(x)$ is four times continuously differentiable on $[a, b]$. Let $s(x)$ be the natural cubic spline interpolating to $f(x)$ at $x_0, \ldots, x_n$. Then there are points x_a and x_b with

$$a \leqq x_a < x_b \leqq b$$

and a constant $K > 0$ such that

$$\underset{x_a \leqq x \leqq x_b}{\text{Max}} |f^{(j)}(x) - s^{(j)}(x)| \leqq Kh^{4-j} \qquad j = 0, 1, 2 \qquad (3.84)$$

The points x_a and x_b satisfy

$$x_a - a = 0(h \; lnh), \quad b - x_b = 0(h \; lnh)$$

If $f''(a) = f''(b) = 0$, then $x_a = a$, $x_b = b$.

In the numerical example given at the end of the section, we will use a uniform grid on $[a, b]$. With $n \geq 2$, choose $h = (b - a)/n$, $x_j = a + jh$ for $j = 0, 1, \ldots, n$. But rather than choose $M_0 = M_n = 0$, we use a special result from Kershaw [K1]. Define

$$x_{-1} = a - h^3 \qquad x_{n+1} = b + h^3 \qquad (3.85)$$

and produce the cubic natural interpolating spline to $f(x)$ at the nodes $x_{-1}, x_0, \ldots, x_{n+1}$. Thus we take $M_{-1} = M_{n+1} = 0$, and we solve the analogue of (3.83) for $M_0, \ldots, M_n$. From Kershaw [K1] we then have the result

$$\underset{a \leqq x \leqq b}{\text{Max}} |f^{(j)}(x) - s^{(j)}(x)| \leqq C_j h^{4-j}$$

$$\times \left[\underset{a \leqq x \leqq b}{\text{Max}} |f^{(4)}(x)| + |f''(a)| + |f''(b)| \right] \qquad (3.86)$$

for suitable constants C_j, $j = 0, 1, 2$, provided $f(x)$ is four times continuously on $[a, b]$. We can ensure a similar convergence result, uniform on $[a, b]$, if other forms of interpolating end conditions are used. But this particular form is closely related to the next formula we take up.

Case 2 Endpoint derivative conditions. To specify M_0 and M_n implicitly, impose conditions on $s'(x)$ at the endpoints:

$$s'(x_0) = y'_0 \qquad s'(x_n) = y'_n \qquad (3.87)$$

This leads to the two additional equations

$$\frac{h_0}{3}M_0 + \frac{h_0}{6}M_1 = \frac{y_1 - y_0}{h_0} - y_0'$$

$$\frac{h_{n-1}}{6}M_{n-1} + \frac{h_{n-1}}{3}M_n = y_n' - \frac{y_n - y_{n-1}}{h_{n-1}} \qquad (3.88)$$

The system composed of (3.80) and (3.88) is of order $n+1$; and it has exactly the same properties as the earlier matrix A in (3.83) of case 1. This interpolating spline can also be obtained as the solution of a minimization problem; see [D2]. The convergence results of (3.84) are valid for this spline, except that now the convergence is uniform on the entire interval $[a,b]$.

There is an interesting connection between the splines of cases 1 and 2. If we use the nodes $x_{-1}, \ldots, x_{n+1}$, as given in (3.85) and its preceding sentence, then the first and last equations of (3.83) (generalized from $n-1$ to $n+1$ equations) are approximations to (3.88). Thus the splines of Cases 1 and 2 are essentially the same on $[a,b]$, for this choice of nodes for Case 1. A numerical example will be given following Case 3.

Case 3 Periodic boundary conditions. Impose the periodic boundary conditions

$$s(x_0) = s(x_n) \qquad s'(x_0) = s'(x_n) \qquad (3.89)$$

This again results in a linear system like that of (3.83), although it is no longer tridiagonal. If $s(x)$ interpolates to a function $f(x)$ on $[a,b]$, if $f(x)$ satisfies the boundary conditions (3.89), and if $f(x)$ is four times continuously differentiable on $[a,b]$, then convergence like (3.84) is valid, uniformly over $[a,b]$. See Kershaw [K1] for much more information.

Example Let $f(x) = \tan^{-1} x, 0 \le x \le 5$. Table 3.13 gives the errors

$$E_i = \underset{0 \le x \le 5}{\text{Max}} |f^{(i)}(x) - f_n^{(i)}(x)| \qquad i = 0, 1, 2, 3 \qquad (3.90)$$

where $f_n(x)$ is the cubic Lagrange function interpolating to $f(x)$ on the nodes $x_j = a + jh, j = 0, 1, \ldots, n, h = (b-a)/n$. The columns labeled Ratio give the rate of decrease in the error when n is doubled. Note that the rate of convergence for $f_n^{(i)}$ to $f^{(i)}$ is proportional to $h^{4-i}, i = 0, 1, 2, 3$. This can be rigorously proven, and an indication of the proof is given in Problem 26.

Table 3.14 gives the analogous errors for the cubic Hermite function $h_n(x)$ interpolating to $f(x)$. And again note that the errors agree with

$$\underset{a \le x \le b}{\text{Max}} |f^{(i)}(x) - h_n^{(i)}(x)| \le ch^{4-i} \qquad i = 0, 1, 2, 3$$

Table 3.13   Cubic Lagrange function interpolation

n	E_0	Ratio	E_1	Ratio	E_2	Ratio	E_3	Ratio
6	1.20E$-$2		1.22E$-$1		7.81E$-$1		2.32	
		3.3		2.1		1.5		1.2
12	3.62E$-$3		5.83E$-$2		5.24E$-$1		1.95	
		11.4		6.1		3.2		1.6
24	3.18E$-$4		9.57E$-$3		1.64E$-$1		1.19	
		16.9		8.1		3.9		1.7
48	1.88E$-$5		1.11E$-$3		4.21E$-$2		.682	
		14.5		7.3		3.7		1.9
96	1.30E$-$6		1.61E$-$4		1.14E$-$2		.359	

Table 3.14   Cubic Hermite function interpolation

n	E_0	Ratio	E_1	Ratio	E_2	Ratio	E_3	Ratio
3	2.64E$-$2		5.18E$-$2		4.92E$-$1		2.06	
		5.6		3.0		2.3		1.5
6	4.73E$-$3		1.74E$-$2		2.14E$-$1		1.33	
		16.0		8.0		3.6		1.5
12	2.95E$-$4		2.17E$-$3		5.91E$-$2		.891	
		13.1		6.7		3.6		1.9
24	2.26E$-$5		3.25E$-$4		1.66E$-$2		.475	
		16.0		8.0		4.0		2.0
48	1.41E$-$6		4.06E$-$5		4.18E$-$3		.241	

which can also be proven, for some $c > 0$. As was stated after (3.73), the functions $f_{2n}(x)$ and $h_n(x)$ are of comparable accuracy, and a comparison of the first two tables confirms this. But the derivative $h_n'(x)$ is a more accurate approximation to $f(x)$ than is $f_{2n}'(x)$. See Problem 26 for an explanation.

Table 3.15 contains the errors on $[0,5]$ for the cubic natural spline interpolation using the nodes

$$x_{-1} = a - h^3, \; x_j = a + jh \quad \text{for} \quad j = 0, 1, \ldots, n \qquad x_{n+1} = b + h^3$$

Table 3.15   Cubic natural spline function interpolation

n	E_0	Ratio	E_1	Ratio	E_2	Ratio	E_3	Ratio
6	1.77E$-$2		8.06E$-$2		2.63E$-$1		1.43	
		36.6		25.6		4.6		1.5
12	4.84E$-$4		3.15E$-$3		5.91E$-$2		.933	
		16.0		7.8		3.6		1.9
24	3.03E$-$5		4.05E$-$4		1.64E$-$2		.486	
		20.5		9.6		3.9		2.0
48	1.48E$-$6		4.22E$-$5		4.23E$-$3		.244	
		16.4		8.1		4.0		2.0
96	9.04E$-$8		5.19E$-$6		1.06E$-$3		.122	

This was discussed above, and the theoretical error results are given in (3.86). The results are better than with the two preceding examples. This was true of all functions $f(x)$ that were tried, although often the difference was not as dramatic as it is with the example given here.

Example Another informative example is to take $f(x) = x^4, 0 \leq x \leq 1$. All of the above interpolation formulas have $f^{(4)}(x)$ as a multiplier in their error formulas. Since $f^{(4)}(x) = 24$, a constant, the error for all three forms of interpolation satisfy

$$\underset{0 \leq x \leq 1}{\text{Max}} \left| \frac{d^j(x^4)}{dx^j} - f_n^{(j)}(x) \right| = C_j h^{4-j} \qquad j = 0, 1, 2, 3 \qquad (3.91)$$

The constants C_j will vary with the form of the interpolation being used. In the actual computations, the errors behaved exactly like (3.91), thus providing another means for comparing the methods. We give the results in Table 3.16 for only the most accurate case. Although the spline is more accurate, it is only marginally so. The above examples should show that all of the methods are adequate in terms of accuracy. Therefore, the decision as to which method of interpolation to use should depend on other factors, usually arising from the intended area of application. Spline functions have proven very useful with data fitting problems; and Lagrange and Hermite functions are more useful for analytic approximations in solving integral and differential equations, respectively.

Table 3.16 Comparison of three forms of piecewise cubic interpolation.

Method	E_0	E_1	E_2	E_3
Lagrange $n = 96$	1.10E−8	6.78E−6	2.93E−3	.375
Hermite $n = 48$	1.18E−8	1.74E−6	8.68E−4	.250
Spline $n = 96$	4.47E−9	2.26E−6	5.41E−4	.216

Discussion of the literature

As noted in the introduction, interpolation theory is a foundation for the development of methods in numerical integration and differentiation, approximation theory, and the numerical solution of differential equations. Each of these topics, except for numerical differentiation, is developed in the following chapters, and the associated literature is discussed at that point. More extensive presentations of interpolation theory are given in the books of Davis [D1, Chapters 2–5] and Hildebrand [H1].

The subject of numerical differentiation is important, although it is not treated in this text. If a low-degree polynomial $p_n(x)$ interpolates to a function $f(x)$ on a set of closely spaced node points, and if the values f_j of f at the node points are known with a high degree of precision, then $p'(x)$ furnishes a good approximation to $f'(x)$ (see Problem 26). But if the function values f_j contain a large amount of noise, then the derivative $p'(x)$ is often a poor estimate for $f'(x)$; the numerical differentiation of noisy data is an example of an ill-posed problem. For a review of the literature on numerical differentiation of empirical data and for a look at some fairly sophisticated solutions to the problem, see the recent papers of Cullum [C2] and Anderssen and Bloomfield [A3].

The introduction of digital computers produced a revolution in numerical analysis, including interpolation theory. Before the use of digital computers, hand computation was necessary; and this meant that numerical methods were used that minimized the need for computation. Such methods were often more complicated than methods now used on computers, taking special advantage of the unique mathematical characteristics of each problem. These methods also used tables extensively, to avoid repeating calculations done by others; and interpolation formulas based on finite differences were used extensively. A large subject was created, called the finite difference calculus, and was used in solving problems in several areas of numerical analysis and applied mathematics. For a general introduction to this approach to numerical analysis, see Hildebrand [H1] and the references contained there.

The use of digital computers has changed the needs of other areas for interpolation theory, vastly reducing the need for finite difference based interpolation formulas. But there is still an important place for both hand computation and the use of mathematical tables, especially for the more complicated functions of mathematical physics. Everyone doing numerical work should possess an elementary book of tables such as the well-known CRC tables [C1]; the NBS tables edited by Abramowitz and Stegun [A1] are also an excellent reference for nonelementary functions. With the increasing sophistication of electronic hand calculators, it is likely that people will use them for more sophisticated calculations and the use of tables may again be important.

Piecewise polynomial approximation theory has been very popular since the early 1960s, and it is finding use in a wide variety of fields, for example, see the applications in Strang and Fix [S3] to the solution of boundary value problems for differential equations. Most of the interest has centered on spline functions. The beginning of the theory of spline functions is generally credited to I. J. Schoenberg in his papers [S1] of 1946; and since that time, Professor Schoenberg and his students have been prominent in developing the area, (e.g., see Schoenberg's monograph [S2]). There is now an extensive literature on spline functions involving many individuals and groups, (e.g., see Ahlberg, Nilson, and Walsh [A2], deBoor [D2], and Kershaw [K1]). Much work now centers on the development of the applications of spline functions, the most promising being in the least squares modeling of empirical data. For a discussion of the recent theory and the practical use of spline functions, see deBoor [D3].

Bibliography

[A1] Abramowitz, M. and I. Stegun, editors, *Handbook of Mathematical Functions*, National Bureau of Standards, 1964.

[A2] Ahlberg, J., E. Nilson, and J Walsh, *The Theory of Splines and Their Applications*, Academic Press, 1967.

[A3] Anderssen, R. and P. Bloomfield, Numerical differentiation procedures for non-exact data, *Numer. Math.*, **22**, 1974, pp. 157–182.

[A4] Atkinson, K. and A. Sharma, A partial characterization of poised Hermite-Birkhoff interpolation problems, *SIAM J. Num. Anal.*, **6** 1969, pp. 230–235.

[C1] *CRC Standard Mathematical Tables*, Chemical Rubber Publishing Co., Cleveland, Ohio, published yearly.

[C2] Cullum, J., Numerical differentiation and regularization, *SIAM J. Num. Anal.*, **8** 1971, pp. 254–265.

[D1] Davis, P., *Interpolation and Approximation*, Blaisdell Pub. Co., New York, 1963.

[D2] deBoor, C., Best approximation properties of spline function of odd degree, *J. Math. and Mech.*, **12** 1963, pp. 747–749.

[D3] deBoor, C., *A Practical Guide to Splines*, Springer-Verlag, New York, 1978.

[H1] Hildebrand, F., *Introduction to Numerical Analysis*, McGraw-Hill Pub. Co., New York, 1956.

[I1] Isaacson, E., and H. Keller, *Analysis of Numerical Methods*, John Wiley and Sons, New York, 1966.

[K1] Kershaw, D., A note on the convergence of interpolatory cubic splines, *SIAM J. Num. Anal.*, **8**, 1971, pp. 67–74.

[S1] Schoenberg, I., Contributions to the problem of approximation of equidistant data by analytic functions, *Quarterly Appl. Math.*, **4**, 1946, Part A, pp. 45–99; Part B, pp. 112–141.

[S2] Schoenberg, I., *Cardinal Spline Interpolation*, SIAM, Philadelphia, 1973.

[S3] Strang, G., and G. Fix, *An Analysis of the Finite Element Method*, Prentice-Hall, Englewood Cliffs, N. J., 1973.

Problems

1. Recall the Vandermonde matrix X given in (3.3), and define

$$V_n(x) = \det \begin{bmatrix} 1 & x_0 & x_0^2 & \cdots & x_0^n \\ \vdots & & & & \\ 1 & x_{n-1} & x_{n-1}^2 & & x_{n-1}^n \\ 1 & x & x^2 & \cdots & x^n \end{bmatrix}$$

(a) Show $V_n(x)$ is a polynomial of degree n, and that its roots are $x_0, \ldots, x_{n-1}$. Obtain the formula

$$V_n(x) = V_{n-1}(x_{n-1})(x-x_0)\ldots(x-x_{n-1})$$

Hint: expand the last row of $V_n(x)$ by minors to show $V_n(x)$ is a polynomial of degree n and to find the coefficient of the term x^n.

(b) Show

$$\det X \equiv V_n(x_n) = \prod_{0 \le j < i \le n} (x_i - x_j)$$

2. For the basis functions $l_{j,n}(x)$ given in (3.5), prove that for any $n \ge 1$,

$$\sum_{j=0}^{n} l_{j,n}(x) = 1 \qquad \text{for all } x$$

3. Consider linear interpolation in a table of values of $e^x, 0 \le x \le 2$, with $h = .01$. Let the table values be given to 5 significant digits, as in [C1]. Bound the error of linear interpolation, including that part due to the rounding errors in the table entries.

4. Consider linear interpolation in a table of $\cos(x)$ with x given in degrees, $0 \le x \le 90°$, with a stepsize $h = 1' = \frac{1}{60}$ degree. Assuming that the table entries are given to five significant digits, bound the total error of interpolation.

5. Consider linear interpolation in a table of square roots, (e.g., in [C1]), with $10 \le x \le 999, h = 1$. Bound the error and relative error. Based on this interpolation error without rounding, to how many significant digits should the table entries be given if the rounding errors are not to significantly worsen the error?

6. Suppose you are to make a table of values of $\sin(x), 0 \le x \le \pi/2$, with a stepsize of h. Assume linear interpolation is to be used with the table; and suppose the total error, including the effects due to rounding in table entries, is to be at most 10^{-6}. What should h equal (choose it in a convenient size for actual use) and to how many significant digits should the table entries be given?

7. Repeat Problem 6, but using e^x on $0 \le x \le 1$.

8. Generalize to quadratic interpolation the material on the effect of rounding errors in table entries, given in Section 3.1. Let $\epsilon_i = f(x_i) - \tilde{f}_i, i = 0, 1, 2$, and $\epsilon \ge \text{Max}\{|\epsilon_0|, |\epsilon_1|, |\epsilon_2|\}$. Show that the effect of these rounding errors on the quadratic interpolation error is bounded by 1.25ϵ, assuming $x_0 \le x \le x_2$ and $x_1 - x_0 = x_2 - x_1 = h$.

9. Repeat Problem 5, but using quadratic interpolation and the result of Problem 8.

10. Consider producing a table of values for $f(x) = \log_{10} x, 1 \le x \le 10$, and assume quadratic interpolation is to be used. Let the total interpolation error, including the effect of rounding in table entries, be less than 10^{-6}. Choose an appropriate grid spacing h and the number of digits to which the entries should be given. Would it be desirable to vary the spacing h as x varies in $[1, 10]$? If so, suggest a suitable partition of $[1, 10]$ with corresponding values of h. Use the result of Problem 8 on rounding error effects.

11. Prove the Newton divided difference recursion relation (3.19).

12. Produce computer subroutine implementations of the algorithms Divdif and Interp given in Section 3.2, and then write a main driver program to use them in doing table interpolation. Choose a table from [A1] to test the program, considering several successive degrees n for the interpolation polynomial.

13. Do an inverse interpolation problem using the table for $J_0(x)$ given in Section 3.2. Find the value of x for which $J_0(x) = 0$, that is, calculate an accurate estimate of the root. Estimate your accuracy, and compare this with the actual value $x = 2.4048255577$.

14. Prove that Volume $(\tau_n) = 1/n!$, where τ_n is the simplex in R^n defined in the (3.25) of Theorem 3.3 in Section 3.2. Hint: use (3.24) and other results on divided differences, along with a special choice for $f(x)$.

15. Prove the relations (3.29), pertaining to the variable divided difference of a polynomial.

16. Consider the following table of values for $j_0(x) = \sqrt{\pi/2x} \, J_{1/2}(x)$, taken from [A1, Chapter 10].

x	$j_0(x)$	x	$j_0(x)$
0.0	1.00000	0.7	.92031
0.1	.99833	0.8	.89670
0.2	.99335	0.9	.87036
0.3	.98507	1.0	.84147
0.4	.97355	1.1	.81019
0.5	.95885	1.2	.77670
0.6	.94107	1.3	.74120

Based on the rounding errors in the table entries, what should be the maximum degree of polynomial interpolation used with the table? Hint: use the forward difference table to detect the influence of the rounding errors.

17. The following data is taken from a polynomial. What is its degree?

x	-2	-1	0	1	2	3
$p(x)$	-5	1	1	1	7	25

18. The following data has noise in it which is large relative to rounding error. Find the noise and change the data appropriately. Only the function values are given, since the node points are unnecessary for computing the forward difference table.

304319	419327	545811	683100
326313	443655	572433	711709
348812	468529	599475	740756
371806	493852	626909	770188
395285	519615	654790	800000

19. Derive the analogue of Lemma (1), given in Section 3.3, for backward differences. Use this and Newton's divided form of the interpolating polynomial (3.20) to derive the backward difference interpolation formula (3.37).

20. Directly verify Everett's formula (3.40) for the cases $n=1$ and $n=3$.

21. Consider the function e^x on $[0,b]$ and its approximation by an interpolating polynomial. For $n \geq 1$, let $h=b/n, x_j=jh, j=0,1,\ldots,n$; and let $p_n(x)$ be the nth degree polynomial interpolating to e^x on the nodes $x_0,\ldots,x_n$. Prove that

$$\underset{0 \leq x \leq b}{\text{Max}} |e^x - p_n(x)| \to 0 \qquad \text{as} \quad n \to \infty$$

22. Let $x_0,\ldots,x_n$ be distinct real points, and consider the following interpolation problem. Choose a function

$$P_n(x)= \sum_{j=0}^{n} c_j e^{jx}$$

such that

$$P_n(x_i)=y_i \qquad i=0,1,\ldots,n$$

with the $\{y_i\}$ given data. Show there is a unique choice of $c_0,\ldots,c_n$. Hint: the problem can be reduced to that of ordinary polynomial interpolation.

23. Prove that the general Hermite problem (3.68) has a unique solution $p(x)$ among all polynomials of degree $\leq N-1$.

24. Consider the problem of finding a quadratic polynomial $p(x)$ for which

$$p(x_0)=y_0 \qquad p'(x_1)=y_1' \qquad p(x_2)=y_2$$

with $x_0 \neq x_2$ and $\{y_0, y_1', y_2\}$ given data. Assuming the nodes x_0, x_1, and x_2 are real, what conditions must be satisfied for such a $p(x)$ to exist and be unique. This and the next problem are examples of Hermite-Birkhoff interpolation problems; see [A4].

25. **(a)** Show there is a unique cubic polynomial $p(x)$ for which

$$p(x_0)=f(x_0) \qquad p(x_2)=f(x_2)$$

$$p'(x_1)=f'(x_1) \qquad p''(x_1)=f''(x_1)$$

where $f(x)$ is a given function and $x_0 \neq x_2$. Derive a formula for $p(x)$, analogous to the Lagrange formula for ordinary interpolation.

(b) Let $x_0=-1, x_1=0, x_2=1$. Assuming $f(x)$ is four times continuously differentiable on $[-1,1]$, show that for $-1 \leq x \leq 1$,

$$f(x)-p(x)= \frac{x^4-1}{4!} f^{(4)}(\xi_x)$$

for some $\xi_x \in [-1,1]$.

26. **(a)** Let $p_3(x)$ denote the cubic polynomial interpolating to $f(x)$ at the evenly spaced points $x_j=x_0+jh, j=0,1,2,3$. Assuming that $f(x)$ is sufficiently differentiable, bound the error in using $p_3'(x)$ as an approximation to $f'(x), x_0 \leq x \leq x_3$.

(b) Let $H_2(x)$ denote the cubic Hermite polynomial interpolating to $f(x)$ and $f'(x)$ at x_0 and $x_1=x_0+h$. Bound the error $f'(x)-H_2'(x)$ for $x_0 \leq x \leq x_1$.

(c) Consider the piecewise polynomial functions $f_{2n}(x)$ and $h_n(x)$ of Section 3.7, and bound the errors $f'(x) - f'_{2n}(x)$ and $f'(x) - h'_n(x)$. Apply it to the specific case of $f(x) = x^4$ and compare your answers with the numerical results given at the end of Section 3.7.

27. For the function $f(x) = \sin(x), 0 \leqslant x \leqslant \pi$, find the piecewise cubic Hermite function $h_n(x)$ and the piecewise cubic Lagrange function $f_{2n}(x)$, for $n = 3, 6, 12$. Evaluate the maximum errors $f(x) - f_{2n}(x), f'(x) - f'_{2n}(x), f(x) - h_n(x)$, and $f'(x) - h'_n(x)$. This can be done with reasonable accuracy by evaluating the errors at $4n$ evenly spaced points on $[0, \pi]$.

FOUR

APPROXIMATION
OF FUNCTIONS

To evaluate most mathematical functions, we must first produce computable approximations to them. Functions are defined in a variety of ways in applications, with integrals and infinite series being the most common type of formula used for the definition. Such a definition is useful in establishing the properties of the function, but it is generally not an efficient way to evaluate the function. In this chapter we examine the use of polynomials as approximations to a given function. Various means of producing polynomial approximations are described, and they are compared as to their relative accuracy.

For evaluating a function $f(x)$ on a computer, it is generally more efficient of space and time to have an analytic approximation to $f(x)$ rather than to store a table and use interpolation. It is also desirable to use the lowest possible degree of polynomial that will give the desired accuracy in approximating $f(x)$. The following sections give a number of methods for producing an approximation, and generally the better approximations are the ones that are more complicated to produce. The amount of time and effort expended on producing an approximation should be directly proportional to how much the approximation will be used. If it is only to be used a few times, a truncated Taylor series will often suffice. But if an approximation is to be used millions of times by many people, then much care should be used in producing the approximation.

There are forms of approximating functions other than polynomials. Rational functions are quotients of polynomials, and they are usually a somewhat more efficient form of approximation. But because polynomials furnish an adequate and efficient form of approximation, and because the theory for rational function approximation is so much more complicated than that of polynomial approximation, we have chosen to consider only polynomials. The results of this chapter can also be used to produce piecewise polynomial approximations, somewhat analogous to the piecewise polynomial interpolating functions of Section 3.7 of the preceding chapter.

4.1   The Weierstrass Theorem and Taylor's Theorem

To justify using polynomials to approximate continuous functions, we present the following theorem.

Theorem 4.1   (Weierstrass)   Let $f(x)$ be continuous for $a \leq x \leq b$ and let $\epsilon > 0$. Then there is a polynomial $p(x)$ for which

$$|f(x) - p(x)| \leq \epsilon \qquad a \leq x \leq b$$

Proof   There are many proofs of this result and of generalizations of it. Since this is not central to our numerical analysis development, we will just indicate a constructive proof.

Assume that $[a,b] = [0,1]$ for simplicity; by an appropriate change of variables, we can always reduce to this case if necessary. Define

$$p_n(x) = \sum_{k=0}^{n} \binom{n}{k} f\left(\frac{k}{n}\right) x^k (1-x)^{n-k} \qquad 0 \leq x \leq 1$$

Let $f(x)$ be bounded on $[0,1]$. Then

$$\operatorname*{Lim}_{n \to \infty} p_n(x) = f(x)$$

at any point x at which f is continuous. If $f(x)$ is continuous at every x in $[0,1]$, then the convergence of p_n to f is uniform on $[0,1]$, that is,

$$\operatorname*{Max}_{0 \leq x \leq 1} |f(x) - p_n(x)| \to 0 \quad \text{as} \quad n \to \infty \tag{4.1}$$

∎

This gives an explicit way to find a polynomial satisfying the conclusion of the Weierstrass theorem. The proof of these results can be found in Davis [D1, pp. 108–118], along with additional properties of the approximating polynomials $p_n(x)$, which are called the Bernstein polynomials. They mimic extremely well the qualitative behavior of the function $f(x)$. For example, if $f(x)$ is r times continuously differentiable on $[0,1]$, then

$$\operatorname*{Max}_{0 \leq x \leq 1} |f^{(r)}(x) - p_n^{(r)}(x)| \to 0 \quad \text{as} \quad n \to \infty$$

But such an overall approximating property has its price, and in this case the convergence in (4.1) is generally very slow. For example, if $f(x) = x^2$, then

$$\operatorname*{Lim}_{n \to \infty} n[p_n(x) - f(x)] = x(1-x)$$

and thus

$$p_n(x) - x^2 \approx \frac{1}{n} x(1-x)$$

for large values of n. The error does not decrease rapidly, even for approximating such a trivial case as $f(x) = x^2$.

Taylor's theorem Taylor's theorem was presented earlier as Theorem 1.4 of Section 1.1, Chapter 1. It is the first important means for the approximation of a function, and it is often used as a preliminary approximation for computing some more efficient approximation. To aid in understanding why the Taylor approximation is not particularly efficient, consider the following example.

Example Find the error of approximating e^x using the fourth-degree Taylor polynomial $p_4(x)$ on the interval $[-1,1]$, expanding about $x=0$. Then

$$p_4(x) = 1 + x + \frac{1}{2}x^2 + \frac{1}{6}x^3 + \frac{1}{24}x^4 \tag{4.2}$$

$$e^x - p_4(x) = R_5(x) = \frac{1}{120} x^5 e^{\xi}$$

with ξ between x and 0.

To examine the error carefully, we bound it from above and below.

$$\frac{1}{120} x^5 \leq e^x - p_4(x) \leq \frac{e}{120} x^5 \qquad 0 \leq x \leq 1$$

$$\frac{e^{-1}}{120} |x|^5 \leq |e^x - p_4(x)| \leq \frac{1}{120} |x|^5 \qquad -1 \leq x \leq 0$$

The error increases for increasing $|x|$, and

$$\underset{-1 \leq x \leq 1}{\text{Max}} |e^x - p_4(x)| = \frac{e}{120} = .0226 \tag{4.3}$$

The error is not distributed evenly through the interval $[-1,1]$; see Figure 4.1. It

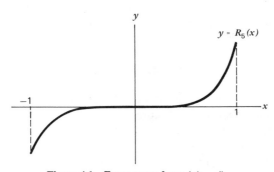

Figure 4.1 Error curve for $p_4(x) \approx e^x$.

is much smaller near the origin than near the endpoints -1 and 1. This uneven distribution of the error, which is typical of the Taylor remainder, means that there are usually much better approximating polynomials of the same degree. Further examples will be given in the next section.

The function space $C[a,b]$ The set $C[a,b]$ of all continuous real-valued functions on the interval $[a,b]$ was introduced earlier in Section 1.1 of Chapter 1. With it we will generally use the norm

$$\|f\|_\infty = \operatorname*{Maximum}_{a \le x \le b} |f(x)| \qquad f \in C[a,b] \tag{4.4}$$

It is called variously the maximum norm, Chebyshev norm, infinity norm, and uniform norm. It is the natural measure to use in approximation theory since we want to determine and compare

$$\|f-p\|_\infty = \operatorname*{Maximum}_{a \le x \le b} |f(x)-p(x)|$$

for various polynomials $p(x)$. Another norm for $C[a,b]$ is introduced in Section 4.3, one which is also quite useful in measuring the size of $f(x)-p(x)$.

As noted earlier in Section 1.1, the maximum norm satisfies the following characteristic properties of a norm:

$$\|f\|=0 \quad \text{if and only if} \quad f \equiv 0 \tag{4.5}$$

$$\|\alpha f\|=|\alpha|\,\|f\|, \quad \text{for all } f \in C[a,b] \text{ and all scalars } \alpha \tag{4.6}$$

$$\|f+g\| \le \|f\|+\|g\|, \quad \text{all } f,g \in C[a,b] \tag{4.7}$$

The proof of these properties for (4.4) is quite straightforward. And they show that the norm should be thought of as a generalization of the absolute value of a number.

Although we will not make any significant use of it, it is often useful to regard $C[a,b]$ as a vector space. The vectors are the functions $f(x)$, $a \le x \le b$. We define the distance from a vector f to a vector g by

$$D(f,g)=\|f-g\| \tag{4.8}$$

which is in keeping with our intuition about the concept of distance in simpler vector spaces. This is illustrated in Figure 4.2.

Using the inequality (4.7),

$$\|f-g\|=\|(f-h)+(h-g)\| \le \|f-h\|+\|h-g\|$$

$$D(f,g) \le D(f,h)+D(h,g) \tag{4.9}$$

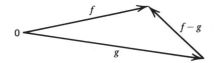

Figure 4.2 Illustration for defining distance $D(f,g)$.

This is called the *triangle inequality*, because of its obvious interpretation in measuring the lengths of sides of a triangle whose vertices are f, g, and h. The name triangle inequality is also applied to the equivalent formulation (4.7).

Another useful result is the *reverse triangle inequality*,

$$\left| \|f\| - \|g\| \right| \leq \|f-g\| \tag{4.10}$$

To prove it, use (4.7) to give

$$\|f\| \leq \|f-g\| + \|g\|$$

$$\|f\| - \|g\| \leq \|f-g\|$$

By the symmetric argument,

$$\|g\| - \|f\| \leq \|g-f\| = \|f-g\|$$

with the last equality following from (4.6) with $\alpha = -1$. Combining these two inequalities proves (4.10).

A more complete introduction to vector spaces (although only finite dimensional) and to vector norms is given in Chapter 7. And some additional geometry for $C[a,b]$ will be given in Sections 4.3 and 4.4 with the introduction of another norm, different from (4.4).

4.2 The Minimax Approximation Problem

Let $f(x)$ be continuous on $[a,b]$. To compare polynomial approximations $p(x)$ to $f(x)$, obtained by various methods, it is natural to ask what is the best possible accuracy that can be attained by using polynomials of each degree $n \geq 0$. Thus we introduce the *minimax error*

$$\rho_n(f) = \underset{\text{degree}(q) \leq n}{\text{Infimum}} \|f-q\|_\infty \tag{4.11}$$

There does not exist a polynomial $q(x)$ of degree $\leq n$ that can approximate $f(x)$ with a smaller maximum error that $\rho_n(f)$.

Having introduced $\rho_n(f)$, we ask whether there is a polynomial $q_n^*(x)$ for which

$$\rho_n(f) = \|f-q_n^*\|_\infty \tag{4.12}$$

And if so, is it unique, what are its characteristics, and how can it be constructed? The approximation $q_n^*(x)$ is called the *minimax approximation* to $f(x)$ on $[a,b]$, and its theory is developed in Section 4.7.

Example Compute the minimax polynomial approximation $q_1^*(x)$ to e^x on $-1 \leqq x \leqq 1$. Let $q_1^*(x) = a_0 + a_1 x$. To find a_0 and a_1, we have to use some geometric insight. Consider the graph of $y = e^x$ with that of a possible approximation $y = q_1(x)$, as in Figure 4.3.

Let

$$\epsilon(x) = e^x - [a_0 + a_1 x] \tag{4.13}$$

Clearly, $q_1^*(x)$ and e^x must be equal at two points in $[-1,1]$, say at $-1 < x_1 < x_2 < 1$. Otherwise we could improve on the approximation by moving the graph of $y = q_1^*(x)$ appropriately. Also,

$$\rho_1 = \underset{-1 \leqq x \leqq 1}{\text{Max}} |\epsilon(x)|$$

and $\epsilon(x_1) = \epsilon(x_2) = 0$. By another argument based on shifting the graph of $y = q_1^*(x)$, we conclude that the maximum error ρ_1 is attained at exactly three points.

$$\epsilon(-1) = \rho_1 \qquad \epsilon(1) = \rho_1 \qquad \epsilon(x_3) = -\rho_1 \tag{4.14}$$

where $x_1 < x_3 < x_2$. Since x_3 is a relative minimum, $\epsilon'(x_3) = 0$. Combining these four equations, we have

$$e^{-1} - [a_0 - a_1] = \rho_1 \qquad e - [a_0 + a_1] = \rho_1$$
$$e^{x_3} - [a_0 + a_1 x_3] = -\rho_1 \qquad e^{x_3} - a_1 = 0 \tag{4.15}$$

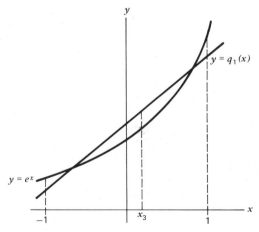

$$y = q_1(x)$$

$$y = e^x$$

Figure 4.3 Linear minimax approximation to e^x.

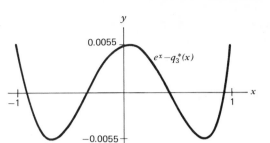

Figure 4.4   Error for cubic minimax approximation to e^x.

These have the solution

$$a_1 = \frac{e - e^{-1}}{2} \doteq 1.1752$$

$$x_3 = \ln_e(a_1) \doteq .1614$$

$$\rho_1 = \frac{1}{2}e^{-1} + \frac{x_3}{4}(e - e^{-1}) \doteq .2788$$

$$a_0 = \rho_1 + (1 - x_3)a_1 \doteq 1.2643 \tag{4.16}$$

Thus

$$q_1^*(x) = 1.2643 + 1.1752x \tag{4.17}$$

and $\rho_1 \doteq .2788$.

By using the numerical method of Section 4.9, we can also construct $q_3^*(x)$ for e^x on $[-1, 1]$.

$$q_3^*(x) = .994579 + .995668x + .542973x^2 + .179533x^3 \tag{4.18}$$

The graph of its error is given in Figure 4.4; and in contrast to the Taylor approximations, the error is evenly distributed throughout the interval of approximation.

We conclude the example with the Table 4.1 of errors for minimax and Taylor approximations for $f(x) = e^x$ on $[-1, 1]$.

Table 4.1   Approximation errors for $f(x) = e^x$

n	$\|f - p_n\|_\infty$	$\|f - q_n^*\|_\infty$
1	.718	.279
2	.218	.0450
3	.0516	.00553

The accuracy of $q_n^*(x)$ is significantly better than that of $p_n(x)$, and the disparity increases as n increases. It should be noted that e^x is a function with a rapidly convergent Taylor series, say in comparison to $\log x$ and $\tan^{-1}x$. With these latter functions, the minimax would look even better in comparison to the Taylor series.

4.3 The Least Squares Approximation Problem

Because of the difficulty in calculating the minimax approximation, we often go to an intermediate approximation called *the least squares approximation*. As notation, introduce

$$\|g\|_2 = \sqrt{\int_a^b |g(x)|^2 dx} \; , \qquad g \in C[a,b] \tag{4.19}$$

This is a function norm, satisfying the same properties (4.5)–(4.7) as the maximum norm of (4.4). It is a generalization of the ordinary Euclidean norm for R^n, defined in (1.20). We return to the proof of the triangle inequality (4.7) in the next section when we generalize the above definition.

For a given $f \in C[a,b]$ and $n \geq 0$, define

$$M_n(f) = \underset{\text{degree}(r) \leq n}{\text{Infimum}} \|f - r\|_2 \tag{4.20}$$

As before, does there exist a polynomial r_n^* that minimizes this expression,

$$M_n(f) = \|f - r_n^*\|_2$$

Is it unique? Can we calculate it?

To further motivate (4.20), consider calculating an average error in the approximation of $f(x)$ by $r(x)$. For an integer $m \geq 1$, define the nodes x_j by

$$x_j = a + \left(j - \frac{1}{2}\right)\left(\frac{b-a}{m}\right) \qquad j = 1, 2, \ldots, m$$

These are the midpoints of m evenly spaced subintervals of $[a,b]$. Then an *average error* of approximating $f(x)$ by $r(x)$ on $[a,b]$ is

$$E = \lim_{m \to \infty} \left\{ \frac{1}{m} \sum_{j=1}^m \left[f(x_j) - r(x_j) \right]^2 \right\}^{1/2}$$

$$= \lim_{m \to \infty} \left\{ \frac{1}{b-a} \sum_{j=1}^m \left[f(x_j) - r(x_j) \right]^2 \left(\frac{b-a}{m}\right) \right\}^{1/2}$$

$$= \frac{1}{\sqrt{b-a}} \sqrt{\int_a^b |f(x) - r(x)|^2 dx}$$

$$E = \frac{\|f - r\|_2}{\sqrt{b-a}} \tag{4.21}$$

Thus the least squares approximation should have a small average error on $[a,b]$. The quantity E is called the *root-mean-square error* in the approximation of $f(x)$ by $r(x)$.

Example Let $f(x)=e^x$, $-1 \le x \le 1$, and let $r_1(x)=b_0+b_1 x$. Minimize

$$\|f-r_1\|_2^2 = \int_{-1}^{1}(e^x - b_0 - b_1 x)^2 dx \equiv F(b_0, b_1) \qquad (4.22)$$

If we expand the integrand and break the integral apart, then $F(b_0, b_1)$ is a quadratic polynomial in the two variables b_0, b_1,

$$F = \int_{-1}^{1}\left(e^{2x} + b_0^2 + b_1^2 x^2 - 2b_0 e^x - 2b_1 x e^x + 2b_0 b_1 x\right)dx$$

To find a minimum, we set

$$\frac{\partial F}{\partial b_0} = 0 \qquad \frac{\partial F}{\partial b_1} = 0$$

which is a necessary condition at a minimal point. Instead of differentiating in the above integral, we merely differentiate through the integral in (4.22),

$$0 = \frac{\partial F}{\partial b_0} = \int_{-1}^{1}\frac{\partial}{\partial b_0}(e^x - b_0 - b_1 x)^2 dx = 2\int_{-1}^{1}[e^x - b_0 - b_1 x](-1)dx$$

$$0 = \frac{\partial F}{\partial b_1} = 2\int_{-1}^{1}(e^x - b_0 - b_1 x)(-x)dx$$

Then

$$b_0 = \tfrac{1}{2}\int_{-1}^{1}e^x dx = \sinh(1) \doteq 1.1752$$

$$b_1 = \tfrac{3}{2}\int_{-1}^{1}xe^x dx = 3e^{-1} \doteq 1.1036$$

$$r_1^*(x) = 1.1752 + 1.1036x \qquad (4.23)$$

By direct examination,

$$\|e^x - r_1^*\|_\infty \doteq 0.44$$

This is intermediate to the approximations q_1^* and $p_1(x)$ derived earlier. Usually the least squares approximation is a fairly good uniform approximation, superior to the Taylor series approximations.

As a further example, the cubic least squares approximation to e^x on $[-1, 1]$ is given by

$$r_3^*(x) = .996294 + .997955x + .536722x^2 + .176139x^3 \qquad (4.24)$$

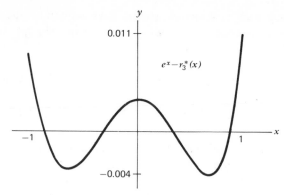

Figure 4.5 Error in cubic least squares approximation to e^x.

and

$$\|e^x - r_3^*\|_\infty = .0112$$

The graph of the error is given in Figure 4.5. Note that the error is not distributed as evenly in $[-1, 1]$ as was true of the minimax $q_3^*(x)$ in Figure 4.4.

The general least squares problem We give a general theory for least squares problems; but we first need to present the theory of orthogonal polynomials, which is given in the next section.

Let $w(x)$ be a nonnegative *weight function* on the interval (a, b), which may be infinite; and assume the following properties:

1. $\displaystyle\int_a^b |x|^n w(x)\,dx$ is integrable and finite for all $n \geq 0$ (4.25)

2. Suppose that   (4.26)

$$\int_a^b w(x)\,g(x)\,dx = 0$$

for some continuous nonnegative continuous function $g(x)$.
Then the function $g(x) \equiv 0$ on (a, b).

Example The following are the weight functions of most interest in the developments of this text:

$$w(x) \equiv 1 \qquad a \leq x \leq b$$

$$w(x) = 1/\sqrt{1 - x^2} \qquad -1 \leq x \leq 1$$

$$w(x) = e^{-x} \qquad 0 \leq x < \infty$$

$$w(x) = e^{-x^2} \qquad -\infty < x < \infty$$

For a finite interval $[a,b]$ the general least squares problem can now be stated. Given $f \in E[a,b]$, does there exist a polynomial $r_n^*(x)$ of degree $\leq n$ that minimizes

$$\int_a^b w(x)[f(x) - r(x)]^2 dx \tag{4.27}$$

among all polynomials $r(x)$ of degree $\leq n$? The function $w(x)$ allows different degrees of importance to be given to the error at different points in the interval $[a,b]$. This will prove useful in developing *near minimax approximations*.

Define

$$F(a_0, a_1, \ldots, a_n) = \int_a^b w(x) \left[f(x) - \sum_{j=0}^{n} a_j x^j \right]^2 dx \tag{4.28}$$

in order to compute (4.27) for an arbitrary polynomial $r(x)$ of degree $\leq n$. We want to minimize F as the coefficients $\{a_i\}$ range over all real numbers. A necessary condition for a point $(a_0, \ldots, a_n)$ to be a minimizing point is

$$\frac{\partial F}{\partial a_i} = 0 \qquad i = 0, 1, \ldots, n \tag{4.29}$$

By differentiating through the integral in (4.28) and using (4.29), we obtain the linear system

$$\sum_{j=0}^{n} a_j \int_a^b w(x) x^{i+j} dx = \int_a^b w(x) f(x) x^i dx \qquad i = 0, 1, \ldots, n \tag{4.30}$$

To see why this solution of the least squares problem is unsatisfactory, consider the special case $w(x) \equiv 1$, $[a,b] = [0,1]$. Then the linear system becomes

$$\sum_{j=0}^{n} \frac{a_j}{i+j+1} = \int_0^1 f(x) x^i dx \qquad i = 0, 1, \ldots, n \tag{4.31}$$

The matrix of coefficients is the Hilbert matrix of order $n+1$, which was introduced in (1.65) of Section 1.5. The solution of linear system (4.31) is extremely sensitive to small changes in the coefficients or right-hand constants. Thus this is not a good way to approach the least squares problem. In single precision arithmetic on an IBM 360 computer, the cases $n \geq 4$ will be unsatisfactory.

4.4 Orthogonal Polynomials

As is evident from graphs of x^n, $n \geq 0$, these monomials are very nearly *linearly dependent*, looking much alike; and this results in the instability of the linear system (4.30). To avoid this problem, we consider an alternative basis for the polynomials, based on polynomials which are *orthogonal* in a function space sense to be given below. These results are of fundamental importance in the approximation of functions and for much work in classical applied mathematics.

Let $w(x)$ be the same as in Section 4.3, and define the *inner product* of two continuous functions f and g by

$$(f,g) = \int_a^b w(x) f(x) g(x) dx \qquad f,g \in C[a,b]. \qquad (4.32)$$

Then the following simple properties are easily shown.

1. $(\alpha f, g) = (f, \alpha g) = \alpha(f, g)$ for all scalars α
2. $(f_1 + f_2, g) = (f_1, g) + (f_2, g)$

    $(f, g_1 + g_2) = (f, g_1) + (f, g_2)$
3. $(f, g) = (g, f)$
4. $(f, f) \geq 0$ for all $f \in C[a,b]$, and $(f, f) = 0$ if and only if $f(x) = 0$, $a \leq x \leq b$

Define the *two-norm* or *Euclidean norm* by

$$\|f\|_2 = \sqrt{\int_a^b w(x) [f(x)]^2 dx} = \sqrt{(f,f)} \qquad (4.33)$$

This definition will satisfy the norm properties (4.5)–(4.7). But the proof of the triangle inequality (4.7) is no longer obvious, and it depends on the following well-known inequality.

Lemma (Cauchy-Schwartz inequality)   For $f, g \in C[a,b]$,

$$|(f,g)| \leq \|f\|_2 \|g\|_2 \qquad (4.34)$$

Proof If $g = 0$, the result is trivially true. For $g \neq 0$, consider the following. For any real number α,

$$0 \leq (f + \alpha g, f + \alpha g) = (f, f) + 2\alpha(f, g) + \alpha^2(g, g)$$

The polynomial on the right has at most one real root, and thus the discriminant cannot be positive,

$$4|(f,g)|^2 - 4(f,f)(g,g) \leq 0$$

and this implies (4.34). Note that we have equality in (4.34) only if the discriminant is exactly zero. But that implies there is an α^* for which the polynomial is zero. Then

$$(f+\alpha^*g, f+\alpha^*g)=0$$

and thus $f=-\alpha^*g$. Thus equality holds in (4.34) if and only if either (1) f is a multiple of g, or (2) f or g is identically zero. ∎

To prove the triangle inequality (4.7),

$$\|f+g\|_2^2=(f+g, f+g)=(f,f)+2(f,g)+(g,g)$$

$$\leqq \|f\|_2^2+2\|f\|_2\|g\|_2+\|g\|_2^2=(\|f\|_2+\|g\|_2)^2$$

Take square roots of each side of the inequality to obtain

$$\|f+g\|_2 \leqq \|f\|_2+\|g\|_2 \tag{4.35}$$

We are interested in obtaining a basis for the polynomials other than the ordinary basis, the monomials $\{1, x, x^2, \dots\}$. We will produce what is called an orthogonal basis, a generalization of orthogonal basis in the space R^n; see Section 7.1. We say that f and g are *orthogonal* if

$$(f,g)=0$$

The following is a constructive existence result for orthogonal polynomials.

Theorem 4.2 (Gram-Schmidt) There exists a sequence of polynomials $\{\varphi_n(x)|$ $n \geqq 0\}$ with degree $(\varphi_n)=n$ for all n, and

$$(\varphi_n, \varphi_m)=0 \quad \text{for all} \quad n \neq m \quad n, m \geqq 0 \tag{4.36}$$

In addition, we can construct the sequence with the additional properties: (1) $(\varphi_n, \varphi_n)=1$ for all n, (2) the coefficient of x^n in $\varphi_n(x)$ is positive. With these additional properties, the sequence $\{\varphi_n\}$ is unique.

Proof We show a constructive and recursive method of obtaining the members of the sequence; it is called the Gram-Schmidt process. Let

$$\varphi_0(x)=c$$

a constant. Pick it such that $\|\varphi_0\|_2=1$ and $c>0$. Then

$$(\varphi_0, \varphi_0)=c^2\int_a^b w(x)\,dx=1$$

$$c=\left[\int_a^b w(x)\,dx\right]^{-1/2}$$

For constructing $\varphi_1(x)$, begin with

$$\psi_1(x) = x + a_{1,0}\varphi_0(x)$$

Then

$(\psi_1, \varphi_0) = 0$ implies $0 = (x, \varphi_0) + a_{1,0}(\varphi_0, \varphi_0)$

$$a_{1,0} = -(x, \varphi_0) = -\int_a^b x w(x) \, dx \Big/ \left[\int_a^b w(x) \, dx\right]^{1/2}$$

Define

$$\varphi_1(x) = \psi_1(x) \Big/ \|\psi_1\|_2$$

and note that

$$\|\varphi_1\|_2 = 1 \qquad (\varphi_1, \varphi_0) = 0$$

To construct $\varphi_n(x)$, first define

$$\psi_n(x) = x^n + a_{n,n-1}\varphi_{n-1}(x) + \ldots + a_{n,0}\varphi_0(x) \tag{4.37}$$

and choose the constants to make ψ_n orthogonal to φ_j for all $0 \leq j < n$. Then

$(\psi_n, \varphi_j) = 0$ implies $a_{n,j} = -(x^n, \varphi_j)$ $\quad j = 0, 1, \ldots, n-1$ (4.38)

The desired $\varphi_n(x)$ is

$$\varphi_n(x) = \psi_n(x) \Big/ \|\psi_n\|_2 \tag{4.39}$$

Continue the derivation inductively. ∎

Example For the special case of $w(x) \equiv 1$, $[a, b] = [-1, 1]$, we have

$$\varphi_0(x) = \sqrt{\tfrac{1}{2}} \qquad \varphi_1(x) = \sqrt{\tfrac{3}{2}}\, x \qquad \varphi_2(x) = \tfrac{1}{2}\sqrt{\tfrac{5}{2}}\, (3x^2 - 1)$$

and further polynomials can be constructed by the same process.

There is a very large literature on these polynomials, including a variety of formulas for them. Of necessity we will only skim the surface. The polynomials are usually given in a form for which $\|\varphi_n\|_2 \neq 1$.

Examples 1 ***The Legendre polynomials*** Let $w(x) \equiv 1$ on $[-1, 1]$. Define

$$P_n(x) = \frac{(-1)^n}{2^n n!} \cdot \frac{d^n}{dx^n}\left[(1 - x^2)^n\right] \qquad n \geq 1 \tag{4.40}$$

with $P_0(x) \equiv 1$. These are orthogonal on $[-1, 1]$, degree $P_n(x) = n$, and $P_n(1) = 1$ for all n. Also,

$$(P_n, P_n) = \frac{2}{2n+1}$$

$$\varphi_n(x) = \sqrt{\frac{2n+1}{2}} \, P_n(x) \tag{4.41}$$

2. The Chebyshev polynomials. Let $w(x) = 1/\sqrt{1-x^2}$, $-1 \le x \le 1$. Then

$$T_n(x) = \cos(n \cos^{-1} x) \qquad n \ge 0 \tag{4.42}$$

is an orthogonal family of polynomials with $\deg(T_n) = n$. To see this more clearly, let $\cos^{-1} x = \theta$, or $x = \cos \theta$, $0 \le \theta \le \pi$. Then

$$T_{n \pm 1}(x) = \cos(n \pm 1)\theta = \cos(n\theta) \cos \theta \mp \sin(n\theta) \sin \theta$$

$$T_{n+1}(x) + T_{n-1}(x) = 2\cos(n\theta) \cos \theta = 2 T_n(x) x$$

$$T_{n+1}(x) = 2x T_n(x) - T_{n-1}(x) \qquad n \ge 1 \tag{4.43}$$

Also by direct calculation in (4.42),

$$T_0(x) \equiv 1 \qquad T_1(x) = x$$

Using (4.43),

$$T_2(x) = 2x^2 - 1, \; T_3(x) = 2x(2x^2 - 1) - x = 4x^3 - 3x$$

The polynomials also satisfy

$$T_n(1) = 1 \qquad (T_n, T_n) = \begin{cases} \pi, & n = 0 \\ \pi/2, & n > 0 \end{cases} \tag{4.44}$$

3. The Laguerre polynomials. Let $w(x) = e^{-x}$, $[a, b] = [0, \infty)$. Then

$$L_n(x) = \frac{1}{n! e^{-x}} \cdot \frac{d^n}{dx^n} \{x^n e^{-x}\} \qquad n \ge 0 \tag{4.45}$$

and $\|L_n\|_2 = 1$ for all n.

We say that a family of functions is an *orthogonal family* if each member is orthogonal to every other member of the family. We call it an *orthonormal family* if it is an orthogonal family and if every member has length one, that is, $\|f\|_2 = 1$. For other examples of orthogonal polynomials, see Abramowitz and Stegun [A1, Chap. 22] Davis [D1, appendix], and Szego [S2].

Some properties of orthogonal polynomials   These results will be useful later in this chapter and in the next chapter.

Theorem 4.3   Let $\{\varphi_n(x)|n \geq 0\}$ be an orthogonal family of polynomials on (a,b) with weight function $w(x)$; and with such a family we always assume implicitly that degree $\varphi_n = n$, $n \geq 0$. If $f(x)$ is a polynomial of degree m, then

$$f(x) = \sum_{n=0}^{m} \frac{(f, \varphi_n)}{(\varphi_n, \varphi_n)} \varphi_n(x) \qquad (4.46)$$

Proof   We begin by showing that every polynomial can be written as a combination of orthogonal polynomials of no greater degree. Since degree $\varphi_0 = 0$, we have $\varphi_0(x) = c$, a constant, and thus

$$1 \equiv \frac{1}{c} \varphi_0(x)$$

Since degree $\varphi_1(x) = 1$, we have, from the construction in the Gram-Schmidt process,

$$\varphi_1(x) = c_{1,1} x + c_{1,0} \varphi_0(x) \qquad c_{1,1} \neq 0$$

$$x = \frac{1}{c_{1,1}} \left[\varphi_1(x) - c_{1,0} \varphi_0(x) \right]$$

By induction in the Gram-Schmidt process,

$$\varphi_r(x) = c_{r,r} x^r + c_{r,r-1} \varphi_{r-1}(x) + \ldots + c_{r,0} \varphi_0(x) \qquad c_{r,r} \neq 0$$

and

$$x^r = \frac{1}{c_{r,r}} \left[\varphi_r(x) - c_{r,r-1} \varphi_{r-1}(x) - \ldots - c_{r,0} \varphi_0(x) \right]$$

Thus every monomial can be rewritten as a combination of orthogonal polynomials of no greater degree. From this it follows easily that an arbitrary polynomial $f(x)$ of degree m can be written as

$$f(x) = b_m \varphi_m(x) + \ldots + b_0 \varphi_0(x)$$

for some choice of $b_0, \ldots, b_m$. To calculate each b_i, multiply both sides by $w(x)$ and $\varphi_i(x)$, and integrate over (a,b). Then

$$(f, \varphi_i) = \sum_{j=0}^{m} b_j(\varphi_j, \varphi_i) = b_i(\varphi_i, \varphi_i)$$

$$b_i = (f, \varphi_i)/(\varphi_i, \varphi_i)$$

which proves (4.46) and the theorem.   ∎

Corollary If $f(x)$ is a polynomial of degree $\leq m-1$, then

$$(f,\varphi_m)=0$$

and $\varphi_m(x)$ is orthogonal to $f(x)$.

Proof It follows easily from (4.46) and the orthogonality of the family $\{\varphi_n(x)\}$.
∎

The following result gives some intuition as to the shape of the graphs of the orthogonal polynomials. It will also be crucial to the work on Gaussian quadrature in Chapter 5.

Theorem 4.4 Let $\{\varphi_n(x)|n\geq 0\}$ be an orthogonal family of polynomials on (a,b) with weight function $w(x)\geq 0$. Then the polynomial $\varphi_n(x)$ has exactly n distinct real roots in the open interval (a,b).

Proof Let $x_1,x_2,\ldots,x_m$ be all zeroes of $\varphi_n(x)$ for which

1. $a<x_i<b$

2. $\varphi_n(x)$ changes sign at x_i

Since degree $\varphi_n=n$, we trivially have $m\leq n$. We will assume $m<n$ and then derive a contradiction.
Define

$$B(x)=(x-x_1)\ldots(x-x_m)$$

By the definition of the points $x_1,\ldots,x_m$, the polynomial

$$\varphi_n(x)B(x)=(x-x_1)\ldots(x-x_m)\varphi_n(x)$$

does not change sign in (a,b). To see this more clearly, the assumptions on $x_1,\ldots,x_m$ imply

$$\varphi_n(x)=h(x)(x-x_1)^{r_1}\ldots(x-x_m)^{r_m}$$

with each r_i odd and with $h(x)$ not changing sign in (a,b). Then

$$\varphi_n(x)B(x)=(x-x_1)^{r_1+1}\ldots(x-x_m)^{r_m+1}h(x)$$

and the conclusion follows. Consequently,

$$\int_a^b w(x)B(x)\varphi_n(x)\,dx\neq 0$$

since clearly $B(x)\varphi_n(x)\not\equiv 0$. But since degree $B=m<n$, the corollary to

Theorem 4.3 implies

$$\int_a^b w(x)B(x)\varphi_n(x)\,dx=(B,\varphi_n)=0$$

This is a contradiction; and thus we must have $m=n$. But then the conclusion of the theorem will follow since $\varphi_n(x)$ can have at most n roots, and the assumptions on $x_1,\dots,x_n$ imply they must all be simple, i.e. $\varphi'(x_i)\neq 0$. ∎

As above, let $\{\varphi_n(x)|n\geq 0\}$ be an orthogonal family on (a,b) with weight function $w(x)\geq 0$. Define A_n and B_n by

$$\varphi_n(x)=A_n x^n + B_n x^{n-1}+\dots. \tag{4.47}$$

Also, write

$$\varphi_n(x)=A_n(x-x_{n,1})(x-x_{n,2})\dots(x-x_{n,n}) \tag{4.48}$$

Let

$$a_n=A_{n+1}/A_n \qquad \gamma_n=(\varphi_n,\varphi_n)>0 \tag{4.49}$$

Theorem 4.5 (Triple recursion relation)   Let $\{\varphi_n\}$ be an orthogonal family of polynomials on (a,b) with weight function $w(x)\geq 0$. Then for $n\geq 1$,

$$\varphi_{n+1}(x)=(a_n x+b_n)\varphi_n(x)-c_n\varphi_{n-1}(x) \tag{4.50}$$

with

$$b_n=a_n\cdot\left[\frac{B_{n+1}}{A_{n+1}}-\frac{B_n}{A_n}\right] \qquad c_n=\frac{A_{n+1}A_{n-1}}{A_n^2}\cdot\frac{\gamma_n}{\gamma_{n-1}} \tag{4.51}$$

Proof   First note that the triple recursion relation (4.43) for Chebyshev polynomials is an example of (4.50). To derive (4.50), we begin by considering the polynomial

$$G(x)=\varphi_{n+1}(x)-a_n x\varphi_n(x)$$

$$=A_{n+1}x^{n+1}+B_{n+1}x^n+\dots-\frac{A_{n+1}}{A_n}x\left(A_n x^n+B_n x^{n-1}+\dots\right)$$

$$=\left(B_{n+1}-\frac{A_{n+1}B_n}{A_n}\right)x^n+\dots,$$

and degree $G \leq n$. By Theorem 4.3, we can write

$$G(x) = d_n \varphi_n(x) + \ldots + d_0 \varphi_0(x)$$

for an appropriate set of constants $d_0, \ldots, d_n$. Calculating d_i,

$$d_i = \frac{(G, \varphi_i)}{(\varphi_i, \varphi_i)} = \frac{1}{\gamma_i} \left[(\varphi_{n+1}, \varphi_i) - a_n (x\varphi_n, \varphi_i) \right]$$

We have $(\varphi_{n+1}, \varphi_i) = 0$ for $i \leq n$; and

$$(x\varphi_n, \varphi_i) = \int_a^b w(x) \varphi_n(x) x\varphi_i(x) \, dx = 0$$

for $i \leq n - 2$, since then degree $x\varphi_i(x) \leq n - 1$. Combining these results,

$$d_i = 0 \qquad 0 \leq i \leq n - 2$$

and therefore

$$G(x) = d_n \varphi_n(x) + d_{n-1} \varphi_{n-1}(x)$$

$$\varphi_{n+1}(x) = (a_n x + d_n) \varphi_n(x) + d_{n-1} \varphi_{n-1}(x)$$

This shows the existence of a triple recursion relation, and the remaining work is manipulating the formulas to obtain d_n and d_{n-1}, given as b_n and c_n in (4.51). These constants will not be derived here, but their values are important for many applications; see Problem 17. ∎

Examples 1. For Laguerre polynomials,

$$L_{n+1}(x) = \frac{1}{n+1}(2n + 1 - x)L_n(x) - \frac{n}{n+1}L_{n-1}(x)$$

2. For Legendre polynomials,

$$P_{n+1}(x) = \frac{2n+1}{n+1} x P_n(x) - \frac{n}{n+1} P_{n-1}(x)$$

Theorem 4.6 (Christoffel-Darboux identity) For $\{\varphi_n\}$ an orthogonal family of polynomials with weight function $w(x) \geq 0$,

$$\sum_{k=0}^n \frac{\varphi_k(x)\varphi_k(y)}{\gamma_k} = \frac{\varphi_{n+1}(x)\varphi_n(y) - \varphi_n(x)\varphi_{n+1}(y)}{a_n \gamma_n (x - y)} \qquad x \neq y \quad (4.52)$$

Proof It is based on manipulating the triple recursion relation, and is not given here. See Szego [S2]. ∎

4.5   The Least Squares Approximation Problem (continued)

Assume that $\{\varphi_n(x)|n \geq 0\}$ is an orthonormal family of polynomials with weight function $w(x) \geq 0$, that is,

$$(\varphi_n, \varphi_m) = \delta_{n,m} = \begin{cases} 1, & n = m \\ 0, & n \neq m. \end{cases}$$

Then an arbitrary polynomial $r(x)$ of degree $\leq n$ can be written as

$$r(x) = b_0 \varphi_0(x) + \ldots + b_n \varphi_n(x)$$

Then for a given $f \in C[a,b]$,

$$\|f - r\|_2^2 = \int_a^b w(x) \left[f(x) - \sum_{j=0}^n b_j \varphi_j(x) \right]^2 dx \equiv G(b_0, \ldots, b_n) \qquad (4.53)$$

We will solve the least squares problem by minimizing G.
As before, we could set

$$\frac{\partial G}{\partial b_i} = 0 \qquad i = 0, 1, \ldots, n$$

But to obtain a more complete result, we proceed in another way. For any choice of $b_0, \ldots, b_n$,

$$0 \leq G(b_0, \ldots, b_n) = \left(f - \sum_{j=0}^n b_j \varphi_j, f - \sum_{i=0}^n b_i \varphi_i \right)$$

$$= (f, f) - 2 \sum_{j=0}^n b_j (f, \varphi_j) + \sum_i \sum_j b_i b_j (\varphi_i, \varphi_j)$$

$$= \|f\|_2^2 - 2 \sum_{j=0}^n b_j (f, \varphi_j) + \sum_{j=0}^n b_j^2$$

$$= \|f\|_2^2 - \sum_{j=0}^n (f, \varphi_j)^2 + \sum_{j=0}^n \left[(f, \varphi_j) - b_j \right]^2$$

which can be checked by expanding the last term. Thus G is a minimum if and only if

$$b_j = (f, \varphi_j) \qquad j = 0, 1, \ldots, n$$

Then the least squares approximation exists, is unique, and is given by

$$r_n^*(x) = \sum_{j=0}^{n} (f, \varphi_j) \varphi_j(x) \tag{4.54}$$

Moreover,

$$\| f - r_n^* \|_2 = \left[\| f \|_2^2 - \sum_{j=0}^{n} (f, \varphi_j)^2 \right]^{1/2}$$

$$= \sqrt{\| f \|_2^2 - \| r_n^* \|_2^2} \tag{4.55}$$

$$\| f \|_2^2 = \| r_n^* \|_2^2 + \| f - r_n^* \|_2^2 \tag{4.56}$$

As a practical note in obtaining $r_{n+1}^*(x)$, use

$$r_{n+1}^*(x) = r_n^*(x) + (f, \varphi_{n+1}) \varphi_{n+1}(x) \tag{4.57}$$

Theorem 4.7 Assuming $[a, b]$ is finite,

$$\lim_{n \to \infty} \| f - r_n^* \|_2 = 0 \tag{4.58}$$

Proof By definition of r_n^* as a minimizing polynomial for $\| f - r_n \|_2$, we have

$$\| f - r_1^* \|_2 \geq \| f - r_2^* \|_2 \geq \cdots \geq \| f - r_n^* \|_2 \geq \cdots \tag{4.59}$$

Let $\epsilon > 0$ be arbitrary. Then by the Weierstrass theorem, there is a polynomial $Q(x)$ of some degree, say m, for which

$$\max_{a \leq x \leq b} |f(x) - Q(x)| \leq \frac{\epsilon}{c} \qquad c = \sqrt{\int_a^b w(x) \, dx}$$

By the definition of $r_m^*(x)$,

$$\| f - r_m^* \|_2 \leq \| f - Q \|_2 = \left\{ \int_a^b w(x) [f(x) - Q(x)]^2 \, dx \right\}^{1/2}$$

$$\leq \left[\int_a^b w(x) \frac{\epsilon^2}{c^2} \, dx \right]^{1/2} = \epsilon$$

Combining this with (4.59),

$$\| f - r_n^* \|_2 \leq \epsilon$$

for all $n \geq m$. Since ϵ was arbitrary, this proves (4.58). ∎

Using (4.56), we have *Bessel's inequality*,

$$\|r_n^*\|_2^2 = \sum_{j=0}^{n} (f, \varphi_j)^2 \le \|f\|_2^2 \tag{4.60}$$

and using (4.58) in (4.56) we obtain *Parseval's equality*,

$$\|f\|_2 = \left[\sum_{j=0}^{\infty} (f, \varphi_j)^2 \right]^{1/2} \tag{4.61}$$

The preceding theorem does not say that $\|f - r_n^*\|_\infty \to 0$. But if additional differentiability assumptions are placed on $f(x)$, results on the uniform convergence of r_n^* to f can be proven. An example is given later.

Legendre polynomial expansions To solve the least squares problem on a finite interval $[a,b]$ with $w(x) \equiv 1$, we can convert it to a problem on $[-1,1]$. The change of variable

$$x = \frac{b + a + (b-a)t}{2} \tag{4.62}$$

converts the interval $-1 \le t \le 1$ to $a \le x \le b$. Define

$$F(t) = f\left(\frac{b + a + (b-a)t}{2} \right) \qquad -1 \le t \le 1 \tag{4.63}$$

for a given $f \in C[a,b]$. Then

$$\int_a^b [f(x) - r_n(x)]^2 dx = \left(\frac{b-a}{2} \right) \int_{-1}^{1} [F(t) - R_n(t)]^2 dt$$

with $R_n(t)$ obtained from $r_n(x)$ using (4.62). The change of variable (4.62) gives a one-to-one correspondence between polynomials of degree m on $[a,b]$ and of degree m on $[-1,1]$, for every $m \ge 0$. Thus minimizing $\|f - r_n\|_2$ on $[a,b]$ is equivalent to minimizing $\|F - R_n\|_2$ on $[-1,1]$. We therefore restrict our interest to the least squares problem on $[-1,1]$.

Given $f \in [-1,1]$, the orthonormal family described in Theorem 4.2 is $\varphi_0(x) \equiv 1/\sqrt{2}$,

$$\varphi_n(x) = \sqrt{\frac{2n+1}{2}} \cdot \frac{(-1)^n}{2^n n!} \cdot \frac{d^n}{dx^n} \{(1-x^2)^n\} \qquad n \ge 1 \tag{4.64}$$

The least squares approximation is

$$r_n^*(x) = \sum_{j=0}^{n} (f, \varphi_j) \varphi_j(x) \qquad (f, \varphi_j) = \int_{-1}^{1} f(x) \varphi_j(x) dx \tag{4.65}$$

the solution of the original least squares problem posed in Section 4.3. The coefficients (f, φ_j) are called Legendre coefficients.

Example For $f(x) = e^x$ on $[-1, 1]$, the expansion coefficients (f, φ_j) of (4.65) are given in Table 4.2. The approximation $r_3^*(x)$ was given earlier in Section 4.3, written in standard polynomial form. For the average error E in $r_3^*(x)$, given in (4.24), combine (4.21), (4.55), and the table coefficients to give

$$E = \frac{1}{\sqrt{2}} \| e^x - r_3^*(x) \|_2 \doteq .0034$$

Table 4.2 Legendre expansion coefficients for e^x.

j	(f, φ_j)	j	(f, φ_j)
0	1.661985	3	.037660
1	.901117	4	.004698
2	.226302	5	.000469

Chebyshev polynomial expansions The weight function is $w(x) = 1/\sqrt{1-x^2}$, and

$$\varphi_0(x) = \frac{1}{\sqrt{\pi}}, \quad \varphi_n(x) = \sqrt{\frac{2}{\pi}} \, T_n(x) \qquad n \geq 1 \qquad (4.66)$$

The least squares solution is

$$C_n(x) = \sum_{j=0}^{n} (f, \varphi_j) \varphi_j(x) \qquad (f, \varphi_j) = \int_{-1}^{1} \frac{f(x)\varphi_j(x)\,dx}{\sqrt{1-x^2}} \qquad (4.67)$$

Using the definition of $\varphi_n(x)$ in terms of $T_n(x)$,

$$C_n(x) = \sum_{j=0}^{n}{}' c_j T_j(x) \qquad c_j = \frac{2}{\pi} \int_{-1}^{1} \frac{f(x)T_j(x)\,dx}{\sqrt{1-x^2}} \qquad (4.68)$$

The prime on the summation symbol means that the first term should be halved before beginning to sum.

The Chebyshev expansion is closely related to Fourier cosine expansions. Using $x = \cos\theta$, $0 \leq \theta \leq \pi$,

$$C_n(\cos\theta) = \sum_{j=0}^{n}{}' c_j \cos(j\theta) \qquad (4.69)$$

$$c_j = \frac{2}{\pi} \int_{0}^{\pi} \cos(j\theta) f(\cos\theta)\,d\theta \qquad (4.70)$$

Thus $C_n(\cos\theta)$ is the truncation after $n+1$ terms of the Fourier cosine expansion

$$f(\cos\theta) = \sum_{0}^{\infty}{}' c_j \cos(j\theta) \tag{4.71}$$

If the Fourier cosine expansion on $[0, \pi]$ of $f(\cos\theta)$ is known, then by substituting $\theta = \cos^{-1}x$, we have the Chebyshev expansion of $f(x)$.

For reasons to be given later, the Chebyshev least squares approximation is more useful than the Legendre least squares approximation. For this reason, we give a more detailed convergence theorem for (4.68).

Theorem 4.8 Let $f(x)$ have r continuous derivatives on $[-1, 1]$, with $r \geq 1$. Then for the Chebyshev least squares approximation $C_n(x)$ defined in (4.68),

$$\|f - C_n\|_\infty \leq \frac{B \ln n}{n^r} \qquad n \geq 0 \tag{4.72}$$

for a constant B dependent on f and r. Thus $C_n(x)$ converges uniformly to $f(x)$ as $n \to \infty$, provided $f(x)$ is continuously differentiable.

Proof Combine Rivlin [R2, Theorem 3.3, p. 134] and Meinardus [M1, Theorem 45, p. 57]. ∎

Example We illustrate the Chebyshev expansion by again considering approximations to $f(x) = e^x$. For the coefficients c_j, from (4.68)

$$c_j = \frac{2}{\pi} \int_{-1}^{1} \frac{e^x T_j(x)\,dx}{\sqrt{1-x^2}}$$

$$= \frac{2}{\pi} \int_{0}^{\pi} e^{\cos\theta} \cos(j\theta)\,d\theta$$

The latter formula is better for numerical integration since there are no singularities in the integrand. Either the midpoint rule or the trapezoidal rule is an excellent integration method because of the periodicity of the integrand; see Corollary (1) to Theorem 5.3 in Section 5.5. Using numerical integration, we obtain the values given in Table 4.3. Using (4.68) and the formulas for $T_j(x)$, we obtain

$$C_1(x) = 1.266 + 1.130x$$

$$C_3(x) = .994571 + .997308x + .542991x^2 + .177347x^3 \tag{4.73}$$

$$\|e^x - C_1(x)\|_\infty = .32 \qquad \|e^x - C_3(x)\|_\infty = .00607$$

Table 4.3   Chebyshev expansion
coefficients for e^x

j	c_j
0	2.53213176
1	1.13031821
2	.27149534
3	.04433685
4	.00547424
5	.00054293

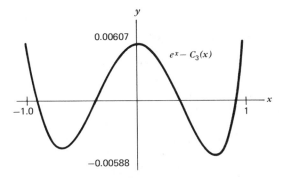

Figure 4.6   Error in cubic Chebyshev least squares approximation to e^x.

The graph of $e^x - C_3(x)$ is given in Figure 4.6, and it is very similar to Figure 4.4, the graph for the minimax error. The maximum errors for these Chebyshev least squares approximations are quite close to the minimax errors, and are generally adequate for most practical purposes.

The polynomial $C_n(x)$ can be evaluated accurately and rapidly using the form (4.68), rather than converting it to the ordinary form using the monomials x^j, as was done above for the example in (4.73). We make use of the triple recursion relation (4.43) for the Chebyshev polynomials $T_n(x)$.

Algorithm   Chebeval (a, n, x, value)

1. Remark: This algorithm evaluates

$$\text{value}:= \sum_{j=0}^{n} a_j T_j(x)$$

2. $v_0:= 1$, $v_1:= x$, sum$:= a_0 v_0 + a_1 v_1$

3. Do through step 6 for $j = 2, 3, \ldots, n$

4. $v_2:= 2xv_1 - v_0$

5. $\text{sum:} = \text{sum} + a_j v_2$

6. $v_0: = v_1, \ v_1: = v_2$

7. Return with value: $= \text{sum}$.

This is almost as efficient as the nested multiplication algorithm (2.64) of Section 2.9; and similar algorithms are available for other orthogonal polynomial expansions, again using their associated triple recursion relation. For an analysis of the effect of rounding errors, see Fox and Parker [F1, p. 57] or Rivlin [R2, p. 126].

4.6 Economization of Taylor Series

Although Taylor series are often very easy to obtain, they are usually inefficient approximations when compared with the minimax approximation of the same degree. Economization is a technique in which first a Taylor series $p_n(x)$ is chosen to yield a certain desired accuracy, say ϵ:

$$|f(x) - p_n(x)| \leq \epsilon \qquad a \leq x \leq b$$

This Taylor series is then changed to a polynomial of lower degree without reducing by much the accuracy of $p_n(x)$. The basis of the process is the following result.

Theorem 4.9 For a fixed integer $n > 0$, consider the minimization problem:

$$\tau_n = \underset{\text{degree}(Q) \leq n-1}{\text{Infimum}} \ \underset{-1 \leq x \leq 1}{\text{Maximum}} |x^n + Q(x)| \qquad (4.74)$$

with $Q(x)$ a polynomial. The minimum τ_n is attained uniquely by letting

$$x^n + Q(x) = \frac{1}{2^{n-1}} T_n(x) \qquad (4.75)$$

defining $Q(x)$ implicitly. The minimum is

$$\tau_n = \frac{1}{2^{n-1}} \qquad (4.76)$$

Proof We begin by considering some facts about the Chebyshev polynomials. From the definition, $T_0(x) \equiv 1$, $T_1(x) = x$. The triple recursion relation

$$T_{n+1}(x) = 2x T_n(x) - T_{n-1}(x)$$

is the basis of an induction proof that

$$T_n(x) = 2^{n-1}x^n + \text{lower-degree terms} \qquad n \geq 1$$

Thus

$$\frac{1}{2^{n-1}} T_n(x) = x^n + \text{lower-degree terms} \qquad n \geq 1 \qquad (4.77)$$

Since $T_n(x) = \cos(n\theta)$, $x = \cos\theta$, $0 \leq \theta \leq \pi$, the polynomial $T_n(x)$ attains relative maxima and minima at $n+1$ points in $[-1, 1]$;

$$\cos(n \cos^{-1} x) = \pm 1$$

$$n \cos^{-1} x = j\pi \quad \text{for all} \quad j$$

$$x_j = \cos\left(\frac{j\pi}{n}\right) \qquad j = 0, 1, \ldots, n \qquad (4.78)$$

Additional values of j do not lead to new values of x_j because of the periodicity of the cosine function. For these points,

$$T_n(x_j) = (-1)^j \qquad j = 0, 1, \ldots, n \qquad (4.79)$$

and

$$-1 = x_n < x_{n-1} < \ldots < x_1 < x_0 = 1$$

The polynomial $T_n(x)/2^{n-1}$ has leading coefficient 1 and

$$\underset{-1 \leq x \leq 1}{\text{Max}} \left| \frac{1}{2^{n-1}} T_n(x) \right| = \frac{1}{2^{n-1}} \qquad (4.80)$$

Thus $\tau_n \leq (\tfrac{1}{2})^{n-1}$. Suppose that

$$\tau_n < \frac{1}{2^{n-1}} \qquad (4.81)$$

We will show that this leads to a contradiction. The assumption (4.81) and the definition (4.74) imply the existence of a polynomial

$$M(x) = x^n + Q(x) \qquad \text{degree}(Q) \leq n - 1$$

with

$$\tau_n \leq \underset{-1 \leq x \leq 1}{\text{Max}} |M(x)| < \left(\tfrac{1}{2}\right)^{n-1} \qquad (4.82)$$

Define

$$R(x) = \frac{1}{2^{n-1}} T_n(x) - M(x)$$

which has degree $\leq n-1$. We will examine the sign of $R(x_j)$ at the points of (4.78). Using (4.79), (4.81), and (4.82),

$$R(x_0) = R(1) = \frac{1}{2^{n-1}} - M(1) > 0$$

$$R(x_1) = -\frac{1}{2^{n-1}} - M(x_1) = -\left[\frac{1}{2^{n-1}} + M(x_1)\right] < 0$$

and the sign of $R(x_j)$ is $(-1)^j$. Since R has $n+1$ changes of sign, R must have at least n zeroes. But then the degree of $R < n$ implies that $R \equiv 0$; and thus $M \equiv (1/2^{n-1})T_n$.

To prove that no polynomial other than $(1/2^{n-1})T_n(x)$ will minimize (4.74), a variation of the above proof is used. We omit it. ∎

Let f be a given function possessing a Taylor series that is convergent on $[-1,1]$, centered at $x=0$. Given an error tolerance $\epsilon > 0$, pick n so that the Taylor series approximation $p_n(x)$ of degree n satisfies

$$\|f - p_n\|_\infty \leq \epsilon_1 \tag{4.83}$$

Then reduce p_n to a lower degree polynomial $M(x)$ with

$$\|p_n - M\|_\infty \leq \epsilon_2$$

making the degree of M as low as practically feasible. Then

$$\|f - M\|_\infty \leq \epsilon_1 + \epsilon_2 \tag{4.84}$$

The process of reducing p_n to a lower-degree polynomial is generally done one degree per step; and we want $\epsilon_1 + \epsilon_2 \leq \epsilon$.

To increase the generality of the following development, let $p_n(x)$ be known, but not necessarily a Taylor approximation. Pick $M(x)$ such that degree $M \leq n-1$ and

$$\operatorname*{Max}_{-1 \leq x \leq 1} |p_n(x) - M(x)|$$

is as small as possible. Then

$$E_n(x) \equiv p_n(x) - M(x) = a_n x^n + p_{n-1}(x) - M(x)$$

$$= a_n\left[x^n + \frac{p_{n-1}(x) - M(x)}{a_n}\right]$$

with $p_n(x) = a_n x^n + p_{n-1}(x)$, $a_n \neq 0$.

$$\|E_n\|_\infty = |a_n| \underset{-1 \leq x \leq 1}{\text{Max}} \left| x^n + \frac{p_{n-1}(x) - M(x)}{a_n} \right|$$

and this is minimized, according to Theorem 4.10, by letting

$$x^n + \frac{p_{n-1}(x) - M(x)}{a_n} = \frac{1}{2^{n-1}} T_n(x)$$

$$M(x) = a_n \left[x^n - \frac{1}{2^{n-1}} T_n(x) \right] + p_{n-1}(x)$$

$$M(x) = p_n(x) - \frac{a_n}{2^{n-1}} T_n(x) \tag{4.85}$$

and the error in the approximation is

$$\|p_n(x) - M(x)\|_\infty = \frac{|a_n|}{2^{n-1}} \tag{4.86}$$

$M(x)$ is the minimax approximation of degree $\leq n-1$ to the function $p_n(x)$.

Example 1.   Let $f(x) = e^x$, $p_2(x) = 1 + x + \frac{1}{2}x^2$. Then

$$\|e^x - p_2\|_\infty = .218$$

Economization to a degree one polynomial gives the new polynomial

$$M_{2,1}(x) = p_2(x) - \frac{1/2}{2} T_2(x) = 1 + x + \frac{x^2}{2} - \frac{1}{4}(2x^2 - 1)$$

$$M_{2,1}(x) = x + \frac{5}{4}$$

$$\|p_2 - M_{2,1}\|_\infty = .25 \qquad \|f - M_{2,1}\|_\infty \leq .218 + .25 = .468$$

By direct calculation, $\|e^x - M_{2,1}\|_\infty = .468$, exactly the bound given above. The approximation $M_{2,1}(x)$ is linear and is a considerable improvement on the Taylor approximation p_1 for which $\|f - p_1\|_\infty = .718$.

We will let $M_{p,q}(x)$ denote the polynomial obtained by reducing a Taylor polynomial of degree p down to a polynomial of degree q, one degree at a time using the above economization procedure.

2.   Again let $f(x) = e^x$, and produce $p_3(x) = 1 + x + \frac{1}{2}x^2 + \frac{1}{6}x^3$. Then

$$\|f - p_3\|_\infty = .0516$$

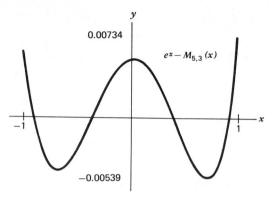

Figure 4.7   Error in economized polynomial $M_{5,3}(x)$ to e^x.

We will perform economization on p_3 twice to obtain a linear approximation.

$$M_{3,2}(x)=p_3(x)-\tfrac{1}{24}T_3(x)=1+\tfrac{9}{8}x+\tfrac{1}{2}x^2$$

$$M_{3,1}(x)=M_{3,2}(x)-\tfrac{1}{4}T_2(x)=\tfrac{5}{4}+\tfrac{9}{8}x.$$

By direct computation,

$$\|f-M_{3,1}(x)\|_\infty=.34$$

a very good linear approximation error.

3.   We give two cubic approximations to e^x, obtained by economization.

$$M_{4,3}(x)=.994792+x+.541667x^2+.166667x^3$$

$$M_{5,3}(x)=.994792+.997396x+.541667x^2+.177083x^3 \quad (4.87)$$

$$\|e^x-M_{4,3}(x)\|_\infty=.0152 \qquad \|e^x-M_{5,3}(x)\|_\infty=.00734$$

The error $e^x-M_{5,3}(x)$ is shown in Figure 4.7, and it is similar to Figure 4.4 for the cubic minimax.

4.7   Minimax Approximations

For a good uniform approximation to a given function $f(x)$, it seems reasonable to expect the error to be fairly uniformly distributed throughout the interval of approximation. The past examples with $f(x)=e^x$ further illustrate this point, and they show that the maximum error will oscillate in sign. Table 4.4 summarizes the statistics for the various forms of approximation to e^x on

Table 4.4 Comparison of various linear and cubic approximations to e^x

Method of approximation	Maximum error	
	Linear	Cubic
Taylor polynomial $p_n(x)$	.718	.0516
Economization with $M_{n+1,n}(x)$	.468	.0152
Economization with $M_{n+2,n}(x)$	.343	.00734
Legendre least squares $r_n^*(x)$	.439	.0112
Chebyshev least squares $C_n(x)$	.322	.00607
Chebyshev node interpolation formula $I_n(x)$	.372	.00666
Chebyshev forced oscillation formula $F_n(x)$	.286	.00558
Minimax $q_n^*(x)$	.279	.00553

$[-1,1]$, including some methods to be given in Section 4.8. To show the importance of a uniform distribution of the error, with the error function oscillating in sign, we present two theorems. The first one is useful for estimating $\rho_n(f)$, the minimax error, without having to calculate the minimax approximation $q_n^*(x)$.

Theorem 4.10 (de la Vallee-Poussin). Let $f \in C[a,b]$ and $n \geq 0$. Suppose we have a polynomial $Q(x)$ of degree $\leq n$ that satisfies

$$f(x_j) - Q(x_j) = (-1)^j e_j \qquad j = 0, 1, \ldots, n+1 \qquad (4.88)$$

with all e_j's of the same sign and nonzero, and with

$$a \leq x_0 < x_1 < \ldots < x_{n+1} \leq b$$

Then

$$\underset{0 \leq j \leq n+1}{\text{Min}} |e_j| \leq \rho_n(f) \equiv \|f - q_n^*\|_\infty \leq \|f - Q\|_\infty \qquad (4.89)$$

Proof The upper inequality follows from the definition of $\rho_n(f)$. To prove the lower bound, we will assume it is false and produce a contradiction. Assume

$$\rho_n(f) < \underset{0 \leq j \leq n+1}{\text{Min}} |e_j|$$

Then by the definition of $\rho_n(f)$, there is a polynomial $P(x)$ of degree $\leq n$ for which

$$\rho_n(f) \leq \|f - P\|_\infty < \text{Min}|e_j| \qquad (4.90)$$

Define

$$R(x) = Q(x) - P(x)$$

a polynomial of degree $\leq n$. For simplicity, let all $e_j > 0$; the same argument works when all $e_j < 0$.

Evaluate $R(x_j)$ for each j and observe the sign of $R(x_j)$. First,

$$R(x_0) = Q(x_0) - P(x_0) = [f(x_0) - P(x_0)] - [f(x_0) - Q(x_0)]$$
$$= [f(x_0) - P(x_0)] - e_0 < 0$$

by using (4.90). Next,

$$R(x_1) = Q(x_1) - P(x_1) = [f(x_1) - P(x_1)] + e_1 > 0$$

Inductively, the sign of $R(x_j)$ is $(-1)^{j+1}$, $j = 0, 1, \ldots, n+1$. This gives $R(x)$ $n+2$ changes of sign and implies $R(x)$ has $n+1$ zeroes. Since degree $R \leq n$, this is not possible unless $R \equiv 0$. Then $P \equiv Q$, contrary to assumption. ∎

Example   Recall the cubic Chebyshev least squares approximation $C_n(x)$ to e^x on $[-1, 1]$. It has the maximum errors on the interval $[-1, 1]$ given in Table 4.5. These errors satisfy the hypotheses of the Theorem 4.10, and thus

$$.00497 \leq \rho_3(f) \leq .00607$$

From this we could conclude that $C_3(x)$ was quite close to the best possible approximation. Note that $\rho_3(f) = .00553$ from direct calculations using (4.18).

Table 4.5   Relative maxima of $|e^x - C_3(x)|$

x	$e^x - C_3(x)$
-1.0	.00497
$-.6919$	$-.00511$
.0310	.00547
.7229	$-.00588$
1.0	.00607

Theorem 4.11   (Chebyshev equioscillation theorem).   Let $f \in C[a, b]$ and $n \geq 0$. Then there exists a unique polynomial $q_n^*(x)$ of degree $\leq n$ for which

$$\rho_n(f) = \|f - q_n^*\|_\infty$$

This polynomial is uniquely characterized by the following property: There are at least $n+2$ points,

$$a \leq x_0 < x_1 < \ldots < x_{n+1} \leq b$$

for which

$$f(x_j) - q_n^*(x_j) = \sigma(-1)^j \rho_n(f) \qquad j = 0, 1, \ldots, n+1 \quad (4.91)$$

with $\sigma = \pm 1$, depending only on f and n.

Proof The proof is quite technical, amounting to a complicated and manipulatory proof by contradiction. For that reason, it is omitted. For a complete development, see Davis [D1, Chapter 7]. A practical method for computing $q_n^*(x)$ will be given in Section 4.9. ∎

Example The cubic minimax approximation $q_3^*(x)$ to e^x, given in (4.18), satisfies the conclusions of this theorem, as can be seen from the graph of the error in Figure 4.4.

From this theorem we can see that the Taylor series is always a poor uniform approximation. The Taylor series error,

$$f(x) - p_n(x) = \frac{(x - x_0)^{n+1}}{(n+1)!} f^{(n+1)}(\xi_x)$$

does not vary uniformly through the interval of approximation, nor does it oscillate much in sign.

To give a better idea of how well $q_n^*(x)$ approximates $f(x)$ with increasing n, we give the following theorem of D. Jackson.

Theorem 4.12 (Jackson) Let $f(x)$ have k continuous derivatives for some $k \geq 0$. Moreover, assume $f^{(k)}(x)$ satisfies

$$\underset{a \leq x, y \leq b}{\mathrm{Sup}} \left| f^{(k)}(x) - f^{(k)}(y) \right| \leq M |x - y|^\alpha \qquad (4.92)$$

for some $M > 0$ and some $0 < \alpha \leq 1$. [We say $f^{(k)}(x)$ satisfies a Lipschitz condition with exponent α.] Then there is a constant d_k, independent of f and n, for which

$$\rho_n(f) \leq \frac{M d_k}{n^{k+\alpha}} \qquad n \geq 1 \quad (4.93)$$

Proof See Meinardus [M1, Theorem 45, p. 57]. Note that if we wish to ignore (4.92) with a k-times continuously differentiable function $f(x)$, then just use $k - 1$ in place of k in the theorem, with $\alpha = 1$ and $M = \| f^{(k)} \|_\infty$. This will then yield

$$\rho_n(f) \leq \frac{d_{k-1}}{n^k} \|f^{(k)}\|_\infty \tag{4.94}$$

Also note that if $f(x)$ is infinitely differentiable, then $q_n^*(x)$ converges to $f(x)$, uniformly on $[a,b]$, faster than any power $1/n^k$, $k \geq 1$. ∎

4.8   Near-Minimax Approximations

In the light of the Chebyshev equioscillation theorem, we can deduce methods that may give a reasonably good approximation to the minimax approximation. We begin by a further examination of the truncated Chebyshev series approximation (4.68), first obtained as a least squares approximation. We repeat the pertinent information:

For $f \in C[-1,1]$,

$$f(x) = \sum_{j=0}^{\infty}{}' c_j T_j(x) \tag{4.95}$$

$$c_j = \frac{2}{\pi} \int_{-1}^{1} \frac{f(x) T_j(x)}{\sqrt{1-x^2}} dx \tag{4.96}$$

with the convergence in (4.95) holding in the sense that

$$\underset{n\to\infty}{\text{Limit}} \int_{-1}^{1} \frac{1}{\sqrt{1-x^2}} \left[f(x) - \sum_{j=0}^{n}{}' c_j T_j(x) \right]^2 dx = 0$$

From Rivlin [R2, p. 134] we have the quite strong result that

$$\rho_n(f) \leq \|f - C_n\|_\infty \leq \left(4 + \frac{4}{\pi^2} \ln n \right) \rho_n(f) \qquad n \geq 1 \tag{4.97}$$

Combining this with Jackson's result (4.94) implies the earlier result (4.72) in Theorem 4.8.

If $f \in C^r[a,b]$, it can be proven that there is a constant c, dependent on f and r, for which

$$|c_j| \leq c/j^r \qquad j \geq 1 \tag{4.98}$$

This is proven by considering the c_j's as the Fourier cosine coefficients of $f(\cos\theta)$, and then by using results from the theory of Fourier series on the rate of decrease of such coefficients. Thus the larger r, the more rapidly the coefficients c_j decrease.

For the truncated expansion $C_n(x)$,

$$f(x) - C_n(x) = \sum_{n+1}^{\infty} c_j T_j(x) \approx c_{n+1} T_{n+1}(x) \tag{4.99}$$

if $c_{n+1} \neq 0$ and if the coefficients c_j are rapidly convergent to zero. Since the term $c_{n+1} T_{n+1}(x)$ has exactly $n+2$ equal relative maxima and minima, this says that C_n should be nearly the minimax approximation $q_n^*(x)$.

Example Recall the example (4.73) for $f(x) = e^x$ near the end of Section 4.5. In it, the coefficients c_j decreased quite rapidly; and

$$\|e^x - C_3\|_\infty = .00607 \qquad c_4 T_4(x) = .00547 T_4(x) \qquad \|e^x - q_3^*\|_\infty = .00553$$

This illustrates the comments of the last paragraph.

Example It can be shown that

$$\tan^{-1} x = 2\left[\alpha T_1(x) - \frac{\alpha^3}{3} T_3(x) + \frac{\alpha^5}{5} T_5(x) - \cdots \right] \tag{4.100}$$

converging uniformly for $-1 \leq x \leq 1$, with $\alpha = \sqrt{2} - 1 \doteq .41$. Then

$$C_{2n+1}(x) = 2\left[\alpha T_1(x) - \frac{\alpha^3}{3} T_3(x) + \cdots + \frac{(-1)^n}{2n+1} \alpha^{2n+1} T_{2n+1}(x) \right] \tag{4.101}$$

For the error,

$$E_{2n+1}(x) \equiv \tan^{-1} x - C_{2n+1}(x) = \frac{2}{2n+3} (-1)^{n+1} \alpha^{2n+3} T_{2n+3}(x)$$

$$+ 2\sum_{j=n+2}^{\infty} \frac{(-1)^j \alpha^{2j+1}}{2j+1} T_{2j+1}(x)$$

We will bound these terms to estimate the error in $C_{2n+1}(x)$ and the minimax error $\rho_{2n+1}(f)$.

By taking upper bounds,

$$\left| 2\sum_{n+2}^{\infty} \frac{(-1)^j \alpha^{2j+1}}{2j+1} T_{2j+1}(x) \right| \leq 2\sum_{n+2}^{\infty} \frac{\alpha^{2j+1}}{2j+1} < \frac{2}{2n+5} \cdot \frac{\alpha^{2n+5}}{1-\alpha^2}$$

Thus we obtain upper and lower bounds for $E_{2n+1}(x)$,

$$E_{2n+1}(x) \leq \frac{2}{2n+3}(-1)^{n+1}\alpha^{2n+3}T_{2n+3}(x) + \frac{2\alpha^{2n+5}}{(2n+5)(1-\alpha^2)}$$

$$\leq \frac{2\alpha^{2n+3}}{2n+3}\left[(-1)^{n+1}T_{2n+3}(x) + .207\right]$$

using $\alpha^2/(1-\alpha^2) \doteq .207$. Similarly,

$$E_{2n+1}(x) \geq \frac{\alpha^{2n+3}}{2n+3}\left[(-1)^{n+1}T_{2n+3}(x) - .207\right]$$

Thus

$$E_{2n+1}(x) \approx \frac{\alpha^{2n+3}}{2n+3}(-1)^{n+1}T_{2n+3}(x)$$

which has $2n+4$ relative maxima and minima, of alternating sign. By Theorem 4.10 and the above inequalities,

$$\frac{2(.793)\alpha^{2n+3}}{2n+3} \leq \rho_{2n+1}(f) = \rho_{2n+2}(f) \leq \frac{2(1.207)\alpha^{2n+3}}{2n+3} \qquad (4.102)$$

Interpolation at the Chebyshev zeroes   If the error in the minimax approximations is nearly $c_{n+1}T_{n+1}(x)$, as derived in (4.99), then the error should be nearly zero at the zeroes of $T_{n+1}(x)$ on $[-1,1]$. These are the points

$$x_j = \cos\left[\frac{2j-1}{2n+2}\cdot\pi\right] \qquad j=1,2,\dots,n+1 \qquad (4.103)$$

Let $I_n(x)$ be the polynomial of degree $\leq n$ that interpolates to $f(x)$ at these nodes $\{x_j\}$. Since it has zero error at the Chebyshev zeroes x_j of $T_{n+1}(x)$, the uniqueness of polynomial interpolation implies that $I_n(x)$ should about equal $q_n^*(x)$. $I_n(x)$ can be calculated using the algorithms Divdif and Interp of Section 3.2. A note of caution: if $c_{n+1}=0$, then the error $f(x)-q_n^*(x)$ is likely to have as its zeroes those of a higher degree Chebyshev polynomial, usually $T_{n+2}(x)$. This is likely to happen if $f(x)$ is either odd or even on $[-1,1]$; see Problem 21. A function $f(x)$ is even (odd) on an interval $[-a,a]$ if $f(-x)=f(x)$ $(f(-x)=-f(x))$ for all x in $[-a,a]$. The above example (4.101) with $f(x)=\tan^{-1}x$ is an apt illustration.

As further motivation for defining $I_n(x)$ using the nodes of (4.103), see Problem 25 in which this choice is shown to be optimal in certain sense. In [P1] Powell has shown that $I_n(x)$ is a good approximation to $f(x)$, regardless of the smoothness of the function $f(x)$. His result is that

$$\|f - I_n\|_\infty \leq v(n)\|f - q_n^*\|_\infty \tag{4.104}$$

where $v(n)$ grows slowly with n. This is reminiscent of (4.97) for $C_n(x)$. For $n \leq 20$, $v(n) \leq 4$, and thus I_n is never a very bad approximation. An illustration can again be given with $f(x) = e^x$; and the maximum errors for the linear and cubic approximations are given in Table 4.4 at the beginning of Section 4.7.

Forced oscillation of the error Let $f(x) \in C[-1, 1]$, and define

$$F_n(x) = \sum_{k=0}^{n}{}' c_{n,k} T_k(x) \qquad -1 \leq x \leq 1 \tag{4.105}$$

Let

$$-1 \leq x_{n+1} < x_n < \ldots < x_1 < x_0 \leq 1 \tag{4.106}$$

be nodes, the choice of which is discussed below. Determine the coefficients $c_{n,k}$ by forcing the error $f(x) - F_n(x)$ to oscillate in the manner specified as necessary by Theorem 4.11:

$$f(x_i) - F_n(x_i) = (-1)^i E_n \qquad i = 0, 1, \ldots, n+1 \tag{4.107}$$

We have introduced another unknown, E_n, which we hope will be nonzero. There are $n+2$ unknowns, the coefficients $c_{n,0}, \ldots, c_{n,n}$ and E_n; and there are $n+2$ equations in (4.107). Thus there is a reasonable chance that a solution exists. If a solution exists, then by Theorem 4.10,

$$|E_n| \leq \rho_n(f) \leq \|f - F_n\|_\infty \tag{4.108}$$

To pick the nodes, note that if $c_{n+1} T_{n+1}(x)$ in (4.99) is nearly the minimax error, then the relative maxima in the minimax error, $f(x) - q_n^*(x)$, occur at the relative maxima of $T_{n+1}(x)$. These relative maxima occur when $T_{n+1}(x) = \pm 1$, and they are

$$x_i = \cos\left(\frac{i\pi}{n+1}\right) \qquad i = 0, 1, \ldots, n+1 \tag{4.109}$$

These seem an excellent choice to use in (4.107) if $F_n(x)$ is to look like the minimax approximation $q_n^*(x)$.

The system (4.107) becomes

$$\sum_{k=0}^{n}{}' c_{n,k} T_k(x_i) + (-1)^i E_n = f(x_i) \qquad i = 0, 1, \ldots, n+1 \tag{4.110}$$

Note that

$$T_k(x_i) = \cos(k \cos^{-1}(x_i)) = \cos\left(\frac{ki\pi}{n+1}\right)$$

Introduce $E_n = c_{n,n+1}/2$. The system (4.110) becomes

$$\sum_{k=0}^{n+1}{}'' c_{n,k} \cos\left(\frac{ki\pi}{n+1}\right) = f(x_i) \qquad i = 0, 1, \ldots, n+1 \qquad (4.111)$$

since

$$\cos\left(\frac{ki\pi}{n+1}\right) = (-1)^i \quad \text{for} \quad k = n+1$$

The notation $\sum''$ means that the first and last terms are to be halved before summation begins.

To solve (4.111), we need the following relations:

$$\sum_{i=0}^{n+1}{}'' \cos\left(\frac{ij\pi}{n+1}\right) \cos\left(\frac{ik\pi}{n+1}\right) = \begin{cases} n+1, & j = k = 0 \text{ or } n+1 \\ \dfrac{n+1}{2}, & 0 < j = k < n+1 \\ 0, & j \neq k, 0 \leq j, k \leq n+1 \end{cases} \qquad (4.112)$$

For a derivation, see Hamming [H1, Chapter 6].

Multiply equation i in (4.111) by $\cos[ij\pi/(n+1)]$ for some $0 \leq j \leq n+1$. Then sum on i, halving the first and last terms. This gives

$$\sum_{k=0}^{n+1}{}'' c_{n,k} \sum_{i=0}^{n+1}{}'' \cos\left(\frac{ij\pi}{n+1}\right) \cos\left(\frac{ik\pi}{n+1}\right) = \sum_{i=0}^{n+1}{}'' f(x_i) \cos\left(\frac{ij\pi}{n+1}\right)$$

Using the identity (4.112), all but one of the terms in the summation on k will be zero. By checking the two cases $j = 0$ or $n+1$ and $0 < j < n+1$, the same formula is obtained:

$$c_{n,j} = \frac{2}{n+1} \sum_{i=0}^{n+1}{}'' f(x_i) \cos\left(\frac{ij\pi}{n+1}\right) \qquad 0 \leq j \leq n+1 \qquad (4.113)$$

The formula for E_n is

$$E_n = \frac{1}{n+1} \sum_{i=0}^{n+1}{}'' (-1)^i f(x_i) \qquad (4.114)$$

There are a number of connections between this approximation $F_n(x)$ and the Chebyshev expansion $C_n(x)$. Most importantly, the coefficients $c_{n,j}$ are approximations to the coefficients c_j in $C_n(x)$.

$$c_j = \frac{2}{\pi} \int_{-1}^{1} \frac{T_j(x)f(x)\,dx}{\sqrt{1-x^2}} = \frac{2}{\pi} \int_{0}^{\pi} f(\cos\theta)\cos(j\theta)\,d\theta$$

$$\approx \frac{2}{\pi} \sum_{i=0}^{n+1} {''} f\left[\cos\left(\frac{i\pi}{n+1}\right)\right] \cdot \cos\left(j\frac{i\pi}{n+1}\right) \cdot \frac{\pi}{n+1} = c_{n,j}$$

The trapezoidal rule (5.13) of numerical integration was used to approximate the integral for c_j. And it is known that for periodic integrands, the trapezoidal rule is especially accurate; see the first corollary to Theorem 5.3 in Section 5.5. Also, the coefficients $c_{n,j}$ can be related to c_j as follows:

$$c_{nj} = c_j + c_{2(n+1)-j} + c_{2(n+1)+j} + c_{4(n+1)-j} + c_{4(n+1)+j} + \cdots \qquad (4.115)$$

If the Chebyshev coefficients c_n in (4.95) decrease rapidly, then the approximation $F_n(x)$ nearly equals $C_n(x)$, and it is easier to calculate.

Example Use $f(x) = e^x$, $-1 \le x \le 1$, as before. For $n = 1$, the nodes are

$$\{x_i\} = \{-1, 0, 1\}$$

then $E_1 = .272$, and

$$F_1(x) = 1.2715 + 1.1752x$$

For the error, the relative maxima for $e^x - F_1(x)$ are given in Table 4.6.

Table 4.6 Relative maxima of $|e^x - F_1(x)|$

x	$e^x - F_1(x)$
-1.0	.272
.1614	$-.286$
1.0	.272

For $n = 3$, $E_3 = .00547$ and

$$F_3(x) = .994526 + .995682x + .543081x^2 + .179519x^3.$$

The points of maximum error are given in Table 4.7. From Theorem 4.10,

$$.00547 \le \rho_3(f) \le .00558$$

and this says that $F_3(x)$ is an excellent approximation to $q_3^*(x)$.

To see that $F_3(x)$ is an approximation to $C_3(x)$, compare the coefficients $c_{3,j}$ with the coefficients c_j given in Table 4.3 in the example (4.73) at the end of

Table 4.7 Relative maxima of $|e^x - F_3(x)|$

x	$e^x - F_3(x)$
-1.0	.00547
$-.6832$	$-.00552$
.0493	.00558
.7324	$-.00554$
1.0	.00547

Table 4.8 Expansion coefficients for $C_3(x)$ and $F_3(x)$ to e^x

j	c_j	$c_{n,j}$
0	2.53213176	2.53213215
1	1.13031821	1.13032142
2	.27149534	.27154032
3	.04433685	.04487978
4	.00547424	$E_4 = .00547424$

Section 4.5. The results are given in Table 4.8.

As with the interpolatory approximation $I_n(x)$, care must be taken if $f(x)$ is odd or even on $[-1, 1]$. In such a case, choose n as follows:

$$\text{If } f \text{ is } \left\{ \begin{matrix} \text{even} \\ \text{odd} \end{matrix} \right\}, \text{ then choose } n \text{ to be } \left\{ \begin{matrix} \text{odd} \\ \text{even} \end{matrix} \right\}.$$

This ensures $c_{n+1} \neq 0$ in (4.99), and thus the nodes chosen will be the correct ones.

Shampine [S1] has given an analysis of the error in $F_n(x)$, comparable to that given for $I_n(x)$ by Powell, summarized in (4.104). Shampine's result is that

$$\|f - F_n\|_\infty \le w(n) \|f - q_n^*\|_\infty \tag{4.116}$$

For n even, the function $w(n) \ge v(n)$, where $v(n)$ is the multiplier in the error result (4.104) for $I_n(x)$. And for arbitrary n, $w(n)$ approaches $v(n)$ rapidly as n increases. In theory, when considering all continuous functions $f(x)$, there is no significant difference between using $I_n(x)$ and $F_n(x)$. But in our experience with smoother functions $f(x)$, $F_n(x)$ has always been superior to $I_n(x)$ as an approximation. We will give an algorithm for computing $F_n(x)$, which can then be evaluated using the algorithm Chebeval given in Section 4.6.

Algorithm Approx (c, E, f, n)

1. Remark: This algorithm calculates the coefficients c_j in

$$F_n(x) = \sum_{j=0}^{n} {}' c_j T_j(x) \qquad -1 \le x \le 1$$

according to the formula (4.113), and E is calculated from (4.114). c_0 should be halved before using algorithm Chebeval.

2. Create $x_i := \cos(i\pi/(n+1))$,
$$f_i := f(x_i), \qquad i = 0, 1, \ldots, n+1$$

3. Do through step 8 for $j = 0, 1, \ldots, n+1$.

4. sum: $= [f_0 + (-1)^j f_{n+1}]/2$

5. Do through step 6 for $i = 1, \ldots, n$.

6. sum: $= \text{sum} + f_i \cos(ij\pi/(n+1))$

7. End loop on i.

8. $c_j := 2 \, \text{sum}/(n+1)$.

9. End loop on j.

10. $E := c_{n+1}/2$ and exit.

4.9 The Remes Algorithm

The minimax approximation $q_n^*(x)$ to $f(x)$ can be found by using an iteration technique called the second algorithm of Remes. This method exploits the oscillation property of the minimax approximation, as described in Theorem 4.11.

We describe one iteration step, dividing it into three parts.

(S1) Given $n+2$ nodes

$$-1 \leqq x_0 < x_1 < \ldots < x_{n+1} \leqq 1 \qquad (4.117)$$

determine the polynomial $q(x)$ for which degree $(q) \leqq n$ and

$$f(x_i) - q(x_i) = (-1)^i E \qquad i = 0, 1, \ldots, n+1 \qquad (4.118)$$

The nodes should be so chosen that $E \neq 0$.

This is a system of $n+2$ linear equations in which the unknowns are E and the $n+1$ coefficients of $q(x)$.

(S2) Having determined $q(x)$, calculate $n+2$ new node points

$$-1 \leqq z_0 < z_1 < \ldots < z_{n+1} \leqq 1 \qquad (4.119)$$

for which $f(x) - q(x)$ is a local optimum at each z_i, and for which $f(x_i) - q(x_i)$ and $f(x_i) - q(z_i)$ have the same sign, $i = 0, 1, \ldots, n+1$. The first requirement implies

$$f'(z_i) - q'(z_i) = 0 \qquad i = 1, 2, \ldots, n \tag{4.120}$$

and also at z_0 and z_{n+1} if they are in $(-1, 1)$. Also, the nodes z_i should be chosen so that

$$\|f - q\|_\infty = |f(z_k) - q(z_k)| \tag{4.121}$$

for some z_k.

(S3)   Using Theorem 4.10 and the properties of the nodes $\{z_i\}$, we have

$$m = \underset{i}{\mathrm{Min}} |f(z_i) - q(z_i)| \leq \rho_n(f) \leq M \equiv \underset{i}{\mathrm{Max}} |f(z_i) - q(z_i)| \tag{4.122}$$

If M/m is sufficiently close to 1, we consider that $q(x)$ is close enough to the minimax for practical purposes; for example, if

$$M/m \leq 1.05 \tag{4.123}$$

If this is not true, then we set the nodes $\{x_i\}$ equal to $\{z_i\}$, and return to (S1). The initial guess of nodes $\{x_i\}$ in (4.117) are usually the points

$$x_i = \cos\left(\frac{i\pi}{n+1}\right) \qquad i = 0, 1, \ldots, n+1$$

discussed in the last section. The first approximation $q(x)$ is therefore the polynomial $F_n(x)$ of Section 4.8, satisfying (4.105), (4.110), and (4.114). The convergence to $q_n^*(x)$ is quite rapid. It is shown in Meinardus [M1, Theorem 84, p. 111] that, subject to mild restrictions on $f(x)$, we have

$$|\rho_n(f) - \|f - q^{(j+1)}\|_\infty| \leq c|\rho_n(f) - \|f - q^{(j)}\|_\infty|^2, \qquad j \geq 1 \tag{4.124}$$

for some $c > 0$. The notation $q^{(j)}(x)$ denotes the $q(x)$ obtained in the jth iteration of the Remes algorithm. A similar quadratic convergence result also holds for $\|q_n^* - q^{(j)}\|_\infty$.

Example   We show the computation of $q_2^*(x)$ for $f(x) = e^x$ on $[-1, 1]$ by the Remes algorithm.
    We begin with

$$x_0 = -1.0 \qquad x_1 = -.5 \qquad x_2 = .5 \qquad x_3 = 1.0$$

and we write $q(x)$ in the standard form

$$q(x) = a_0 + a_1 x + a_2 x^2$$

Table 4.9 Remes iterate $q^{(1)}(x)$ for $q_2^*(x) \approx e^x$

a_i, E	z_i	$f(z_i) - q(z_i)$
.989141	-1.0	$-.0443369$
1.130864	$-.438621$	.0452334
.553940	.560939	$-.0454683$
$E = .0443369$	1.0	.0443369

Table 4.10 Remes iterate $q^{(2)}(x)$ for $q_2^*(x) \approx e^x$

a_i, E	z_i	$f(z_i) - q(z_i)$
.989039	-1.0	$-.0450171$
1.130184	$-.436958$	.0450177
.554041	.560059	$-.0450174$
$E = .0450171$	1.0	.0450171

The iterates $q^{(1)}(x)$ and $q^{(2)}(x)$ are summarized in Tables 4.9 and 4.10. For this first iterate, the ratio $M/m = 1.026$; and if the test (4.123) was being used, the first iterate would be adequate. But for purposes of illustration, we give the second iterate. The third iterate agrees with the second one to the number of places shown, and the values of the error are all

$$f(z_i) - q(z_i) = \pm .045017388403$$

This is far more accuracy than is needed, based on the limited accuracy of $\rho_2(f) \doteq .045$ in the minimax approximation. But it illustrates the rapid convergence of the Remes algorithm.

Discussion of the literature

Approximation theory is a very popular subject within mathematics, and it continues to attract people interested in both its classical and modern aspects. The variety of approaches and topics can be seen in the books by Achieser [A2], Davis [D1], Lorentz [L1], Rice [R1], and Meinardus [M1]. For detailed information on the special functions of mathematical physics, see the handbook by Abramowitz and Stegun [A1] and the two volume set of Luke [L2]. The classic work on orthogonal polynomials is that of Szego [S2]; and useful presentations are also given in Abramowitz and Stegun [A1] and Davis [D1]. For Chebyshev polynomials and their many uses throughout numerical analysis and mathematics, see Fox and Parker [F1] and Rivlin [R2]. This includes a further discussion of the near-minimax methods of Section 4.8. For a discussion of approximation by functions other than polynomials, especially by rational functions, see Meinardus [M1, Part II] and Rice [R1, Vol. II].

With the advent of large-scale computing, there has been a need for polynomial approximations of the common mathematical functions. The derivation of these approximations depends on the classical approximation theory, but the actual computer implementation requires additional knowledge involving computing. A general survey of numerical methods is given in Fraser [F2], and it has influenced the organization of this chapter. The best source of information on the computer implementation of polynomial approximations is the book by Hart et al. [H2]. A successful and widely used program for generating minimax approximations is given in Cody et al. [C1].

Bibliography

[A1] Abramowitz, M. and I. Stegun, editors, *Handbook of Mathematical Functions*, National Bureau of Standards, U.S. Gov't Printing Office, 1964.

[A2] Achieser, N., *Theory of Approximation*, transl. by C. Hyman, F. Ungar Pub., New York, 1956.

[C1] Cody, W., W. Fraser, and J. Hart, Rational Chebyshev approximation using linear equations, *Numer. Math.*, **12**, 1968, pp. 242–251.

[D1] Davis, P., *Interpolation and Approximation*, Blaisdell Pub., New York, 1963.

[F1] Fox, L. and I. Parker, *Chebyshev Polynomials in Numerical Analysis*, Oxford Press, London, 1968.

[F2] Fraser, W., A survey of methods of computing minimax and near-minimax polynomial approximations for functions of a single independent variable, *Jour. Assoc. Comput. Mach.*, **12**, 1965, pp. 295–314.

[H1] Hamming, R., *Numerical Methods for Scientists and Engineers*, McGraw-Hill, New York, 1962.

[H2] Hart, J. et al., *Computer Approximations*, John Wiley, New York, 1968.

[I1] Isaacson, E. and H. Keller, *Analysis of Numerical Methods*, John Wiley, New York, 1966.

[L1] Lorentz, G. G., *Approximation of Functions*, Holt, Rinehart and Winston, New York, 1966.

[L2] Luke, Y., *The Special Functions and Their Applications*, Vol. I and II, Academic Press, New York, 1969.

[M1] Meinardus, G., *Approximation of Functions: Theory and Numerical Methods*, transl. by L. Schumaker, Springer-Verlag, New York, 1967.

[P1] Powell, M., On the maximum errors of polynomial approximations defined by interpolation and by least squares criteria, *The Comput. Jour.*, **9** 1967, pp. 404–407.

[R1] Rice, J., *The Approximation of Functions*, Vol. I: *Linear Theory*, Vol. II: *Advanced Topics*, Addison-Wesley, Reading, Mass., 1964 and 1968.

[R2] Rivlin, T., *The Chebyshev Polynomials*, John Wiley, New York, 1974.

[S1] Shampine, L., Efficiency of a procedure for near-minimax approximation, *Jour. Assoc. Comput. Mach.*, **17**, 1970, pp. 655–660.

[S2] Szego, G., *Orthogonal Polynomials*, 3rd edition, Amer. Math. Soc., Providence, R.I., 1967.

Problems

1. To illustrate that the Bernstein polynomials $p_n(x)$ in Theorem 1.1 are poor approximations, calculate the fourth-degree approximation $p_4(x)$ for $f(x) = \sin(\pi x)$, $0 \leqslant x \leqslant 1$. Compare it with the fourth-degree Taylor series approximation, expanded about $x = \frac{1}{2}$.

2. **(a)** Let $f(x)$ be continuously differentiable on $[a,b]$. Let $p(x)$ be a polynomial for which

$$\|f' - p\|_\infty \leq \epsilon$$

and define

$$q(x) = f(a) + \int_a^x p(t)\,dt \quad a \leq x \leq b$$

Show that $q(x)$ is a polynomial and satisfies

$$\|f - q\|_\infty \leq \epsilon(b-a)$$

(b) Extend part (a) to the case where $f(x)$ is N times continuously differentiable on $[a,b]$, $N \geq 2$, and $p(x)$ is a polynomial satisfying

$$\|f^{(N)} - p\|_\infty \leq \epsilon$$

Obtain a formula for a polynomial $q(x)$ approximating $f(x)$, with the formula for $q(x)$ involving a single integral.

(c) Assume that $f(x)$ is infinitely differentiable on $[a,b]$, that is, $f^{(j)}(x)$ exists and is continuous on $[a,b]$, for all $j \geq 0$. [This does not imply that $f(x)$ has a convergent Taylor series on $[a,b]$.] Prove there exists a sequence of polynomials $\{p_n(x)|n \geq 1\}$ for which

$$\operatorname*{Limit}_{n \to \infty} \|f^{(j)} - p_n^{(j)}\|_\infty = 0$$

for all $j \geq 0$. Hint: use the Weierstrass theorem and part (b).

3. Let $S = \sum_1^\infty (-1)^j a_j$ be a convergent series; and assume that all $a_j \geq 0$ and

$$a_1 \geq a_2 \geq \ldots \geq a_n \geq \ldots .$$

Prove that

$$\left| S - \sum_1^n (-1)^j a_j \right| \leq a_{n+1}$$

4. Using Problem 3, examine the convergence of the following series. Bound the error when truncating after n terms, and note the dependence on x. Find the value of n for which the error is less than 10^{-5}. This problem illustrates another common technique for bounding the error in Taylor series.

(a) $J_0(x) = \sum_0^\infty \dfrac{(-1)^j \left(\frac{1}{4}x^2\right)^j}{(j!)^2}$

(b) $\sum_{j=1}^\infty \dfrac{(-1)^j x^{2j}}{j^2}$

5. Graph the errors of the Taylor series approximations $p_n(x)$ to $f(x) = \sin[(\pi/2)x]$ on $-1 \leq x \leq 1$, for $n = 1, 3, 5$. Note the behavior of the error both near the origin and near the endpoints.

6. Let $f(x)$ be three times continuously differentiable on $[-\alpha, \alpha]$ for some $\alpha > 0$, and consider approximating it by the rational function

$$R(x) = \frac{a + bx}{1 + cx}$$

To generalize the idea of Taylor series, choose the constants a, b, and c so that

$$R^{(j)}(0) = f^{(j)}(0) \qquad j = 0, 1, 2$$

Is it always possible to find such an approximation $R(x)$? The function $R(x)$ is an example of a *Pade approximation* to $f(x)$.

7. Apply the results of Problem 6 to the case $f(x)=e^x$, and give the resulting approximation $R(x)$. Analyze its error on $[-1,1]$; and compare it with the error for the quadratic Taylor polynomial.

8. (a) Produce the linear Taylor polynomial to $f(x)=\ln x$ on $1 \leq x \leq e$, expanding about $x_0=(1+e)/2$. Graph the error.

   (b) Produce the linear minimax approximation to $f(x)=\ln x$ on $[1,e]$. Graph the error, and compare it with the Taylor approximation.

9. Prove the following result. Let $f \in C^2[a,b]$ with $f''(x)>0$ for $a \leq x \leq b$. If $q_1^*(x)=a_0+a_1 x$ is the linear minimax approximation to $f(x)$ on $[a,b]$, then

$$a_1 = \frac{f(b)-f(a)}{b-a} \qquad a_0 = \frac{f(a)+f(c)}{2} - \left(\frac{a+c}{2}\right)\left[\frac{f(b)-f(a)}{b-a}\right]$$

   where c is the unique solution of

$$f'(c) = \frac{f(b)-f(a)}{b-a}$$

   What is ρ?

10. (a) Show that the linear minimax approximation to $\sqrt{1+x^2}$ on $[0,1]$ is

$$q_1^*(x)=.955+.414x$$

   (b) Using (a), derive the approximation

$$\sqrt{x^2+y^2} \approx .955y + .414x \qquad 0 \leq x \leq y$$

   and determine the error.

11. Find the linear least squares approximation to $f(x)=\ln x$ on $[1,e]$. Compare the error with the results of Problem 8.

12. Find the value of α that minimizes

$$\int_0^1 |e^x - \alpha| dx$$

   What is the minimum? This is a trivial illustration of yet another way to measure the error of an approximation and of the resulting best approximation.

13. Solve the following minimization problems, and determine whether there is a unique value of α which gives the minimum. In each case, α is allowed to range over all real numbers.

(a) Minimize$_\alpha$ $\int_{-1}^{1} [x - \alpha x^2]^2 dx$

(b) Minimize$_\alpha$ $\int_{-1}^{1} |x - \alpha x^2| dx$

(c) Minimize$_\alpha$ $\underset{-1 \leq x \leq 1}{\text{Max}} |x - \alpha x^2|$

14. Let $f \in C[0, \pi]$. Find the least squares approximation to $f(x)$ on $[0, \pi]$ by a trigonometric polynomial of the form

$$p(x) = \sum_{j=0}^{n} a_j \cos(jx)$$

Use the weight function $w(x) \equiv 1$. There is a large literature on approximation by trigonometric polynomials, including the entire area of Fourier series; see [D1] or [M1].

15. Use the Gram-Schmidt method to determine the first three orthonormal polynomials for $[0, 1]$ with weight function $w(x) = \ln(1/x)$.

16. Define $S_n(x) = (1/(n+1)) T'_{n+1}(x)$, $n \geq 0$, with $T_{n+1}(x)$ the Chebyshev polynomial of degree $n+1$. The polynomials $S_n(x)$ are called Chebyshev polynomials of the second kind.

(a) Show that $\{S_n(x) | n \geq 0\}$ is an orthogonal family on $[-1, 1]$ with respect to the weight function $w(x) = \sqrt{1 - x^2}$.

(b) Show that the family $\{S_n(x)\}$ satisfies the same triple recursion relation (4.43) as the family $\{T_n(x)\}$.

(c) Given $f \in C[-1, 1]$, solve the problem

$$\text{Minimize} \int_{-1}^{1} \sqrt{1 - x^2} \left[f(x) - p_n(x) \right]^2 dx$$

where $p_n(x)$ is allowed to range over all polynomials of degree $\leq n$.

17. Derive the formulas for b_n and c_n given in the triple recursion relation in (4.50) of Theorem 4.5.

18. Let $\{\varphi_n(x) | n \geq 1\}$ be orthogonal on (a, b) with weight function $w(x) \geq 0$. Denote the zeroes of $\varphi_n(x)$ by

$$a < z_{n,n} < z_{n-1,n} < \cdots < z_{1,n} < b$$

Prove that the zeroes of $\varphi_{n+1}(x)$ are separated by those of $\varphi_n(x)$, that is,

$$a < z_{n+1,n+1} < z_{n,n} < z_{n,n+1} < \cdots < z_{2,n+1} < z_{1,n} < z_{1,n+1} < b$$

Hint: Use induction on the degree n. Write $\varphi_n(x) = A_n x^n + \ldots$ with $A_n > 0$, and use the triple recursion relation (4.50) to evaluate the polynomials at the zeroes of $\varphi_n(x)$. Observe the sign changes for $\varphi_{n-1}(x)$ and $\varphi_{n+1}(x)$.

19. For $f(x) = \sin[(\pi/2)x]$, $-1 \le x \le 1$, find both the Legendre and Chebyshev least squares approximations of degree 3 to $f(x)$. Determine the error in each approximation and graph them. Use Theorem 4.10 to bound the minimax error $\rho_3(f)$.

20. Using economization find a polynomial of degree 4 that approximates e^x on $[-1,1]$ with an error $\le .005$.

21. Let $f(x)$ be a continuous even (odd) function on $[-a,a]$. Show that the minimax approximation $q_n^*(x)$ to $f(x)$ will be an even (odd) function on $[-a,a]$, regardless of whether n is even or odd. Hint: use Theorem 4.11, including the uniqueness result.

22. Using your own program of the Remes algorithm of Section 4.9 or else a package program for producing minimax approximations, produce the minimax approximations $q_n^*(x)$ to $J_0(x)$ on $[0,3]$, with $n=2$ thru 9. You will first need to write a program to evaluate $J_0(x)$ using

$$J_0(x) = \sum_0^\infty \frac{(-1)^j \left(\frac{1}{4}x^2\right)^j}{(j!)^2}$$

23. For $f(x) = \tan^{-1} x$, $-1 \le x \le 1$, produce the cubic near-minimax approximations $C_n(x)$, $I_n(x)$, and $F_n(x)$, $n=4$, of Section 4.8. Evaluate the error and compare the results.

24. Produce the near-minimax approximation $F_n(x)$ for

$$f(x) = \frac{1}{x} \int_0^x e^{-t^2} dt \qquad -1 \le x \le 1$$

for $n=1,3,5,7,9$. Bound the error $\|f - F_n\|_\infty$, and use Theorem 4.10 to bound $\rho_n(f)$.

25. To further motivate the definition of $I_n(x)$ in Section 4.8, consider the polynomial $p(x)$ interpolating to $f(x)$ at the $n+1$ nodes

$$-1 \le x_n < x_{n-1} < \cdots < x_0 \le 1$$

The error formula is

$$f(x) - p(x) = \frac{(x - x_0)\ldots(x - x_n)}{(n+1)!} f^{(n+1)}(\xi_x) \qquad -1 \le x \le 1$$

for some ξ_x in $[-1, 1]$. Since ξ_x is unknown, our only way to decrease the error is to minimize the multiplying polynomial. Pick the nodes $\{x_i\}$ to minimize

$$\underset{-1 \le x \le 1}{\text{Max}} |(x - x_0)\ldots(x - x_n)|$$

Hint: use Theorem 4.9.

26. For $f(x) = e^x$ on $[-1, 1]$, consider constructing $I_n(x)$. Bound the error $\|f - I_n\|_\infty$, for any $n \ge 1$. Prove

$$\alpha_n |T_{n+1}(x)| \le |f(x) - I_n(x)| \le \beta_n |T_{n+1}(x)| \qquad -1 \le x \le 1$$

for appropriate constants α_n, β_n.

FIVE

NUMERICAL INTEGRATION

In this chapter we derive and analyze numerical methods for evaluating definite integrals. The integrals will be mainly of the form

$$I(f) = \int_a^b f(x)\, dx \tag{5.1}$$

with $[a, b]$ finite. Most such integrals cannot be evaluated explicitly; and with many others it is often faster to integrate them numerically rather than evaluating them exactly using a complicated antiderivative of $f(x)$.

There are many numerical methods for evaluating (5.1), but most can be made to fit within the following simple framework. For the integrand $f(x)$, find an approximating family $\{f_n(x) | n \geq 1\}$ and define

$$I_n(f) = \int_a^b f_n(x)\, dx = I(f_n) \tag{5.2}$$

We usually require the approximations $f_n(x)$ to satisfy

$$\|f - f_n\|_\infty \to 0 \quad \text{as} \quad n \to \infty \tag{5.3}$$

And the form of each $f_n(x)$ should be chosen such that $I(f_n)$ can be evaluated easily. For the error,

$$E_n(f) = I(f) - I_n(f) = \int_a^b \left[f(x) - f_n(x) \right] dx$$

$$|E_n(f)| \leq \int_a^b |f(x) - f_n(x)|\, dx \leq (b - a)\|f - f_n\|_\infty \tag{5.4}$$

Most numerical integration methods can be viewed within this framework, although some of them are better studied from some other perspective. The one class of methods that don't fit within the framework are those based on extrapolation using asymptotic estimates of the error. These are examined in Section 5.5.

Most numerical integrals $I_n(f)$ will have the following form when they are evaluated:

$$I_n(f) = \sum_{j=1}^{n} w_{j,n} f(x_{j,n}) \qquad n \geq 1 \tag{5.5}$$

The coefficients $w_{j,n}$ are called the integration weights or quadrature weights; and the points $x_{j,n}$ are the integration nodes, usually chosen in $[a,b]$. The dependence on n is usually suppressed, writing w_j and x_j, although it will be understood implicitly. Standard methods have nodes and weights that have simple formulas or else they are tabulated in tables that are readily available. Thus there is usually no need to explicitly construct the functions $f_n(x)$ of (5.2), although their role in defining $I_n(f)$ may be useful to keep in mind.

The following example is a simple illustration of (5.2) to (5.4), but it is not of the form (5.5).

Example Evaluate

$$I = \int_0^1 \frac{e^x - 1}{x} \, dx$$

This integrand has a removable singularity at the origin. Use a Taylor series for e^x [see (1.8) of Chapter 1] to define $f_n(x)$, and then define

$$I_n = \int_0^1 \sum_{j=1}^{n} \frac{x^{j-1}}{j!} \, dx$$

$$= \sum_{j=1}^{n} \frac{1}{(j!)(j)} \tag{5.6}$$

For the error in I_n, use the Taylor series (1.8) to obtain

$$f(x) - f_n(x) = \frac{x^n}{(n+1)!} e^{\xi_x}$$

for some $0 \leq \xi_x \leq x$. Then

$$I - I_n = \int_0^1 \frac{x^n}{(n+1)!} e^{\xi_x} dx$$

$$\frac{1}{(n+1)!(n+1)} \leq I - I_n \leq \frac{e}{(n+1)!(n+1)} \tag{5.7}$$

The sequence in (5.6) is rapidly convergent, and (5.7) allows us to estimate the error very accurately. For example, with $n=6$

$$I_6 = 1.31787037$$

and from (5.7),

$$2.83 \times 10^{-5} \leq I - I_6 \leq 7.70 \times 10^{-5}$$

The true error is 3.18×10^{-5}.

For integrals in which the integrand has some kind of bad behavior, for example an infinite value at some point, we often will consider the integrand in the form

$$I(f) = \int_a^b w(x) f(x) dx \qquad (5.8)$$

The bad behavior is assumed to be located in $w(x)$, called the weight function, and the function $f(x)$ will be assumed well behaved. For example, consider evaluating

$$\int_0^1 (\ln x) f(x) dx$$

for arbitrary continuous functions $f(x)$. The framework (5.2) to (5.4) generalizes easily to the treatment of (5.8). Methods for such integrals will be considered in Sections 5.3 and 5.7.

Most numerical integration formulas are based on defining $f_n(x)$ in (5.2) by using polynomial or piecewise polynomial interpolation. Formulas using such interpolation with evenly spaced node points are derived and discussed in Sections 5.1 and 5.2. Optimal formulas with very rapid convergence are given in Sections 5.3 and 5.4; and they are based on defining $f_n(x)$ using polynomial interpolation at carefully selected node points that need not be evenly spaced.

Asymptotic error formulas for the methods of Sections 5.1 and 5.2 are given and discussed in Section 5.5; and some new formulas are derived based on extrapolation with these error formulas. Some methods that control the integration error in an automatic way, while remaining efficient, are given in Section 5.6. The final section (5.7) discusses methods for integrals that are singular or ill behaved in some sense.

5.1 The Trapezoidal Rule and Simpson's Rule

We begin our development of numerical integration by giving two well-known numerical methods for evaluating

$$I(f) = \int_a^b f(x) dx \qquad (5.9)$$

We will analyze and illustrate these methods very completely, and they will serve as an introduction to the material of later sections. The interval [a, b] will always be finite in this section.

The trapezoidal rule The simple trapezoidal rule is based on approximating $f(x)$ by the straight line joining $(a, f(a))$ and $(b, f(b))$. By integrating the formula for this straight line, we obtain the approximation

$$I_1(f) = \left(\frac{b-a}{2}\right)\left[f(a) + f(b)\right] \tag{5.10}$$

This is of course the area of the trapezoid shown in Figure 5.1. To obtain an error formula, we use the interpolation error formula (3.22),

$$f(x) - \frac{(b-x)f(a) + (x-a)f(b)}{b-a} = (x-a)(x-b)f[a,b,x]$$

We also assume for all work with the error for the trapezoidal rule in this section that $f(x)$ is twice continuously differentiable on $[a,b]$. Then

$$E_1(f) = \int_a^b f(x)dx - \frac{(b-a)}{2}\left[f(a) + f(b)\right]$$

$$= \int_a^b (x-a)(x-b)f[a,b,x]\,dx \tag{5.11}$$

Using the integral mean-value theorem (Theorem 1.3 of Chapter 1),

$$E_1(f) = f[a,b,\xi]\int_a^b (x-a)(x-b)\,dx \qquad \text{some } a \le \xi \le b$$

$$= \left[\tfrac{1}{2}f''(\eta)\right]\left[-\tfrac{1}{6}(b-a)^3\right] \qquad \text{some } \eta \in [a,b]$$

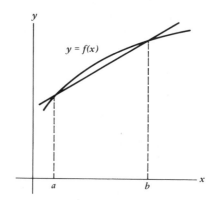

Figure 5.1 Illustration of trapezoidal rule (5.10).

using (3.23). Thus

$$E_1(f) = -\frac{(b-a)^3}{12} f''(\eta) \qquad \eta \in [a,b] \tag{5.12}$$

If $b-a$ is not sufficiently small, the trapezoidal rule (5.10) is not of much use. For such an integral, we break it into a sum of integrals over small subintervals, and then we apply (5.10) to each of these smaller integrals. Let $n \geq 1$, $h = (b-a)/n$, and $x_j = a + jh$ for $j = 0, 1, \ldots, n$. Then

$$I(f) = \int_a^b f(x)dx = \sum_{j=1}^n \int_{x_{j-1}}^{x_j} f(x)dx$$

$$= \sum_{j=1}^n \left\{ \left(\frac{h}{2}\right)[f(x_{j-1}) + f(x_j)] - \frac{h^3}{12} f''(\eta_j) \right\}$$

with $x_{j-1} \leq \eta_j \leq x_j$. There is no reason why the subintervals $[x_{j-1}, x_j]$ must all have equal length, but it is customary to first introduce the general principles involved in this way. Although this is also the customary way in which the method is applied, there are situations in which it is desirable to vary the spacing of the nodes.

The first terms in the sum can be combined to give the composite trapezoidal rule,

$$I_n(f) = h\left\{ \tfrac{1}{2}f_0 + f_1 + f_2 + \cdots + f_{n-1} + \tfrac{1}{2}f_n \right\} \qquad n \geq 1 \tag{5.13}$$

with $f(x_j) \equiv f_j$. For the error in $I_n(f)$,

$$E_n(f) = I(f) - I_n(f) = \sum_{j=1}^n -\frac{h^3}{12} f''(\eta_j)$$

$$= -\frac{h^3 n}{12} \left[\frac{1}{n} \sum_{j=1}^n f''(\eta_j) \right] \tag{5.14}$$

For the term in brackets,

$$\underset{a \leq x \leq b}{\text{Min}} f''(x) \leq M \equiv \frac{1}{n} \sum_{j=1}^n f''(\eta_j) \leq \underset{a \leq x \leq b}{\text{Max}} f''(x)$$

Since $f''(x)$ is continuous for $a \leq x \leq b$, it must attain all values between its minimum and maximum at some point of $[a,b]$; and thus $f''(\eta) = M$ for some $\eta \in [a,b]$. Thus we can write

$$E_n(f) = -\frac{(b-a)h^2}{12} f''(\eta) \qquad \text{some } \eta \in [a,b] \tag{5.15}$$

Another error estimate can be derived using this analysis. From (5.14),

$$\underset{n\to\infty}{\text{Limit}} \frac{E_n(f)}{h^2} = \underset{n\to\infty}{\text{Limit}} \left[-\frac{b-a}{12n} \sum_{j=1}^{n} f''(\eta_j) \right]$$

$$= -\tfrac{1}{12} \underset{n\to\infty}{\text{Limit}} \sum_{j=1}^{n} f''(\eta_j)h$$

Since $x_{j-1} \le \eta_j \le x_j, j = 1,\ldots,n$, the last sum is a Riemann sum, and thus

$$\underset{n\to\infty}{\text{Limit}} \frac{E_n(f)}{h^2} = -\tfrac{1}{12} \int_a^b f''(x)\,dx = -\tfrac{1}{12} \left[f'(b) - f'(a) \right] \tag{5.16}$$

$$E_n(f) \approx -\frac{h^2}{12} \left[f'(b) - f'(a) \right] \equiv \tilde{E}_n(f) \tag{5.17}$$

The term $\tilde{E}_n(f)$ is called an asymptotic error estimate for $E_n(f)$, valid in the sense of (5.16).

Definition Let $E_n(f)$ be an exact error formula, and let $\tilde{E}_n(f)$ be an estimate of it. We say that $\tilde{E}_n(f)$ is an asymptotic error estimate for $E_n(f)$ if

$$\underset{n\to\infty}{\text{Limit}} \frac{\tilde{E}_n(f)}{E_n(f)} = 1 \tag{5.18}$$

or equivalently,

$$\underset{n\to\infty}{\text{Limit}} \frac{E_n(f) - \tilde{E}_n(f)}{E_n(f)} = 0$$

The estimate in (5.17) meets this criteria, based on (5.16).

The composite trapezoidal rule (5.13) could also have been obtained by replacing $f(x)$ by a piecewise linear interpolating function $f_n(x)$ interpolating to $f(x)$ at the nodes $x_0, x_1, \ldots, x_n$. We will generally refer to the composite trapezoidal rule as simply the trapezoidal rule.

Example We will use the trapezoidal rule (5.13) to calculate

$$I = \int_0^\pi e^x \cos(x)\,dx \tag{5.19}$$

The true value is $I = -12.0703463164$. The values of I_n are given in Table 5.1, along with the true errors E_n and the asymptotic estimates $\tilde{E}_n$, obtained from (5.17). Note that the errors decrease by a factor of four when n is doubled (and

Table 5.1   Trapezoidal rule for evaluating (5.19)

n	I_n	E_n	Ratio	$\tilde{E}_n$
2	-17.389259	5.32		4.96
			4.20	
4	-13.336023	1.27		1.24
			4.06	
8	-12.382162	$3.12\mathrm{E}-1$		$3.10\mathrm{E}-1$
			4.02	
16	-12.148004	$7.77\mathrm{E}-2$		$7.76\mathrm{E}-2$
			4.00	
32	-12.089742	$1.94\mathrm{E}-2$		$1.94\mathrm{E}-2$
			4.00	
64	-12.075194	$4.85\mathrm{E}-3$		$4.85\mathrm{E}-3$
			4.00	
128	-12.071558	$1.21\mathrm{E}-3$		$1.21\mathrm{E}-3$
			4.00	
256	-12.070649	$3.03\mathrm{E}-4$		$3.03\mathrm{E}-4$
			4.00	
512	-12.070422	$7.57\mathrm{E}-5$		$7.57\mathrm{E}-5$

hence h is halved). This result was predictable from the multiplying factor of h^2 present in (5.15) and (5.17). This example will also show that the trapezoidal rule is relatively inefficient when compared with other methods to be developed in this chapter.

Simpson's rule   To improve on the simple trapezoidal rule (5.10), we use a quadratic interpolating polynomial to approximate $f(x)$ on $[a,b]$. Let $c = (a+b)/2$, and define

$$I_2(f) = \int_a^b \left[\frac{(x-c)(x-b)}{(a-c)(a-b)} f(a) + \frac{(x-a)(x-b)}{(c-a)(c-b)} f(c) + \frac{(x-a)(x-c)}{(b-a)(b-c)} f(b) \right] dx$$

$$= \frac{h}{3}\left[f(a) + 4f(c) + f(b) \right] \qquad h = \frac{b-a}{2}$$

Simpson's rule is

$$I_2(f) = \frac{h}{3}\left[f(a) + 4f\left(\frac{a+b}{2}\right) + f(b) \right] \qquad h = \frac{b-a}{2} \qquad (5.20)$$

For the error, we begin with the interpolation error formula (3.22) to obtain

$$E_2(f) = I(f) - I_2(f)$$

$$= \int_a^b (x-a)(x-c)(x-b) f[a,b,c,x] dx$$

We cannot apply the integral mean-value theorem since the polynomial in the integrand changes sign at $x=c=(a+b)/2$. We assume that $f(x)$ is four times continuously differentiable on $[a,b]$ for the work of this section on Simpson's rule. Define

$$w(x)=\int_a^x (t-a)(t-c)(t-b)\,dt$$

It is not hard to show that

$$w(a)=w(b)=0 \qquad w(x)>0 \quad \text{for} \quad a<x<b$$

Integrating by parts,

$$E_2(f)=\int_a^b w'(x)f[a,b,c,x]\,dx$$

$$=\left[w(x)f[a,b,c,x]\right]_{x=a}^{x=b}-\int_a^b w(x)\frac{d}{dx}f[a,b,c,x]\,dx$$

$$=-\int_a^b w(x)f[a,b,c,x,x]\,dx$$

The last equality used (3.28). Applying the integral mean-value theorem and (3.23),

$$E_2(f)=-f[a,b,c,\xi,\xi]\int_a^b w(x)\,dx \qquad \text{some } a\le\xi\le b$$

$$=-\frac{f^{(4)}(\eta)}{24}\left[\frac{4}{15}h^5\right] \qquad h=\frac{b-a}{2} \qquad \text{some } \eta\in[a,b]$$

Thus

$$E_2(f)=-\frac{h^5}{90}f^{(4)}(\eta) \qquad \eta\in[a,b] \tag{5.21}$$

From this we see that $E_2(f)=0$ if $f(x)$ is a polynomial of degree ≤ 3, even though quadratic interpolation is exact only if $f(x)$ is a polynomial of degree at most two. This results in Simpson's rule being much more accurate than the trapezoidal rule.

Again we create a composite rule. For $n\ge 2$ and even, define $h=(b-a)/n$, $x_j=a+jh$ for $j=0,1,\ldots,n$. Then

$$I(f)=\int_a^b f(x)\,dx=\sum_{j=1}^{n/2}\int_{x_{2j-2}}^{x_{2j}}f(x)\,dx$$

$$=\sum_{j=1}^{n/2}\left[\frac{h}{3}(f_{2j-2}+4f_{2j-1}+f_{2j})-\frac{h^5}{90}f^{(4)}(\eta_j)\right]$$

with $x_{2j-2} \leq \eta_j \leq x_{2j}$. Simplifying the first terms in the sum, we obtain the composite Simpson rule:

$$I_n(f) = \frac{h}{3}(f_0 + 4f_1 + 2f_2 + 4f_3 + 2f_4 + \ldots + 2f_{n-2} + 4f_{n-1} + f_n) \qquad (5.22)$$

For the error, as with the trapezoidal rule,

$$E_n(f) = I(f) - I_n(f) = -\frac{h^5(n/2)}{90} \cdot \frac{2}{n} \sum_{j=1}^{n/2} f^{(4)}(\eta_j)$$

$$E_n(f) = -\frac{h^4(b-a)}{180} f^{(4)}(\eta) \qquad \text{some } \eta \in [a,b] \qquad (5.23)$$

We can also derive the asymptotic error formula

$$E_n(f) \approx -\frac{h^4}{180}\left[f^{(3)}(b) - f^{(3)}(a)\right] \equiv \tilde{E}_n(f) \qquad (5.24)$$

The proof is essentially the same as was used to obtain (5.17).

Example   We use Simpson's rule (5.22) to evaluate the integral (5.19),

$$I = \int_0^\pi e^x \cos(x)\,dx$$

used earlier as an example for the trapezoidal rule. The numerical results are given in Table 5.2. Again, the rate of decrease in the error confirms the results given by (5.23) and (5.24). Comparing with the earlier results in Table 5.1 for the trapezoidal rule, it is clear that Simpson's rule is superior.

Table 5.2   Simpson rule for evaluating (5.19)

n	I_n	E_n	Ratio	$\tilde{E}_n$
2	−11.5928395534	−4.78E−1		−1.63
			5.59	
4	−11.9849440198	−8.54E−2		−1.02E−1
			14.9	
8	−12.0642089572	−6.14E−3		−6.38E−3
			15.5	
16	−12.0699513233	−3.95E−4		−3.99E−4
			15.9	
32	−12.0703214561	−2.49E−5		−2.49E−5
			16.0	
64	−12.0703447599	−1.56E−6		−1.56E−6
			16.0	
128	−12.0703462191	−9.73E−8		−9.73E−8
			16.0	
256	−12.0703463103	−6.08E−9		−6.08E−9

Peano kernel error formulas   There is another approach to deriving the error formulas that does not result in the derivative being evaluated at an unknown point η. We first consider the trapezoidal rule. Assume $f' \in C[a,b]$ and that $f''(x)$ is integrable on $[a,b]$. Then using Taylor's theorem (Theorem 1.4 in Chapter 1),

$$f(x) = p_1(x) + R_2(x) \qquad p_1(x) = f(a) + (x-a)f'(a)$$

$$R_2(x) = \int_a^x (x-t) f''(t)\, dt$$

Note that from substitution into (5.11),

$$E_1(F+G) = E_1(F) + E_1(G) \tag{5.25}$$

for any two functions F, $G \in C[a,b]$. Thus

$$E_1(f) = E_1(p_1) + E_1(R_2) = E_1(R_2) \tag{5.26}$$

since $E_1(p_1) = 0$ from (5.12). Substituting,

$$E_1(R_2) = \int_a^b R_2(x)\, dx - \left(\frac{b-a}{2}\right)\left[R_2(a) + R_2(b)\right]$$

$$= \int_a^b \int_a^x (x-t) f''(t)\, dt\, dx - \left(\frac{b-a}{2}\right)\int_a^b (b-t)f''(t)\, dt$$

In general for any integrable function $G(x,t)$,

$$\int_a^b \int_a^x G(x,t)\, dt\, dx = \int_a^b \int_t^b G(x,t)\, dx\, dt$$

Thus

$$E_1(R_2) = \int_a^b f''(t) \int_t^b (x-t)\, dx\, dt - \left(\frac{b-a}{2}\right)\int_a^b (b-t)f''(t)\, dt$$

Combining integrals and simplifying the results,

$$E_1(f) = \tfrac{1}{2}\int_a^b f''(t)(t-a)(t-b)\, dt \tag{5.27}$$

For the composite trapezoidal rule (5.13),

$$E_n(f) = \int_a^b K(t) f''(t)\, dt \tag{5.28}$$

$$K(t) = \tfrac{1}{2}(t - t_{j-1})(t - t_j) \qquad t_{j-1} \le t \le t_j \qquad j = 1, 2, \ldots, n \tag{5.29}$$

The formulas (5.27) and (5.28) are called the Peano kernel formulation of the error, and $K(t)$ is called the *Peano kernel*. For a more general presentation, see Davis [D1, Chapter 3].

As a simple illustration of its use, take bounds in (5.28) to obtain

$$|E_n(f)| \le \|K\|_\infty \int_a^b |f''(t)| dt = \frac{h^2}{8} \int_a^b |f''(t)| dt \tag{5.30}$$

If $f''(t)$ is very peaked, this may give a better bound on the error than (5.15), because in (5.15) we generally must replace $|f''(\eta)|$ by $\|f''\|_\infty$.

For Simpson's rule, write

$$f(x) = p_3(x) + R_4(x)$$

$$R_4(x) = \frac{1}{6} \int_a^x (x - t)^3 f^{(4)}(t) dt$$

As before

$$E_2(f) = E_2(p_3) + E_2(R_4) = E_2(R_4)$$

and we then calculate $E_2(R_4)$ by direct substitution and simplification:

$$E_2(f) = \int_a^b R_4(x) dx - \frac{h}{3} \left[R_4(a) + 4R_4\left(\frac{a+b}{2}\right) + R_4(b) \right]$$

This yields

$$E_2(f) = \int_a^b K(t) f^{(4)}(t) dt \tag{5.31}$$

$$K(t) = \begin{cases} \dfrac{1}{72}(t-a)^3(3t-a-2b), & a \le t \le \dfrac{a+b}{2} \\[2mm] \dfrac{1}{72}(b-t)^3(b+2a-3t), & \dfrac{a+b}{2} \le t \le b \end{cases} \tag{5.32}$$

A graph of $K(t)$ is given in Figure 5.2. By direct evaluation,

$$\|K\|_\infty = \frac{h^4}{72}, \quad \int_a^b K(t) dt = -\frac{h^5}{90} \quad h = \frac{b-a}{2}$$

As with the trapezoidal method, these results extend easily to the composite Simpson rule.

The following examples are intended to describe more fully the behavior of the Simpson and trapezoidal rules. Example (3) does not satisfy the previous assumptions on differentiability of the integrand, and Example (2) is peaked on the interval of integration.

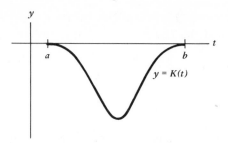

Figure 5.2 The Peano kernel for Simpson's rule.

Examples 1. $f(x)=x^3\sqrt{x}$, $[a,b]=[0,1]$, $I=\frac{2}{9}$. Table 5.3 gives the error for increasing values of n. Although $f^{(4)}(x)$ is singular at $x=0$, the generalization of (5.31) to the composite Simpson rule gives

$$|E_n(f)| \leq \|K\|_\infty \int_a^b |f^{(4)}(t)| dt = \left(\frac{h^4}{72}\right)\left(\frac{105}{8}\right) = \frac{35}{192} h^4$$

Thus the error should decrease by a factor of 16 when h is halved (i.e., n is doubled). This also gives a fairly realistic bound on the error. Note the close agreement of ratios with the theoretically predicted values of 4 and 16, respectively.

Table 5.3 Trapezoidal, Simpson integration: case 1

	Trapezoidal Rule		Simpson's Rule	
n	Error	Ratio	Error	Ratio
2	$-7.197-2$		$-3.370-3$	
4	$-1.817-2$	3.96	$-2.315-4$	14.6
8	$-4.553-3$	3.99	$-1.543-5$	15.0
16	$-1.139-3$	4.00	$-1.008-6$	15.3
32	$-2.848-4$	4.00	$-6.489-8$	15.5
64	$-7.121-5$	4.00	$-4.141-9$	15.7
128	$-1.780-5$	4.00	$-2.626-10$	15.8
256	$-4.450-6$	4.00	$-1.658-11$	15.8
512	$-1.113-6$	4.00	$-1.043-12$	15.9
1024	$-2.782-7$	4.00	$-6.489-14$	16.1

2. $f(x)=\dfrac{1}{1+(x-\pi)^2}$ $[a,b]=[0,5]$ $I \doteq 2.33976628367$

According to theory, the infinite differentiability of $f(x)$ implies a value for ratios of 4.0 and 16.0 for the trapezoidal and Simpson rules, respectively. But these need not hold for the first several values of I_n, as Table 5.4 shows.

Table 5.4   Trapezoidal, Simpson integration: case 2

	Trapezoidal rule		Simpson's rule	
n	Error	Ratio	Error	Ratio
2	1.731 − 1		− 2.853 − 1	
4	7.110 − 2	2.43	3.709 − 2	− 7.69
8	7.496 − 3	9.48	− 1.371 − 2	− 2.71
16	1.953 − 3	3.84	1.059 − 4	− 130
32	4.892 − 4	3.99	1.080 − 6	9.81
64	1.223 − 4	4.00	6.743 − 8	16.0
128	3.059 − 5	4.00	4.217 − 9	16.0
256	7.647 − 6	4.00	2.636 − 10	16.0
512	1.912 − 6	4.00	1.648 − 11	16.0
1024	4.780 − 7	4.00	1.037 − 12	15.9

3. $f(x) = \sqrt{x}$, $[a,b] = [0, 1]$, $I = 2/3$.

Since $f'(x)$ has an infinite value at $x = 0$, none of the theoretical results given previously apply to this case. The numerical results are in Table 5.5; and note that there is still a regular behavior to the error. In fact, the errors of the two methods decrease by the same ratio as n is doubled. This ratio of $2^{1.5} \doteq 2.83$ will be explained in Section 5.5, formula (5.98).

Table 5.5   Trapezoidal, Simpson integration: case 3

	Trapezoidal rule		Simpson's rule	
n	Error	Ratio	Error	Ratio
2	6.311 − 2		2.860 − 2	
4	2.338 − 2	2.70	1.012 − 2	2.82
8	8.536 − 3	2.74	3.587 − 3	2.83
16	3.085 − 3	2.77	1.268 − 3	2.83
32	1.108 − 3	2.78	4.485 − 4	2.83
64	3.959 − 4	2.80	1.586 − 4	2.83
128	1.410 − 4	2.81	5.606 − 5	2.83

5.2   Newton-Cotes Integration Formulas

The simple trapezoidal rule (5.10) and Simpson rule (5.20) are the first two cases of the Newton-Cotes integration formula. For $n \geq 1$, let $h = (b - a)/n$, $x_j = a + jh$ for $j = 0, 1, \ldots, n$. Define $I_n(f)$ by replacing $f(x)$ by its interpolating polynomial $P_n(x)$ on the nodes $x_0, x_1, \ldots, x_n$:

$$I(f) = \int_a^b f(x)dx \approx I_n(f) = \int_a^b P_n(x)dx$$

Using the Lagrange formula (3.7) for $P_n(x)$,

$$I_n(f) = \int_a^b \sum_{j=0}^{n} l_{j,n}(x) f(x_j) dx = \sum_{j=0}^{n} w_{j,n} f(x_j) \tag{5.33}$$

with

$$w_{j,n} = \int_a^b l_{j,n}(x) dx \qquad j = 0, 1, \ldots, n \tag{5.34}$$

Usually we suppress the subscript n and write just w_j. We have already calculated the cases $n=1$ and 2. To illustrate the calculation of the weights, we give the case of w_0 for $n=3$.

$$w_0 = \int_a^b l_0(x) dx = \int_{x_0}^{x_3} \frac{(x-x_1)(x-x_2)(x-x_3) dx}{(x_0-x_1)(x_0-x_2)(x_0-x_3)}$$

A change of variable simplifies the calculations. Let $x = x_0 + \mu h$, $0 \le \mu \le 3$. Then

$$w_0 = -\frac{1}{6h^3} \int_{x_0}^{x_3} (x-x_1)(x-x_2)(x-x_3) dx$$

$$= -\frac{1}{6h^3} \int_0^3 (\mu-1)h(\mu-2)h(\mu-3)h \cdot h d\mu$$

$$= -\frac{h}{6} \int_0^3 (\mu-1)(\mu-2)(\mu-3) d\mu$$

$$w_0 = \frac{3h}{8}$$

The complete formula for $n=3$ is

$$I_3(f) = \frac{3h}{8} \left[f(x_0) + 3f(x_1) + 3f(x_2) + f(x_3) \right]$$

and is called the three-eights rule.

For the error, we give the following theorem.

Theorem 5.1 **1.** For n even, assume $f(x)$ is $n+2$ times continuously differentiable on $[a,b]$. Then

$$I(f) - I_n(f) = C_n h^{n+3} f^{(n+2)}(\eta) \qquad \text{some } \eta \in [a,b] \tag{5.35}$$

with

$$C_n = \frac{1}{(n+2)!} \int_0^n \mu^2(\mu-1) \ldots (\mu-n) d\mu$$

2. For n odd, assume $f(x)$ is $n+1$ times continuously differentiable on $[a,b]$. Then

$$I(f) - I_n(f) = C_n h^{n+2} f^{(n+1)}(\eta) \qquad \text{some } \eta \in [a,b] \quad (5.36)$$

with

$$C_n = \frac{1}{(n+1)!} \int_0^n \mu(\mu-1)\dots(\mu-n)d\mu$$

Proof We will sketch the proof for part (1), the most important case. For complete proofs of both cases, see Isaacson and Keller [I2, pp. 308–314]. From (3.22),

$$E_n(f) = I(f) - I_n(f)$$

$$= \int_a^b (x-x_0)(x-x_1)\dots(x-x_n)f[x_0, x_1, \dots, x_n, x]dx$$

Define

$$w(x) = \int_a^x (t-x_0)\dots(t-x_n)dt$$

Then

$$w(a) = w(b) = 0 \qquad w(x) > 0 \qquad \text{for } a < x < b$$

The proof that $w(x) > 0$ can be found in [I2, p. 309]. It is easy to show $w(b) = 0$ since the integrand $(t-x_0)\dots(t-x_n)$ is an odd function with respect to the middle node $x_{n/2} = (a+b)/2$.

Using integration by parts and (3.28),

$$E_n(f) = \int_a^b w'(x)f[x_0, \dots, x_n, x]dx$$

$$= \left[w(x)f[x_0, \dots, x_n, x] \right]_a^b - \int_a^b w(x)\frac{d}{dx}f[x_0, \dots, x_n, x]dx$$

$$E_n(f) = -\int_a^b w(x)f[x_0, \dots, x_n, x, x]dx \qquad (5.37)$$

Using the integral mean-value theorem and (3.23),

$$E_n(f) = f[x_0, \dots, x_n, \xi, \xi] \int_a^b w(x)dx$$

$$= \frac{f^{(n+2)}(\eta)}{(n+2)!} \int_a^b \int_a^x (t-x_0)\dots(t-x_n)dt\, dx \qquad (5.38)$$

We change the order of integration and then use the change of variable $t = x_0 + \mu h$, $0 \le \mu \le n$.

$$\int_a^b w(x)dx = \int_a^b \int_t^b (t-x_0)\ldots(t-x_n)dx\,dt$$

$$= \int_{x_0}^{x_n} (t-x_0)\ldots(t-x_{n-1})(t-x_n)(x_n-t)dt$$

$$= -h^{n+3} \int_0^n \mu(\mu-1)\ldots(\mu-n+1)(\mu-n)^2 d\mu$$

Use the change of variable $\nu = n - \mu$ to give the final result

$$\int_a^b w(x)dx = -h^{n+3} \int_0^n \nu^2(\nu-1)\ldots(\nu-n)d\nu$$

Combining this with (5.38) completes the proof.   ■

For easy reference, the most commonly used Newton-Cotes formulas are given in Table 5.6. For $n=4$, $I_4(f)$ is often called *Boole's rule*. As above, let $h = (b-a)/n$ in the table.

Table 5.6   Commonly used Newton-Cotes formulas

$n=1$	$\int_a^b f(x)dx = \dfrac{h}{2}[f(a)+f(b)] - \dfrac{h^3}{12}f''(\xi),$ trapezoidal rule
$n=2$	$\int_a^b f(x)dx = \dfrac{h}{3}\left[f(a)+4f\left(\dfrac{a+b}{2}\right)+f(b)\right] - \dfrac{h^5}{90}f^{(4)}(\xi),$ Simpson rule
$n=3$	$\int_a^b f(x)dx = \dfrac{3h}{8}[f(a)+3f(a+h)+3f(b-h)+f(b)] - \dfrac{3h^5}{80}f^{(4)}(\xi)$
$n=4$	$\int_a^b f(x)dx = \dfrac{2h}{45}\left[7f(a)+32f(a+h)+12f\left(\dfrac{a+b}{2}\right)+32f(b-h)+7f(b)\right] - \dfrac{8h^7}{945}f^{(6)}(\xi)$

Definition   A numerical integration formula $\tilde{I}(f)$ that approximates $I(f)$ is said to have degree of precision m if (1) $\tilde{I}(f) = I(f)$ for all polynomials $f(x)$ of degree $\le m$ and (2) $\tilde{I}(f) \ne I(f)$ for some polynomial f of degree $m+1$.

Example   With $n=1,3$ in the Table 5.6, the degrees of precision are also $n=1,3$ respectively. But with $n=2,4$, the degrees of precision are $n+1=3,5$ respectively. This illustrates the general result that Newton-Cotes formulas with an even index n gain an extra degree of precision as compared with those of an odd index. See formulas (5.35) and (5.36).

Each Newton-Cotes formula can be used to construct a composite rule. The most useful remaining one is probably that based on Boole's rule; see Problem 8. We omit any further details.

Convergence discussion   The next question of interest is whether $I_n(f)$ converges to $I(f)$ as $n \to \infty$. Given the lack of convergence of the interpolatory polynomials on evenly spaced nodes for some choices of $f(x)$ [see Section 3.5, formula (3.52)], we should expect some difficulties. Table 5.7 gives the results for a well-known example,

$$I = \int_{-4}^{4} \frac{dx}{1+x^2} = 2\tan^{-1}(4) \doteq 2.6516$$

These numerical integrals are diverging; and this illustrates the fact that the Newton-Cotes integration formulas $I_n(f)$, given in (5.33), need not converge to $I(f)$.

Table 5.7   Newton-Cotes example

n	I_n
2	5.4902
4	2.2776
6	3.3288
8	1.9411
10	3.5956

Using the lack of convergence of the Newton-Cotes formula (5.33) for this integral, it can be shown that

$$\operatorname*{Sup}_{n} \sum_{j=0}^{n} |w_{j,n}| = \infty \qquad (5.39)$$

Supposing the contrary, there is $B < \infty$ with

$$\sum_{j=0}^{n} |w_{j,n}| \le B \qquad n \ge 1$$

We will show that $I_n(f)$ will then converge to $I(f)$ for any continuous function $f(x)$ on $[a,b]$. Let $\epsilon > 0$ be given. Choose a polynomial $q(x)$, say of degree m, for which

$$\operatorname*{Max}_{a \le x \le b} |f(x) - q(x)| \le \operatorname{Min}\left\{ \frac{\epsilon}{2B}, \frac{\epsilon}{2(b-a)} \right\}$$

Such a polynomial exists by the Weierstrass Theorem 4.1 of Chapter 4. Now consider the error $I(f) - I_n(f)$ for all $n \geq m$;

$$I(f) - I_n(f) = [I(f) - I(q)] + [I(q) - I_n(q)] + [I_n(q) - I_n(f)]$$

Since the Newton-Cotes integration formula I_n is exact when integrating polynomials of degree $\leq n$, we have

$$I_n(q) = I(q) \qquad \text{all } n \geq m$$

Thus

$$|I(f) - I_n(f)| \leq |I(f - q)| + |I_n(q - f)|$$

$$\leq \frac{\epsilon}{2} + \frac{\epsilon}{2} = \epsilon \qquad n \geq m$$

using the definition of $q(x)$ and the formulas for I and I_n. This shows $I_n(f) \rightarrow I(f)$ as $n \rightarrow \infty$. But this is contrary to the lack of convergence for the case of

$$f(x) = \frac{1}{1 + x^2} \qquad -4 \leq x \leq 4$$

This proves (5.39).

Using $f(x) \equiv 1$ implies $I(f) = I_n(f)$, all n; and thus

$$\sum_{1}^{n} w_{j,n} = b - a \qquad n \geq 1$$

Combining this with (5.39), we see that the integration weights must vary in sign for sufficiently large n. For example, using $n = 8$

$$\int_{x_0}^{x_8} f(x)dx \approx I_8(f) = \frac{4h}{14175} \left[989(f_0 + f_8) + 5888(f_1 + f_7) - 928(f_2 + f_6) \right.$$

$$\left. + 10496(f_3 + f_5) - 4540f_4 \right]$$

Such formulas can cause loss of significance errors, although it is unlikely to be a problem until n becomes larger. But because of this problem, people have generally avoided using Newton-Cotes formulas for $n \geq 8$, even in forming composite formulas.

The midpoint rule   There are additional Newton-Cotes formulas in which one or both of the endpoints of integration are deleted from the interpolation (and integration) node points. The best known of these is also the simplest, the midpoint rule:

$$\int_a^b f(x)\,dx = (b-a)f\left(\frac{a+b}{2}\right) + \frac{(b-a)^3}{24}f''(\eta) \qquad \text{some } \eta \in [a,b] \quad (5.40)$$

For its composite form, define

$$x_j = a + \left(j - \tfrac{1}{2}\right)h \qquad j = 1,2,\ldots,n$$

the midpoints of the intervals $[a+(j-1)h, a+jh]$. Then

$$\int_a^b f(x)\,dx = I_n(f) + E_n(f)$$

$$I_n(f) = h(f_1 + f_2 + \ldots + f_n)$$

$$E_n(f) = \frac{h^2(b-a)}{24}f''(\eta) \qquad \text{some } \eta \in [a,b] \qquad (5.41)$$

The proof of these results is left as Problem 7.

These integration formulas in which one or both endpoints are missing are called open Newton-Cotes formulas, and the previous formulas are called closed formulas. The open formulas of higher order are used mainly for deriving formulas for the solution of ordinary differential equations.

5.3 Gaussian Quadrature

The idea behind Gaussian quadrature is to find an integration formula

$$I_n(f) = \sum_{j=1}^n w_j f(x_j) \approx \int_a^b w(x)f(x)\,dx = I(f) \qquad (5.42)$$

which is exact for all polynomials $f(x)$ of as large a degree as possible. The weights and nodes are restricted to be real, and the nodes must belong to the interval of integration. The weight function $w(x)$ should be nonnegative and satisfy the hypotheses (4.25) and (4.26) of Section 4.3. The resulting integration formulas are extremely accurate; and they should be considered seriously by anyone faced with many integrals to evaluate. This will be discussed in greater detail near the end of the section.

To obtain some intuition for the problem, consider the special case

$$\int_{-1}^1 f(x)\,dx \approx \sum_{j=1}^n w_j f(x_j) \qquad (5.43)$$

The weights $\{w_j\}$ and nodes $\{x_j\}$ are to be determined to make the error

$$E_n(f) = \int_{-1}^{1} f(x)dx - \sum_{j=1}^{n} w_j f(x_j)$$

equal zero for as high a degree polynomial $f(x)$ as possible. To derive equations for the nodes and weights, we first note that

$$E_n(a_0 + a_1 x + \ldots + a_m x^m) = a_0 E_n(1) + a_1 E_n(x) + \ldots + a_m E_n(x^m)$$

Thus $E_n(f) = 0$ for every polynomial of degree $\leq m$ if and only if

$$E_n(x^i) = 0 \qquad i = 0, 1, \ldots, m$$

Case 1 $n = 1$. Since there are two parameters, w_1 and x_1, we consider requiring

$$E_1(1) = 0 \qquad E_1(x) = 0$$

This gives

$$\int_{-1}^{1} 1\,dx - w_1 = 0 \qquad \int_{-1}^{1} x\,dx - w_1 x_1 = 0$$

This implies $w_1 = 2$ and $x_1 = 0$. Thus the formula (5.43) becomes

$$\int_{-1}^{1} f(x)dx \approx 2f(0)$$

the midpoint rule (5.40).

Case 2 $n = 2$. There are four parameters, w_1, w_2, x_1, x_2, and thus we put four constraints on these parameters:

$$E_2(x^i) = \int_{-1}^{1} x^i dx - \left(w_1 x_1^i + w_2 x_2^i\right) = 0 \qquad i = 0, 1, 2, 3$$

or

$$w_1 + w_2 = 2$$

$$w_1 x_1 + w_2 x_2 = 0$$

$$w_1 x_1^2 + w_2 x_2^2 = \tfrac{2}{3}$$

$$w_1 x_1^3 + w_2 x_2^3 = 0$$

These nonlinear equations have the unique solution

$$w_1 = w_2 = 1 \qquad x_2 = -x_1 = \frac{\sqrt{3}}{3}$$

and the rule

$$\int_{-1}^{1} f(x)dx \approx f\left(-\frac{\sqrt{3}}{3}\right)+f\left(\frac{\sqrt{3}}{3}\right) \tag{5.44}$$

has degree of precision three. Compare this with Simpson's rule (5.20), which uses three nodes to attain the same precision.

Case 3 For a general n there are $2n$ free parameters $\{x_i\}$ and $\{w_i\}$; and we guess that there is a formula (5.42) that uses n nodes and gives a degree of precision of $2n-1$. The equations to be solved are

$$E_n(x^i)=0 \qquad i=0,1,\ldots,2n-1$$

or

$$\sum_{j=1}^{n} w_j x_j^i = \begin{cases} 0 & i=1,3,\ldots,2n-1 \\ \dfrac{2}{i+1} & i=0,2,\ldots,2n-2 \end{cases} \tag{5.45}$$

These are nonlinear equations, and their solvability is not at all obvious. Because of the difficulty in working with this nonlinear system, we use another approach to the theory for (5.42), one which is somewhat circuitous.

Let $\{\varphi_n(x)|n \geq 0\}$ be the orthogonal polynomials on (a,b) with respect to the weight function $w(x) \geq 0$. Denote the zeroes of $\varphi_n(x)$ by

$$a < x_1 < \ldots < x_n < b \tag{5.46}$$

Also, recall the notation from (4.49),

$$\varphi_n(x)=A_n x^n + \ldots \qquad a_n = A_{n+1}/A_n$$

$$\gamma_n = \int_a^b w(x)\left[\varphi_n(x)\right]^2 dx$$

Theorem 5.2 For each $n \geq 1$, there is a unique numerical integration formula (5.42), which is exact for all polynomials of degree $\leq 2n-1$. Assuming $f(x)$ is $2n$ times continuously differentiable on $[a,b]$, the formula plus its error is given by

$$\int_a^b w(x)f(x)dx = \sum_{j=1}^{n} w_j f(x_j) + \frac{\gamma_n}{A_n^2(2n)!} f^{(2n)}(\eta) \quad a < \eta < b \tag{5.47}$$

The nodes $\{x_j\}$ are the zeroes of $\varphi_n(x)$; and the weights $\{w_j\}$ are given by

$$w_j = \frac{-a_n \gamma_n}{\varphi'_n(x_j)\varphi_{n+1}(x_j)} \qquad j=1,\ldots,n \qquad (5.48)$$

Proof The proof is divided into three parts. We first obtain a formula with degree of precision $2n-1$, using the nodes (5.46). We then show it is unique, and finally we sketch the derivation of the error formula and (5.48).

1. Construction of the formula. Hermite interpolation is used as the vehicle for the construction; see Section 3.6 to review the notation and results. For the nodes in (5.46), the Hermite polynomial interpolating to $f(x)$ and $f'(x)$ is

$$H_n(x) = \sum_{j=1}^{n} f(x_j)h_j(x) + \sum_{j=1}^{n} f'(x_j)\tilde{h}_j(x) \qquad (5.49)$$

with $h_j(x)$ and $\tilde{h}_j(x)$ defined in (3.55) of Section 3.6. The interpolation error is given by

$$\mathcal{E}_n(x) = f(x) - H_n(x) = [\psi_n(x)]^2 f[x_1, x_1, \ldots, x_n, x_n, x]$$

$$= \frac{[\psi_n(x)]^2}{(2n)!} f^{(2n)}(\xi) \qquad \xi \in [a,b] \qquad (5.50)$$

with

$$\psi_n(x) = (x-x_1)\ldots(x-x_n)$$

Note that

$$\psi_n(x) = \varphi_n(x)/A_n \qquad (5.51)$$

since both $\varphi_n(x)$ and $\psi_n(x)$ are degree n and have the same zeroes.

Using (5.49), if $f(x)$ is continuously differentiable, then

$$\int_a^b w(x)f(x)\,dx = \int_a^b w(x)H_n(x)\,dx + \int_a^b w(x)\mathcal{E}_n(x)\,dx$$

$$\equiv I_n(f) + E_n(f) \qquad (5.52)$$

The degree of precision is at least $2n-1$ since $\mathcal{E}_n(x)=0$ if $f(x)$ is a polynomial of degree $<2n$, from (5.50). Also from (5.50),

$$E_n(x^{2n}) = \int_a^b w(x)\mathcal{E}_n(x)\,dx = \int_a^b w(x)[\psi_n(x)]^2\,dx > 0 \qquad (5.53)$$

Thus the degree of precision of $I_n(f)$ is exactly $2n-1$.

To derive a simpler formula for $I_n(f)$,

$$I_n(f) = \sum_{j=1}^{n} f(x_j) \int_a^b w(x) h_j(x)\, dx + \sum_{j=1}^{n} f'(x_j) \int_a^b w(x) \tilde{h}_j(x)\, dx$$

We will show that all of the integrals in the second sum are zero. Recall that from (3.55),

$$\tilde{h}_j(x) = (x - x_j)[l_j(x)]^2$$

$$l_j(x) = \frac{\psi_n(x)}{(x - x_j)\psi_n'(x_j)} = \frac{\varphi_n(x)}{(x - x_j)\varphi_n'(x_j)}$$

The last step uses (5.51). Thus

$$\tilde{h}_j(x) = (x - x_j) l_j(x) l_j(x) = \frac{\varphi_n(x) l_j(x)}{\varphi_n'(x_j)}$$

Since degree $(l_j) = n - 1$, and since $\varphi_n(x)$ is orthogonal to all polynomials of degree $< n$, we have

$$\int_a^b w(x) \tilde{h}_j(x)\, dx = \frac{1}{\varphi_n'(x_j)} \int_a^b w(x) \varphi_n(x) l_j(x)\, dx = 0 \qquad j = 1, \dots, n \quad (5.54)$$

The integration formula (5.52) becomes

$$\int_a^b w(x) f(x)\, dx = \sum_{j=1}^{n} w_j f(x_j) + E_n(f)$$

$$w_j = \int_a^b w(x) h_j(x)\, dx, \qquad j = 1, \dots, n \qquad (5.55)$$

2. Uniqueness of formula (5.55). Suppose that we have a numerical integration formula

$$\int_a^b w(x) f(x)\, dx \approx \sum_{j=1}^{n} v_j f(z_j) \qquad (5.56)$$

which has degree of precision $\geq 2n - 1$. Construct the Hermite interpolation formula to $f(x)$ at the nodes $z_1, \dots, z_n$. Then for any polynomial $f(x)$ of degree $\leq 2n - 1$,

$$f(x) = \sum_{j=1}^{n} f(z_j) h_j(x) + \sum_{j=1}^{n} f'(z_j) \tilde{h}_j(x) \qquad \text{degree } (f) \leq 2n - 1 \quad (5.57)$$

where $h_j(x)$ and $\tilde{h}_j(x)$ are defined using $\{z_j\}$. Multiply by $w(x)$, use the assumption on the degree of precision of (5.56), and integrate in (5.57) to get

$$\sum_{j=1}^n v_j f(z_j) = \sum_{j=1}^n f(z_j) \int_a^b w(x) h_j(x)\, dx + \sum_{j=1}^n f'(z_j) \int_a^b w(x) \tilde{h}_j(x)\, dx \quad (5.58)$$

for any polynomial $f(x)$ of degree $\leq 2n-1$.

Let $f(x) = \tilde{h}_i(x)$ and use the properties of $\tilde{h}_i(x)$ to obtain from (5.58) that

$$0 = \int_a^b w(x) \tilde{h}_i(x)\, dx \quad i = 1,\ldots,n \quad (5.59)$$

As before, following (5.54), we can write

$$\tilde{h}_i(x) = (x - z_i)\left[l_i(x) \right]^2 = \frac{l_i(x)\omega_n(x)}{\omega_n'(z_i)}$$

$$\omega(x) = (x - z_1)\ldots(x - z_n)$$

Then (5.59) becomes

$$\int_a^b w(x)\omega_n(x) l_i(x)\, dx = 0 \quad i = 1,2,\ldots,n$$

Since all polynomials of degree $\leq n-1$ can be written as a combination of $l_1(x),\ldots,l_n(x)$, we have that $\omega_n(x)$ is orthogonal to every polynomial of degree $\leq n-1$. Using the uniqueness of the orthogonal polynomials (from Theorem 4.2), $\omega_n(x)$ must be a constant multiple of $\varphi_n(x)$. Thus they must have the same zeroes, and

$$z_i = x_i \quad i = 1,\ldots,n$$

To complete the proof of the uniqueness, we must show that $w_i = v_i$, where v_i is the weight in (5.56) and w_i in (5.55). Use (5.58) with (5.59) and $f(x) = h_i(x)$. The result will follow immediately, since the $h_i(x)$ is now constructed using $\{x_i\}$,

3. **The error formula.** We begin by deducing some further useful properties about the weights $\{w_i\}$ in (5.55).

$$w_i = \int_a^b w(x) h_i(x)\, dx = \int_a^b w(x)\left[1 - 2l_i'(x_i)(x - x_i)\right]\left[l_i(x) \right]^2 dx$$

$$= \int_a^b w(x)\left[l_i(x) \right]^2 dx - 2l_i'(x_i)\int_a^b w(x)(x - x_i)\left[l_i(x) \right]^2 dx$$

The last integral is zero from (5.54), since $\tilde{h}_i(x) = (x - x_i)[l_i(x)]^2$. Thus

$$w_i = \int_a^b w(x)\left[l_i(x)\right]^2 dx > 0 \qquad i = 1, 2, \dots, n \qquad (5.59)$$

and all the weights are positive, for all n.

To construct w_i, begin by substituting $f(x) = l_i(x)$ into (5.55), and note that $E_n(f) = 0$ since degree $(l_i) = n - 1$. Then using $l_i(x_j) = \delta_{ij}$, we have

$$w_i = \int_a^b w(x)l_i(x)\, dx \qquad i = 1, \dots, n$$

To further simplify the formula, the Christoffel-Darboux identity (Theorem 4.6) can be used, followed by much manipulation, to give the formula (5.48). For the details, see Isaacson and Keller [I2, pp. 333–334].

For the integration error, if $f(x)$ is $2n$ times continuously differentiable on $[a, b]$, then

$$E_n(f) = \int_a^b w(x)\mathcal{E}_n(x)\, dx = \int_a^b w(x)\left[\psi_n(x)\right]^2 f[x_1, x_1, \dots, x_n, x_n, x]\, dx$$

$$= f[x_1, x_1, \dots, x_n, x_n, \xi]\int_a^b w(x)\left[\psi_n(x)\right]^2 dx \qquad \text{some } \xi \in [a, b]$$

the last step using the integral mean-value theorem. Using (5.51) in the last integral, and replacing the divided difference by a derivative, we have

$$E_n(f) = \frac{f^{(2n)}(\eta)}{(2n)!}\int_a^b w(x)\frac{\left[\varphi_n(x)\right]^2}{A_n^2}\, dx$$

which gives the error formula in (5.47).   ■

Gauss-Legendre quadrature   For $w(x) \equiv 1$, on $[-1, 1]$ the Gaussian integration formula is

$$\int_{-1}^1 f(x)\, dx \approx \sum_{j=1}^n w_j f(x_j) \qquad (5.60)$$

with the nodes the zeroes of the degree n Legendre polynomial $P_n(x)$ on $[-1, 1]$. The weights are

$$w_i = \frac{-2}{(n+1)P_n'(x_i)P_{n+1}(x_i)} \qquad i = 1, 2, \dots, n$$

and

$$E_n(f) = \frac{2^{2n+1}(n!)^4}{(2n+1)\left[(2n)!\right]^2}\cdot\frac{f^{(2n)}(\eta)}{(2n)!} \equiv e_n\frac{f^{(2n)}(\eta)}{(2n)!} \qquad (5.61)$$

Table 5.8 Gauss-Legendre nodes and weights

n	x_i	w_i
2	$\pm.5773502692$	1.0
3	$\pm.7745966692$	.5555555556
	0.0	.8888888889
4	$\pm.8611363116$	.3478548451
	$\pm.3399810436$	.6521451549
5	$\pm.9061798459$	.2369268851
	$\pm.5384693101$	.4786286705
	0.0	.5688888889
6	$\pm.9324695142$	.1713244924
	$\pm.6612093865$	.3607615730
	$\pm.2386191861$	.4679139346
7	$\pm.9491079123$	.1294849662
	$\pm.7415311856$	.2797053915
	$\pm.4058451514$	.3818300505
	0.0	.4179591837
8	$\pm.9602898565$	.1012285363
	$\pm.7966664774$	.2223810345
	$\pm.5255324099$	.3137066459
	$\pm.1834346425$	.3626837834

for some $-1 < \eta < 1$. For integrals on other finite intervals with weight function $w(x) \equiv 1$, use the following linear change of variables:

$$\int_a^b f(t)\,dt = \left(\frac{b-a}{2}\right)\int_{-1}^1 f\left(\frac{a+b+x(b-a)}{2}\right)dx \qquad (5.62)$$

reducing the integral to the standard interval $[-1, 1]$.

For convenience, we enclose Table 5.8 giving the nodes and weights for formula (5.60) with small values of n. For larger values of n, see the very complete tables in Stroud and Secrest [S3, pp. 100–151], which go up to $n = 512$.

Example Evaluate the integral (5.19), $I = \int_0^\pi e^x \cos(x)\,dx = -12.0703463164$, which was used previously in Section 5.1 as an example for the trapezoidal rule (see Table 5.1) and Simpson's rule (see Table 5.2). The results given in Table 5.9 show the marked superiority of Gaussian quadrature.

Table 5.9 Gaussian quadrature for (5.19)

n	I_n	$I - I_n$
2	-12.33621046570	$2.66E-1$
3	-12.12742045017	$5.71E-2$
4	-12.07018949029	$-1.57E-4$
5	-12.07032853589	$-1.78E-5$
6	-12.07034633110	$1.47E-8$
7	-12.07034631753	$1.14E-9$
8	-12.07034631639	$<5.0E-12$

Discussion of the error For the remainder of this section, we restrict the discussion to the Gauss-Legendre integration formula (5.60), which will be referred to as simply Gaussian quadrature. Other forms of Gaussian quadrature will be discussed in Section 5.7 on singular integrals.

We begin by attempting to make the error term (5.61) more understandable. First define

$$M_m = \underset{-1 \leq x \leq 1}{\text{Maximum}} \frac{|f^{(m)}(x)|}{m!} \qquad m \geq 0 \qquad (5.63)$$

For a large class of infinitely differentiable functions f on $[-1,1]$, we have $\text{Sup}_{m \geq 0} M_m < \infty$. For example, this will be true if $f(z)$ is analytic on the region R of the complex plane defined by

$$R = \{ z : |z - x| \leq 1 \text{ for some } x, \ -1 \leq x \leq 1 \}$$

With many functions, $M_m \to 0$ as $m \to \infty$, for example, $f(x) = e^x$ or $\cos(x)$.

Combining (5.61) and (5.63), we obtain

$$|E_n(f)| \leq e_n M_{2n} \qquad n \geq 1 \qquad (5.64)$$

and the size of e_n is essential in examining the speed of convergence.

The term e_n can be made more understandable by estimating it using Stirling's formula,

$$n! \approx e^{-n} n^n \sqrt{2\pi n}$$

which is true in a relative error sense as $n \to \infty$. Then we obtain

$$e_n \approx \frac{\pi}{4^n} \qquad \text{as } n \to \infty \qquad (5.65)$$

This is quite a good estimate. For example, $e_5 = 0.00293$, and (5.65) gives the estimate 0.00307. Combined with (5.64), this implies

$$|E_n(f)| \leq \frac{\pi}{4^n} \cdot M_{2n} \qquad (5.66)$$

which is a correct bound in an asymptotic sense as $n \to \infty$. This shows that $E_n(f) \to 0$ with an exponential rate of decrease as a function of n. Compare this with the polynomial rates of $1/n^2$ and $1/n^4$ for the trapezoidal and Simpson rules, respectively.

In order to obtain convergence, it is not necessary to have $f(x)$ be infinitely differentiable on $[-1,1]$. It can be shown that if $f(x)$ is continuous on $[-1,1]$, then $E_n(f) \to 0$ as $n \to \infty$; see Problem 16. In response to a need for error formulas in which the order of the derivative does not grow, we can show the following. If $f(x)$ is r times continuously differentiable on $[-1,1]$, then

$$E_n(f) = \int_{-1}^{1} K_{n,r}(t) f^{(r)}(t) \, dt \qquad n > r/2 \qquad (5.67)$$

for an appropriate Peano kernel $K_{n,r}(t)$. The procedure for constructing $K_{n,r}(t)$ is exactly the same as with the Peano kernels (5.29) and (5.32) in Section 5.1. From (5.67),

$$|E_n(f)| \leq e_{n,r} M_r$$

$$e_{n,r} = r! \int_{-1}^{1} |K_{n,r}(t)| \, dt \qquad (5.68)$$

The values of $e_{n,r}$ given in Table 5.10 are taken from Stroud and Secrest [S3, pp. 152–153]. The table shows that for f twice continuously differentiable, Gaussian quadrature converges at least as rapidly as the trapezoidal rule (5.17). Using (5.68) and the table, we can construct the empirical bound

$$|E_n(f)| \leq \frac{.42}{n^2} \operatorname*{Max}_{-1 \leq x \leq 1} |f''(x)| \qquad (5.69)$$

The corresponding formula (5.18) for the trapezoidal rule on $[-1, 1]$ gives

$$|\text{Trapezoidal error}| \leq \frac{.67}{n^2} \operatorname*{Max}_{-1 \leq x \leq 1} |f''(x)|$$

which is slightly worse than (5.69). And in actual computation, Gaussian quadrature appears to always be superior to the trapezoidal rule; see the below example for the integral $I^{(1)}$. An analogous discussion, using Table 5.10 with $e_{n,4}$, can be carried out for integrands $f(x)$ that are four times differentiable; see Problem 15.

Table 5.10   Error constants $e_{n,r}$ for (5.68)

n	$e_{n,2}$	Ratio	$e_{n,4}$	Ratio
2	.162		.178	
		3.7		27.5
4	.437E−1		.647E−2	
		3.7		15.5
8	.118E−1		.417E−3	
		3.8		14.9
16	.311E−2		.279E−4	
		3.9		15.3
32	.800E−3		.183E−5	
		3.9		
64	.203E−3			
		4.0		
128	.511E−4			

Example   We give three further examples that are not as well behaved as the one in Table 5.9. Consider

$$I^{(1)} = \int_0^1 \sqrt{x}\ dx = \frac{2}{3}$$

$$I^{(2)} = \int_0^1 \frac{4dx}{1 + 256(x - .375)^2} = .719193830921$$

$$I^{(3)} = \int_0^{2\pi} e^{-x} \sin(50x)\ dx = .019954669278$$

The values in Table 5.11 show that Gaussian quadrature is still very effective in spite of the bad behavior of the integrand.

Table 5.11   Gaussian quadrature examples

n	$I^{(1)} - I_n^{(1)}$	Ratio	$I^{(2)} - I_n^{(2)}$	$I^{(3)} - I_n^{(3)}$
2	$-7.22\text{E}-3$		$4.20\text{E}-1$	$3.48\text{E}-1$
		6.2		
4	$-1.16\text{E}-3$		$-2.33\text{E}-1$	$-1.04\text{E}-1$
		6.9		
8	$-1.69\text{E}-4$		$-6.12\text{E}-2$	$-1.80\text{E}-2$
		7.4		
16	$-2.30\text{E}-5$		$-8.90\text{E}-3$	$-3.34\text{E}-1$
		7.6		
32	$-3.00\text{E}-6$		$-3.06\text{E}-4$	$1.16\text{E}-1$
		7.8		
64	$-3.84\text{E}-7$		$5.98\text{E}-8$	$1.53\text{E}-1$
		7.9		
128	$-4.85\text{E}-8$		$3.33\text{E}-16$	$6.69\text{E}-15$

Compare the results for $I^{(1)}$ with those in Table 5.5 for the trapezoidal and Simpson rules. Gaussian quadrature is converging with an error proportional to $1/n^3$, whereas in Table 5.5, the errors converged with a rate proportional to $1/n\sqrt{n}$. The initial convergence of $I_n^{(2)}$ to $I^{(2)}$ is quite slow, but as n increases the speed also increases dramatically. The approximations for $I^{(3)}$ are quite poor because it is so oscillatory; it has 101 zeroes in the interval of integration. The approximating polynomial (implicit in defining Gaussian quadrature) must have degree at least 101, and this explains why the approximates do poorly until $n = 128$.

General comments   Gaussian quadrature has a number of strengths and weaknesses.

1.   Because of form of the nodes and weights and the resulting need to use a table, many people prefer a simpler formula such as Simpson's rule. This shouldn't be a problem when doing integration using a computer. Programs should be written containing these weights and nodes for standard values of n, for example, $n = 2, 4, 8, 16, \ldots, 512$, taken from Stroud and Secrest [S3]. Although initially a problem to keypunch, the effort repays itself many times over with the much greater speed of numerical integration.

2.   It is difficult to estimate the error, and thus we usually take

$$I - I_n \approx I_m - I_n \tag{5.70}$$

for some $m > n$, for example, $m = n + 2$ with well-behaved integrands and $m = 2n$ otherwise. This results in greater accuracy than necessary; but even with the increased number of function evaluations, Gaussian quadrature is still faster than most other methods.

3.   The nodes for each formula I_n are distinct from those of preceding formulas I_m, and this results in some inefficiency. If I_n is not sufficiently accurate, based on an error estimate like (5.70), then we must compute a new value of I_n. But none of the previous values of the integrand can be reused, resulting in wasted effort. This problem is treated in the next section, leading to a modification of Gaussian quadrature. But in many situations, the resulting inefficiency in Gaussian quadrature is not significant because of its rapid rate of convergence.

4.   If a large class of integrals of a similar nature are to be evaluated, then proceed as follows. Pick a few representative integrals, including some with the worst behavior that is likely to occur in the integrand. Determine a value of n for which $I_n(f)$ will have sufficient accuracy among the representative set. Then fix that value of n, and use $I_n(f)$ as the numerical integral for all members of the original class of integrals.

5.   Gaussian quadrature can handle many near-singular integrands very effectively, as with example $I^{(1)}$ in Table 5.11. But all points of singular behavior should occur as endpoints of the integration interval. Gaussian quadrature is very poor on an integral such as

$$\int_0^1 \sqrt{|x - .7|} \, dx$$

containing a singular point internal to the interval of integration (most other numerical integration methods will also perform poorly on this integral). The integral should be decomposed and evaluated in the form

$$\int_0^{.7} \sqrt{.7-x}\ dx + \int_{.7}^1 \sqrt{x-.7}\ dx$$

5.4   Patterson's Method

The general philosophy of the method is as follows. Given a numerical integration formula $I_n(f)$ using n node points and given an integer $p \geq 1$, calculate a new formula $I_{n+p}(f)$ using $n+p$ node points. Choose these $n+p$ points such that (1) the n original node points in $I_n(f)$ are included, and (2) the degree of precision of $I_{n+p}(f)$ is as large as possible. We will limit the discussion to

$$I(f) = \int_{-1}^1 f(x)\,dx$$

Patterson derived a number of general results within the above framework (see [P1]), but we will concentrate on the best known numerical integration method from his work. Begin with the three-point Gauss-Legendre formula

$$I_3(f) = \frac{8}{9} f(0) + \frac{5}{9}\left[f\left(-\sqrt{\frac{3}{5}}\,\right) + f\left(\sqrt{\frac{3}{5}}\,\right)\right] \tag{5.71}$$

and add $p=4$ new points to obtain a formula $I_7(f)$. At each stage of the construction, we will have a formula $I_n(f)$ with n odd, and we will add $n+1$ new node points to obtain a formula $I_{2n+1}(f)$ with a higher degree of precision.

The node points will always be chosen to be symmetrically distributed about the origin in $[-1, 1]$, and the weights will also be symmetric. Without any further requirements, this will guarantee the integration formulas will integrate all odd polynomials [i.e., $p(-x) \equiv -p(x)$] exactly, giving an integral of zero. Suppose we have the formula $I_n(f)$ and we wish to create $I_{2n+1}(f)$. Let $m = (n-1)/2$, and denote the nodes for $I_n(f)$ by

$$x_0 = 0 \qquad x_{\pm 1}, \ldots, x_{\pm m} \tag{5.72}$$

Denote the new node points by

$$x_{\pm(m+1)}, \ldots, x_{\pm(2m+1)} \tag{5.73}$$

where $x_{-i} = -x_i$, $2m+1 = n$. Also define

$$\omega(x) = (x - x_{-n}) \ldots (x - x_0) \ldots (x - x_n)$$

We will show that

$$\omega(x) = \sum_{l=1}^{m+1} C_l P_{n+2l}(x) \tag{5.74}$$

for an appropriate choice of constants $\{C_l\}$, given in (5.82). The polynomial $P_j(x)$ is the Legendre polynomial of degree j; see (4.40) in Section 4.4 for the definition of $P_j(x)$. The new node points (5.73) can then be calculated by finding the roots of (5.74).

Given the integrand $f(x)$, let $p_{2n}(x;f)$ denote the polynomial interpolating to $f(x)$ at the $2n+1$ nodes $\{x_{-n}, \ldots, x_n\}$. Define

$$I_{2n+1}(f) = \int_{-1}^{1} p_{2n}(x;f) \, dx \tag{5.75}$$

Use the interpolation error formula in (3.22) to obtain

$$I(f) - I_{2n+1}(f) = \int_{-1}^{1} \omega(x) f[x_{-n}, \ldots, x_n, x] \, dx \tag{5.76}$$

The divided difference is of order $2n+1$, and thus the degree of precision of $I_{2n+1}(f)$ is at least $2n$, regardless of the choice of the new nodes.

To increase the degree of precision of $I_{2n+1}(f)$, note that if $f(x)$ is a polynomial, then $f[x_{-n}, \ldots, x_n, x]$ is also a polynomial. From (3.29) of Section 3.2, we have

$$\text{degree}(f) = 2n + q + 1 \Rightarrow \text{degree } f[x_{-n}, \ldots, x_n, x] = q, \quad q \geq 0. \tag{5.77}$$

To make $I_{2n+1}(f) = I(f)$ for polynomials f of degree greater than $2n$, (5.76) and (5.77) imply that $\omega(x)$ must be orthogonal to polynomials of as large a degree as possible. We impose conditions on $\omega(x)$, and implicitly on the new nodes in (5.73), by requiring

$$\int_{-1}^{1} \omega(x) g(x) \, dx = 0 \quad \text{degree } (g) \leq n \tag{5.78}$$

This imposes $n+1$ constraints on $\omega(x)$, and there are $n+1$ nodes in (5.73), roots of $\omega(x)$, to be determined. With (5.76)–(5.78), the degree of precision of $I_{2n+1}(f)$ will be at least $3n+1$; because if $f(x)$ is a polynomial of degree $\leq 3n+1$, then degree $f[x_{-n}, \ldots, x_n, x] \leq n$, and (5.78) will then show $I(f) - I_n(f) = 0$ in (5.76).

Since the formula $I_{2n+1}(f)$ will be exact for all odd polynomials, it follows that the degree of precision will be at least $3n+2$, and this is what is attained. Table 5.12 gives the degree of precision of $I_r(f)$ for smaller values of r, and the general formula is

$$\text{Degree of precision of } I_r(f) = \frac{1}{2}(3r+1) \quad r = 3, 7, 15, 31, \ldots$$

Table 5.12 Degree of Precision

n	Degree
3	5
7	11
15	23
31	47

Using the orthogonal family $\{P_j(x)|j \geq 0\}$, we can rewrite (5.78) in the equivalent form

$$\int_{-1}^{1} \omega(x)P_i(x)\,dx = 0 \qquad i = 0, 1, \ldots, n \tag{5.79}$$

Also, since degree $(\omega) = 2n + 1$, $\omega(x)$ can be written

$$\omega(x) = \sum_{j=0}^{2n+1} t_j P_j(x) \tag{5.80}$$

for some constants $\{t_j\}$. Since $\omega(x) = x^{2n+1} + $ lower-degree terms, the coefficient t_{2n+1} should equal $1/A_{2n+1}$, where A_{2n+1} is the coefficient of x^{2n+1} in $P_{2n+1}(x)$. But since we will calculate the new nodes by finding the roots of $\omega(x)$, it does not matter if $\omega(x)$ is changed by a constant multiple. Nor does it matter for (5.79). Thus choose $t_{2n+1} = 1$ for convenience.

By combining (5.79) and (5.80), and using the orthogonality of $\{P_j(x)\}$, it follows that

$$t_0 = t_1 = \ldots = t_n = 0$$

To further simplify the formula (5.80), note that the symmetry of the nodes $\{x_j\}$ about the origin (including $x_0 = 0$) forces $\omega(x)$ to be an odd function. And this will then imply that

$$t_{n+1} = t_{n+3} = \ldots = t_{2n} = 0$$

Thus

$$\omega(x) = \sum_{l=1}^{m+1} C_l P_{n+2l}(x) \qquad C_l = t_{n+2l} \tag{5.81}$$

with $C_{m+1} = 1$. To obtain the remaining coefficients, use the conditions

$$\omega(x_i) = 0 \qquad i = 1, \ldots, m$$

for the nonzero nodes of (5.72) that are known. This implies

$$\sum_{l=1}^{m} C_l P_{n+2l}(x_i) = -P_{2n+1}(x_i) \qquad i = 1, \ldots, m \tag{5.82}$$

and this determines the constants $\{C_l\}$ in (5.81). This completes the derivation.

Example Take $n=3$ and use (5.71). Thus $m=1$, $x_0=0$, $x_1=\sqrt{3/5}$. Equation (5.81) becomes

$$\omega(x)=P_7(x)+C_1P_5(x) \tag{5.83}$$

and C_1 is determined from (5.82) by

$$C_1P_5(\sqrt{3/5})=-P_7(\sqrt{3/5})$$

The roots of $\omega(x)$ are $x_0=0$, $\pm x_1$, $\pm x_2$, $\pm x_3$, with

$$x_1=.774597 \qquad x_2=.434244 \qquad x_3=.960491$$

This leads to the formula

$$I_7(f)=w_0f(0)+w_1\big[f(x_1)+f(-x_1)\big]$$
$$+w_2\big[f(x_2)+f(-x_2)\big]+w_3\big[f(x_3)+f(-x_3)\big]$$

which has degree of precision 11, compared with 13 for the seven-point Gauss-Legendre formula. The weights are

$$w_0=.450917 \qquad w_1=.268488 \qquad w_2=.401397 \qquad w_3=.104656$$

and can be obtained using (5.85).

The weights in the integration formula

$$I_{2n+1}(f)=w_0f(0)+\sum_{j=1}^{n}w_j\big[f(x_j)+f(-x_j)\big] \tag{5.84}$$

can be calculated using (5.75). Write

$$P_{2n}(x;f)=\sum_{j=-n}^{n}f(x_j)l_j(x) \qquad l_j(x)=\frac{\omega(x)}{(x-x_j)\omega'(x_j)}$$

the Lagrange form of the interpolating formula. Substituting into (5.75), we see that

$$w_j=\int_{-1}^{1}l_j(x)\,dx \qquad -n\le j\le n \tag{5.85}$$

It can easily be shown that $w_{-j}=w_j$ because of the summetry of the integration nodes.

The nodes and weights have been computed through $n=255$. All nodes lie in $[-1,1]$, and all weights are positive. But no general theory is yet known that guarantees the existence of the nodes $\{x_i\}$ for larger n; nor is it known that the weights will all be positive. The calculation of the nodes $\{x_j\}$ from (5.81) and (5.82) can be an unstable and difficult process, and a great deal of attention is given to the problem in Patterson [P1], along with tables of nodes and weights.

Patterson has used this numerical integration rule as the basis of an automatic numerical integration computer program, called QSUB. The rules $I_3(f), I_7(f), \ldots$ are computed successively until $|I_{2n+1}(f) - I_n(f)|$ is sufficiently small. If this is not satisfied and $I_{255}(f)$ has been computed, then the interval of integration is divided into smaller subintervals, for example,

$$\int_{-1}^{1} f(x)\, dx = \int_{-1}^{0} f(x)\, dx + \int_{0}^{1} f(x)\, dx$$

The numerical integration rule is then applied to each new integral in succession. This is continued, perhaps with additional interval halvings, until either the original integral is successfully evaluated or else the number of integrand evaluations exceeds some preset upper limit. For additional details, see Patterson [P2], which contains a Fortran program QSUB and additional discussion.

Empirically, QSUB performs very well on smooth integrands. It is generally competitive with Gaussian quadrature, often faster. And its characteristic behavior over a range of integrand types is essentially the same as that of Gaussian quadrature. See the following example for additional insight.

Example QSUB was used to evaluate the following integrals. Three different relative error tolerances were requested of QSUB; and the actual resulting relative errors (RE) and number of integrand evaluations (IE) are given in Table 5.13. An asterisk (*) entry indicates that convergence was not attained by the program.

Most of the integrals are chosen to exhibit some type of poor behavior in the integrand; and under the circumstances most of the results are quite good.

$$I_1 = \int_0^1 \frac{4dx}{1 + 256(x - .375)^2} = .71919383092$$

$$I_2 = \int_0^1 x^2 \sqrt{x}\ dx = .28571428571$$

$$I_3 = \int_0^1 \sqrt{x}\ dx = 2/3$$

$$I_4 = \int_0^1 \ln(x)\, dx = -1$$

$$I_5 = \int_0^1 \ln|x - .7|\, dx = -1.6108643021$$

$$I_6 = \int_0^{2\pi} e^{-x} \sin(50x)\, dx = .019954669278$$

Table 5.13 Examples of Patterson's QSUB Program

| Integral | Desired relative error | | | | | |
| | 5.0E−2 | | 5.0E−5 | | 5.0E−8 | |
	RE	IE	RE	IE	RE	IE
I_1	2.79E−4	31	5.68E−7	127	6.07E−12	650
I_2	2.55E−8	7	2.88E−10	15	2.88E−10	15
I_3	2.13E−4	7	5.03E−7	31	4.63E−8	2937
I_4	8.05E−3	7	2.74E−6	127	6.06E−8	19216
I_5	1.79E−2	31	1.72E−5	2633	*	*
I_6	2.36E−6	509	2.36E−6	509	4.84E−10	1269

The main weakness of QSUB is its inability to integrate efficiently a singular or near-singular integrand with high accuracy, for example I_3, I_4, and I_5 with desired error 5.0E−8. Also, example I_5 shows the much poorer behavior resulting when the singular point is interior to the integration interval rather than at an endpoint. Integral I_5 should be written as

$$I_5 = \int_0^{.7} \ln(.7-x)\,dx + \int_{.7}^{1.0} \ln(x-.7)\,dx$$

and each integral evaluated separately; many fewer evaluations will be required. Also note that the singularity in the third derivative of the integrand $x^2\sqrt{x}$ of I_2 does not present any difficulty. The program QSUB is especially recommended for smooth integrands, with possible oscillations of a less severe type than in I_6. Patterson's article [P2] contains another program QSUBA that is more efficient for singular integrands.

5.5 Asymptotic Error Formulas and Their Application

Recall the definition (5.18) of an asymptotic error formula for a numerical integration formula $I_n(f) \approx I(f)$. We say $\tilde{E}_n(f)$ is an asymptotic error formula for $E_n(f) = I(f) - I_n(f)$ if

$$\lim_{n\to\infty} \frac{\tilde{E}_n(f)}{E_n(f)} = 1 \tag{5.86}$$

or equivalently

$$\lim_{n\to\infty} \frac{E_n(f) - \tilde{E}_n(f)}{E_n(f)} = 0$$

Examples are (5.17) and (5.24) in Section 5.1.

By obtaining an asymptotic error formula, we are obtaining the form or structure of the error. And with this information, we can either develop new and more accurate integration formulas or we can accurately estimate the error in the present formula $I_n(f)$. Both ideas are illustrated in this section; and we conclude the section with the well-known Romberg integration method, which is rapidly convergent. But first we develop additional asymptotic error formulas.

The Bernoulli polynomials For use in the next theorem, we introduce the Bernoulli polynomials $B_n(x)$, $n \geq 0$. These are defined implicitly by the *generating function*

$$\frac{t(e^{xt}-1)}{e^t-1} = \sum_{j=1}^{\infty} B_j(x)\frac{t^j}{j!} \tag{5.87}$$

The first few polynomials are

$$B_0(x) \equiv 1 \qquad B_1(x) = x \qquad B_2(x) = x^2 - x$$

$$B_3(x) = x^3 - \frac{3x^2}{2} + \frac{x}{2} \qquad B_4(x) = x^2(1-x)^2$$

With these polynomials,

$$B_k(0) = 0 \qquad k \geq 1$$

There are easily computable recursion relations for calculating these polynomials; see Problem 18.

Also of interest are the Bernoulli numbers, defined implicitly by

$$\frac{t}{e^t-1} = \sum_{j=0}^{\infty} B_j \frac{t^j}{j!} \tag{5.88}$$

The first few numbers are

$$B_0 = 1 \qquad B_1 = -\tfrac{1}{2} \qquad B_2 = \tfrac{1}{6} \qquad B_4 = -\tfrac{1}{30} \qquad B_6 = \tfrac{1}{42} \qquad B_8 = \tfrac{1}{30}$$

and for all odd integers $j \geq 3$, $B_j = 0$. To obtain a relation to the Bernoulli polynomials $B_j(x)$, integrate (5.87) with respect to x on $[0, 1]$. Then

$$1 - \frac{t}{e^t-1} = \sum_{1}^{\infty} \frac{t^j}{j!} \int_0^1 B_j(x)\,dx$$

and thus

$$B_j = -\int_0^1 B_j(x)\,dx \qquad j \geq 1 \tag{5.89}$$

We will also need to define a periodic extension of $B_j(x)$,

$$\bar{B}_j(x) = \begin{cases} B_j(x) & 0 \le x < 1 \\ \bar{B}_j(x-1) & x \ge 1 \end{cases}$$

The Euler-MacLaurin formula The following theorem gives a very detailed asymptotic error formula for the trapezoidal rule; and it is at the heart of much of the asymptotic error analysis of this section. The connection with some other integration formulas appears later in the section.

Theorem 5.3 (Euler-MacLaurin formula) Let $m \ge 0$, $n \ge 1$, and define $h = (b-a)/n$, $x_j = a + jh$ for $j = 0, 1, \ldots, n$. Further assume that $f(x)$ is $2m+2$ times continuously differentiable on $[a,b]$. Then for the error in the trapezoidal rule,

$$E_n(f) = \int_a^b f(x)\,dx - h \sum_{j=0}^n {}'' f(x_j)$$

$$= - \sum_{i=1}^m \frac{B_{2i}}{(2i)!} h^{2i} \left[f^{(2i-1)}(b) - f^{(2i-1)}(a) \right]$$

$$+ \frac{h^{2m+2}}{(2m+2)!} \int_a^b \bar{B}_{2m+2}\left(\frac{x-a}{h}\right) f^{(2m+2)}(x)\,dx \qquad (5.90)$$

Proof A complete proof is given in Ralston [R1, pp. 131–133]; and a more general development is given in Steffensen [S1, Chapter 14]. The proof in [R1] is short and correct, making full use of the special properties of the Bernoulli polynomials. Here we give a simpler, but less general, version of that proof, showing it to be based on integration by parts with a bit of clever algebraic manipulation.

The proof of (5.90) for general $n \ge 1$ is based on first proving the result for $n = 1$. Thus we will concentrate on

$$E_1(f) = \int_0^h f(x)\,dx - \frac{h}{2}\left[f(0) + f(h) \right]$$

$$= \tfrac{1}{2} \int_0^h f''(x) x(x-h)\,dx \qquad (5.91)$$

the latter formula coming from (5.27). Since we know the asymptotic formula

$$E_1(f) \approx - \frac{h^2}{12}\left[f'(h) - f'(0) \right]$$

we attempt to manipulate (5.91) to obtain this. Write

$$E_1(f) = \int_0^h f''(x)\left[-\frac{h^2}{12}\right]dx + \int_0^h f''(x)\left[\frac{x(x-h)}{2} + \frac{h^2}{12}\right]dx$$

Then

$$E_1(f) = -\frac{h^2}{12}[f'(h) - f'(0)] + \int_0^h f''(x)\left[\frac{x^2}{2} - \frac{xh}{2} + \frac{h^2}{12}\right]dx$$

and using integration by parts,

$$E_1(f) = -\frac{h^2}{12}[f'(h) - f'(0)] + \left[f''(x)\left(\frac{x^3}{6} - \frac{x^2h}{4} + \frac{h^2x}{12}\right)\right]_0^h$$

$$-\int_0^h f'''(x)\left(\frac{x^3}{6} - \frac{x^2h}{4} + \frac{h^2x}{12}\right)dx$$

The evaluation of the quantity in brackets at $x=0$ and $x=h$ gives zero. Integrate by parts again; the parts outside the integral will again be zero. The result will be

$$E_1(f) = -\frac{h^2}{12}[f'(h) - f'(0)] + \frac{1}{24}\int_0^h f^{(4)}(x)x^2(x-h)^2 dx \quad (5.92)$$

which is (5.90) with $m=1$. To obtain the $m=2$ case, first note that

$$\frac{1}{24}\int_0^h x^2(x-h)^2 dx = \frac{h^5}{720}$$

Then as before, write

$$\frac{1}{24}\int_0^h f^{(4)}(x)x^2(x-h)^2 dx = \int_0^h f^{(4)}(x)\left[\frac{h^4}{720}\right]dx$$

$$+ \int_0^h f^{(4)}(x)\left[\frac{x^2(x-h)^2}{24} - \frac{h^4}{720}\right]dx$$

$$= \frac{h^4}{720}[f'''(h) - f'''(0)]$$

$$+ \int_0^h f^{(4)}(x)\left(\frac{x^4 - 2x^3h + x^2h^2}{24} - \frac{h^4}{720}\right)dx$$

Integrate by parts twice to obtain the $m=2$ case of (5.90). This can be continued indefinitely. The proof in [R1] uses integration by parts, taking advantage of special relations for the Bernoulli polynomials; see Problem 18.

For the proof for general $n > 1$, write

$$E_n(f) = \sum_{j=1}^{n} \left\{ \int_{x_{j-1}}^{x_j} f(x)\,dx - \frac{h}{2} \left[f(x_{j-1}) + f(x_j) \right] \right\}$$

For the $m = 1$ case, using (5.92),

$$E_n(f) = \sum_{j=1}^{n} \left\{ -\frac{h^2}{12} \left[f'(x_j) - f'(x_{j-1}) \right] \right\}$$

$$+ \sum_{j=1}^{n} \frac{1}{24} \int_{x_{j-1}}^{x_j} f^{(4)}(x)(x - x_{j-1})^2 (x - x_j)^2\,dx$$

$$= -\frac{h^2}{12} \left[f'(b) - f'(a) \right] + \frac{h^4}{24} \int_a^b f^{(4)}(x) \bar{B}_4 \left(\frac{x-a}{h} \right) dx$$

The proof for $m > 1$ is essentially the same. ∎

The error term in (5.90) can be simplified using the integral mean value theorem. It can be shown that

$$B_{2j}(x) > 0 \qquad 0 < x < 1$$

and consequently the error term satisfies

$$\int_a^b \bar{B}_{2m+2} \left(\frac{x-a}{h} \right) f^{(2m+2)}(x)\,dx = f^{(2m+2)}(\xi) \int_a^b \bar{B}_{2m+2} \left(\frac{x-a}{h} \right) dx$$

$$= n f^{(2m+2)}(\xi) \int_a^{a+h} B_{2m+2} \left(\frac{x-a}{h} \right) dx$$

$$= nh f^{(2m+2)}(\xi) \int_0^1 B_{2m+2}(u)\,du$$

$$= -(b-a) B_{2m+2} f^{(2m+2)}(\xi) \qquad \text{some } a \le \xi \le b$$

Thus (5.90) becomes

$$E_n(f) = -\sum_{i=1}^{n} \frac{B_{2i}}{(2i)!} h^{2i} \left[f^{(2i-1)}(b) - f^{(2i-1)}(a) \right] - \frac{h^{2m+2}(b-a) B_{2m+2}}{(2m+2)!} f^{(2m+2)}(\xi)$$

$$\tag{5.93}$$

for some $a \le \xi \le b$.

As an important corollary of (5.90), we can show that the trapezoidal rule performs especially well when applied to periodic functions.

Corollary 1 Suppose $f(x)$ is infinitely differentiable for $a \leq x \leq b$, and suppose that all of its odd ordered derivatives are periodic with $b - a$ an integer multiple of the period. Then the order of convergence of the trapezoidal rule $I_n(f)$ applied to

$$I(f) = \int_a^b f(x)\,dx$$

is greater than any power of h.

Proof Directly from the assumptions on $f(x)$,

$$f^{(2j-1)}(b) = f^{(2j-1)}(a) \qquad j \geq 1$$

Consequently for any $m \geq 0$, with $h = (b-a)/n$, (5.93) implies

$$I(f) - I_n(f) = -\frac{h^{2m+2}}{(2m+2)!}(b-a)B_{2m+2}f^{(2m+2)}(\xi) \qquad a \leq \xi \leq b \quad (5.94)$$

Thus as $n \to \infty$ (and $h \to 0$) the rate of convergence is proportional to h^{2m+2}. But m was arbitrary, which shows the desired result. ∎

The Euler-MacLaurin summation formula Although it doesn't involve numerical integration, an important application of (5.90) or (5.93) is the summation of series.

Corollary 2 (Euler-MacLaurin summation formula) Assume $f(x)$ is $2m+2$ times continuously differentiable for $0 \leq x < \infty$, for some $m \geq 0$. Then for all $n \geq 1$,

$$\sum_{j=0}^{n} f(j) = \int_0^n f(x)\,dx + \tfrac{1}{2}\left[f(0) + f(n) \right]$$

$$+ \sum_{i=1}^{m} \frac{B_{2i}}{(2i)!}\left[f^{(2i-1)}(n) - f^{(2i-1)}(0) \right]$$

$$- \frac{1}{(2m+2)!} \int_0^n \overline{B}_{2m+2}(x) f^{(2m+2)}(x)\,dx \qquad (5.95)$$

Proof Merely substitute $a = 0$, $b = n$ into (5.90), and then rearrange the terms appropriately. ∎

Example. In a later example we will need the sum

$$S=\sum_1^\infty \frac{1}{n^{3/2}}$$

If we use the obvious choice $f(x)=(x+1)^{-3/2}$ in (5.95), the results are disappointing. By letting $n\to\infty$, we obtain

$$S=\int_0^\infty \frac{dx}{(x+1)^{3/2}}+\frac{1}{2}-\sum_1^m \frac{B_{2i}}{(2i)!}f^{(2i-1)}(0)$$

$$-\frac{1}{(2m+2)!}\int_0^\infty \overline{B}_{2m+2}(x)f^{(2m+2)}(x)\,dx$$

and the error term does not become small for any choice of m. But if we divide the series S into two parts, we are able to treat it very accurately.

Let $f(x)=(x+10)^{-3/2}$. Then with $m=1$,

$$\sum_{10}^\infty \frac{1}{n^{3/2}}=\sum_0^\infty f(j)=\int_0^\infty \frac{dx}{(x+10)^{3/2}}+\frac{1}{2(10)^{3/2}}-\frac{(1/6)}{2}\frac{(-3/2)}{(10)^{5/2}}+E$$

$$E=-\tfrac{1}{24}\int_0^\infty \overline{B}_4(x)f^{(4)}(x)\,dx$$

Since $\overline{B}_4(x)\ge 0$, $f^{(4)}>0$, we have $E<0$. Also

$$0<-E<\frac{1}{24}\int_0^\infty \left(\frac{1}{16}\right)f^{(4)}(x)\,dx=\frac{35}{(1024)(10)^{9/2}}\doteq 1.08\times 10^{-6}$$

Thus

$$\sum_{10}^\infty \frac{1}{n^{3/2}}=.648662205+E$$

By directly summing $\sum_1^9(1/n^{3/2})=1.963713717$, we obtain

$$\sum_1^\infty \frac{1}{n^{3/2}}=2.6123759+E \qquad 0<-E<1.08\times 10^{-6} \qquad (5.96)$$

See Ralston [R1, pp. 134–138] and Steffensen [S1] for more information on summation techniques. To appreciate the importance of the above summation method, it would have been necessary to have added 3.43×10^{12} terms in S to have obtained comparable accuracy.

A generalized Euler-MacLaurin formula For integrals in which the integrand is not differentiable at some points, it is still often possible to obtain an asymptotic error expansion. For the trapezoidal rule and other numerical integration rules applied to integrands with algebraic and/or logarithmic singularities, see the article by Lyness and Ninham [L2, Section 7]. We specialize their results to the integral

$$I = \int_0^1 x^\alpha f(x) \, dx \qquad \alpha > 0$$

with $f \in C^{m+1}[0,1]$, with the trapezoidal numerical integration rule.

Assuming α is not an integer,

$$E_n(f) = \sum_{j=0}^{m-1} \frac{c_j}{n^{\alpha+j+1}} + \sum_{j=1}^{m-1} \frac{d_j}{n^{j+1}} + O\left(\frac{1}{n^{m+1}}\right) \tag{5.97}$$

The term $O(1/n^{m+1})$ denotes a quantity whose size is proportional to $1/n^{m+1}$, or possibly smaller. The constants are given by

$$c_j = \frac{2\Gamma(\alpha+j+1)\sin\left[\frac{\pi}{2}(\alpha+j)\right]\zeta(\alpha+j+1)}{(2\pi)^{\alpha+j+1}j!} f^{(j)}(0)$$

$$d_j = 0, \quad \text{for } j \text{ even}$$

$$d_j = (-1)^{(j-1)/2} \frac{2\zeta(j+1)}{(2\pi)^{j+1}} g^{(j)}(1) \quad j \text{ odd}$$

with $g(x) = x^\alpha f(x)$, $\Gamma(x) =$ gamma function, and $\zeta(p)$ is the zeta function,

$$\zeta(p) = \sum_{j=1}^{\infty} \frac{1}{j^p} \qquad p > 1$$

For $0 < \alpha < 1$ with $m = 1$, we obtain the asymptotic error estimate

$$E_n(f) = \frac{2\Gamma(\alpha+1)\sin\left(\frac{\pi}{2}\alpha\right)\zeta(\alpha+1)f(0)}{(2\pi)^{\alpha+1}n^{\alpha+1}} + O\left(\frac{1}{n^2}\right), \quad f \in C^2[0,1] \tag{5.98}$$

For example with $\int_0^1 \sqrt{x}\, f(x) \, dx$, and using (5.96) for evaluating $\zeta(3/2)$,

$$E_n(f) = \frac{c}{n\sqrt{n}} f(0) + O\left(\frac{1}{n^2}\right) \qquad c = \frac{\zeta(3/2)}{4\pi} \doteq .208$$

This is confirmed numerically in the example given in Table 5.5 in Section 5.1.

For logarithmic singularities, the results of [L2] imply an asymptotic error formula

$$E_n(f) \approx \frac{c \log n}{n^p} \tag{5.99}$$

for some $p > 0$ and some constant c. For numerical purposes, this is essentially $O(1/n^p)$. To justify this, calculate the following limit using L'Hospital's rule, with $p > q$.

$$\underset{n \to \infty}{\text{Limit}} \frac{\log(n)/n^p}{1/n^q} = \underset{n \to \infty}{\text{Limit}} \frac{\log(n)}{n^{p-q}} = 0$$

This means that $\log(n)/n^p$ decreases more rapidly than $1/n^q$ for any $q < p$. And it clearly decreases less rapidly than $1/n^p$, although not by much. For practical computation, (5.99) is essentially $O(1/n^p)$. For example, calculate the limit of successive errors

$$\underset{n \to \infty}{\text{Limit}} \frac{I - I_n}{I - I_{2n}} = \underset{n \to \infty}{\text{Limit}} \frac{c \log(n)/n^P}{c \log(2n)/2^P n^P} = \underset{n \to \infty}{\text{Limit}} 2^P \frac{\log(n)}{\log(2n)}$$

$$= \underset{n \to \infty}{\text{Limit}} 2^P \frac{1}{1 + \dfrac{\log 2}{\log n}} = 2^P$$

This is the same limiting ratio as would occur if the error were just $O(1/n^p)$.

Aitken extrapolation Motivated by the preceding, we assume that the integration formula has an asymptotic error formula

$$I - I_n \approx c/n^p \qquad p > 0 \tag{5.100}$$

Much as in Section 2.5, we estimate I using the successive iterates I_n, I_{2n}, and I_{4n}. But first we estimate p.

$$\frac{I_{2n} - I_n}{I_{4n} - I_{2n}} = \frac{(I - I_n) - (I - I_{2n})}{(I - I_{2n}) - (I - I_{4n})} \approx \frac{\dfrac{c}{n^p} - \dfrac{c}{2^p n^p}}{\dfrac{c}{2^p n^p} - \dfrac{c}{4^p n^p}} = 2^p$$

$$R_n \equiv \frac{I_{2n} - I_n}{I_{4n} - I_{2n}} \approx 2^p \tag{5.101}$$

and this gives a simple way of computing p.

Example. Consider the use of Simpson's rule with $\int_0^1 x \sqrt{x} \, dx = 0.4$. In Table 5.14, the column R_n should approach $2^{2.5} \doteq 5.66$, a theoretical result from [L2] for the order of convergence. Clearly the results confirm the theory.

Table 5.14   Simpson integration errors for $\int_0^1 x\sqrt{x}\,dx$

n	I_n	$I-I_n$	$I_{2n}-I_n$	R_n
2	.402368927062	$-2.369-3$	$-1.937-3$	
				5.46
4	.400431916045	$-4.319-4$	$-3.547-4$	
				5.58
8	.400077249447	$-7.725-5$	$-6.354-5$	
				5.63
16	.400013713469	$-1.371-5$	$-1.129-5$	
				5.65
32	.400002427846	$-2.428-6$	$-1.998-6$	
				5.65
64	.400000429413	$-4.294-7$	$-3.535-7$	
				5.66
128	.400000075924	$-7.592-8$	$-6.250-8$	
				5.66
256	.400000013423	$-1.342-8$	$-1.105-8$	
512	.400000002373	$-2.373-9$		

To estimate the integral I with increased accuracy, suppose that I_n, I_{2n}, and I_{4n} have been computed. Using (5.100),

$$\frac{I-I_n}{I-I_{2n}}\approx 2^p\approx\frac{I-I_{2n}}{I-I_{4n}}$$

and thus

$$(I-I_n)(I-I_{4n})\approx(I-I_{2n})^2$$

Solving for I, and manipulating to obtain a desirable form,

$$I\approx\tilde{I}_n\equiv I_{4n}-\frac{(I_{4n}-I_{2n})^2}{(I_{4n}-I_{2n})-(I_{2n}-I_n)}\tag{5.102}$$

Example   Using the previous example for $f(x)=x\sqrt{x}$ and Table 5.14, we obtain the difference table in Table 5.15.

Table 5.15   Difference table for Simpson integration

m	I_m	$\Delta I_m = I_{2m}-I_m$	$\Delta^2 I_m$
16	.400013713469		
		$-1.1285623-5$	
32	.400002427846		$9.28719-6$
		$-1.998433-6$	
64	.400000429413		

Then

$$I \approx \tilde{I}_{16} = .399999999387$$

$$I - \tilde{I}_{16} = 6.13 \times 10^{-10} \qquad I - I_{64} = -4.29 \times 10^{-7}$$

Thus $\tilde{I}_{16}$ is a considerable improvement on I_{64}. Also note that $\tilde{I}_{16} - I_{64}$ is an excellent approximation to $I - I_{64}$.

Summing up, given a numerical integration rule satisfying (5.100) and given three values I_n, I_{2n}, I_{4n}, calculate the Aitken extrapolate $\tilde{I}_n$ in (5.102). It is usually a significant improvement on I_{4n} as an approximation to I; and based on this,

$$I - I_{4n} \approx \tilde{I}_n - I_{4n}$$

With Simpson's rule, or any other composite closed Newton-Cotes formula, the expense of evaluating I_n, I_{2n}, I_{4n} is no more than that of I_{4n} alone, that is, $4n + 1$ function evaluations. And when the algorithm is designed correctly, there is no need for temporary storage of large numbers of function values $f(x_j)$. For that reason, one should never use Simpson's rule with just one value of the index n. With no extra expenditure of time, and with only a slightly more complicated algorithm, an Aitken extrapolate and an error estimate can be produced. The following is such an algorithm.

Algorithm Simpson $(f, a, b, m, I_m, A, p, \text{ier})$

1. Remark: This algorithm calculates the Simpson numerical integral,

$$I_m \approx I = \int_a^b f(x)\, dx$$

with m evenly divisible by 8. If $\text{ier} = 0$ on exit, then A contains the Aitken extrapolate to I, based on $I_m, I_{m/2}, I_{m/4}$, and p contains the order based on (5.101). If $\text{ier} \neq 0$, then A is not calculated. $\text{ier} = 1$ means that the ratio R in (5.101) was negative; and $\text{ier} = 2$ means the denominator in (5.101) or in (5.102) was zero.

2. $n := m/4, \quad h := (b - a)/n$

3. $\text{sum}(1) := f(a) + f(b)$

4. $\text{sum}(2) := \displaystyle\sum_{j=1}^{n/2-1} f(a + 2jh)$

5. Do through step 10 for $i = 0, 1, 2$

6. $N:=2^i n,\ h:=(b-a)/N$

7. $\text{sum}(3):=\displaystyle\sum_{j=1}^{N/2} f[a+(2j-1)h]$

8. $\text{Int}(i):=\dfrac{h}{3}[\text{sum}(1)+2\text{sum}(2)+4\text{sum}(3)]$

9. $\text{sum}(2):=\text{sum}(2)+\text{sum}(3)$

10. End loop on i

11. $I_m:=\text{Int}(2)$

12. $D(1):=\text{Int}(1)-\text{Int}(0),\ D(2):=\text{Int}(2)-\text{Int}(1)$
    $D(3):=D(2)-D(1)$

13. If $D(2)=0$ or $D(3)=0$, then set ier$=2$ and return.

14. $R:=D(1)/D(2)$

15. If $R<0$, then set ier$=1$ and return.

16. $p:=\log_2(R)$, ier:$=0$.

17. $A:=\text{Int}(2)-D(2)^2/D(3)$

18. End Simpson and return.

Richardson extrapolation   If we assume sufficient smoothness for the integrand $f(x)$ in our integral $I(f)$, then we can write the trapezoidal error term (5.90) as

$$I-I_n=\frac{d_2^{(0)}}{n^2}+\frac{d_4^{(0)}}{n^4}+\cdots+\frac{d_{2m}^{(0)}}{n^{2m}}+F_{n,m}$$

$$F_{n,m}=\frac{(b-a)^{2m+2}}{(2m+2)!\,n^{2m+2}}\int_a^b \bar{B}_{2m+2}\left(\frac{x-a}{h}\right)f^{(2m+2)}(x)\,dx$$

$$d_{2j}^{(0)}=-\frac{B_{2j}}{(2j)!}(b-a)^{2j}\left[f^{(2j-1)}(b)-f^{(2j-1)}(a)\right].\tag{5.103}$$

Although the series dealt with are always finite and have an error term, we will usually not concern ourselves directly with it.
    For n even,

$$I-I_{n/2}=\frac{4d_2^{(0)}}{n^2}+\frac{16d_4^{(0)}}{n^4}+\frac{64d_6^{(0)}}{n^6}+\cdots$$

Multiply (5.103) by 4 and subtract from it the last equation,

$$4(I - I_n) - (I - I_{n/2}) = \frac{-12d_4^{(0)}}{n^4} - \frac{60d_6^{(0)}}{n^6} - \cdots$$

$$I = \frac{4I_n - I_{n/2}}{3} - \frac{4d_4^{(0)}}{n^4} - \frac{20d_6^{(0)}}{n^6} - \cdots$$

Define

$$I_n^{(1)} = \tfrac{1}{3}\left[4I_n^{(0)} - I_{n/2}^{(0)}\right] \qquad n \text{ even, } n \geq 2 \tag{5.104}$$

and $I_m^{(0)} \equiv I_m$.

The sequence

$$I_2^{(1)}, I_4^{(1)}, I_6^{(1)}, \ldots$$

is a new numerical integration rule. For the error,

$$I - I_n^{(1)} = \frac{d_4^{(1)}}{n^4} + \frac{d_6^{(1)}}{n^6} + \cdots$$

$$d_4^{(1)} = -4d_4^{(0)} \qquad d_6^{(1)} = -20d_6^{(0)}, \ldots \tag{5.105}$$

To see the explicit formula for $I_n^{(1)}$, let $h = (b - a)/n$ and $x_j = a + jh$ for $j = 0, 1, \ldots, n$. Then

$$I_n^{(1)} = \frac{4h}{3}\left(\tfrac{1}{2}f_0 + f_1 + f_2 + f_3 + \cdots + f_{n-1} + \tfrac{1}{2}f_n\right)$$

$$- \frac{2h}{3}\left(\tfrac{1}{2}f_0 + f_2 + f_4 + f_6 + \cdots + f_{n-2} + \tfrac{1}{2}f_n\right)$$

$$= \frac{h}{3}\left(f_0 + 4f_1 + 2f_2 + 4f_3 + \cdots + 2f_{n-2} + 4f_{n-1} + f_n\right)$$

which is Simpson's rule with n subdivisions. For the error, using (5.105) and the previous definitions of $d_j^{(0)}$,

$$I - I_n^{(1)} = -\frac{h^4}{180}\left[f'''(b) - f'''(a)\right] + \frac{h^6}{1512}\left[f^{(5)}(b) - f^{(5)}(a)\right] + \cdots \tag{5.106}$$

This means that the work on the Euler-MacLaurin formula transfers to Simpson's rule by means of some simple algebraic manipulations.

The above argument which led to $I_n^{(1)}$ can be continued to produce other new formulas. The use of the asymptotic error formula in this way is referred to as Richardson extrapolation. As before, if n is a multiple of 4, then

$$I - I_{n/2}^{(1)} = \frac{16d_4^{(1)}}{n^4} + \frac{64d_6^{(1)}}{n^6} + \dots$$

Then

$$16\left(I - I_n^{(1)}\right) - \left(I - I_{n/2}^{(1)}\right) = \frac{-48d_6^{(1)}}{n^6} + \dots$$

$$I = \frac{16I_n^{(1)} - I_{n/2}^{(1)}}{15} - \frac{48d_6^{(1)}}{15n^6} \dots$$

$$I - I_n^{(2)} = \frac{d_6^{(2)}}{n^6} + \frac{d_8^{(2)}}{n^8} + \dots$$

with

$$I_n^{(2)} = \frac{16I_n^{(1)} - I_{n/2}^{(1)}}{15} \qquad n \geq 4 \tag{5.107}$$

and n divisible by 4.

Using the above formulas, we can obtain useful estimates of the error.

$$I - I_n^{(1)} = \frac{16I_n^{(1)} - I_{n/2}^{(1)}}{15} - I_n^{(1)} + \frac{d_6^{(2)}}{n^6} - \dots$$

$$= \frac{I_n^{(1)} - I_{n/2}^{(1)}}{15} + \frac{d_6^{(2)}}{n^6} - \dots$$

Using $h = (b - a)/n$,

$$I - I_n^{(1)} = \tfrac{1}{15}\left(I_n^{(1)} - I_{n/2}^{(1)}\right) + O\left(h^6\right) \tag{5.108}$$

and thus

$$I - I_n^{(1)} \approx \tfrac{1}{15}\left(I_n^{(1)} - I_{n/2}^{(1)}\right) \tag{5.109}$$

since both terms are $O(h^4)$ and the remainder term is $O(h^6)$.

Richardson extrapolation can be continued inductively. Define

$$I_n^{(k)} = \frac{4^k I_n^{(k-1)} - I_{n/2}^{(k-1)}}{4^k - 1} \qquad n \geq 2^k \tag{5.110}$$

with n a multiple of 2^k, $k \geq 1$. It can be shown that the error has the form

$$I - I_n^{(k)} = \frac{d_{2k+2}^{(k)}}{n^{2k+2}} + \cdots$$

$$= \beta_k (b-a) h^{2k+2} f^{(2k+2)}(\zeta_n) \qquad a < \zeta_n < b$$

with β_k a constant independent of f and h, and

$$d_{2k+2}^{(k)} = \beta_k (b-a)^{2k+2} \left[f^{(2k+1)}(b) - f^{(2k+1)}(a) \right]$$

Finally, it can be shown that for any $f \in C[a,b]$,

$$\underset{n \to \infty}{\text{Limit}}\, I_n^{(k)}(f) = I(f) \tag{5.111}$$

The rules $I_n^{(k)}(f)$ for $k > 2$ bear no direct relation to the composite Newton-Cotes rules. See Bauer, Rutishauser, and Stiefel [B1] for complete details.

Romberg integration   Define

$$J_k(f) = I_{2^k}^{(k)} \qquad k = 0, 1, 2, \dots \tag{5.112}$$

This is the Romberg integration rule. Consider the diagram in Figure 5.3 for the Richardson extrapolates of the trapezoidal rule, with the number of subdivisions a power of two. The first column denotes the trapezoidal rule, the second Simpson's rule, and so forth. By (5.111), each column converges to $I(f)$. The Romberg integration is the rule of taking the diagonal. Since each column converges more rapidly than the preceding column, assuming $f(x)$ is infinitely differentiable, it could be expected that $J_k(f)$ would converge more rapidly than $\{I_n^{(k)}\}$ for any k. This is indeed the case, and consequently the method has been very popular since the late 1950s. Compared with Gaussian quadrature and Patterson's method, it has the advantage of using evenly spaced abscissas.

$$
\begin{array}{ccccc}
I_1^{(0)} & & & & \\[4pt]
I_2^{(0)} & I_2^{(1)} & & & \\[4pt]
I_4^{(0)} & I_4^{(1)} & I_4^{(2)} & & \\[4pt]
I_8^{(0)} & I_8^{(1)} & I_8^{(2)} & I_8^{(3)} & \\[4pt]
I_{16}^{(0)} & I_{16}^{(1)} & I_{16}^{(2)} & I_{16}^{(3)} & I_{16}^{(4)} \\[4pt]
\vdots & \vdots & \vdots & \vdots & \vdots
\end{array}
$$

Figure 5.3   Romberg integration table.

To compute $J_k(f)$ for a particular k, having already computed $J_1(f), \ldots, J_{k-1}(f)$, the row

$$I_n^{(0)}, I_n^{(1)}, \ldots, I_n^{(k-1)} \qquad n = 2^{k-1} \tag{5.113}$$

should have been saved in temporary storage. Also the values $f(a)$ and $f(b)$ need to have been saved. Then compute $I_{2^k}^{(0)}$ from $I_{2^{k-1}}^{(0)}$ and 2^{k-1} new function values. Using (5.110), (5.112), and (5.113), compute the next row in the table, including $J_k(f)$. Compare $J_k(f)$ and $J_{k-1}(f)$ to see if there is sufficient accuracy to accept $J_k(f)$ as an accurate approximation to $I(f)$.

For an analysis of Romberg integration, see Bauer, Rutishauser, and Stiefel [B1]. A further discussion is given in Davis and Rabinowitz [D2], including some variations and a Fortran program. Like Gaussian quadrature and Patterson's method, Romberg integration is rapidly convergent; although it is not as fast as the former methods. It is often more convenient because of using evenly spaced node points.

Example Evaluate the integral (5.19), used previously in Sections 5.1 and 5.3, using the Romberg integration. The results are given in Table 5.16; compare them with the earlier results in Tables 5.1, 5.2, and 5.9. The Romberg integration is much superior to Simpson's rule, Table 5.2; but Gaussian quadrature is still the most rapidly convergent method.

Table 5.16 Example of Romberg integration for (5.19)

k	Nodes	$J_k(f)$	Error
1	3	-11.59283955342	$-4.8E-1$
2	5	-12.01108431754	$-5.9E-2$
3	9	-12.07042041287	$7.4E-5$
4	17	-12.07034720873	$8.9E-7$
5	33	-12.07034631632	$-7.0E-11$
6	65	-12.07034631639	$< 5.0E-12$

5.6. Adaptive Numerical Integration

Many integrands vary in their smoothness at different points of the interval of integration $[a,b]$. For example, with

$$I = \int_0^1 \sqrt{x} \; dx$$

the integrand has infinite slope at $x = 0$, but the function is well behaved at points x near 1. Most numerical integration methods use a uniform grid of node points, that is, the density of node points is about equal throughout the integration interval. This includes composite Newton-Cotes formulas, Gaussian quadrature, Patterson's method, and Romberg integration. When the integrand is badly behaved at some point α in the interval $[a,b]$, many node points must be placed near α to compensate for this. But this forces many more node points than necessary to be used at all other parts of $[a,b]$. Adaptive integration attempts to place node points according to the behavior of the integrand, with the density of node points being greater near points of bad behavior.

We will explain the basic concept of adaptive integration using a simplified *adaptive Simpson's rule*. To see more precisely why variable spacing is necessary, consider Simpson's rule with such spacing of the nodes:

$$I(f) = \sum_{j=1}^{n/2} \int_{x_{2j-2}}^{x_{2j}} f(x)\,dx \approx I_n(f) = \sum_{j=1}^{n/2} \left(\frac{x_{2j} - x_{2j-2}}{6} \right)(f_{2j-2} + 4f_{2j-1} + f_{2j})$$

with $x_{2j-1} = (x_{2j-2} + x_{2j})/2$. Using (5.24),

$$I(f) - I_n(f) = -\frac{1}{2880} \sum_{j=1}^{n/2} (x_{2j} - x_{2j-2})^5 f^{(4)}(\xi_j) \tag{5.114}$$

with $x_{2j-2} \leq \xi_j \leq x_{2j}$. Clearly, you want to choose $x_{2j} - x_{2j-2}$ according to the size of $f^{(4)}(\xi_j)$, which is unknown in general. If $f^{(4)}(x)$ varies greatly in magnitude, you do not want even spacing on $[a,b]$.

As notation, introduce

$$I_{\alpha,\beta} = \int_\alpha^\beta f(x)\,dx$$

$$I_{\alpha,\beta}^{(1)} = \frac{h}{3}\left[f(\alpha) + 4f\left(\frac{\alpha+\beta}{2}\right) + f(\beta)\right] \qquad h = \frac{\beta-\alpha}{2}$$

$$I_{\alpha,\beta}^{(2)} = I_{\alpha,\gamma}^{(1)} + I_{\gamma,\beta}^{(1)} \qquad \gamma = \frac{\beta+\alpha}{2}$$

To describe the adaptive algorithm for computing

$$I = \int_a^b f(x)\,dx$$

we use a recursive definition. Suppose that $\epsilon > 0$ is given, and that we want to find an approximate integral $\tilde{I}$ for which

$$|I - \tilde{I}| \leq \epsilon$$

Begin by setting $\alpha = a$, $\beta = b$. Compute $I_{\alpha,\beta}^{(1)}$ and $I_{\alpha,\beta}^{(2)}$. If

$$\left|I_{\alpha,\beta}^{(2)} - I_{\alpha,\beta}^{(1)}\right| \leq \epsilon \tag{5.115}$$

then accept $I_{\alpha,\beta}^{(2)}$ as the adaptive integral approximation to $I_{\alpha,\beta}$. Otherwise set the adaptive integral for $I_{\alpha,\beta}$ equal to the sum of the adaptive integrals for $I_{\alpha,\gamma}$ and $I_{\gamma,\beta}$, $\gamma=(\alpha+\beta)/2$, each computed with an error tolerance of $\epsilon/2$.

In an actual implementation as a computer program, many extra limitations are included as safeguards; and the error estimation is usually much more sophisticated. All function evaluations are handled carefully in order to ensure that the integrand is never evaluated twice at the same point. This requires a clever stacking procedure for those values of $f(x)$ that must be temporarily stored because they will be needed again later in the computation. There are many small modifications that can be made to improve the performance of the program; but generally a great deal of experience and empirical investigation is necessary first. And for that and other reasons, it is recommended that standard well-tested adaptive procedures be used, for example, De Boor [D3], Hillstrom [H1], and Robinson [R2], in preference to users trying to write their own. This will be discussed further at the end of the section.

Example Consider using the above simple-minded adaptive Simpson procedure to evaluate

$$I=\int_0^1 \sqrt{x}\ dx$$

with $\epsilon=.005$ on $[0,1]$. The final intervals $[\alpha,\beta]$ and integrals $I_{\alpha,\beta}^{(2)}$ are given in Table 5.17. The column labeled "ϵ" gives the error tolerance used in the test (5.115), which bounds the error in $I_{\alpha,\beta}^{(1)}$. The error estimated for $I_{\alpha,\beta}^{(1)}$ on $[0,.0625]$ was incorrectly too small; but it was accurate for the remaining subintervals. The value used to estimate $I_{\alpha,\beta}$ is actually $I_{\alpha,\beta'}^{(2)}$, and it is sufficiently accurate on all subintervals. The total integral, obtained by summing over all $I_{\alpha,\beta}^{(2)}$, is

$$\tilde{I}=.666505 \qquad I-\tilde{I}=1.6E-4$$

and the calculated bound is

$$|I-\tilde{I}|\leq 3.3E-4$$

obtained by summing the column labeled $|I^{(2)}-I^{(1)}|$. Note that the error is concentrated on the first subinterval, as could have been predicted from the behaviour of the integrand near $x=0$.

Table 5.17 Adaptive Simpson's method example

$[\alpha,\beta]$	$I^{(2)}$	$I-I^{(2)}$	$I-I^{(1)}$	$|I^{(2)}-I^{(1)}|$	ϵ
$[0,.0625]$	.010258	$1.6E-4$	$4.5E-4$	$2.9E-4$	.0003125
$[.0625,.0125]$	.019046	$1.2E-7$	$1.1E-6$	$1.0E-6$	.0003125
$[.125,.25]$	.053871	$4.5E-7$	$3.6E-6$	$4.0E-6$	.000625
$[.25,.5]$	.152368	$9.3E-7$	$1.1E-5$	$1.0E-5$	.00125
$[.5,1.0]$	.430962	$2.4E-6$	$3.0E-5$	$2.8E-5$	.0025

Example   Consider evaluating

$$I = \int_0^1 \ln(x)\,dx = -1 \tag{5.116}$$

in which the integrand is set equal to zero at $x=0$, somewhat arbitrarily. In the Table 5.18, the adaptive programs were requested to obtain an accuracy of $\epsilon = 1.0E-6$. Clearly, the adaptive programs are superior to the standard Simpson rule for this singular integral. Also, the adaptive Boole's rule is usually superior to the adaptive Simpson's rule, except when the error tolerance ϵ is fairly large, for example, $\epsilon = .01$. The adaptive programs were taken from Hillstrom [H1].

Table 5.18   Integration of (5.116)

Numerical method	Error attained	Integral evaluations
Simpson rule (5.25)	3.1E−3	1025
Adaptive Simpson rule	1.6E−8	341
Adaptive Boole's rule	5.0E−8	241

CADRE   One of the most widely used adaptive integration programs is CADRE (Cautious Adaptive Romberg Extrapolation), written by DeBoor [D3]. It includes a means of recognizing algebraic singularities at the endpoints of the integration interval. And the asymptotic formulas of Lyness and Ninham [L2], given in (5.97) in a special case, are used to produce a more rapidly convergent integration method, again based on repeated Richardson extrapolation. The routine has been found empirically to be both quite reliable and efficient.

Example   We use the six integrals used in the example for QSUB, given at the end of Section 5.4. Compare the results for CADRE, given in Table 5.19, with the earlier results for QSUB. An extra integral is included,

$$I_7 = \int_{-9}^{10000} \frac{dx}{\sqrt{|x|}} = 206$$

showing CADRE to be unreliable for an integrand with a singularity internal to the integration interval. In the table, the error preceded by A is the actual relative error, and that preceded by P is the relative error predicted by CADRE.

The entries marked with (*) are cases in which the predicted error was greater than the requested error, although the actual error still fell within the desired error tolerance. Those entries marked by (**), all for I_7, are cases where the predicted error was smaller than the actual error, sometimes significantly so. But for two of these cases (requested errors of $5.0E-5$ and $5.0E-8$), the program CADRE noted in an error message that the predicted errors were

Table 5.19 Examples of DeBoor's CADRE program

			Desired relative error			
	5.0E−2		5.0E−5		5.0E−8	
	RE	IE	RE	IE	RE	IE
I_1	$A = 3.47E−6$	75	$A = 2.64E−7$	73	$A = 4.94E−10$	193
	$P = 7.37E−4$		$P = 1.29E−5$		$P = 1.89E−8$	
I_2	$A = 4.13E−5$	9	$A = 1.39E−6$	17	$A = 9.57E−10$	129
	$P = 1.15E−2$		$P = 1.25E−5$		$P = 9.85E−9$	
I_3	$A = 5.43E−4$	9	$A = 4.85E−8$	33	$A = 2.97E−9$	65
	$P = 1.23E−2$		$P = 6.65E−8$		$P = 4.30E−9$	
I_4	$A = 3.16E−4$	81	$A = 2.14E−7$	185	$A = 5.32E−9$	257
	$P = 1.53E−2$		$P = 3.97E−5$		$P = 6.10E−8*$	
I_5	$A = 5.69E−5$	76	$A = 2.48E−6$	180	$A = 3.81E−9$	516
	$P = 1.87E−2$		$P = 5.90E−5*$		$P = 1.11E−7*$	
I_6	$A = 3.56E−6$	1054	$A = 2.42E−7$	1441	$A = 1.78E−10$	3313
	$P = 3.10E−2$		$P = 1.28E−4*$		$P = 1.05E−7*$	
I_7	$A = 3.87E−2**$	65	$A = 4.75E−4**$	269	$A = 4.74E−4**$	529
	$P = 1.79E−2$		$P = 2.23E−5$		$P = 4.77E−8$	

probably wrong and that the user should be cautious in accepting the results.

The results for I_5 and I_7 reinforce a comment made earlier in connection with both Gaussian quadrature and Patterson's method. If $x = c$ is a point of singular behavior of the integrand $f(x)$, and if it is interior to the integration interval $[a,b]$, then the integral should be split into integrals over smaller intervals with $x = c$ an endpoint,

$$\int_a^b f(x)\,dx = \int_a^c f(x)\,dx + \int_c^b f(x)\,dx$$

This will improve greatly the performance of most numerical integration methods, including CADRE. The results for I_6 show the superiority of QSUB (and Gaussian quadrature) for highly oscillatory integrands.

Automatic programs There are now many automatic programs that have been published, both adaptive and nonadaptive. See [Davis and Rabinowitz, D2, Chapter 6] and its references for a much more detailed introduction to automatic numerical integration. One of the most promising adaptive routines is being developed by Krogh and Snyder [K3]. It is based on Patterson's method, but is more sophisticated in its adaptive procedure than that of QSUBA in Patterson [P2]. Preliminary results indicate it is somewhat superior to CADRE, both in efficiency and reliability. The program by Robinson [R2] is based on a unique adaptive rule using three point Gauss-Legendre quadrature. It is very reliable, although somewhat less efficient than CADRE. Its program size is smaller than CADRE.

With the growth of automatic numerical integration, and more generally of automatic programs in all areas of numerical analysis, people have begun to study how best to compare and certify algorithms. For numerical integration, see Keast [K2], Lyness and Kaganove [L1], and Dixon [D6], and the references contained therein. The area of certification is very important, especially the determination of the class of problems for which an algorithm is reliable. Given any program for numerical integration, a person knowledgeable about it can create an integral for which the program will fail. The certification process should establish the class of problems for which the program is reliable and communicate that information to the potential user of the routine. Efficiency is also important, but there is a tradeoff with reliability, and more importance is usually given to reliability.

5.7 Singular Integrals

The term *singular integral* generally refers to any integral for which the standard methods of Sections 5.1 through 5.5 either do not apply or else lead to very slow convergence. This is usually caused by (1) an infinite interval of integration, and/or (2) a singularity in either the integrand or one of its low-order derivatives. Examples of the slower convergence caused by (2) have been given in previous sections. In this section, we consider a variety of techniques for treating these singular integrals.

Change of the variable of integration   We illustrate the importance of this idea with several examples. For

$$I = \int_0^b \frac{f(x)\,dx}{\sqrt{x}}$$

with $f(x)$ a function with several continuous derivatives, let $x = u^2$, $0 \le u \le \sqrt{b}$. Then

$$I = 2\int_0^{\sqrt{b}} f(u^2)\,du$$

This integral has a smooth integrand and standard techniques can be applied to it.

Similarly,

$$\int_0^1 \sin(x)\sqrt{1-x^2}\,dx = 2\int_0^1 u^2 \sqrt{2-u^2}\,\sin(1-u^2)\,du$$

using the change of variable $u = \sqrt{1-x}$. The second integrand has an infinite number of continuous derivatives on $[0, 1]$, whereas the derivative of the first integrand was singular at $x = 1$.

For an infinite interval of integration, the change of variable technique is occasionally useful. Suppose

$$I = \int_1^\infty \frac{f(x)}{x^p} dx \qquad p > 1 \qquad (5.117)$$

with $f(x) \to c \neq 0$ as $x \to \infty$. Also assume that $f(x)$ is smooth on $[1, \infty)$. Then use the change of variable

$$x = \frac{1}{u^\alpha} \qquad dx = \frac{-\alpha}{u^{1+\alpha}} du \qquad \text{for some } \alpha > 0$$

Then

$$I = \alpha \int_0^1 u^{p\alpha} f\left(\frac{1}{u^\alpha}\right) \frac{du}{u^{1+\alpha}} = \alpha \int_0^1 u^{(p-1)\alpha - 1} f\left(\frac{1}{u^\alpha}\right) du$$

Maximize the smoothness of the new integrand at $u = 0$ by picking α to produce a large value for the exponent $(p-1)\alpha - 1$. For example, with

$$I = \int_1^\infty \frac{f(x) dx}{x\sqrt{x}}$$

the change of variable $x = 1/u^6$ leads to

$$I = 6 \int_0^1 u^2 f\left(\frac{1}{u^6}\right) du$$

If we assume a behaviour at $x = \infty$ of

$$f(x) = c_0 + \frac{c_1}{x} + \frac{c_2}{x^2} + \cdots$$

then

$$u^2 f\left(\frac{1}{u^6}\right) = c_0 u^2 + c_1 u^8 + c_2 u^{14} + \cdots$$

and I has a smooth integrand at $u = 0$.

An interesting idea has been given in Iri, Moriguti, and Takasawa [11] (see DeDoncker and Piessens [D4] for an English language description, including programs) to deal with endpoint singularities in the integral

$$I = \int_a^b f(x) dx \qquad (5.118)$$

Define

$$\psi(t) = \exp\left(-\frac{c}{1-t^2}\right) \tag{5.119}$$

$$\varphi(t) = a + \frac{b-a}{\gamma} \int_{-1}^{t} \psi(u)\,du \qquad -1 \le t \le 1$$

where c is a positive constant and

$$\gamma = \int_{-1}^{1} \psi(u)\,du$$

In DeDoncker and Piessens [D4], $c=4$ gives very satisfactory results for all examples. Note that as t varies from -1 to 1, $\varphi(t)$ varies from a to b. Using $x = \varphi(t)$ as a change of variable in (5.118), we obtain

$$I = \int_{-1}^{1} f(\varphi(t))\varphi'(t)\,dt \tag{5.120}$$

The function $\varphi'(t) = \psi(t)$ is infinitely differentiable on $[-1, 1]$, and it and all of derivatives are zero at $t = \pm 1$. In (5.120), the integrand and all of derivatives will vanish at $t = \pm 1$ for virtually all functions of interest. Using the error formula (5.93) for the trapezoidal rule on $[-1, 1]$, it can be seen that the trapezoidal rule will converge very rapidly when applied to (5.120): This is the basic idea used in [D4], and it is quite successful when applied with some care. Integrals over $[0, \infty)$ are treated by first using the change of variable $x = (1+u)/(1-u)$, $-1 \le u \le 1$, followed by the change of variable $u = \varphi(t)$. See [D4] for additional details, programs, and numerical examples.

Gaussian quadrature In Section 5.3, a general theory was given for the existence of Gaussian quadrature formulas

$$\int_{a}^{b} w(x)f(x)\,dx \approx \sum_{1}^{n} w_{j,n}f(x_{j,n}) \qquad n \ge 1 \tag{5.121}$$

with $w(x)$ a nonnegative weight function. The weights and nodes for many different forms of $w(x)$ are given in Stroud and Secrest [S3], and other tables cover additional weight functions. The tables of [S3] cover integrals of the form

$$\int_{0}^{\infty} e^{-x}f(x)\,dx \qquad \int_{-\infty}^{\infty} e^{-x^2}f(x)\,dx \qquad \int_{0}^{1} \ln\left(\frac{1}{x}\right)f(x)\,dx$$

and many others.

One especially easy case is for the singular integral

$$I(f) = \int_{-1}^{1} \frac{f(x)\,dx}{\sqrt{1-x^2}} \tag{5.122}$$

With this weight function, the orthogonal polynomials are the Chebyshev polynomials $\{T_n(x), n \geq 0\}$. Thus the integration nodes in (5.121) are given by

$$x_{j,n} = \cos\left(\frac{2j-1}{2n}\pi\right) \qquad j = 1, \ldots, n \tag{5.123}$$

and from (5.48), the weights are

$$w_{j,n} = \frac{\pi}{n} \qquad j = 1, \ldots, n$$

Using the formula for the error in (5.47), the Gaussian quadrature formula for (5.122) is given by

$$\int_{-1}^{1} \frac{f(x)\,dx}{\sqrt{1-x^2}} = \frac{\pi}{n} \sum_{j=1}^{n} f(x_{j,n}) + \frac{2\pi}{2^{2n}(2n)!} f^{(2n)}(\eta) \tag{5.124}$$

for some $-1 < \eta < 1$.

This formula is related to the composite midpoint rule (5.41). Make the change of variable $x = \cos\theta$ in (5.124) to obtain

$$\int_0^{\pi} f(\cos\theta)\,d\theta = \frac{\pi}{n} \sum_{j=1}^{n} f(\cos\theta_{j,n}) + E \tag{5.125}$$

where $\theta_{j,n} = (2j-1)\pi/2n$. Thus Gaussian quadrature for (5.122) is equivalent to the composite midpoint rule applied to integral on the left in (5.125). Like the trapezoidal rule, the midpoint rule has an error expansion very similar to that given in (5.90) using the Euler-MacLaurin formula. Corollary 1 to Theorem 5.3 also is valid, showing the composite midpoint rule to be highly accurate for periodic functions. This is reflected in the high accuracy of (5.125). Thus Gaussian quadrature for (5.122) results in a formula that would have been reasonable from the asymptotic error expansion for the composite midpoint rule applied to the integral on the left of (5.125).

Analytic treatment of singularity   Divide the interval of integration into two parts, one containing the singular point, which is to be treated analytically. For example, consider

$$I = \int_0^b f(x)\ln x\,dx = \int_0^{\epsilon} + \int_{\epsilon}^b f(x)\ln x\,dx \equiv I_1 + I_2$$

Assuming $f(x)$ is smooth on $[\epsilon, b]$, apply a standard technique to the evaluation of I_2. For $f(x)$ about zero, assume it has a convergent Taylor series on $[0, \epsilon]$. Then

$$I_1 = \int_0^\epsilon f(x) \ln x \, dx = \int_0^\epsilon \left(\sum_0^\infty a_j x^j \right) \ln x \, dx$$

$$= \sum_0^\infty a_j \frac{\epsilon^{j+1}}{j+1} \left(\ln \epsilon - \frac{1}{j+1} \right)$$

For example, with

$$I = \int_0^{4\pi} \cos(x) \ln x \, dx$$

define

$$I_1 = \int_0^{.1} \cos(x) \ln x \, dx \qquad \epsilon = .1$$

$$= \epsilon(\ln \epsilon - 1) - \frac{\epsilon^3}{6} \left(\ln \epsilon - \frac{1}{3} \right) + \frac{\epsilon^5}{600} \left(\ln \epsilon - \frac{1}{5} \right) - \cdots$$

This is an alternating series, and thus it is clear that using the first three terms will give a very accurate value for I_1. A standard method can be applied to I_2 on $[.1, 4\pi]$.

Similar techniques can be used with infinite intervals of integration, truncating the integral over $[b, \infty)$ for some large value of b. It will not be developed here.

Product integration   Let $I(f) = \int_a^b w(x) f(x) \, dx$ with a near-singular or singular weight function $w(x)$ and a smooth function $f(x)$. The main idea is to produce a sequence of functions $f_n(x)$ for which

1. $\|f - f_n\|_\infty = \underset{a \le x \le b}{\text{Max}} |f(x) - f_n(x)| \to 0$ as $n \to \infty$

2.   The integrals

$$I_n(f) \equiv \int_a^b w(x) f_n(x) \, dx \qquad (5.126)$$

can be fairly easily evaluated.

This generalizes the schema (5.2) to (5.4) of the introduction. For the error,

$$|I(f) - I_n(f)| \leq \int_a^b |w(x)||f(x) - f_n(x)| dx$$

$$\leq \|f - f_n\|_\infty \int_a^b |w(x)| dx \tag{5.127}$$

Thus $I_n(f) \to I(f)$ as $n \to \infty$, and the rate of convergence is at least as rapid as that of $f_n(x)$ to $f(x)$ on $[a,b]$.

Within the above framework, the product integration methods are usually defined by using piecewise polynomial interpolation to define $f_n(x)$ from $f(x)$. To illustrate the main ideas, while keeping the algebra simple, we will define the product trapezoidal method for evaluating

$$I(f) = \int_0^b f(x) \ln x \, dx \tag{5.128}$$

Let $n \geq 1$, $h = b/n$, $x_j = jh$ for $j = 0, 1, \ldots, n$. Define $f_n(x)$ as the piecewise linear function interpolating to $f(x)$ on the nodes $x_0, x_1, \ldots, x_n$. For $x_{j-1} \leq x \leq x_j$, define

$$f_n(x) = \frac{1}{h} \left[(x_j - x) f(x_{j-1}) + (x - x_{j-1}) f(x_j) \right] \tag{5.129}$$

for $j = 1, 2, \ldots, n$. From (3.9) it is straightforward to show

$$\|f - f_n\|_\infty \leq \frac{h^2}{8} \|f''\|_\infty$$

provided $f(x)$ is twice continuously differentiable for $0 \leq x \leq b$. From (5.127) we obtain the error bound

$$|I(f) - I_n(f)| \leq \frac{h^2}{8} \|f''\|_\infty \int_0^b |\ln(x)| dx \tag{5.130}$$

This method of defining $I_n(f)$ is similar to the regular trapezoidal rule (5.13). The rule (5.13) could also have been obtained by integrating the above function $f_n(x)$, but the weight function would have been simply $w(x) \equiv 1$. We can easily generalize the above by using higher-degree piecewise polynomial interpolation. The use of piecewise quadratic interpolation to define $f_n(x)$ leads to a formula $I_n(f)$ called the product Simpson's rule. And using the same reasoning as led to (5.30), it can be shown that

$$|I(f) - I_n(f)| \leq \frac{\sqrt{3}}{27} h^3 \|f'''\|_\infty \int_0^b |\ln(x)| dx \tag{5.131}$$

Higher-order formulas can be obtained by using even higher-degree interpolation.

For the computation of $I_n(f)$ using (5.129),

$$I_n(f) = \sum_{j=1}^{n} \int_{x_{j-1}}^{x_j} (\ln x) \left[\frac{(x_j - x)f(x_{j-1}) + (x - x_{j-1})f(x_j)}{h} \right] dx$$

$$= \sum_{k=0}^{n} w_k f(x_k)$$

$$w_0 = \frac{1}{h} \int_{x_0}^{x_1} (x_1 - x) \ln x \, dx \qquad w_n = \frac{1}{h} \int_{x_{n-1}}^{x_n} (x - x_{n-1}) \ln x \, dx$$

$$w_j = \frac{1}{h} \int_{x_{j-1}}^{x_j} (x - x_{j-1}) \ln x \, dx + \frac{1}{h} \int_{x_j}^{x_{j+1}} (x_{j+1} - x) \ln x \, dx \qquad j = 1, \ldots, n-1$$

$$(5.132)$$

The calculation of these weights can be simplified considerably. Making the change of variable $x - x_{j-1} = uh$, $0 \le u \le 1$, we have

$$\frac{1}{h} \int_{x_{j-1}}^{x_j} (x - x_{j-1}) \ln x \, dx = h \int_0^1 u \ln\left[(j - 1 + u)h \right] du$$

$$= \frac{h}{2} \ln h + h \int_0^1 u \ln(j - 1 + u) \, du$$

and

$$\frac{1}{h} \int_{x_{j-1}}^{x_j} (x_j - x) \ln x \, dx = h \int_0^1 (1 - u) \ln\left[(j - 1 + u)h \right] du$$

$$= \frac{h}{2} \ln h + h \int_0^1 (1 - u) \ln(j - 1 + u) \, du$$

Define

$$\psi_1(k) = \int_0^1 u \ln(u + k) \, du \qquad \psi_2(k) = \int_0^1 (1 - u) \ln(u + k) \, du \qquad (5.133)$$

for $k = 0, 1, 2, \ldots$. Then

$$w_0 = \frac{h}{2} \ln h + h\psi_2(0) \qquad w_n = \frac{h}{2} \ln h + h\psi_1(n - 1)$$

$$w_j = h \ln h + h\left[\psi_1(j - 1) + \psi_2(j) \right] \qquad j = 1, 2, \ldots, n-1 \qquad (5.134)$$

The functions $\psi_1(k)$ and $\psi_2(k)$ do not depend on h, b, or n. They can be calculated and stored in a table for use with a variety of values of b and n. For example, a table of $\psi_1(k)$ and $\psi_2(k)$ for $k=0,1,\ldots,99$ can be used with any $b>0$ and with any $n\leq 100$. Once the table of the values $\psi_1(k)$ and $\psi_2(k)$ has been calculated, the cost of using the product trapezoidal rule is no greater than the cost of any other integration rule.

The integrals $\psi_1(k)$ and $\psi_2(k)$ in (5.133) can be evaluated explicitly; some values are given in Table 5.20.

Table 5.20 Weights for Product Trapezoidal Rule

k	$\psi_1(k)$	$\psi_2(k)$
0	$-.250$	$-.750$
1	.250	.1362943611
2	.4883759281	.4211665768
3	.6485778545	.6007627239
4	.7695705457	.7324415720
5	.8668602747	.8365069785
6	.9482428376	.9225713904
7	1.018201652	.9959596385

Example Compute $I=\int_0^1 (1/(x+2))\ln x\,dx \doteq -.4484137$. The computed values are given in Table 5.21. The computed rate of convergence is in agreement with the order of convergence of (5.130).

Many types of interpolation may be used to define $f_n(x)$, but most applications to date have used piecewise polynomial interpolation on evenly spaced node points. Other weight functions may be used, for example

$$w(x)=x^\alpha \qquad \alpha>-1 \qquad x\geq 0$$

and again the weights can be reduced to a fairly simple formula similar to

Table 5.21 Example of Product Trapezoidal rule

n	I_n	$I-I_n$	**Ratio**
1	$-.4583333$	.00992	
			3.10
2	$-.4516096$	.00320	
			3.61
4	$-.4493011$	.000887	
			3.82
8	$-.4486460$	.000232	

(5.134). For an irrational value of α, say

$$w(x) = \frac{1}{x^{\sqrt{2}} - 1}$$

a change of variables can no longer be used to remove the singularity in the integral. Also, one of the major applications of product integration is to integral equations in which the kernel function has an algebraic and/or logarithmic singularity. For such equations, changes of variable are no longer possible, even with square root singularities. For example, consider the equation

$$\lambda \varphi(x) - \int_a^b \frac{\varphi(y) \, dy}{|x - y|^{1/2}} = f(x) \qquad a \leq x \leq b$$

with λ, a, b, and f given and φ the desired unknown function. Product integration leads to efficient procedures for such equations, provided $\varphi(y)$ is a smooth function; see Atkinson [A2].

For complicated weight functions in which the weights w_j can no longer be calculated, it is often possible to modify the problem to one in which product integration is still easily applicable. This will be examined using an example.

Example   Consider $I = \int_0^\pi e^x \ln(\sin x) \, dx$. The integrand has a singularity at both $x = 0$ and $x = \pi$. Use

$$\ln(\sin x) = \ln\left[\frac{\sin(x)}{x(\pi - x)}\right] + \ln(x) + \ln(\pi - x)$$

and this gives

$$I = \int_0^\pi e^x \ln\left[\frac{\sin x}{x(\pi - x)}\right] dx + \int_0^\pi e^x \ln x \, dx + \int_0^\pi e^x \ln(\pi - x) \, dx$$

$$\equiv I_1 + I_2 + I_3$$

I_1 has an infinitely differentiable integrand, and any standard numerical method will perform well. I_2 has already been discussed, with $w(x) = \ln x$. For I_3, use a change of variable to write

$$I_3 = \int_0^\pi e^x \ln(\pi - x) \, dx = \int_0^\pi e^{\pi - z} \ln z \, dz$$

Combining with I_2,

$$I_2 + I_3 = \int_0^\pi (\ln x)\left[e^x + e^{\pi - x}\right] dx$$

to which the preceding work applies. By such manipulations, the applicability of cases the $w(x) = \ln x$ and $w(x) = x^\alpha$ is much greater than might first be imagined.

For an asymptotic error analysis of product integration, see the work of deHoog and Weiss [D5], in which some generalizations of the Euler-MacLaurin expansion are derived. Using their results, it can be shown that for the case of $w(x) = \log(x)$, the error in the product Simpson rule is $O(h^4 \ln h)$. Thus the bound (5.131) based on the interpolation error $f(x) - f_n(x)$ does not predict the correct rate of convergence. This is similar to the result (5.21) for the Simpson rule error, in which the error was smaller than the use of quadratic interpolation would lead us to believe.

Discussion of the literature

The topic of numerical integration is one of the oldest in numerical analysis, and there is a very large literature. Nonetheless, new papers continue to appear with increasing frequency. Many of the results give methods for special classes of problems, for example, oscillatory integrands and others are a response to changes in the computing area. The best survey of numerical integration is the large and detailed work of Davis and Rabinowitz [D2]. It contains a very extensive bibliography, a set of computer programs, and many methods for quite special forms of integrals. The special area of multidimensional integration is much less developed, and is presently an active area of research. The best general reference is Stroud [S2]. A very general program embodying several new programming ideas is given in Kahaner and Wells [K1].

One of the important current trends in numerical analysis is towards automatic computer programs, including monitoring of the error. Automatic numerical integration is a very active area, and a survey of popular techniques is given in Dixon [D6] and [D2, Chapter 6]. The popular automatic programs QSUB of Patterson [P2] and CADRE of DeBoor [D3] were discussed earlier in Sections 5.4 and 5.6, respectively. Because of the increasing numbers of such programs, work is beginning on standardized testing and certification procedures; see the ideas in Keast [K2] and Lyness and Kaganove [L1].

For hand computation, Simpson's rule is still popular because of its simplicity. But serious consideration should also be given to Gaussian quadrature because of its much greater accuracy. And the nodes and weights can be found conveniently in either Stroud and Secrest [S3] or Abramowitz and Stegun [A1, Chapter 25].

Bibliography

[A1]   Abramowitz, M. and I. Stegun, editors, *Handbook of Mathematical Functions*, National Bureau of Standards, U. S. Government Printing Office, 1964.

[A2]   Atkinson, K., The numerical solution of Fredholm integral equations of the second kind, *SIAM J. Num. Anal.*, **4**, 1967, pp. 337-348.

[B1]   Bauer, F., H. Rutishauser, and E. Stiefel, New aspects in numerical quadrature in *Experimental Arithmetic, High Speed Computing, and Mathematics*, pp. 199-218, Amer. Math. Soc., Providence, R. I., 1963.

[C1]   Clenshaw, C. and A. Curtis, A method for numerical integration on an automatic computer, *Numer. Math.*, **2**, 1960, pp. 197-205.

[D1]   Davis, P., *Interpolation and Approximation*, Blaisdell, New York, 1963.

[D2]   Davis, P. and P. Rabinowitz, *Methods of Numerical Integration*, 2nd ed., Academic Press, New York, 1975.

[D3]   DeBoor, C., CADRE: An algorithm for numerical quadrature in *Mathematical Software*, John Rice, ed., pp. 201-209, Academic Press, New York, 1971.

[D4]   De Doncker, E. and R. Piessens, Automatic computation of integrals with singular integrand over a finite or infinite range, *Tech. Rep. TW22*, Applied Mathematics Division, Catholic University of Leuven, Belgium, 1975.

[D5]   De Hoog, F. and R. Weiss, Asymptotic expansions for product integration, *Math. Comp.*, **27**, 1973, pp. 295-306.

[D6]   Dixon, V., Numerical quadrature: A survey of the available algorithms in *Software for Numerical Mathematics*, D. Evans, ed., pp. 105-137, Academic Press, New York, 1974.

[H1]   Hillstrom, K. E., Comparison of several adaptive Newton-Cotes quadrature routines in evaluating definite integrals with peaked integrands, *Tech. Rep. ANL-7511*, Argonne National Lab., 1968.

[I1]   Iri, M., S. Moriguti, and Y. Takasawa, On a numerical integration formula (in Japanese), *Journal of Research Inst. for Math. Sci.*, Kyoto Univ., **91**, 1970, p. 82.

[I2]   Isaacson, E., and H. Keller, *Analysis of Numerical Methods*, John Wiley, New York, 1966.

[K1]   Kahaner, D. and M. Wells, An algorithm for N-dimensional adaptive quadrature using advanced programming techniques, *Los Alamos Laboratory Tech. Rep. 76-2310*, 1976.

[K2]   Keast, P., The evaluation of one-dimensional quadrature routines, *Tech. Rep. 83*, Computer Science Dept., University of Toronto, Canada, 1975.

[**K3**] Krogh, F. and W. Snyder, Test results with quadrature software, Tech. Rep. CL77-576, Computer Laboratory, Jet Propulsion Laboratory, California Institute of Technology, Pasadena, 1977.

[**L1**] Lyness, J. and J. Kaganove, Comments on the nature of automatic quadrature routines, *ACM Trans. Math. Software*, pp. 65-81.

[**L2**] Lyness, J. and B. Ninham, Numerical quadrature and asymptotic expansions, *Math. Comp.*, **21**, 1967, pp. 162-178.

[**P1**] Patterson, T., The optimum addition of points to quadrature formulae, *Math. Comp.*, **22**, 1968, pp. 847-856 plus a microfiche. Microfiche correction in Vol. 23, 1969.

[**P2**] Patterson, T., Algorithm 468: Algorithm for numerical integration over a finite interval, *Comm. ACM*, **16**, 1973, pp. 694-699.

[**R1**] Ralston, A., *A First Course in Numerical Analysis*, McGraw-Hill, New York, 1965.

[**R2**] Robinson, I., An algorithm for automatic integration using the adaptive Gaussian technique, *Australian Computer Jour.*, **8**, 1976, pp. 106-115.

[**S1**] Steffensen, J. F., *Interpolation*, Chelsea, New York, 1950.

[**S2**] Stroud, A., *Approximate Calculation of Multiple Integrals*, Prentice-Hall, Englewood Cliffs, N.J., 1971.

[**S3**] Stroud, A., and D. Secrest, *Gaussian Quadrature Formulas*, Prentice-Hall, Englewood Cliffs, N.J., 1966.

Problems

1. Write a program to evaluate $I = \int_a^b f(x)\,dx$ using the trapezoidal rule with n subdivisions, calling the result I_n. Use the program to calculate the following integrals with $n = 2, 4, 8, 16, \ldots, 512$.

   (a) $\displaystyle\int_0^1 e^{-x^2}\,dx$ (b) $\displaystyle\int_0^1 x^{5/2}\,dx$ (c) $\displaystyle\int_{-4}^4 \frac{dx}{1+x^2}$

   (d) $\displaystyle\int_0^{2\pi} \frac{dx}{2+\cos(x)}$ (e) $\displaystyle\int_0^\pi e^x \cos(4x)\,dx$

   Analyze empirically the rate of convergence of I_n to I by calculating the ratios

$$R_{4n} = \frac{I_{2n} - I_n}{I_{4n} - I_{2n}}$$

As justification, see (5.101).

2. Repeat Problem 1 using Simpson's rule.

3. Using the program of Problem 1, determine empirically the rate of convergence of the trapezoidal rule applied to $\int_0^1 x^\alpha \ln x \, dx$, $0 \le \alpha \le 1$, with a range of values of α, say $\alpha = .25, .5, .75,$ and 1.0.

4. (a) Assume that $f(x)$ is continuous and that $f'(x)$ is integrable on $[0, 1]$. Show that the error in the Trapezoidal rule for calculating $\int_0^1 f(x) \, dx$ has the form

$$E_n(f) = \int_0^1 K(t) f'(t) \, dt$$

$$K(t) = (t_{j-1} + t_j)/2 - t \qquad t_{j-1} \le t \le t_j \qquad j = 1, \ldots, n$$

This contrasts with (5.28) in which $f'(x)$ is continuous and $f''(x)$ is integrable.

(b) Apply the result to $f(x) = x^\alpha$ and to $f(x) = x^\alpha \ln x$, $0 < \alpha < 1$. This gives an order of convergence, although it is less than the true order. See Problem 3 and (5.97).

5. Show that when Simpson's rule is applied to $\int_0^{2\pi} \sin(x) \, dx$, there is zero error, assuming no rounding error. Obtain a generalization to other integrals $\int_a^b f(x) \, dx$. Apply it to show that Simpson's rule is exact for $\int_0^{\pi/2} \cos^2(x) \, dx$.

6. Let $P_2(x)$ be the quadratic polynomial interpolating to $f(x)$ at $x = 0, h, 2h$. Use this to derive a numerical integration formula I_h for $I = \int_0^{3h} f(x) \, dx$. Use a Taylor series expansion of $f(x)$ to show

$$I - I_h = \tfrac{3}{8} h^4 f'''(0) + O(h^5)$$

7. For the midpoint integration formula (5.40), derive the Peano kernel error

formula

$$\int_0^h f(x)\,dx - hf\left(\frac{h}{2}\right) = \int_0^h K(t) f''(t)\,dt$$

$$K(t) = \begin{cases} \frac{1}{2}t^2, & 0 \le t \le h/2 \\ \frac{1}{2}(h-t)^2, & h/2 \le t \le h \end{cases}$$

Use this to derive the error term in (5.40).

8. Derive the composite form of Boole's rule, which is given as the $n=4$ entry in Table 5.6. Develop error formulas analogous to those given in (5.23) and (5.24) for Simpson's rule.

9. Repeat Problem 1 using the composite Boole's rule, obtained in Problem 8.

10. Prove the Gauss-Legendre nodes and weights on $[-1,1]$ are symmetrically distributed about the origin $x=0$.

11. Prove that for Gaussian quadrature with weight function $w(x)$ on $[a,b]$,

$$\sum_{j=1}^{n} w_{j,n} = \int_a^b w(x)\,dx$$

12. Derive the two point Gaussian quadrature formula for

$$I(f) = \int_0^1 f(x) \ln\left(\frac{1}{x}\right) dx$$

in which the weight function is $w(x) = \ln(1/x)$.

13. Use Gauss-Legendre quadrature to evaluate $\int_{-4}^{4} dx/(1+x^2)$ with the $n=2,4,6,8$ node-point formulas. Compare with the results in Table 5.7, obtained using the Newton-Cotes formula.

14. For the integral $I = \int_{-1}^{1} \sqrt{1-x^2}\, f(x)\,dx$ with weight $w(x) = \sqrt{1-x^2}$, find explicit formulas for the nodes and weights of the Gaussian quadrature formula. Also give the error formula. Hint: see Problem 16 of Chapter 4.

15. Using the column $e_{n,4}$ of Table 5.10, produce the fourth-order error formula analogous to the second order formula (5.69) for $e_{n,2}$. Compare it with the fourth-order Simpson error formula (5.23).

16. Let $I_n(f)$ denote the n point Gauss-Legendre formula for integrating $I(f) = \int_{-1}^{1} f(x)\,dx$. Show that $I_n(f) \to I(f)$ as $n \to \infty$, for any continuous function $f(x)$. Hint: use the Weierstrass theorem 4.1 to approximate $f(x)$ by a polynomial $p(x)$, with any desired error tolerance. Write $f(x) = p(x) + r(x)$, and say $\|r\|_\infty \le E$. Then analyze the error in integrating $p(x)$ and $r(x)$, using Problem 11 and the positivity of the weights, given in (5.59).

17. Implement the program QSUB or QSUBA of [P2] if your computer center does not have it. Use it to calculate the integrals of Problem 1.

18. (a)   Derive the relation (5.89) for the Bernoulli polynomials $B_j(x)$ and Bernoulli numbers B_j. Show that $B_j = 0$ for all odd integers $j \ge 3$.

     (b)   Derive the identities

$$B_j'(x) = jB_{j-1}(x) \qquad j \ge 4 \text{ and even}$$

$$B_j'(x) = j[B_{j-1}(x) + B_{j-1}] \qquad j \ge 3 \text{ and odd}$$

These can be used to give a general proof of the Euler-MacLaurin formula (5.90).

19. Using the Euler-MacLaurin summation formula (5.95), obtain an estimate of $\zeta(5/4)$, accurate to three decimal places. The Zeta function $\zeta(p)$ is defined preceding (5.98).

20. Obtain the asymptotic error formula for the trapezoidal rule applied to $\int_{0}^{1} \sqrt[4]{x}\, f(x)\,dx$. Use the estimate from Problem 19.

21. Consider the following table of approximate integrals I_n produced using Simpson's rule. Predict the order of convergence of I_n to I,

n	I_n
2	.28451779686
4	.28559254576
8	.28570248748
16	.28571317731
32	.28571418363
64	.28571427643

That is, if $I - I_n \approx c/n^p$, then what is p? Does this appear to be a valid form for the error for this data? Predict a value of c and the error in I_{64}. How large should n be chosen if I_n is to be in error by less than 10^{-11}?

22. For the trapezoidal rule [denoted by $I_n^{(T)}$] for evaluating $I = \int_a^b f(x)\,dx$ we have the asymptotic error formula

$$I - I_n^{(T)} = -\frac{h^2}{12}\left[f'(b) - f'(a)\right] + O(h^4)$$

and for the midpoint formula $I_n^{(M)}$, we have

$$I - I_n^{(M)} = \frac{h^2}{24}\left[f'(b) - f'(a)\right] + O(h^4)$$

provided f is sufficiently differentiable on $[a,b]$. Using these results obtain a new numerical integration formula $\tilde{I}_n$ combining $I_n^{(T)}$ and $I_n^{(M)}$, with a higher order of convergence. Write out the weights to the new formula $\tilde{I}_n$.

23. Assume that the error in an integration formula has the asymptotic expansion

$$I - I_n = \frac{C_1}{n\sqrt{n}} + \frac{C_2}{n^2} + \frac{C_3}{n^2\sqrt{n}} + \frac{C_4}{n^3} + \cdots$$

Generalize the Richardson extrapolation process of Section 5.5. Assume that three values I_n, I_{2n}, and I_{4n} have been computed, and use these to compute an estimate of I with an error of order $1/n^2\sqrt{n}$.

24. Obtain an asymptotic error formula for Simpson's rule, comparable to the Euler-MacLaurin formula (5.90) for the trapezoidal rule. Use (5.90), (5.103), and (5.104).

25. Show that the formula (5.107) for $I_n^{(2)}$ is the composite Boole's rule. See Problem 8.

26. Use an adaptive integration program (e.g., CADRE in [D3]) to calculate the integrals in Problem 1. Compare the results with those of Problems 1, 2, 9, and 17.

27. Evaluate the following singular integrals within the prescribed error tolerance. Do not use an automatic program.

   (a) $\displaystyle\int_0^{4\pi} \cos(u)\ln(u)\,du$ $\epsilon = 10^{-8}$

   (b) $\displaystyle\int_0^{2/\pi} x^2 \sin\left(\frac{1}{x}\right) dx$ $\epsilon = 10^{-3}$

   (c) $\displaystyle\int_0^1 \frac{\cos(x)\,dx}{\sqrt[3]{x^2}}$ $\epsilon = 10^{-8}$

   (d) $\displaystyle\int_0^1 \frac{\sqrt{1 - x^4}\,dx}{x^\alpha}$ $\epsilon = 10^{-5}$ $\alpha = 1 - \dfrac{1}{\pi}$

28. Use an adaptive program (e.g., CADRE) to evaluate the integrals of Problem 27.

29. Use Gauss-Laguerre quadrature with $n=2,4,6,$ and 8 node points to evaluate the following integrals.

(a) $\displaystyle\int_0^\infty e^{-x^2}dx=\frac{\sqrt{\pi}}{2}$ (b) $\displaystyle\int_0^\infty \frac{dx}{1+x^2}=\frac{\pi}{2}$

(c) $\displaystyle\int_0^\infty \frac{\sin(x)}{x}dx=\frac{\pi}{2}$

Hint: for the integral $I=\displaystyle\int_0^\infty g(x)dx$, write it as

$$I=\int_0^\infty e^{-x}f(x)dx \qquad f(x)=e^x g(x)$$

and use the standard Gauss-Laguerre formula,

$$\int_0^\infty e^{-x}f(x)dx\approx \sum_{j=1}^n w_{j,n}f(x_{j,n})$$

Find the necessary nodes and weights from [A1] or [S3].

30. (a) Develop the product trapezoidal integration rule for calculating

$$I(f)=\int_0^b \frac{f(x)\,dx}{x^\alpha} \qquad 0<\alpha<1$$

Put the weights into a convenient form, analogous to (5.133), (5.134) or the product trapezoidal rule when $w(x)=\ln(x)$. Also, give an error formula analogous to (5.130).

(b) Write a simple program to evaluate the following integrals using the results of part (a).

(i) $\displaystyle\int_0^1 \frac{e^x}{x^{1/\pi}}dx$ (ii) $\displaystyle\int_0^1 \frac{dx}{\sin(x^{1/\pi})}$

31. (a) Consider the following two-dimensional numerical integration formula,

$$I=\int_{-h}^h\int_{-h}^h f(x,y)\,dx\,dy$$

$$\approx I_h=\frac{h^2}{3}\left[8f(0,0)+f(h,h)+f(h,-h)+f(-h,h)+f(-h,-h)\right]$$

Show that $I_h=I$ if $f(x,y)$ is any polynomial in x and y of degree ≤ 3. Also, use a Taylor series expansion of $f(x,y)$ to show

$$I - I_h = -\frac{8h^6}{72}\left\{\frac{1}{5}\left[\frac{\partial^4 f(0,0)}{\partial x^4} + \frac{\partial^4 f(0,0)}{\partial y^4}\right] + 2\frac{\partial^4 f(0,0)}{\partial x^2 \partial y^2}\right\} + O(h^8)$$

(b) Give a composite formula for evaluating

$$I = \int_0^1 \int_0^1 f(x,y)\,dx\,dy$$

using part (a). Also, bound the error.

SIX

NUMERICAL METHODS FOR DIFFERENTIAL EQUATIONS

Differential equations are one of the most important mathematical tools used in modeling problems in the physical sciences. In this chapter we derive and analyze numerical methods for solving the initial value problem for ordinary differential equations. The specific form of problem considered is

$$y' = f(x,y) \qquad y(x_0) = Y_0 \tag{6.1}$$

The function $f(x,y)$ is to be continuous for all (x,y) in some domain D of the xy-plane; and (x_0, Y_0) is a point in D. The results obtained for (6.1) will generalize in a straightforward way to both systems of differential equations and higher-order equations, provided appropriate vector and matrix notation is used. These generalizations will be discussed and illustrated in the next two sections.

We say that a function $Y(x)$ is a solution on $[a,b]$ of (6.1) if for all $a \leqslant x \leqslant b$,

1. $(x, Y(x)) \in D$.

2. $Y(x_0) = Y_0$.

3. $Y'(x)$ exists and $Y'(x) = f(x, Y(x))$.

As notation throughout this chapter, $Y(x)$ denotes the true solution of whatever differential equation problem is being considered.

Examples 1.   The general first-order linear differential equation is

$$y' = a_0(x)y + g(x) \qquad a \leqslant x \leqslant b$$

in which the coefficients $a_0(x)$ and $g(x)$ are assumed to be continuous on $[a,b]$. The domain D for this problem is

$$D = \{(x,y) | a \leqslant x \leqslant b, -\infty < y < \infty\}$$

The exact solution of this equation can be found in any elementary differential equations textbook, for example, Boyce and Diprima [B1, pp. 11-23]. As a special case, consider

$$y' = \lambda y + g(x) \qquad 0 \leqslant x < \infty \tag{6.2}$$

with $g(x)$ continuous on $[0, \infty)$. The solution that satisfies $Y(0) = Y_0$ is given by

$$Y(x) = Y_0 e^{\lambda x} + \int_0^x e^{\lambda(x-t)} g(t)dt \qquad 0 \leqslant x < \infty \tag{6.3}$$

This is used later to illustrate several theoretical points.

2.   The equation $y' = -y^2$ is nonlinear. Its general solution is

$$Y(x) = \frac{1}{x+c}$$

with c an arbitrary constant. Note that $|Y(-c)| = \infty$. Thus the global smoothness of $f(x,y) = -y^2$ does not guarantee a similar behavior in the solutions.

To obtain some geometric insight for the solutions of a single first-order differential equation, we can look at the *vector field* induced by the equation on the xy-plane. If $Y(x)$ is a solution that passes through (x_0, Y_0), then the slope of $Y(x)$ at (x_0, Y_0) is $Y'(x_0) = f(x_0, Y_0)$. Within the domain D of $f(x,y)$, pick a representative set of points (x,y) and then draw a short line segment with slope $f(x,y)$ through (x,y).

Example   Consider the equation $y' = -y$. The vector field is given in Figure 6.1, and several representative solutions have been drawn in. It is clear from the

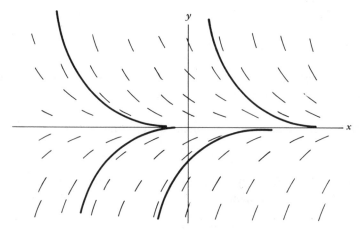

Figure 6.1   Vector field for $y' = -y$.

graph that all solutions $Y(x)$ satisfy

$$\underset{x \to \infty}{\text{Limit}} \, Y(x) = 0$$

To make it easier to draw the vector field of $y' = f(x,y)$, look for those curves in the xy-plane along which $f(x,y)$ is constant. Solve

$$f(x,y) = c$$

for various choices of c. Then each solution $Y(x)$ of $y' = f(x,y)$ that intersects the curve $f(x,y) = c$ satisfies $Y'(x) = c$ at the point of intersection. The curves $f(x,y) = c$ are called *level curves* of the differential equation.

Example Sketch the qualitative behavior of solutions of

$$y' = x^2 + y^2$$

The level curves of the equation are given by

$$x^2 + y^2 = c \qquad c > 0$$

which are circles with center $(0,0)$ and radius $\sqrt{c}$. The vector field with some

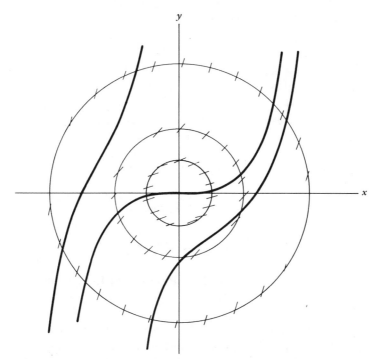

Figure 6.2   Vector field for $y' = x^2 + y^2$.

representative solutions is given in Figure 6.2. Three circles are drawn, with $c = \frac{1}{4}, 1, 4$. This gives some qualitative behavior for the solutions $Y(x)$, and occasionally this can be useful.

6.1 Existence, Uniqueness, and Stability Theory

From the examples of the introduction it should be intuitive that in most cases the initial value problem (6.1) has a unique solution. Before beginning the numerical analysis for (6.1), we present some theoretical results for it. Conditions are given to ensure that it has a unique solution; and we consider the stability of the solution when the initial data Y_0 and the derivative $f(x,y)$ are changed by small amounts. These results are necessary in order to better understand the numerical methods presented later. The domain D will henceforth be assumed to satisfy the following minor technical requirement: if two points (x,y_1) and (x,y_2) both belong to D, then the vertical line segment joining them is also contained in D.

Theorem 6.1 Let $f(x,y)$ be a continuous function of x and y, for all (x,y) in D; and let (x_0, Y_0) be an interior point of D. Assume $f(x,y)$ satisfies the *Lipschitz condition*

$$|f(x,y_1) - f(x,y_2)| \leqslant K|y_1 - y_2|, \quad \text{all} \quad (x,y_1),(x,y_2) \quad \text{in} \quad D, \quad (6.4)$$

for some $K \geqslant 0$. Then for a suitably chosen interval $I = (x_0 - \alpha, x_0 + \alpha)$, there is a unique solution $Y(x)$ on I of (6.1).

Remark If $\partial f(x,y)/\partial y$ exists and is bounded on D, then (6.4) is satisfied. Simply let

$$K = \underset{(x,y) \in D}{\text{Max}} \left| \frac{\partial f(x,y)}{\partial y} \right| \qquad (6.5)$$

Then using the mean value theorem 1.2,

$$f(x,y_1) - f(x,y_2) = \frac{\partial f(x,\zeta)}{\partial y}(y_1 - y_2)$$

for some ζ between y_1 and y_2. Result (6.4) follows immediately using (6.5).

Proof Only a sketch will be given. Complete details are given in Boyce and Diprima [B1, pp. 70–79] or Coddington and Levinson [C2] or any number of other textbooks on differential equations. The proof gives a constructive means to obtain the solution $Y(x)$, although it is generally quite impractical as a numerical method. It is known variously as Picard iteration, successive approximations, and other names.

Begin by reformulating (6.1) as an equivalent integral equation. Suppose $y(t)$ is a solution of (6.1). Then

$$\int_{x_0}^{x} y'(t)dt = \int_{x_0}^{x} f(t, y(t))dt$$

$$y(x) = Y_0 + \int_{x_0}^{x} f(t, y(t))dt \qquad (6.6)$$

This process is reversible. If $y(t)$ is a continuous solution of (6.6), then $f(t, y(t))$ is continuous. But then the integral in (6.6) is differentiable, and this says that the left side of (6.6) must also be differentiable. If we differentiate (6.6), then $y'(x) = f(x, y(x))$ by the fundamental theorem of calculus. Also, any solution of (6.6) satisfies $y(x_0) = Y_0$.

To show (6.6) has a unique solution, we define a sequence of functions $Y_n(x)$: for $x_0 - \alpha \leqslant x \leqslant x_0 + \alpha$,

$$Y_0(x) = Y_0$$

$$Y_{n+1}(x) = Y_0 + \int_{x_0}^{x} f(t, Y_n(t))dt \qquad n \geqslant 0 \qquad (6.7)$$

The constant α is chosen so that

$$\alpha K < 1 \qquad (6.8)$$

In addition, it can be shown that if α is chosen sufficiently small, then all of the functions $Y_n(x)$ will remain in D and they will converge to a function $Y(x)$, uniformly on $[x_0 - \alpha, x_0 + \alpha]$. Figure 6.3 shows the kind of behavior to be expected. It is then straightforward to take limits in (6.7) to obtain

$$Y(x) = Y_0 + \int_{x_0}^{x} f(t, Y(t))dt \qquad x_0 - \alpha \leqslant x \leqslant x_0 + \alpha \qquad (6.9)$$

To see the rate of convergence, subtract (6.7) from (6.9) and take bounds. For $x_0 - \alpha \leqslant x \leqslant x_0 + \alpha$,

$$Y(x) - Y_{n+1}(x) = \int_{x_0}^{x} \left[f(t, Y(t)) - f(t, Y_n(t)) \right] dt,$$

$$|Y(x) - Y_{n+1}(x)| \leqslant \int_{x_0}^{x} |f(t, Y(t)) - f(t, Y_n(t))| dt$$

$$\leqslant K \int_{x_0}^{x} |Y(t) - Y_n(t)| dt$$

$$\leqslant \alpha K \| Y - Y_n \|_{\infty}.$$

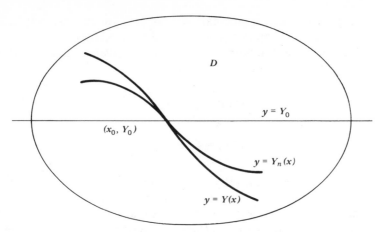

Figure 6.3 Schematic for possible convergence of Picard iteration.

Since the right-hand side is independent of x, we take the maximum of the left side to obtain

$$\|Y - Y_{n+1}\|_\infty \leqslant \alpha K \|Y - Y_n\|_\infty \qquad n \geqslant 0 \qquad (6.10)$$

The maximum norm is over the interval $[x_0 - \alpha, x_0 + \alpha]$. This inequality and the requirement (6.8) indicate a linear order of convergence with a decrease in the error of αK per iterate. [See (2.11) and what follows it in Chapter 2 for the definition of linear order of convergence.]

The proof of the uniqueness of $Y(x)$ is similar to the derivation of (6.10). Let $\tilde{Y}(x)$ be a second solution of (6.6). Then

$$Y(x) - \tilde{Y}(x) = \int_{x_0}^{x} \left[f(t, Y(t)) - f(t, \tilde{Y}(t)) \right] dt$$

and just as with (6.10),

$$\|Y - \tilde{Y}\|_\infty \leqslant \alpha K \|Y - \tilde{Y}\|_\infty$$

Since $\alpha K < 1$, this implies that $\tilde{Y} = Y$, completing the proof. ∎

Examples These two examples illustrate the theorem.

1. Consider $y' = 1 + \sin(xy)$ with

$$D = \{(x,y)|0 \leqslant x \leqslant 1, -\infty < y < \infty\}$$

To compute the Lipschitz constant K of (6.4), use (6.5). Then

$$\frac{\partial f(x,y)}{\partial y} = x\cos(xy) \qquad K = 1$$

Thus for any (x_0, Y_0) with $0 < x_0 < 1$, there is a solution $Y(x)$ to the associated initial value problem on some interval $[x_0 - \alpha, x_0 + \alpha] \subset [0, 1]$.

2. Consider the problem

$$y' = \frac{2x}{a^2} y^2 \qquad y(0) = 1$$

for any constant $a > 0$. The solution is

$$Y(x) = \frac{a^2}{a^2 - x^2} \qquad -a < x < a$$

The solution exists on only a small interval if a is small. To determine the Lipschitz constant, we must calculate

$$\frac{\partial f(x, y,)}{\partial y} = \frac{4xy}{a^2}$$

If we choose a domain D that is bounded in x and y, in order to have a bounded Lipschitz constant as in (6.5), we see that the solution $Y(x)$ will leave the region D. Thus the interval $I = [-\alpha, \alpha]$ of the theorem will be contained in $(-a, a)$ for any D. Lest it be thought that we can evaluate the partial derivative at the initial point $(x_0, Y_0) = (0, 0)$ and obtain sufficient information to estimate the Lipschitz constant K, note that

$$\frac{\partial f(0, 1)}{\partial y} = 0$$

for any $a > 0$.

Example (Picard iteration)   Consider $y' = -xy, y(1) = 2$. The true solution is

$$Y(x) = 2e^{.5(1 - x^2)}$$

The integral equation reformulation is

$$y(x) = 2 - \int_1^x ty(t)dt$$

The first few iterates are

$$Y_0(x) \equiv 2 \qquad Y_1(x) = 3 - x^2 \qquad Y_2(x) = \frac{13}{4} - \frac{3}{2}x^2 + \frac{1}{4}x^4$$

$$Y_3(x) = \frac{79}{24} - \frac{13}{8}x^2 + \frac{3}{8}x^4 - \frac{x^6}{24}$$

A table of values is given in Table 6.1. In general, the Picard iteration method is

Table 6.1   Values of Picard iterates

x	$Y_0(x)$	$Y_1(x)$	$Y_2(x)$	$Y_3(x)$	$Y(x)$
.9	2.0	2.19	2.1990	2.199311	2.199318
1.0	2.0	2.0	2.0	2.0	2.0
1.1	2.0	1.79	1.8010	1.800639	1.800649
1.5	2.0	2.18	1.1406	1.05924	1.070523
2.0	2.0	-1.0	1.25	.125	.446260

not a practical numerical method. But there are special situations, often of a theoretical nature, in which the technique is quite useful.

Stability of the solution   The stability of the solution $Y(x)$ will be examined when the initial value problem is changed by a small amount. This is related to the discussion of stability in Section 1.5 of Chapter 1. We consider the perturbed problem

$$y' = f(x,y) + \delta(x)$$

$$y(x_0) = Y_0 + \epsilon \tag{6.11}$$

with the same hypotheses for $f(x,y)$ as in Theorem 6.1. Furthermore, we assume that $\delta(x)$ is continuous for all x such that $(x,y) \in D$ for some y. The problem (6.11) can then be shown to have a unique solution, denoted by $Y(x; \delta, \epsilon)$. The results of Theorem 6.1 will be valid for one fixed interval $[x_0 - \alpha, x_0 + \alpha]$, uniformly for all perturbations ϵ and $\delta(x)$ provided they satisfy

$$|\epsilon| \leqslant \epsilon_0 \qquad \|\delta\|_\infty \leqslant \epsilon_0 \tag{6.12}$$

for some sufficiently small ϵ_0.

The function $Y(x; \delta, \epsilon)$ satisfies

$$Y(x; \delta, \epsilon) = Y_0 + \epsilon + \int_{x_0}^x \left[f(t, Y(t; \delta, \epsilon)) + \delta(t) \right] dt \tag{6.13}$$

Subtracting (6.9),

$$Y(x; \delta, \epsilon) - Y(x) = \epsilon + \int_{x_0}^x \left[f(t, Y(t; \delta, \epsilon)) - f(t, Y(t)) \right] dt$$

$$+ \int_{x_0}^x \delta(t) dt \tag{6.14}$$

Using this, we can prove the following.

Theorem 6.2   Consider the initial value problem (6.1) and the perturbed problem (6.11). Assuming that $\delta(x)$ is continuous on D and that ϵ_0 in (6.12) is sufficiently small, the perturbed solution $Y'(x; \delta, \epsilon)$ satisfies

$$\| Y(\cdot; \delta, \epsilon) - Y \|_\infty \leqslant k \left[|\epsilon| + \alpha \| \delta \|_\infty \right] \tag{6.15}$$

with $k = 1/(1 - \alpha K)$. We say the solution of the initial value problem (6.1) is stable with respect to perturbations in its data.

Proof Proceed in a manner similar to that used to derive (6.10). Then from (6.14),

$$\| Y(\cdot; \delta, \epsilon) - Y \|_\infty \leqslant |\epsilon| + \alpha K \| Y(\cdot; \delta, \epsilon) - Y \|_\infty + \alpha \| \delta \|_\infty$$

Collect terms and divide by $1 - \alpha K$ to complete the proof. Note that $\alpha K < 1$ was assumed in (6.8). ∎

Although this is stated somewhat differently than the concept of stability in Section 1.5, it does show that small changes in the data lead to small changes in the answer (although k may be quite large). Thus the solution Y depends continuously on the data; and this means the initial value problem is well posed or stable.

In Section 1.5, it was pointed out that a problem could be well posed, but ill conditioned with respect to numerical computation. The same is true with differential equations, although it does not occur often in practice. To better understand when this might happen, we can use (6.14) to estimate the perturbation in Y due to the perturbations ϵ and $\delta(x)$. Because ϵ and $\delta(x)$ enter into the final results in much the same way, as indicated in (6.15), we simplify the discussion by letting $\delta = 0$.

Let $Y(x; \epsilon)$ denote the solution of the perturbed problem

$$y' = f(x, y) \qquad y(x_0) = Y_0 + \epsilon$$

From (6.14),

$$Y(x; \epsilon) - Y(x) = \epsilon + \int_{x_0}^x \left[f(t, Y(t; \epsilon)) - f(t, Y(t)) \right] dt$$

Apply the mean value theorem 1.2 to the integrand to obtain the approximation

$$Z(x; \epsilon) \doteq \epsilon + \int_{x_0}^x \frac{\partial f(t, Y(t))}{\partial y} Z(t; \epsilon) dt \tag{6.17}$$

with $Z(x; \epsilon) = Y(x; \epsilon) - Y(x)$. This can be converted back to a linear differential equation problem, and it can be solved to give

$$Z(x; \epsilon) \doteq \epsilon \exp \left[\int_{x_0}^x \frac{\partial f(t, Y(t))}{\partial y} dt \right] \tag{6.18}$$

This is a valid approximation as long as $Z(x;\epsilon)$ does not become too large, since the approximation in (6.17) would then no longer be valid.

Example   As a simple illustration of (6.18), suppose that $\partial f(x, Y(x))/\partial y$ is positive and that the solution $Y(x)$ is a decreasing function, for $x \geqslant x_0$ and increasing. Then the perturbations $Z(x;\epsilon)$ will be rapidly increasing with x, as well as the relative error in $Y(x;\epsilon)$. More concretely, consider the problem

$$y' = 100y - 101e^{-x} \qquad y(0) = 1 \tag{6.19}$$

Its solution is $Y(x) = e^{-x}$. Perturbing the initial value by ϵ, the solution will be exactly

$$Y(x;\epsilon) = e^{-x} + \epsilon e^{100x}$$

This also agrees with (6.18). It is clear that for any $\epsilon \neq 0$, the perturbed solution will depart rapidly from the true solution. Thus (6.19) will be an ill-conditioned problem when solved numerically.

The problems that are most well-conditioned are those for which $\partial f(x, Y(x))/\partial y$ is negative for $x \geqslant x_0$. Then from (6.18), the effect of the perturbation ϵ will decrease with increasing x. Those equations for which $\partial f(x, Y(x))/\partial y$ is negative but very large in magnitude will be well-conditioned problems; but they will also be troublesome for most of the numerical methods of this chapter. Such equations are called *stiff differential equations*, and we return later to some discussion of them in Section 6.8.

Systems of differential equations   The material of this chapter will generalize to a system of n first-order equations, written

$$y_1' = f_1(x, y_1, y_2, \ldots, y_n) \qquad y_1(x_0) = Y_{1,0}$$
$$\vdots \tag{6.20}$$
$$y_n' = f_n(x, y_1, \ldots, y_n) \qquad y_n(x_0) = Y_{n,0}$$

This is often made to look like a first-order equation by introducing vector notation. Let

$$\mathbf{y}(x) = \begin{bmatrix} y_1(x) \\ \vdots \\ y_n(x) \end{bmatrix} \qquad \mathbf{f}(x, \mathbf{y}) = \begin{bmatrix} f_1(x, \mathbf{y}) \\ \vdots \\ f_n(x, \mathbf{y}) \end{bmatrix} \qquad \mathbf{Y}_0 = \begin{bmatrix} Y_{1,0} \\ \vdots \\ Y_{n,0} \end{bmatrix}$$

The system (6.20) can then be written as

$$\mathbf{y}' = \mathbf{f}(x, \mathbf{y}), \quad \mathbf{y}(x_0) = \mathbf{Y}_0 \tag{6.21}$$

By introducing vector norms to replace absolute values, virtually all of the preceding results generalize to this vector initial value problem.

For higher-order equations, the initial value problem

$$y^{(m)} = f(x, y, y', \dots, y^{(m-1)})$$

$$y(x_0) = Y_0, \dots, y^{(m-1)}(x_0) = Y_0^{(m-1)}$$

can be converted to a first-order system. Introduce the new unknown functions

$$y_1 = y, y_2 = y', \dots, y_m = y^{(m-1)}$$

These functions satisfy the system

$$
\begin{aligned}
y_1' &= y_2 & y_1(x_0) &= Y_0 \\
y_2' &= y_3 & y_2(x_0) &= Y_0' \\
&\;\;\vdots & &\;\;\vdots \\
y_{m-1}' &= y_m & & \\
y_m' &= f(x, y_1, \dots, y_m) & y_m(x_0) &= Y_0^{(m-1)}
\end{aligned}
$$

Example The linear second-order initial value problem

$$y'' = a_1(x)y' + a_0(x)y + g(x), \qquad y(x_0) = \alpha, \qquad y'(x_0) = \beta$$

becomes

$$
\begin{bmatrix} y_1 \\ y_2 \end{bmatrix}' =
\begin{bmatrix} 0 & 1 \\ a_0(x) & a_1(x) \end{bmatrix}
\begin{bmatrix} y_1 \\ y_2 \end{bmatrix} +
\begin{bmatrix} 0 \\ g(x) \end{bmatrix}, \qquad
\begin{bmatrix} y_1(x_0) \\ y_2(x_0) \end{bmatrix} =
\begin{bmatrix} \alpha \\ \beta \end{bmatrix}
$$

In vector form with $A(x)$ denoting the coefficient matrix,

$$\mathbf{y}' = A(x)\mathbf{y} + \mathbf{G}(x) \qquad \mathbf{y}(x_0) = \mathbf{Y}_0 = \begin{bmatrix} \alpha \\ \beta \end{bmatrix}$$

a linear system of first-order differential equations.

There are special numerical methods for mth order equations; but these have been developed to a significant extent only for $m = 2$, which arises in applications of Newtonian mechanics due to Newton's second law. Most high-order equations are solved by first converting them to an equivalent first-order system, as was described above.

6.2 Euler's Method

The most popular numerical methods for solving (6.1) are called finite difference methods. Approximate values are obtained for the solution at a set of grid points

$$x_0 < x_1 < x_2 < \cdots < x_n < \cdots$$

and the approximate value at each x_n is obtained by using some of the values obtained in previous steps. We begin with a simple but computationally inefficient method attributed to Euler. The analysis of it has many of the features of the analyses of the more efficient finite difference methods, but without their additional complexity. First we give several derivations of Euler's method, and follow with a complete convergence and stability analysis for it. We give an asymptotic error formula, and conclude the section by generalizing the earlier results to systems of equations.

As before, $Y(x)$ denotes the true solution to (6.1),

$$Y'(x) = f(x, Y(x)) \qquad Y(x_0) = Y_0 \tag{6.22}$$

The approximate solution is denoted by $y(x)$, and the values $y(x_0), y(x_1), \cdots$, $y(x_n), \cdots$ are often denoted by $y_0, y_1, \cdots, y_n, \cdots$. An equal grid size $h > 0$ defines the node points,

$$x_j = x_0 + jh \qquad j = 0, 1, \cdots$$

In later sections with more practical methods, we discuss varying the grid size to control the error. The problem (6.1) is solved on a fixed finite interval, which is always denoted by $[x_0, b]$. The notation $N(h)$ denotes the largest index N for which

$$x_N \leq b \qquad x_{N+1} > b$$

Derivation of Euler's method Euler's method is defined by

$$y_{n+1} = y_n + hf(x_n, y_n) \qquad n = 0, 1, 2, \cdots \tag{6.23}$$

with $y_0 \approx Y_0$. Four viewpoints of it are given.

1. A geometric viewpoint. Consider the graph of the solution $Y(x)$ in Figure 6.4. Form the tangent line to the graph of $Y(x)$ at x_0, and use this line as an approximation to the curve for $x_0 \leq x \leq x_1$. Then

$$\frac{\Delta y}{h} = Y'(x_0) = f(x_0, Y_0)$$

$$Y(x_1) - Y(x_0) \approx \Delta y = hY'(x_0)$$

$$Y(x_1) \approx Y(x_0) + hf(x_0, Y(x_0))$$

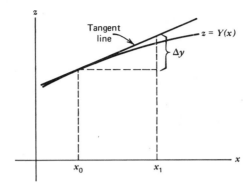

Figure 6.4   Geometric interpretation of Euler's method.

By repeating this argument on $[x_1, x_2], [x_2, x_3], \cdots$, we obtain the general formula (6.23).

2. Taylor series. Expand $Y(x_{n+1})$ about x_n,

$$Y(x_{n+1}) = Y(x_n) + hY'(x_n) + \frac{h^2}{2} Y''(\xi_n) \qquad x_n \leqslant \xi_n \leqslant x_{n+1} \qquad (6.24)$$

By dropping the error term, we obtain the Euler method (6.23). The term

$$T_n = \frac{h^2}{2} Y''(\xi_n) \qquad (6.25)$$

is called the truncation error or discretization error at x_{n+1}. We use the former name in this text.

3. Numerical differentiation. From the definition of a derivative,

$$\frac{Y(x_{n+1}) - Y(x_n)}{h} \approx Y'(x_n) = f(x_n, Y(x_n))$$

$$Y(x_{n+1}) \approx Y(x_n) + hf(x_n, Y(x_n))$$

4. Numerical integration. Integrate $Y'(t) = f(t, Y(t))$ over $[x_n, x_{n+1}]$:

$$Y(x_{n+1}) = Y(x_n) + \int_{x_n}^{x_{n+1}} f(t, Y(t)) dt \qquad (6.26)$$

Consider the simple numerical integration method

$$\int_a^{a+h} g(t) dt \approx hg(a) \qquad (6.27)$$

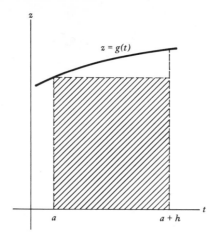

Figure 6.5 Illustration of integration rule (6.27).

called the *left-hand rectangular rule*. Figure 6.5 shows the numerical integral as the cross-hatched area. Applying this to (6.26), we obtain

$$Y(x_{n+1}) \approx Y(x_n) + hf(x_n, Y(x_n))$$

as before.

Of the three analytical derivations (2) to (4), both (2) and (4) are the simplest cases of a set of increasingly accurate methods. Approach (2) leads to the *single-step methods*, particularly the Runge-Kutta formulas. Approach (4) leads to *multistep methods*, especially the predictor-corrector methods. Perhaps surprisingly, method (3) does not lead to other successful and more accurate methods. The one example given of (3) is the midpoint method in Section 6.4, and it leads to problems of numerical instability. The bulk of this chapter is devoted to multistep methods, partly because they are generally the most efficient class of methods and partly because they are more complex to analyze than are the Runge-Kutta methods. The latter are taken up in Section 6.9. If desired, Section 6.9 can be covered following the present section, with only minimal inconvenience.

Before analyzing the Euler method, we give some numerical examples. They also serve as illustrations for some of the theory that will be presented.

Examples 1. Consider the equation $y' = y, y(0) = 1$. Its true solution is $Y(x) = e^x$. Numerical results are given in Table 6.2 for several values of h. The answers $y(x_n)$ are given at only a few points, rather than at all points at which they were calculated. Note that the error at each point x decreases by about half when h is halved.

Table 6.2 Euler's method, example 1

	x	$y(x)$	$Y(x)$	$Y(x)-y(x)$
$h=.2$	.40	1.44000	1.49182	.05182
	.80	2.07360	2.22554	.15194
	1.20	2.98598	3.32012	.33413
	1.60	4.29982	4.95303	.65321
	2.00	6.19174	7.38906	1.19732
$h=.1$	.40	1.46410	1.49182	.02772
	.80	2.14359	2.22554	.08195
	1.20	3.13843	3.32012	.18169
	1.60	4.59497	4.95303	.35806
	2.00	6.72750	7.38906	.66156
$h=.05$	.40	1.47746	1.49182	.01437
	.80	2.18287	2.22554	.04267
	1.20	3.22510	3.32012	.09502
	1.60	4.76494	4.95303	.18809
	2.00	7.03999	7.38906	.34907

2. Consider the equation

$$y' = \frac{1}{1+x^2} - 2y^2 \qquad y(0)=0$$

The true solution is

$$Y(x) = x/(1+x^2)$$

and the results are given in Table 6.3. Again note the behavior of the error as h is decreased.

Table 6.3 Euler's method, example 2

	x	$y(x)$	$Y(x)$	$Y(x)-y(x)$
$h=.2$	.00	0.0	0.0	0.0
	.40	.37631	.34483	$-.03148$
	.80	.54228	.48780	$-.05448$
	1.20	.52709	.49180	$-.03529$
	1.60	.46632	.44944	$-.01689$
	2.00	.40682	.40000	$-.00682$
$h=.1$	.40	.36085	.34483	$-.01603$
	.80	.51371	.48780	$-.02590$
	1.20	.50961	.49180	$-.01781$
	1.60	.45872	.44944	$-.00928$
	2.00	.40419	.40000	$-.00419$
$h=.05$	.40	.35287	.34483	$-.00804$
	.80	.50049	.48780	$-.01268$
	1.20	.50073	.49180	$-.00892$
	1.60	.45425	.44944	$-.00481$
	2.00	.40227	.40000	$-.00227$

Convergence analysis At each step of the Euler method, an additional truncation error (6.25) is introduced. We will analyze the cumulative effect of these errors. The truncation error is often referred to as *local error*, and the cumulative error is called *global error*. The last column in Tables 6.2 and 6.3 is the global error.

We begin with the following lemma, which is quite useful in the analysis of finite difference methods.

Lemma 1 For any real x,

$$1 + x \leqslant e^x$$

and for any $x \geqslant -1$,

$$0 \leqslant (1+x)^m \leqslant e^{mx} \tag{6.28}$$

Proof Using Taylor's theorem,

$$e^x = 1 + x + \frac{x^2}{2} e^\xi$$

with ξ between 0 and x. Since the remainder is always positive, the first result is proven. (6.28) follows easily. ∎

For the remainder of this chapter it is assumed that the function $f(x,y)$ satisfies the following stronger Lipschitz condition:

$$|f(x,y_1) - f(x,y_2)| \leqslant K|y_1 - y_2|, \ -\infty < y_1, y_2 < \infty, \ x_0 \leqslant x \leqslant b \tag{6.29}$$

for some $K \geqslant 0$. Although stronger than necessary, it will simplify the proofs. And given a function $f(x,y)$ satisfying the weaker condition (6.4) and a solution $Y(x)$ to the initial value problem (6.1), the function f can be modified to satisfy (6.29) without changing the solution $Y(x)$ or the essential character of the problem (6.1) and its numerical solution. See Shampine and Gordon [S2, p. 24] for the details.

Theorem 6.3 Assume that the solution $Y(x)$ of (6.1) has a bounded second derivative on $[x_0, b]$. Then the solution $\{y_h(x_n) | x_0 \leqslant x_n \leqslant b\}$ obtained by Euler's method (6.23) satisfies

$$\underset{x_0 \leqslant x_n \leqslant b}{\text{Max}} |Y(x_n) - y_h(x_n)| \leqslant e^{(b-x_0)K}|e_0| + \left[\frac{e^{(b-x_0)K} - 1}{K}\right]\tau(h) \tag{6.30}$$

where

$$\tau(h) = \frac{h}{2}\|Y''\|_\infty \tag{6.31}$$

and $e_0 = Y_0 - y_h(x_0)$. The subscript h on the approximate solution $y_h(x)$ is used to note explicitly the dependence of y on h; usually it is suppressed, although implicitly understood.

If, in addition,

$$|Y_0 - y_h(x_0)| \leqslant c_1 h \qquad \text{as} \quad h \to 0 \qquad (6.32)$$

for some $c_1 \geqslant 0$ (e.g., if $Y_0 = y_0$ for all h), then there is a constant $B \geqslant 0$ for which

$$\underset{x_0 \leqslant x_n \leqslant b}{\text{Max}} |Y(x_n) - y_h(x_n)| \leqslant Bh \qquad (6.33)$$

Proof Let $e_n = Y(x_n) - y(x_n), n \geqslant 0$, and recall the definition of $N(h)$ from the beginning of the section. Define

$$\tau_n = \frac{h}{2} Y''(\xi_n) \qquad 0 \leqslant n \leqslant N(h)$$

based on the truncation error in (6.25). Easily

$$\underset{0 \leqslant n \leqslant N(h)}{\text{Max}} |\tau_n| \leqslant \tau(h)$$

using (6.31). From (6.24), (6.22), and (6.23), we have

$$Y_{n+1} = Y_n + hf(x_n, Y_n) + h\tau_n \qquad (6.34)$$

$$y_{n+1} = y_n + hf(x_n, y_n) \qquad 0 \leqslant n \leqslant N(h) \qquad (6.35)$$

We are using the common notation $Y_n \equiv Y(x_n)$. Subtracting (6.35) from (6.34),

$$e_{n+1} = e_n + h[f(x_n, Y_n) - f(x_n, y_n)] + h\tau_n \qquad (6.36)$$

and taking bounds using (6.29),

$$|e_{n+1}| \leqslant |e_n| + hK|Y_n - y_n| + h|\tau_n|$$

$$|e_{n+1}| \leqslant (1 + hK)|e_n| + h\tau(h) \qquad (6.37)$$

Apply (6.37) recursively to obtain

$$|e_n| \leqslant (1 + hK)^n |e_0| + \left[1 + (1 + hK) + \cdots + (1 + hK)^{n-1} \right] h\tau(h)$$

Using the formula for the sum of a finite geometric series,

$$1 + r + r^2 + \cdots + r^{n-1} = \frac{r^n - 1}{r - 1} \qquad r \neq 1$$

we obtain

$$|e_n| \leqslant (1+hK)^n |e_0| + \left[\frac{(1+hK)^n - 1}{K} \right] \tau(h) \tag{6.38}$$

Using Lemma 1,

$$(1+hK)^n \leqslant e^{nhK} = e^{(x_n - x_0)K} \leqslant e^{(b-x_0)K}$$

and this with (6.38) implies the main result (6.30).

The remaining result (6.33) is a trivial corollary of (6.30), with the constant B given by

$$B = c_1 e^{(b-x_0)K} + \left[\frac{e^{(b-x_0)K} - 1}{K} \right] \frac{\|Y''\|_\infty}{2} \tag{6.39}$$

This completes the proof. ∎

The result (6.33) implies that the error should decrease by at least one-half when the stepsize h is halved. This is confirmed in the examples of Tables 6.2 and 6.3. It is shown later in the section that (6.33) gives the correct rate of convergence. But for now, a simple example is carried out in detail.

Example Consider $y' = y, y(0) = 1$. The true solution is $Y(x) = e^x$. Euler's method is

$$y_{n+1} = y_n + hy_n = (1+h)y_n \qquad y_0 = 1$$

Inductively,

$$y_n = (1+h)^n = \left[(1+h)^{1/h} \right]^{nh} \equiv c(h)^{x_n}$$

For $c(h)$, with $0 < h < 1$,

$$c(h) = (1+h)^{1/h} = \exp\left[\frac{1}{h} \ln(1+h) \right] = e^{1-(h/2)+(h^2/3)-(h^3/4)+\cdots} < e.$$

Consequently, $c(h)^{x_n}$ increases less rapidly than e^{x_n}; and thus

$$\underset{0 \leqslant x \leqslant 1}{\text{Max}} |e^x - c(h)^x| = e - c(h)$$

Using the above formula,

$$e - c(h) = e\{1 - e^{-(h/2)+(h^2/3)-(h^3/4)+\cdots}\}$$

$$= e\left\{ 1 - \left(1 - \frac{h}{2} + \frac{11}{24}h^2 + \cdots \right) \right\}$$

$$= e\left\{ \frac{h}{2} - \frac{11}{24}h^2 + \cdots \right\}$$

and we obtain the asymptotic error estimate

$$e - c(h) \approx \frac{h}{2} e$$

Summarizing, for $y' = y$, $y(0) = 1$,

$$\underset{0 \leqslant x_n \leqslant 1}{\text{Max}} |Y(x_n) - y_n| \approx \frac{e}{2} h \tag{6.40}$$

This shows that (6.33) is qualitatively correct.

It is of some interest to note that the bound (6.33) is worse than that given by (6.40), and this is a case in which (6.33) should perform at its best. The Lipschitz constant $K = 1$, and the error bound is

$$\underset{0 \leqslant x_n \leqslant 1}{\text{Max}} |e^{x_n} - y_n| \leqslant (e - 1) \frac{e}{2} h$$

somewhat worse than the actual error (6.40).

The bound (6.33) gives the correct speed of convergence for Euler's method, but the multiplying constant B is much too large for most equations. For example, with the earlier example (2), $y' = 1/(1 + x^2) - 2y^2$, (6.33) will predict that the error grows with b. But, clearly from Table 6.3, the error decreases with increasing x.

Stability analysis   Recall the stability analysis for the initial value problem, given in Theorem 6.2. To consider a similar idea for Euler's method, we consider the numerical method

$$z_{n+1} = z_n + h[f(x_n, z_n) + \delta(x_n)] \qquad 0 \leqslant n \leqslant N(h) - 1 \tag{6.41}$$

with $z_0 = y_0 + \epsilon$. This is in analog to the comparison of (6.11) with (6.1). Let the perturbations satisfy (6.12), as before. We compare the two numerical solutions $\{z_n\}$ and $\{y_n\}$ as $h \to 0$.

Let $e_n = z_n - y_n$, $n \geqslant 0$. Then $e_0 = \epsilon$, and subtracting (6.23) from (6.41),

$$e_{n+1} = e_n + h[f(x_n, z_n) - f(x_n, y_n)] + h\delta(x_n)$$

This has exactly the same form as (6.36). Using the same procedure as that following (6.36), we have

$$\underset{0 \leqslant n \leqslant N(h)}{\text{Max}} |z_n - y_n| \leqslant e^{(b - x_0)K} |\epsilon| + \left[\frac{e^{(b - x_0)K} - 1}{K} \right] \|\delta\|_\infty$$

Consequently, there are constants $\tilde{k}_1, \tilde{k}_2$, independent of h, with

$$\underset{0 \leqslant n \leqslant N(h)}{\text{Max}} |z_n - y_n| \leqslant \tilde{k}_1 |\epsilon| + \tilde{k}_2 \|\delta\|_\infty \tag{6.42}$$

This is the analog to the result (6.15) for the original problem (6.1). This says that Euler's method is a stable numerical method for the solution of the initial value problem (6.1). We insist that all numerical methods for initial value problems possess this form of stability; and usually we require other forms of stability as well. In the future we take $\delta(x) \equiv 0$ and consider only the effect of perturbing the initial value Y_0. This simplifies the analysis, and the results are still equally useful.

Rounding error analysis Introduce an error into each step of the Euler method, with each error derived from the rounding errors of the operations being performed. This number, denoted by ρ_n, is called the local rounding error. Calling the resultant numerical values $\tilde{y}_n$, we have

$$\tilde{y}_{n+1} = \tilde{y}_n + hf(x_n, \tilde{y}_n) + \rho_n \qquad n = 0, 1, \cdots, N(h) - 1 \qquad (6.43)$$

The values $\tilde{y}_n$ are the finite place numbers actually obtained in the computer, and y_n is the value that would be obtained if exact arithmetic was being used. Let $\rho(h)$ be a bound on the rounding errors,

$$\rho(h) = \underset{0 \leqslant n \leqslant N(h) - 1}{\text{Max}} |\rho_n|$$

In a practical situation, using a fixed-word-length computer, the bound $\rho(h)$ does not decrease as $h \to 0$. Instead it remains approximately constant, and $\rho(h)/\|Y\|_\infty$ is proportional to the unit roundoff u on the computer, that is, the smallest number u for which $1 + u > 1$ (see Problem 7, Chapter 1).

To see the effect of the roundoff errors in (6.43), subtract it from the true solution in (6.34), to obtain

$$\tilde{e}_{n+1} = \tilde{e}_n + h \left[f(x_n, Y_n) - f(x_n, \tilde{y}_n) \right] + h\tau_n - \rho_n$$

where $\tilde{e}_n = Y(x_n) - \tilde{y}_n$. Proceed as in the proof of Theorem 6.3, but identify $\tau_n - \rho_n/h$ with τ_n in the earlier proof. Then we obtain

$$|\tilde{e}_n| \leqslant e^{(b-x_0)K} |Y_0 - \tilde{y}_0| + \left[\frac{e^{(b-x_0)K} - 1}{K} \right] \left[\tau(h) + \frac{\rho(h)}{h} \right] \qquad (6.44)$$

For simplicity, let $\rho(h)/\|Y\|_\infty \doteq u$, as discussed above. Then

$$|\tilde{e}_n| \leqslant c \left(\frac{h}{2} \|Y''\|_\infty + \frac{u\|Y\|_\infty}{h} \right) \equiv E(h)$$

The qualitative behavior of $E(h)$ is shown in the graph of Figure 6.6. At h^*, $E(h)$ is a minimum; and any further decrease will cause an increased error bound $E(h)$.

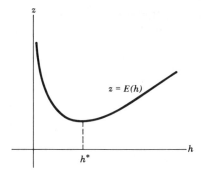

Figure 6.6 Error curve for Euler's method including rounding error effect.

This derivation gives the worst possible case for the effect of the rounding errors. In practice the rounding errors vary in both size and sign. The resultant cancellation will cause $\tilde{e}_n$ to increase less rapidly than is implied by the $1/h$ term in $E(h)$ and (6.44). But there will still be an optimum value of h^*, and below it the error will again increase. For a more complete analysis, see Henrici [H1, pp. 35–59].

Usually rounding error is not a serious problem. However, if the desired accuracy is close to the best that can be attained because of the computer word length, then greater attention must be given to the effects due to rounding. On an IBM 360 model computer with double-precision arithmetic, if $h \geqslant .001$, then the maximum of u/h is 2.2×10^{-13}, where u is the unit rounding error. Thus the rounding error will usually not present a significant problem unless very small error tolerances are desired. But in single precision with the same restriction on h, the maximum of u/h is 9.5×10^{-4}; and with error tolerances of this magnitude (not an unreasonable one), the rounding error will be a more significant factor.

Asymptotic error analysis An asymptotic estimate of the error in Euler's method will be derived, ignoring any effects due to rounding. Before beginning, some special notation is necessary to simplify the algebra in the analysis. If $B(x,h)$ is a function defined for $x_0 \leqslant x \leqslant b$ and for all sufficiently small h, then the notation

$$B(x,h) = O(h^p)$$

for some $p > 0$ means there is a constant c such that

$$|B(x,h)| \leqslant ch^p \qquad x_0 \leqslant x \leqslant b$$

for all sufficiently small h. If B depends on h only, the same kind of bound is implied.

Theorem 6.4 Assume that $Y(x)$ is the solution of the initial value problem (6.1) and that it is three times continuously differentiable. Assume that

$$f_y(x,y) \equiv \frac{\partial f(x,y)}{\partial y} \qquad f_{yy}(x,y) \equiv \frac{\partial^2 f(x,y)}{\partial y^2}$$

are continuous and bounded for $x_0 \leqslant x \leqslant b$, $-\infty < y < \infty$. Let the initial value $y_h(x_0)$ satisfy

$$Y_0 - y_h(x_0) = \delta_0 h + O(h^2) \tag{6.45}$$

Usually this error is zero and thus $\delta_0 = 0$.

Then the error in Euler's method (6.23) satisfies

$$Y(x_n) - y_h(x_n) = D(x_n)h + O(h^2) \tag{6.46}$$

where $D(x)$ is the solution of the linear initial value problem

$$D'(x) = f_y(x, Y(x))D(x) + \tfrac{1}{2}Y''(x) \qquad D(x_0) = \delta_0 \tag{6.47}$$

Proof Using Taylor's theorem,

$$Y(x_{n+1}) = Y(x_n) + hY'(x_n) + \frac{h^2}{2}Y''(x_n) + \frac{h^3}{6}Y'''(\xi_n)$$

for some $x_n \leqslant \xi_n \leqslant x_{n+1}$. Subtract (6.23) and use (6.22) to obtain

$$e_{n+1} = e_n + h[f(x_n, Y_n) - f(x_n, y_n)] + \frac{h^2}{2}Y''(x_n) + \frac{h^3}{6}Y'''(\xi_n) \tag{6.48}$$

Using Taylor's theorem on $f(x_n, y_n)$, regarded as a function of y_n,

$$f(x_n, y_n) = f(x_n, Y_n) + (y_n - Y_n)f_y(x_n, Y_n) + \tfrac{1}{2}(y_n - Y_n)^2 f_{yy}(x_n, \zeta_n)$$

for some ζ_n between y_n and Y_n. Using this in (6.48),

$$e_{n+1} = [1 + hf_y(x_n, Y_n)]e_n + \frac{h^2}{2}Y''(x_n) + B_n$$

$$\tag{6.49}$$

$$B_n = \frac{h^3}{6}Y'''(\xi_n) - \tfrac{1}{2}hf_{yy}(x_n, \zeta_n)e_n^2$$

Using (6.33),

$$B_n = O(h^3) \tag{6.50}$$

Because B_n is small relative to the remaining terms in (6.49), we find the dominant part of the error by neglecting B_n. Let g_n represent the dominant part of the error. It is defined implicitly by

$$g_{n+1} = \left[1 + hf_y(x_n, Y_n)\right]g_n + \frac{h^2}{2}Y''(x_n) \qquad (6.51)$$

with

$$g_0 = \delta_0 h \qquad (6.52)$$

the dominant part of the initial error (6.45). Since we expect $g_n \approx e_n$, and since $e_n = O(h)$, we introduce the sequence $\{\delta_n\}$ implicitly by

$$g_n = h\delta_n \qquad (6.53)$$

Substituting into (6.51), canceling h and rearranging, we obtain

$$\delta_{n+1} = \delta_n + h\left[f_y(x_n, Y_n)\delta_n + \tfrac{1}{2}Y''(x_n)\right] \qquad x_0 \leqslant x_n \leqslant b \qquad (6.54)$$

The initial value δ_0 from (6.45) was defined as independent of h.

The equation (6.54) is Euler's method for solving the initial value problem (6.47); and thus from Theorem 6.3,

$$D(x_n) - \delta_n = O(h) \qquad x_0 \leqslant x_n \leqslant b$$

Using (6.53),

$$g_n = D(x_n)h + O(h^2) \qquad (6.55)$$

To complete the proof, it must be shown that g_n is indeed the principal part of the error e_n. Introduce

$$k_n = e_n - g_n$$

Then $k_0 = e_0 - g_0 = O(h^2)$ from (6.52) and (6.45). Subtracting (6.51) from (6.49), and using (6.50),

$$k_{n+1} = \left[1 + hf_y(x_n, Y_n)\right]k_n + B_n$$

$$|k_{n+1}| \leqslant (1 + hK)|k_n| + O(h^3)$$

This is of the form of (6.37) in the proof of Theorem 6.3, with the term $h\tau(h)$ replaced by $O(h^3)$. Using the same derivation,

$$|k_n| = O(h^2)$$

Combining the preceding results,

$$e_n = g_n + k_n = \left[hD(x_n) + O(h^2) \right] + O(h^2)$$

which proves (6.46).   ∎

The function $D(x)$ is rarely produced explicitly, but the form of the error (6.46) furnishes useful qualitative information. It is often used as the basis for extrapolation procedures, some of which will be discussed in later sections.

Example   Consider the problem

$$y' = -y \qquad y(0) = 1$$

with the solution $Y(x) = e^{-x}$. The equation for $D(x)$ is

$$D'(x) = -D(x) + \tfrac{1}{2}e^{-x} \qquad D(0) = 0$$

and its solution is

$$D(x) = \tfrac{1}{2}xe^{-x}$$

This gives the following asymptotic formula for the error in Euler's method.

$$Y(x_n) - y_h(x_n) \approx \frac{h}{2} x_n e^{-x_n} \tag{6.56}$$

Table 6.4 contains the actual errors and the errors predicted by (6.56) for $h = .05$. Note then the error decreases with increasing x, just as with the solution $Y(x)$. But the relative error increases linearly with x,

$$\frac{Y(x_n) - y_h(x_n)}{Y(x_n)} \approx \frac{h}{2} x_n$$

Also, the estimate (6.56) is much better than the bound given by (6.30) in Theorem 6.3.

Table 6.4   Example (6.56) of Theorem 6.4.

x_n	$Y_n - y_n$	$hD(x_n)$
.4	.00689	.00670
.8	.00920	.00899
1.2	.00921	.00904
1.6	.00819	.00808
2.0	.00682	.00677

Systems of equations To simplify the presentation, we consider only the following system of order two,

$$y_1' = f_1(x, y_1, y_2) \quad y_1(x_0) = Y_{1,0}$$
$$y_2' = f_2(x, y_1, y_2) \quad y_2(x_0) = Y_{2,0} \tag{6.57}$$

The generalization to the higher-order systems is straightforward. The Euler's method is

$$y_{1,n+1} = y_{1,n} + hf_1(x_n, y_{1,n}, y_{2,n})$$

$$y_{2,n+1} = y_{2,n} + hf_2(x_n, y_{1,n}, y_{2,n}) \tag{6.58}$$

a clear generalization of (6.23).

All of the preceding results of this section generalize to (6.58); and the form of the generalization is clear if we use the vector notation for (6.57) and (6.58), as given in (6.21) in the last section. Use

$$\mathbf{y}' = \mathbf{f}(x, \mathbf{y}) \quad \mathbf{y}(x_0) = \mathbf{Y}_0$$

In place of absolute values, use the norm (1.19) from Chapter 1,

$$\|\mathbf{y}\| = \underset{i}{\text{Max}} |y_i|$$

To generalize the Lipschitz condition (6.29), Taylor's Theorem 1.5 for functions of several variables can be used to obtain

$$\|\mathbf{f}(x, \mathbf{z}) - \mathbf{f}(x, \mathbf{y})\| \leqslant K \|\mathbf{z} - \mathbf{y}\| \tag{6.59}$$

$$K = \underset{i}{\text{Max}} \sum_j \underset{\substack{x_0 \leqslant x \leqslant b \\ -\infty < w_1, w_2 < \infty}}{\text{Maximum}} \left| \frac{\partial f_i(x, w_1, w_2)}{\partial w_j} \right| \tag{6.60}$$

The role of $\partial f(x, y) / \partial y$ is replaced by the Jacobian matrix

$$\mathbf{f_y}(x, \mathbf{y}) = \begin{bmatrix} \dfrac{\partial f_1}{\partial y_1} & \dfrac{\partial f_1}{\partial y_2} \\ \dfrac{\partial f_2}{\partial y_1} & \dfrac{\partial f_2}{\partial y_2} \end{bmatrix}$$

As an example, the asymptotic error formula (6.46) becomes

$$\mathbf{Y}(x_n) - \mathbf{y}_h(x_n) = h\mathbf{D}(x_n) + \mathbf{R}_n \quad \|\mathbf{R}_n\| = O(h^2) \tag{6.61}$$

with $\mathbf{D}(x)$ the solution of the linear system

$$\mathbf{D}'(x)=\mathbf{f}_y(x,\mathbf{Y}(x))\mathbf{D}(x)+\tfrac{1}{2}\mathbf{Y}''(x) \qquad \mathbf{D}(x_0)=\boldsymbol{\delta}_0$$

using the above matrix $\mathbf{f}_y(x,\mathbf{y})$.

Example Solve the pendulum equation,

$$\ddot{\theta}(t)=-\sin(\theta(t)) \qquad \theta(0)=\pi/2 \qquad \dot{\theta}(0)=0$$

Convert this to a system by letting $y_1=\theta$, $y_2=\dot{\theta}$, and replace the variable t by x. Then

$$\begin{aligned} y_1'&=y_2 & y_1(0)&=\pi/2 \\ y_2'&=-\sin(y_1) & y_2(0)&=0 \end{aligned} \tag{6.62}$$

The numerical results are given in Table 6.5. Note that the error decreases by about half when h is halved.

Table 6.5 Euler's method for example (6.62)

h	x_n	$y_{1,n}$	$Y_1(x_n)$	Error	$y_{2,n}$	$Y_2(x_n)$	Error
.2	.2	1.5708	1.5508	−.0200	−.20000	−.199999	.000001
	.6	1.4508	1.3910	−.0598	−.59984	−.59806	.00178
	1.0	1.1711	1.0749	−.0962	−.99267	−.97550	.01717
.1	.2	1.5608	1.5508	−.0100	−.20000	−.199999	.000001
	.6	1.4208	1.3910	−.0298	−.59927	−.59806	.00121
	1.0	1.1223	1.0749	−.0474	−.98568	−.97550	.01018

6.3 Multistep Methods

This section contains an introduction to the theory of multistep methods. Some particular methods are examined in greater detail in the following sections. A more complete theory is given in Section 6.8.

As before, let $h>0$ and define the nodes by $x_n=x_0+nh, n \geqslant 0$. The general form of the multistep methods to be considered is

$$y_{n+1}=\sum_{j=0}^{p} a_j y_{n-j}+h\sum_{j=-1}^{p} b_j f(x_{n-j},y_{n-j}) \qquad n=p,p+1,\cdots \tag{6.63}$$

The coefficients $a_0,\cdots,a_p,b_{-1},b_0,\cdots,b_p$ are constants, and $p \geqslant 0$. If either $a_p \neq 0$ or $b_p \neq 0$, the method is called a $p+1$ step method, because $p+1$ previous solution values are being used to compute y_{n+1}. The values $y_1, \cdots, y_p$ must be obtained by other means; and this is discussed in later sections. Euler's method is an example of a one-step method, with $p=0$ and

$$a_0=1 \qquad b_0=1 \qquad b_{-1}=0$$

If $b_{-1}=0$, then y_{n+1} occurs on only the left side of equation (6.63). Such formulas are called *explicit methods*. If $b_{-1} \neq 0$, then y_{n+1} is present on both sides of (6.63), and the formula is called an *implicit method*. Implicit methods are generally solved by iteration, which will be discussed in detail for the Trape zoidal method in Section 6.5.

Example 1. The midpoint method is defined by

$$y_{n+1}=y_{n-1}+2hf(x_n,y_n) \qquad n \geqslant 1$$

and it is an explicit two-step method. It is examined in Section 6.4.

2. The trapezoidal method is defined by

$$y_{n+1}=y_n+\frac{h}{2}\left[f(x_n,y_n)+f(x_{n+1},y_{n+1})\right] \qquad n \geqslant 0$$

It is an implicit one-step method; and it is discussed in Sections 6.5 and 6.6.

For any differentiable function $Y(x)$, define the truncation error for integrating $Y'(x)$ by

$$T_{n+1}(Y)=Y(x_{n+1})-\left[\sum_{j=0}^{p} a_j Y(x_{n-j})+h \sum_{j=-1}^{p} b_j Y'(x_{n-j})\right] \qquad n \geqslant p \quad (6.64)$$

Define the function $\tau_{n+1}(Y)$ by

$$\tau_{n+1}(Y)=\frac{1}{h}T_{n+1}(Y) \tag{6.65}$$

In order to prove the convergence of the approximate solution $\{y_n|x_0 \leqslant x_n \leqslant b\}$ of (6.63) to the solution $Y(x)$ of the initial value problem (6.1), it is necessary to have

$$\tau(h) \equiv \operatorname*{Max}_{x_p \leqslant x_n \leqslant b} |\tau_{n+1}(Y)| \to 0 \qquad \text{as} \qquad h \to 0 \tag{6.66}$$

This is often called the *consistency condition* for the method (6.63). The speed of convergence of the solution $\{y_n\}$ to the true solution $Y(x)$ is related to the speed of convergence in (6.66), and thus we need to know the conditions under which

$$\tau(h) = O(h^m) \tag{6.67}$$

for some desired choice of $m \geqslant 1$. We now examine the implications of (6.66) and (6.67) for the coefficients in (6.63). The convergence result for (6.63) is given as Theorem 6.6.

Theorem 6.5 Let $m \geqslant 1$ be a given integer. In order that (6.66) hold for all continuously differentiable functions $Y(x)$, that is, that the method (6.63) be consistent, it is necessary and sufficient that

$$\sum_{j=0}^{p} a_j = 1 \qquad - \sum_{j=0}^{p} ja_j + \sum_{j=-1}^{p} b_j = 1 \tag{6.68}$$

And for (6.67) to be valid for all functions $Y(x)$ that are $m+1$ times continuously differentiable, it is necessary and sufficient that (6.68) hold and that

$$\sum_{j=0}^{p} (-j)^i a_j + i \sum_{j=-1}^{p} (-j)^{i-1} b_j = 1 \qquad i = 2, \ldots, m \tag{6.69}$$

Proof Note that

$$T_{n+1}(\alpha Y + \beta W) = \alpha T_{n+1}(Y) + \beta T_{n+1}(W) \tag{6.70}$$

for all constants α, β and all differentiable functions Y, W. To examine the consequences of (6.66) and (6.67), expand $Y(x)$ about x_n, using Taylor's Theorem 1.4, to obtain

$$Y(x) = \sum_{i=0}^{m} \frac{1}{i!} (x - x_n)^i Y^{(i)}(x_n) + R_{m+1}(x) \tag{6.71}$$

assuming $Y(x)$ is $m+1$ times continuously differentiable. Substituting into (6.64) and using (6.70),

$$T_{n+1}(Y) = \sum_{i=0}^{m} \frac{1}{i!} Y^{(i)}(x_n) T_{n+1}\big((x - x_n)^i\big) + T_{n+1}(R_{m+1}). \tag{6.72}$$

It is necessary to calculate $T_{n+1}((x - x_n)^i)$ for $i \geqslant 0$.
For $i = 0$,

$$T_{n+1}(1) = c_0 \equiv 1 - \sum_{j=0}^{p} a_j \tag{6.73}$$

For $i \geqslant 1$,

$$T_{n+1}\big((x-x_n)^i\big)$$

$$= (x_{n+1} - x_n)^i - \left[\sum_{j=0}^{p} a_j(x_{n-j} - x_n)^i + h \sum_{j=-1}^{p} b_j i (x_{n-j} - x_n)^{i-1} \right]$$

$$\tag{6.74}$$

$$= c_i h^i$$

$$c_i = 1 - \left[\sum_{j=0}^{p} (-j)^i a_j + i \sum_{j=-1}^{p} (-j)^{i-1} b_j \right] \qquad i \geqslant 1$$

This gives

$$T_{n+1}(Y) = \sum_{i=0}^{m} \frac{c_i}{i!} h^i Y^{(i)}(x_n) + T_{n+1}(R_{m+1}) \tag{6.75}$$

And if we write the remainder $R_{m+1}(x)$ as

$$R_{m+1}(x) = \frac{1}{(m+1)!} (x - x_n)^{m+1} Y^{(m+1)}(x_n) + \cdots$$

then

$$T_{n+1}(R_{m+1}) = \frac{c_{m+1}}{(m+1)!} h^{m+1} Y^{(m+1)}(x_n) + O(h^{m+2}) \tag{6.76}$$

To obtain the consistency condition (6.66), we need $\tau(h) = O(h)$; and this requires $T_{n+1}(Y) = O(h^2)$. Using (6.75) with $m=1$, we must have $c_0, c_1 = 0$, which gives the set of equations (6.68). In some texts, these are referred to as the consistency conditions. To obtain (6.67) for some $m \geqslant 1$, we must have $T_{n+1}(Y) = O(h^{m+1})$. From (6.75) and (6.76), this will be true if and only if $c_i = 0, i = 0, 1, \cdots, m$. This proves the conditions (6.69) and completes the proof. ∎

The largest value of m for which (6.67) holds is called the *order* or *order of convergence* of the method (6.63). In Section 6.7 procedures are given for deriving methods of any desired order.

We now give a convergence result for the solution of (6.63). Although the theorem will not include all methods (6.63) which are convergent, it does include all methods of current interest. Moreover, the proof is much easier than that of the more general Theorem 6.8 of Section 6.8.

Theorem 6.6 Consider solving the initial value problem

$$y' = f(x,y) \qquad y(x_0) = Y_0 \qquad x_0 \leqslant x \leqslant b$$

using the multistep method (6.63). Let the initial errors satisfy

$$\eta(h) = \underset{0 \leqslant i \leqslant p}{\text{Max}} |Y(x_i) - y_i| \to 0 \qquad \text{as} \qquad h \to 0 \qquad (6.77)$$

Assume the method is consistent, that is, it satisfies (6.66). And finally, assume that the coefficients a_j are all nonnegative,

$$a_j \geqslant 0 \qquad j = 0, 1, \cdots, p \qquad (6.78)$$

Then the method (6.63) is convergent, and

$$\underset{x_0 \leqslant x_n \leqslant b}{\text{Max}} |Y(x_n) - y_n| \leqslant c_1 \eta(h) + c_2 \tau(h) \qquad (6.79)$$

for suitable constants c_1, c_2. If the method (6.63) is of order m, and if the initial errors satisfy $\eta(h) = O(h^m)$, then the speed of convergence of the method (6.63) is $O(h^m)$.

Proof Rewrite (6.64) and use $Y'(x) = f(x, Y(x))$ to get

$$Y(x_{n+1}) = \sum_{j=0}^{p} a_j Y(x_{n-j}) + h \sum_{j=-1}^{p} b_j f(x_{n-j}, Y(x_{n-j})) + h\tau_{n+1}(Y).$$

Subtracting (6.63), and using the notation $e_i = Y(x_i) - y_i$,

$$e_{n+1} = \sum_{j=0}^{p} a_j e_{n-j} + h \sum_{j=-1}^{p} b_j \big[f(x_{n-j}, Y_{n-j})$$

$$-f(x_{n-j}, y_{n-j}) \big] + h\tau_{n+1}(Y).$$

Apply the Lipschitz condition and the assumption (6.78) to obtain

$$|e_{n+1}| \leqslant \sum_{j=0}^{p} a_j |e_{n-j}| + hK \sum_{j=-1}^{p} |b_j| |e_{n-j}| + h\tau(h)$$

Introduce the following error bounding function,

$$f_n = \underset{0 \leqslant i \leqslant n}{\text{Max}} |e_i| \qquad n = 0, 1, \cdots, N(h)$$

Using this function,

$$|e_{n+1}| \leqslant \sum_{j=0}^{p} a_j f_n + hK \sum_{j=-1}^{p} |b_j| f_{n+1} + h\tau(h)$$

and applying (6.68),

$$|e_{n+1}| \leqslant f_n + hcf_{n+1} + h\tau(h) \qquad c = K \sum_{j=-1}^{p} |b_j|$$

The right side is trivially a bound for f_n, and thus

$$f_{n+1} \leqslant f_n + hcf_{n+1} + h\tau(h)$$

For $hc \leqslant \frac{1}{2}$, which must be true as $h \rightarrow 0$,

$$f_{n+1} \leqslant \frac{f_n}{1-hc} + \frac{h}{1-hc}\tau(h)$$

$$\leqslant (1+2hc)f_n + 2h\tau(h)$$

Noting that $f_p = \eta(h)$, proceed as in the proof of Theorem 6.3 following (6.37). Then

$$f_n \leqslant e^{2c(b-x_0)}\eta(h) + \left[\frac{e^{2c(b-x_0)}-1}{c}\right]\tau(h) \qquad x_0 \leqslant x_n \leqslant b \qquad (6.80)$$

This completes the proof. ∎

The conclusions of the theorem can be proven under weaker assumptions; and, in particular, (6.78) can be replaced by a much weaker assumption. These results are given in Section 6.8. To obtain a rate of convergence of $O(h^m)$ for the method (6.63), it is necessary that each step have an error

$$T_{n+1}(Y) = O(h^{m+1})$$

But the initial values $y_0, \cdots, y_p$ need to be computed with only an accuracy of $O(h^m)$, since $\eta(h) = O(h^m)$ is sufficient in (6.79).

The result (6.80) can be improved somewhat for particular cases, but the speed of convergence remains the same. Examples of the theorem are given in following sections. As with Euler's method, a complete stability analysis can be given, including a result of the form (6.42). The proof is a straight-forward modification of the above proof of Theorem 6.6. Also, an asymptotic error analysis can be given; this is illustrated in the next two sections.

6.4 The Midpoint Method

As with Euler's method, the Midpoint method can be derived in several ways. Numerical differentiation will be used here.

Lemma 2 Assume $g(x)$ is three times continuously differentiable for $a - h \leqslant x \leqslant a + h$. Then

$$g'(a) = \frac{g(a+h) - g(a-h)}{2h} - \frac{h^2}{6} g'''(\zeta) \qquad (6.81)$$

for some $a - h \leqslant \zeta \leqslant a + h$

Proof Use Taylor's Theorem 1.4 to expand $g(a+h)$ and $g(a-h)$. Then

$$g(a+h) = g(a) + hg'(a) + \frac{h^2}{2} g''(a) + \frac{h^3}{6} g'''(\xi_1) \quad a \leqslant \xi_1 \leqslant a + h$$

$$g(a-h) = g(a) - hg'(a) + \frac{h^2}{2} g''(a) - \frac{h^3}{6} g'''(\xi_2) \quad a - h \leqslant \xi_2 \leqslant a$$

Subtracting and dividing by $2h$,

$$\frac{g(a+h) - g(a-h)}{2h} = g'(a) + \frac{h^2}{6} \left[\frac{g'''(\xi_1) + g'''(\xi_2)}{2} \right]$$

Since $g'''(x)$ is continuous on $[a - h, a + h]$, and since

$$\frac{g'''(\xi_1) + g'''(\xi_2)}{2} \qquad (6.82)$$

lies between the maximum and minimum of $g'''(x)$ on $[a - h, a + h]$, there is a point ζ in the interval at which $g'''(\zeta)$ equals (6.82). This completes the proof. ∎

Approximate $Y'(x_n) = f(x_n, Y(x_n))$ using (6.81):

$$\frac{Y(x_{n+1}) - Y(x_{n-1})}{2h} - \frac{h^2}{6} Y'''(\xi_n) = f(x_n, Y(x_n)),$$

$$Y(x_{n+1}) = Y(x_{n-1}) + 2hf(x_n, Y(x_n)) + \tfrac{1}{3} h^3 Y'''(\xi_n) \qquad (6.83)$$

for some $x_{n-1} \leqslant \xi_n \leqslant x_{n+1}$. The midpoint method for solving the initial value problem (6.1) is defined by

$$y_{n+1} = y_{n-1} + 2hf(x_n, y_n) \qquad n \geqslant 1 \tag{6.84}$$

It is a two-step explicit method, and the order of convergence is also two. The value of y_1 must be obtained by another method.

The midpoint method could have been obtained by applying the midpoint numerical integration rule (5.40) to the integral in the reformulation of (6.22),

$$Y(x_{n+1}) = Y(x_{n-1}) + \int_{x_{n-1}}^{x_{n+1}} f(t, Y(t)) dt$$

This approach is used in Section 6.7 to derive some higher-order formulas.

To use Theorem 6.6, note that

$$\tau_{n+1}(Y) = \tfrac{1}{3} h^2 Y'''(\xi_n) \qquad x_{n-1} \leqslant \xi_n \leqslant x_{n+1} \tag{6.85}$$

A slight modification of the proof of Theorem 6.6 yields

$$\underset{x_0 \leqslant x_n \leqslant b}{\text{Max}} |Y(x_n) - y_n| \leqslant e^{2K(b-x_0)} \eta(h) + \left[\frac{e^{2K(b-x_0)} - 1}{2K} \right] \left[\frac{1}{3} h^2 \| Y'' \|_\infty \right] \tag{6.86}$$

$$\eta(h) = \text{Max}\{|Y_0 - y_0|, |Y(x_1) - y_1|\}$$

Assuming $y_0 = Y_0$, we need to have $Y(x_1) - y_1 = O(h^2)$ in order to have a global order of convergence $O(h^2)$ in (6.86). From (6.24) with $n=0$, the Euler method with a single step has the desired property,

$$y_1 = y_0 + hf(x_0, y_0)$$

$$Y(x_1) - y_1 = \frac{h^2}{2} Y''(\xi) \qquad x_0 \leqslant \xi \leqslant x_1$$

With this initial guess, we have

$$\underset{x_0 \leqslant x_n \leqslant b}{\text{Max}} |Y(x_n) - y_n| = O(h^2)$$

A complete stability analysis can be given for the midpoint method, paralleling that given for Euler's method. And if we assume for simplicity that $\eta(h) = O(h^3)$, then we have the following asymptotic formula:

$$Y(x_n) - y_n = D(x_n) h^2 + O(h^3)$$

$$D' = f_y(x, Y(x)) D + \tfrac{1}{6} Y'''(x) \qquad D(x_0) = 0$$

There is little that is different in the proofs of these results, and we dispense with them.

Weak stability   In contrast to the earlier stability analysis that considered solutions on a fixed interval $[x_0, b]$, we are interested in the stability of the solution $\{y_n\}$ with respect to perturbations in the data as $x_n \to \infty$ and h remains fixed. We examine this for the midpoint method by considering the equation

$$y' = \lambda y \qquad y(0) = 1 \tag{6.87}$$

which has the solution $Y(x) = e^{\lambda x}$. This model problem enables us to better understand the midpoint method and how it will perform on other equations. A further discussion of the usefulness of (6.87) as a model for (6.1) is given in Section 6.8 following (6.169).

The midpoint method for (6.87) is

$$y_{n+1} = y_{n-1} + 2h\lambda y_n \qquad n \geqslant 1 \tag{6.88}$$

We calculate the exact solution of this equation and compare it to the true solution $e^{\lambda x}$. Equation (6.88) is an example of a linear difference equation of the second order. There is a general theory for pth-order linear difference equations that parallels the theory for pth-order linear differential equations. Most methods for solving the differential equation have an analog in solving the difference equation, and this is a guide in solving (6.88). We begin by looking for linearly independent solutions of the difference equation, and then these are combined to form the general solution. See Henrici [H1, pp. 210–215] for a general theory for linear difference equations.

In analogy with the exponential solutions of linear differential equations, we assume a solution for (6.88) of the form

$$y_n = r^n \qquad n \geqslant 0 \tag{6.89}$$

for some unknown r. Substituting into (6.88) to find necessary conditions on r,

$$r^{n+1} = r^{n-1} + 2h\lambda r^n$$

and canceling r^{n-1},

$$r^2 = 1 + 2h\lambda r \tag{6.90}$$

This argument is reversible. If r satisfies the quadratic equation (6.90), then (6.89) satisfies (6.88).

The equation (6.90) is called the *characteristic equation* for the midpoint method. Its roots are

$$r_1 = h\lambda + \sqrt{1 + h^2\lambda^2} \qquad r_2 = h\lambda - \sqrt{1 + h^2\lambda^2} \tag{6.91}$$

The general solution to (6.88) is then

$$y_n = \beta_1 r_1^n + \beta_2 r_2^n \qquad n \geqslant 0 \tag{6.92}$$

The coefficients β_1 and β_2 are chosen so that the values of y_0 and y_1, which were given originally, will agree with the values calculated using (6.92),

$$\beta_1 + \beta_2 = y_0$$

$$\beta_1 r_1 + \beta_2 r_2 = y_1$$

The general solution is

$$\beta_1 = \frac{y_1 - r_2 y_0}{r_1 - r_2} \qquad \beta_2 = \frac{y_0 r_1 - y_1}{r_1 - r_2}$$

To gain some intuition for these formulas, consider taking the exact initial values

$$y_0 = 1 \qquad y_1 = e^{\lambda h}$$

Then

$$\beta_1 = \frac{e^{\lambda h} - r_2}{2\sqrt{1 + h^2 \lambda^2}} = 1 + O(h^2 \lambda^2)$$

$$\beta_2 = \frac{r_1 - e^{\lambda h}}{2\sqrt{1 + h^2 \lambda^2}} = O(h^3 \lambda^3)$$

(6.93)

The estimates necessary for the proof of these estimates for β_1 and β_2 are discussed in the following paragraphs.

From these coefficients and (6.92), the term $\beta_1 r_1^n$ should correspond to the true solution $e^{\lambda x}$ and the term $\beta_2 r_2^n$ should usually be quite small in comparison. We need to obtain detailed estimates of r_1 and r_2 before concluding anything further.

Introduce the function

$$g(w) = w + \sqrt{1 + w^2}$$

Then

$$r_1 = g(\lambda h) \qquad r_2 = -g(-\lambda h)$$

(6.94)

Using the Taylor series for $\sqrt{1 + w^2}$,

$$g(w) = w + \left(1 + \tfrac{1}{2}w^2 - \tfrac{1}{8}w^4 + \cdots\right)$$

$$= \left(1 + w + \tfrac{1}{2}w^2 + \tfrac{1}{6}w^3 + \tfrac{1}{24}w^4 + \cdots\right)$$

$$- \left(\tfrac{1}{6}w^3 + \tfrac{1}{24}w^4 + \cdots\right) + \left(-\tfrac{1}{8}w^4 + \cdots\right),$$

$$g(w) = e^w + w^3 B(w)$$

(6.95)

with $B(w)$ bounded and continuous about zero.

Since we want to examine r_1^n and r_2^n, we first examine

$$g(w)^n = \left[e^w + w^3 B(w) \right]^n = e^{wn} \left[1 + w^3 e^{-w} B(w) \right]^n. \tag{6.96}$$

To avoid a more detailed analysis than is necessary, we restrict ourselves to a large fixed interval $[0, b]$. Then from (6.94) and (6.96),

$$r_1^n = e^{\lambda x_n} \left[1 + O(h^2) \right] \qquad r_2^n = (-1)^n e^{-\lambda x_n} \left[1 + O(h^2) \right]$$

and a general solution to (6.88) is given by

$$y_n = \beta_1 e^{\lambda x_n} \left[1 + O(h^2) \right] + \beta_2 (-1)^n e^{-\lambda x_n} \left[1 + O(h^2) \right] \tag{6.97}$$

Using the coefficients β_1 and β_2 from (6.93), it can be seen that the solution $\{ y_n \}$ converges to the true solution with an order of convergence $O(h^2)$.

The term $\beta_2 r_2^n$ is called a *parasitic solution* of (6.88) since it does not correspond to any solution of the original differential equation $y' = \lambda y$. The original differential equation $y' = \lambda y$ has a one-parameter family of solutions, but its approximation (6.88) has the two-parameter family (6.92). The new solution $\beta_2 r_2^n$ is a creation of the numerical method; and for the problem (6.87), it can cause difficulty.

If $\lambda > 0$, there is no difficulty since the parasitic solution decreases in magnitude with increasing x_n. But for $\lambda < 0$, $e^{\lambda x_n}$ decreases with increasing x_n, and $e^{-\lambda x_n}$ increases. Even though the coefficient β_2 is small, the term $\beta_2 r_2^n$ will increase as n increases, and it will eventually dominate the term $\beta_1 r_1^n$ corresponding to the true solution. The oscillation in sign of the parasitic solution is characteristic of this form of instability. Because of this instability, we say the midpoint method is only *weakly stable*.

This topic will be returned to in Section 6.8, after some necessary theory has been introduced. We generalize the applicability of the model problem (6.87) by considering the sign of $\partial f(x, Y(x))/\partial y$. If it is negative, then the weak instability of the midpoint method will usually appear in solving the associated initial value problem. This is illustrated in the second example below.

Table 6.6 Example (1) of midpoint method instability

x_n	y_n	$Y(x_n)$	Error
.25	.7500	.7788	.0288
.50	.6250	.6065	−.0185
.75	.4375	.4724	.0349
1.00	.4063	.3679	−.0384
1.25	.2344	.2865	.0521
1.50	.2891	.2231	−.0659
1.75	.0898	.1738	.0839
2.00	.2441	.1353	−.1088
2.25	−.0322	.1054	.1376

Example 1. Consider the model problem (6.87) with $\lambda = -1$. The numerical results are given in Table 6.6 for $h = 0.25$. The value y_1 was obtained using Euler's method, as discussed following (6.86). From the values in the table, the parasitic solution is clearly growing in magnitude. For $x_n = 2.25$, the numerical solution y_n becomes negative; and it alternates in sign with each succeeding step.

2.   Consider the problem

$$y' = x - y^2 \qquad y(0) = 0$$

The solution $Y(x)$ is strictly increasing for $x \geqslant 0$; and for large x, $Y(x) \approx \sqrt{x}$. Although it is increasing,

$$\frac{\partial f(x,y)}{\partial y} = -2y < 0 \qquad \text{for} \quad y > 0$$

Therefore, we expect that the midpoint method would exhibit some instability. This is confirmed with the results given in Table 6.7, obtained with a stepsize of $h = 0.25$. By $x_n = 2.25$, the numerical solution begins to decrease; and at $x_n = 3.25$, the solution y_n becomes negative.

Table 6.7   Example (2) of midpoint method instability

x_n	y_n	$Y(x_n)$	Error
.25	0.0	.0312	.0312
.50	.1250	.1235	−.0015
.75	.2422	.2700	.0278
1.00	.4707	.4555	−.0151
1.25	.6314	.6585	.0271
1.50	.8963	.8574	−.0389

6.5   The Trapezoidal Method

Integrate the equation (6.22) to obtain the reformulation

$$Y(x_{n+1}) = Y(x_n) + \int_{x_n}^{x_{n+1}} f(t, Y(t)) \, dt$$

Apply the simple trapezoidal method (5.10) and (5.12) to obtain

$$Y(x_{n+1}) = Y(x_n) + \frac{h}{2} \left[f(x_n, Y(x_n)) + f(x_{n+1}, Y(x_{n+1})) \right]$$

$$- \frac{h^3}{12} Y'''(\xi_n)$$

(6.98)

for some $x_n \leqslant \xi_n \leqslant x_{n+1}$. By dropping the remainder term, we obtain the trapezoidal method,

$$y_{n+1} = y_n + \frac{h}{2} \left[f(x_n, y_n) + f(x_{n+1}, y_{n+1}) \right] \qquad n \geqslant 0 \qquad (6.99)$$

It is a one-step method with an $O(h^2)$ order of convergence. It is also a simple example of an implicit method, since y_{n+1} occurs on both sides of (6.99). A numerical example is given at the end of this section.

Iterative solution The formula (6.99) is a nonlinear equation with root y_{n+1}, and any of the techniques of Chapter 2 could be used to solve it. Simple linear iteration is sufficient and most convenient; see Section 2.4. Let $y_{n+1}^{(0)}$ be a good initial guess of the solution y_{n+1}, and define

$$y_{n+1}^{(j+1)} = y_n + \frac{h}{2} \left[f(x_n, y_n) + f(x_{n+1}, y_{n+1}^{(j)}) \right] \qquad j = 0, 1, \dots \qquad (6.100)$$

The initial guess is usually obtained using an explicit method.

To analyze the iteration and to determine conditions under which it will converge, subtract (6.100) from (6.99). Take bounds and use the Lipschitz condition (6.29) to obtain

$$y_{n+1} - y_{n+1}^{(j+1)} = \frac{h}{2} \left[f(x_{n+1}, y_{n+1}) - f(x_{n+1}, y_{n+1}^{(j)}) \right]$$

$$\left| y_{n+1} - y_{n+1}^{(j+1)} \right| \leqslant \frac{hK}{2} \left| y_{n+1} - y_{n+1}^{(j)} \right| \qquad j \geqslant 0 \qquad (6.101)$$

If we assume that h is so small that

$$\frac{hK}{2} < 1 \qquad (6.102)$$

then $y_{n+1}^{(j)} \to y_{n+1}$ as $j \to \infty$. In practice, the initial guess $y_{n+1}^{(0)}$ and the stepsize h are generally chosen to insure that only one iterate need be computed. And then we take $y_{n+1} \doteq y_{n+1}^{(1)}$.

The computation of y_{n+1} from y_n contains a truncation error which is $O(h^3)$ (see (6.104) below for a precise statement). To maintain this order of accuracy, the eventual iterate $y_{n+1}^{(i)}$ which is chosen to represent y_{n+1} should satisfy $|y_{n+1} - y_{n+1}^{(i)}| = O(h^3)$. And if we want the iteration error to be less significant (as we will in the next section), then $y_{n+1}^{(i)}$ should be chosen to satisfy

$$\left| y_{n+1} - y_{n+1}^{(i)} \right| = O(h^4) \qquad (6.103)$$

To analyze the error in choosing an initial guess $y_{n+1}^{(0)}$, let $u_n(x)$ denote the solution of $y' = f(x, y)$ which passes through (x_n, y_n). We need to derive a more detailed statement on the truncation error. To begin,

$$u_n(x_{n+1}) = y_n + \frac{h}{2}\left[f(x_n, y_n) + f(x_{n+1}, u_n(x_{n+1}))\right] - \frac{h^3}{12}u_n'''(\xi_n)$$

for some $x_n \leqslant \xi_n \leqslant x_{n+1}$. Let $\tilde{e}_{n+1} = u_n(x_{n+1}) - y_{n+1}$, the local error in computing y_{n+1} from y_n. Subtracting (6.99),

$$\tilde{e}_{n+1} = \frac{h}{2}\left[f(x_{n+1}, u_n(x_{n+1})) - f(x_{n+1}, y_{n+1})\right] - \frac{h^3}{12}u_n'''(\xi_n)$$

$$= \frac{h}{2}f_y(x_{n+1}, y_{n+1})\tilde{e}_{n+1} + O\left(h\tilde{e}_{n+1}^2\right) - \frac{h^3}{12}u_n'''(\xi_n).$$

Solving for $\tilde{e}_{n+1}$,

$$\tilde{e}_{n+1} = \left[1 - \frac{h}{2}f_y(x_{n+1}, y_{n+1})\right]^{-1}\left[-\frac{h^3}{12}u_n'''(\xi_n) + O(h^4)\right]$$

and noting that $u_n'''(\xi_n) = u_n'''(x_n) + O(h)$,

$$u_n(x_{n+1}) - y_{n+1} = -\frac{h^3}{12}u_n'''(x_n) + O(h^4) \tag{6.104}$$

This shows the local error is essentially the truncation error, for one step of (6.99).

If Euler's method is used to compute $y_{n+1}^{(0)}$,

$$y_{n+1}^{(0)} = y_n + hf(x_n, y_n) \tag{6.105}$$

then $u_n(x_{n+1})$ can be expanded to show that

$$u_n(x_{n+1}) - y_{n+1}^{(0)} = \frac{h^2}{2}u_n''(\zeta_n) \qquad x_n \leqslant \zeta_n \leqslant x_{n+1}$$

Combined with (6.104),

$$y_{n+1} - y_{n+1}^{(0)} = O(h^2)$$

To satisfy (6.103), the bound (6.101) implies that two iterates will have to be computed; and then we use $y_{n+1}^{(2)}$ to represent y_{n+1}.

Using the midpoint method, we can obtain the more accurate initial guess

$$y_{n+1}^{(0)} = y_{n-1} + 2hf(x_n, y_n) \tag{6.106}$$

To estimate the error, begin with the identity

$$u_n(x_{n+1}) = u_n(x_{n-1}) + 2hf(x_n, u_n(x_n)) + \frac{h^3}{3}u_n'''(\eta_n)$$

for some $x_{n-1} \leqslant \eta_n \leqslant x_{n+1}$. Subtracting (6.106),

$$u_n(x_{n+1}) - y_{n+1}^{(0)} = u_n(x_{n-1}) - y_{n-1} + \frac{h^3}{3} u_n'''(\eta_n)$$

The quantity $u_n(x_{n-1}) - y_{n-1}$ can be computed in a manner similar to that used for (6.104) with about the same result,

$$u_n(x_{n-1}) - y_{n-1} = \frac{h^3}{12} u_n'''(x_n) + O(h^4)$$

Then

$$u_n(x_{n+1}) - y_{n+1}^{(0)} = \frac{5h^3}{12} u_n'''(x_n) + O(h^4)$$

And combining this with (6.104),

$$y_{n+1} - y_{n+1}^{(0)} = \frac{h^3}{2} u_n'''(x_n) + O(h^4) \tag{6.107}$$

With the initial guess (6.106), one iterate from (6.100) will be sufficient to satisfy (6.103), based on the bound in (6.101).

The formulas (6.105) and (6.106) are often called *predictor formulas*, and the trapezoidal iteration formula (6.100) is called a *corrector formula*. Together they form a predictor-corrector method, and they are the basis of a method that can be used to control the size of the truncation error. This is illustrated in the next section.

Convergence and stability results The convergence of the trapezoidal method (6.99) is assured by Theorem 6.6. Assuming $hK \leqslant 1$ [although $(hK/2) < 1$ is sufficient],

$$\underset{x_0 \leqslant x_n \leqslant b}{\text{Max}} |Y(x_n) - y_n| \leqslant e^{2K(b-x_0)} |e_0|$$

$$+ \left[\frac{e^{2K(b-x_0)} - 1}{K} \right] \left[\frac{h^2}{12} \| Y''' \|_\infty \right] \tag{6.108}$$

The derivation of an asymptotic error formula is similar to that for Euler's method. Assuming $e_0 = \delta_0 h^2 + O(h^3)$, we can show

$$Y(x_n) - y_n = D(x_n) h^2 + O(h^3) \tag{6.109}$$

$$D'(x) = f_y(x, Y(x)) D(x) - \tfrac{1}{12} Y'''(x) \qquad D(x_0) = \delta_0$$

As with the midpoint method, we consider the behavior of the trapezoidal rule applied to $y' = \lambda y$, $y(0) = 1$. The comparison with the earlier work is somewhat poor since the trapezoidal rule is a one-step method and thus has no parasitic solution. Nonetheless, it has interesting properties which are worth examining. Applying the trapezoidal method,

$$y_{n+1} = y_n + \frac{h}{2}\lambda(y_n + y_{n+1}) \qquad y_0 = 1$$

$$y_{n+1} = \left[\frac{1 + \frac{h}{2}\lambda}{1 - \frac{h}{2}\lambda} \right] y_n \qquad n \geqslant 0$$

Assuming

$$h\lambda/2 < 1$$

we have inductively

$$y_n = \left[\frac{1 + \frac{1}{2}h\lambda}{1 - \frac{1}{2}h\lambda} \right]^n \qquad n \geqslant 0$$

If $\lambda < 0$, then for any $h > 0$ we have $y_n \to 0$ as $x_n \to \infty$. This is an especially good form of stability, which most methods do not have. It is called *A-stability*, and it is a desirable property if the differential equations being solved are stiff. For more discussion of A-stability, see Gear [G1, p. 43].

Richardson extrapolation   This was introduced in Section 5.5 of Chapter 5, and it was used both to predict the error [as in (5.109)] and to produce more rapidly convergent numerical integration methods [as in (5.104)]. It can be used in both ways in solving differential equations, although we use it mainly to predict the error.

   Let $y_h(x)$ and $y_{h/2}(x)$ denote numerical solutions to (6.1) on $[x_0, b]$, obtained using the trapezoidal method with stepsizes h and $h/2$, respectively. Using (6.109),

$$Y(x_n) - y_h(x_n) = D(x_n)h^2 + O(h^3)$$

$$Y(x_n) - y_{h/2}(x_n) = D(x_n)\frac{h^2}{4} + O(h^3)$$

Multiply the second equation by four, subtract the first, and solve for $Y(x_n)$ to obtain

$$Y(x_n) = \tfrac{1}{3}\left[4y_{h/2}(x_n) - y_h(x_n) \right] + O(h^3) \qquad (6.110)$$

The formula on the right has a higher order of convergence than the trapezoidal method; but it requires computation of both $y_h(x_n)$ and $y_{h/2}(x_n)$ for all x_n in $[x_0, b]$.

The formula (6.110) is of greater use in predicting the global error in $y_{h/2}(x_n)$. Using (6.110),

$$Y(x_n) - y_{h/2}(x_n) = \tfrac{1}{3}[y_{h/2}(x_n) - y_h(x_n)] + O(h^3)$$

$$Y(x_n) - y_{h/2}(x_n) \approx \tfrac{1}{3}[y_{h/2}(x_n) - y_h(x_n)] \tag{6.111}$$

This is a correct asymptotic estimate since $Y(x_n) - y_{h/2}(x_n)$ and $\tfrac{1}{3}[y_{h/2}(x_n) - y_h(x_n)]$ are both $O(h^2)$, and their difference is $O(h^3)$. This is a practical way to estimate the global error; although we have not derived it so as to allow for a variable stepsize.

Example   Consider the problem

$$y' = -y^2 \qquad y(0) = 1$$

which has the solution $Y(x) = 1/(1+x)$. The results in Table 6.8 are for stepsizes of $h = .5$ and $h/2 = .25$. The last column is the error estimate (6.111); and it is an accurate estimate for the true error $Y(x) - y_{h/2}(x)$.

Table 6.8   Trapezoidal method and Richardson extrapolation

x	$y_h(x)$	$Y(x) - y_h(x)$	$y_{h/2}(x)$	$Y(x) - y_{h/2}(x)$	$\tfrac{1}{3}[y_{h/2}(x) - y_h(x)]$
1.0	.483144	.016856	.496021	.003979	.004292
2.0	.323610	.009723	.330991	.002342	.002460
3.0	.243890	.006110	.248521	.001479	.001543
4.0	.195838	.004162	.198991	.001009	.001051
5.0	.163658	.003008	.165937	.000730	.000759

6.6   A Low-Order Predictor-Corrector Algorithm

In this section, a fairly simple algorithm is described for solving the initial value problem (6.1). It uses the trapezoidal method (6.99), and it controls the size of the truncation error by varying the stepsize h. The method is not practical because of its low order of convergence; but it demonstrates some of the ideas and techniques involved in constructing a variable-stepsize predictor-corrector algorithm. It is also simpler to understand than algorithms based on higher order methods.

Each step from x_n to x_{n+1} consists of constructing y_{n+1} from y_n and y_{n-1}; and y_{n+1} is an approximate solution of (6.99) based on using some iterate from

(6.100). A regular step has $x_{n+1} - x_n = x_n - x_{n-1} = h$, the midpoint predictor (6.106) is used, and the truncation error is predicted using the difference of the predictor and corrector formulas. When the stepsize is being changed, the Euler predictor (6.105) is used.

The user of the algorithm will have to specify several parameters in addition to those defining the differential equation problem (6.1). The stepsize h will vary, and the user must specify values h_{min} and h_{max}, which limit the size of h. The user should also specify an initial value for h; and the value should be one for which $\frac{1}{2} h f_y(x_0, y_0)$ is sufficiently less than one in magnitude, say less than .1. This quantity will determine the speed of convergence of the iteration in (6.100) and will be discussed later in the section, following the numerical example. An error tolerance ϵ must be given, and the stepsize h is so chosen that the truncation error *trunc* satisfies

$$\tfrac{1}{4}\epsilon h \leqslant |\text{trunc}| \leqslant \epsilon h \qquad (6.112)$$

at each step. This is called controlling the *error per unit stepsize*. Its significance is discussed near the end of the section.

The notation of the preceding section is continued. The function $u_n(x)$ will be the solution of $y' = f(x, y)$, which passes through (x_n, y_n). The truncation error to be estimated and controlled is (6.104), which is the error in obtaining $u_n(x_{n+1})$ using the trapezoidal method,

$$u_n(x_{n+1}) - y_{n+1} = -\frac{h^3}{12} u_n'''(x_n) + O(h^4) \qquad h = x_{n+1} - x_n \qquad (6.113)$$

If y_n is sufficiently close to $Y(x_n)$, then this is a good approximation to the closely related truncation error in (6.98).

$$-\frac{h^3}{12} Y'''(\xi_n)$$

And (6.113) is the only quantity for which we have the information needed to control it.

Choosing the initial stepsize The problem is to find an initial value of h and node $x_1 = x_0 + h$, for which $|y_1 - Y(x_1)|$ satisfies the bounds in (6.112). With the initial h supplied by the user, the value $y_h(x_1)$ is obtained by using the Euler predictor (6.105) and iterating twice in (6.100). Using the same procedure, the values $y_{h/2}(x_0 + h/2)$ and $y_{h/2}(x_1)$ are also calculated. The Richardson extrapolation procedure is used to predict the error in $y_h(x_1)$,

$$Y(x_1) - y_h(x_1) \approx \tfrac{4}{3} [y_{h/2}(x_1) - y_h(x_1)] \qquad (6.114)$$

If this error satisfies the bounds of (6.112), then the value of h is accepted; and the regular trapezoidal step using the midpoint predictor (6.106) is begun. But if (6.112) is not satisfied by (6.114), then a new value of h is chosen.

Using the values

$$f_0 = f(x_0, y_0), \qquad f_1 = f(x_0 + h/2, y_{h/2}(x_0 + h/2))$$

$$f_2 = f(x_1, y_{h/2}(x_1))$$

obtain the approximation

$$y'''(x_0) \approx D_3 y \equiv (f_2 - 2f_1 + f_0)/(h^2/4) \tag{6.115}$$

This is an approximation using the second-order divided difference of $Y' = f(x, y)$; for example, apply Lemma 2 in Section 3.4, Chapter 3. For any small stepsize h, the truncation error at $x_0 + h$ is well approximated by

$$-\frac{h^3}{12} Y'''(\xi_0) \approx -\frac{h^3}{12} D_3 y$$

The new stepsize h is chosen so that

$$\left| \frac{h^3}{12} D_3 y \right| = \tfrac{1}{2} \epsilon h$$

$$h = \sqrt{\frac{6\epsilon}{|D_3 y|}} \tag{6.116}$$

This should place the initial truncation error in approximately the middle of the range (6.112) for the error per unit step criteria. With this new value of h, the test (6.114) is again repeated, as a safety check.

By choosing h so that the truncation error will satisfy the bound (6.112) when it is doubled or halved, we ensure that the stepsize will not have to be changed for several steps, provided the derivative $Y'''(x)$ is not changing rapidly. Changing the stepsize will be more expensive than a normal step, and we want to minimize the need for such changes.

The regular predictor-corrector step   The stepsize h satisfies $x_n - x_{n-1} = x_{n+1} - x_n = h$. To solve for the value y_{n+1}, use the midpoint predictor (6.106) and iterate once in (6.100). The truncation error is estimated from (6.113); and combining with (6.107),

$$-\frac{1}{6}(y_{n+1} - y_{n+1}^{(0)}) = -\frac{h^3}{12} u_n'''(x_n) + O(h^4) \tag{6.117}$$

Thus we measure the truncation error (6.113) using

$$\text{trunc} = -\tfrac{1}{6}(y_{n+1} - y_{n+1}^{(0)}) \tag{6.118}$$

If trunc satisfies (6.112), then the value of h is not changed and calculation continues with this regular step procedure. But when (6.112) is not satisfied, the values y_{n+1} and x_{n+1} are discarded; and a new stepsize is chosen based on the value of trunc.

Changing the stepsize   Using (6.117),

$$-\frac{u_n'''(x_n)}{12} \approx \frac{\text{trunc}}{h_0^3}$$

where h_0 denotes the stepsize used in obtaining trunc. For an arbitrary stepsize h, the truncation error in obtaining y_{n+1} is estimated by

$$-\frac{h^3}{12} u_n'''(x_n) \approx \left(\frac{h}{h_0}\right)^3 \text{trunc}$$

Choose h so that

$$\left(\frac{h}{h_0}\right)^3 |\text{trunc}| = \tfrac{1}{2}\epsilon h$$

$$h = \sqrt{\frac{\epsilon h_0^3}{2 \cdot |\text{trunc}|}} \tag{6.119}$$

Calculate y_{n+1} by using the Euler predictor and iterating twice in (6.100). Then return to the regular predictor-corrector step. To avoid rapid changes in h, which can lead to significant errors, the new value of h is never allowed to be more than twice the previous value. If the new value of h is less than $h_{\min}$, then calculation is terminated. But if the new value of h is greater than $h_{\max}$, we just let $h = h_{\max}$ and proceed with the calculation. This has possible problems, which will be discussed following the numerical example.

Algorithm   Detrap $(f, x_0, y_0, x_{\text{end}}, \epsilon, h, h_{\min}, h_{\max}, \text{ier})$

1. **Remark:**   the problem being solved is $Y' = f(x, Y)$, $Y(x_0) = y_0$, for $x_0 \leqslant x \leqslant x_{\text{end}}$, using the method described earlier in the section. The approximate solution values are printed at each node point. The error parameter ϵ and the stepsize parameters were discussed earlier in the section. The variable *ier* is an error indicator, output when exiting the algorithm. $\text{ier} = 0$ means a normal return, $\text{ier} = 1$ means that $h = h_{\max}$ at some node points, and $\text{ier} = 2$ means that the integration was terminated due to a necessary $h < h_{\min}$.

2. Initialize loop: $=1$, ier: $=0$.

3. Remark: choose an initial value of h.

4. Calculate $y_h(x_0+h)$, $y_{h/2}(x_0+\frac{1}{2}h)$, $y_{h/2}(x_0+h)$ using the method (6.99). In each case, use the Euler predictor (6.105) and follow it by two iterations of (6.100).

5. For the error in $y_h(x_0+h)$, use

$$\text{trunc:} = \tfrac{4}{3}\left[y_{h/2}(x_0+h)-y_h(x_0+h)\right]$$

6. If $\frac{1}{4}\epsilon h \leqslant |\text{trunc}| \leqslant \epsilon h$, or if loop$=2$, then $x_1:=x_0+h$, $y_1:=y_h(x_0+h)$, print x_1,y_1, and go to step 10.

7. Calculate $D_3 y \approx Y'''(x_0)$ from (6.115). Let

$$h: = \left[6\epsilon/|D_3 y|\right]^{1/2}$$

If $D_3 y=0$, then $h:=h_{\max}$ and loop$:=2$.

8. If $h<h_{\min}$, then ier$:=2$ and exit. If $h>h_{\max}$, then $h:=h_{\max}$, ier$:=1$, loop$:=2$.

9. Go to step 4.

10. Remark: this portion of the algorithm contains the regular predictor-corrector step with error control.

11. Let $x_2:=x_1+h$, and $y_2^{(0)}:=y_0+2hf(x_1,y_1)$. Iterate (6.100) once to obtain y_2.

12. $\text{trunc:} = -\tfrac{1}{6}(y_2-y_2^{(0)})$.

13. If $|\text{trunc}|>\epsilon h$ or $|\text{trunc}|<\frac{1}{4}\epsilon h$, then go to step 16.

14. Print x_2,y_2.

15. $x_0:=x_1$, $x_1:=x_2$, $y_0:=y_1$, $y_1:=y_2$. If $x_1<x_{\text{end}}$, then go to step 11. Otherwise exit.

16. Remark: change the stepsize.

17. $x_0:=x_1$, $y_0:=y_1$, $h_0:=h$, and calculate h using (6.119).

18. $h:=\text{Minimum }\{h,2h_0\}$.

19. If $h<h_{\min}$, then ier$:=2$ and exit. If $h>h_{\max}$, then ier$:=1$ and $h:=h_{\max}$.

20. $y_1^{(0)}:=y_0+hf(x_0,y_0)$, and iterate twice in (6.100) to calculate y_1.

21. Print x_1,y_1.

22. If $x_1<x_{\text{end}}$, then go to step 10. Otherwise exit.

The following example uses an implementation of Detrap that also prints trunc; and a section was added to predict the truncation error in y_1 of step 20.

Example   Consider the problem

$$y' = \frac{1}{1+x^2} - 2y^2 \qquad y(0) = 0 \tag{6.120}$$

which has the solution

$$Y(x) = \frac{x}{1+x^2}$$

This is an interesting problem for testing Detrap, and it performs quite well. The equation was solved on $[0, 7]$ with $h_{min} = .001$, $h_{max} = 1.0$, $h = 0.2$, and $\epsilon = .0005$. Table 6.9 contains some of the results, including the true global error. Only selected sections of output are shown because of space.

Several points are illustrated using the example. First, step 18 is necessary in order to avoid stepsizes that are far too large. For the numerical example (6.120), we have

$$Y'''(x) = \frac{-6(x^4 - 6x^2 + 1)}{(1+x^2)^4}$$

which is zero at $x = \pm .414, \pm 2.414$. Thus the truncation error in solving the

Table 6.9   Example of algorithm Detrap

x_n	h	y_n	$Y(x_n) - y_n$	trunc
.02376	.02376	.02376	6.7E−6	6.7E−6
.04751	.02376	.047393	1.3E−5	6.7E−6
.07127	.02376	.070889	2.0E−5	6.6E−6
.09503	.02376	.094149	2.6E−5	6.4E−6
.28508	.02376	.263594	5.7E−5	3.1E−6
.32083	.03576	.290830	6.2E−5	6.8E−6
.35659	.03576	.316299	6.5E−5	6.2E−6
.42811	.07151	.361738	5.9E−5	1.7E−6
.48640	.05830	.393303	4.0E−5	1.5E−5
.54470	.05830	.420054	1.2E−5	1.7E−5
.60299	.05830	.442230	−2.3E−5	2.6E−5
.64179	.03880	.454592	−3.1E−5	1.0E−5
1.98527	.11853	.401876	−1.0E−4	1.8E−5
2.22232	.23705	.374289	−7.9E−5	1.2E−6
2.69642	.47410	.325595	4.3E−4	5.5E−4
2.99928	.30286	.299680	3.8E−4	9.5E−5
3.30214	.30286	.277047	3.5E−4	5.8E−5
3.60500	.30286	.257247	3.3E−4	5.5E−5
3.90786	.30286	.239863	3.0E−4	4.9E−5
4.21072	.30286	.224528	2.8E−4	4.1E−5
4.66700	.45628	.204557	3.1E−4	1.3E−4

differential equation (6.120) will be very small near these points, as can be seen by combining (6.117), (6.118), and (6.119). This leads to a prediction of a very large h, one that will be too large for following points x_n. At both $x_n = .35659$ and $x_n = 2.22232$, the step 18 was needed to avoid a misleadingly large value of h.

In all of the derivations of this and the preceding section, estimates were made that were accurate if h was sufficiently small. In most cases, the crucial quantity is actually $hf_y(x_n, y_n)$, as in (6.102) when analyzing the rate of convergence of the iteration (6.100). The rate of convergence for solving (6.99) by the iteration (6.100) is

$$\tfrac{1}{2} hf_y(x_n, y_n)$$

and for the problem (6.120) the rate of convergence is

$$\text{rate} = -2hy_n$$

If this is near one, then several iterations are necessary to obtain an accurate estimate of y_{n+1}. From the table, the rate is roughly increasing in size as h increases. At $x_n = 4.66700$, this rate is about .19. This seems small enough, but the effect shows itself in the inaccuracy of the prediction of the truncation error. The truncation error in going from $x_n = 4.21072$ to $x_n = 4.66700$ should be about

$$\frac{h^3}{12} Y'''(4.44) = .000024$$

whereas the algorithm predicts .00013. By inference, the value of y_n at $x_n = 4.66700$ was not obtained with sufficient accuracy. The algorithm can be made more sophisticated in order to detect the problems of too large an h, but setting a reasonably sized h_{max} will also help.

The global error We begin by giving an error bound analogous to the bound (6.108) for a fixed stepsize. Write

$$Y(x_{n+1}) - y_{n+1} = \left[Y(x_{n+1}) - u_n(x_{n+1}) \right] + \left[u_n(x_{n+1}) - y_{n+1} \right] \quad (6.121)$$

For the last term, we assume the error per unit step criteria (6.112) is satisfied,

$$|u_n(x_{n+1}) - y_{n+1}| \leqslant \epsilon(x_{n+1} - x_n). \quad (6.122)$$

For the other term in (6.121), introduce the integral equation reformulations

$$Y(x) = Y(x_n) + \int_{x_n}^{x} f(t, Y(t)) dt$$

$$u_n(x) = y_n + \int_{x_n}^{x} f(t, u_n(t)) dt \qquad x \geqslant x_n \quad (6.123)$$

Subtract and take bounds using the Lipschitz condition (6.29), to obtain

$$|Y(x) - u_n(x)| \leqslant |e_n| + K(x - x_n) \underset{x_n \leqslant t \leqslant x}{\text{Max}} |Y(t) - u_n(t)| dt$$

with $e_n = Y(x_n) - y_n$. Using this,

$$|Y(x_{n+1}) - u_n(x_{n+1})| \leqslant \frac{1}{1 - K(x_{n+1} - x_n)} |e_n| \qquad (6.124)$$

Introduce

$$H = \underset{x_0 \leqslant x_n \leqslant b}{\text{Maximum}} (x_{n+1} - x_n) = \text{Max } h$$

and assume that

$$HK < 1$$

Combining (6.124), (6.122), and (6.121), we obtain

$$|e_{n+1}| \leqslant \frac{1}{1 - HK} |e_n| + \epsilon H \qquad x_0 \leqslant x_n \leqslant b$$

This is easily solved just as in Section 6.2; and we obtain

$$|Y(x_n) - y_n| \leqslant e^{c(b - x_0)K} |Y(x_0) - y_0| + \left[\frac{e^{c(b - x_0)} - 1}{c} \right] \epsilon, \qquad (6.125)$$

for an appropriate constant $c > 0$. This is the basic error result when using a variable stepsize, and it is a partial justification of the error criteria (6.112).

In some situations, we can obtain a more realistic bound. For simplicity, assume $f_y(x,y) < 0$ for all (x,y) of interest. Subtracting in (6.123),

$$Y(x) - u_n(x) = e_n + \int_{x_n}^x \left[f(t, Y(t)) - f(t, u_n(t)) \right] dt$$

$$= e_n + \int_{x_n}^x \frac{\partial f(t, \zeta(t))}{\partial y} \left[Y(t) - u_n(t) \right] dt$$

The last step uses the mean-value theorem (1.2); and it can be shown that $\partial f(t, \zeta(t))/\partial y$ is a continuous function of t. This shows that $v(x) \equiv Y(x) - u_n(x)$ is a solution of the linear problem

$$v' = \frac{\partial f(t, \zeta(t))}{\partial y} v \qquad v(x_n) = e_n$$

The solution of this linear problem, along with the assumption $f_y(x,y) < 0$, implies

$$|Y(x_{n+1}) - u_n(x_{n+1})| \leqslant |e_n|$$

Combining with (6.121) and (6.122),

$$|Y(x_{n+1}) - y_{n+1}| \leqslant |Y(x_n) - y_n| + \epsilon(x_{n+1} - x_n)$$

Solving the inequality, we obtain the more realistic bound

$$|Y(x_n) - y_n| \leqslant |Y(x_0) - y_0| + \epsilon(x_n - x_0) \tag{6.126}$$

This partially explains the good behavior of the example in Table 6.9; and even better theoretical results are possible. But the results (6.125) and (6.126) are sufficient justification for the use of the test (6.122), which controls the error per unit step. For systems of equations $y' = f(x, y)$, the condition $f_y(x, y) < 0$ is replaced by requiring that all eigenvalues of the Jacobian matrix $f_y(x, y)$ have real parts that are negative.

The Detrap algorithm could be improved in a number of ways. But it illustrates the construction of a predictor-corrector algorithm with variable stepsize. The output is printed at an inconvenient set of node points x_n, but a simple interpolation algorithm can take care of this. The predictors can be improved, but that too would make the algorithm more complicated. In the next section, we return to a discussion of currently available practical predictor-corrector algorithms, most of which also vary the order.

6.7   Derivation of Higher-Order Multistep Methods

Recall from Section 6.3 the general formula for a $p + 1$ step method for solving (6.1).

$$y_{n+1} = \sum_{j=0}^{p} a_j y_{n-j} + h \sum_{j=-1}^{p} b_j f(x_{n-j}, y_{n-j}) \qquad n \geqslant p \tag{6.127}$$

A theory was given for these methods in Section 6.3, and some specific higher order methods will now be derived. There are two principal means of deriving higher order formulas: (1) The method of undetermined coefficients, and (2) numerical integration. The methods based on numerical integration are currently the most popular, but the method of undetermined coefficients is still important.

The implicit formulas can be solved by iteration, in complete analogy with (6.99) and (6.100) for the trapezoidal method. If $b_{-1} \neq 0$ in (6.127), the iteration is defined by

$$y_{n+1}^{(l+1)} = \sum_{j=0}^{p} \left[a_j y_{n-j} + h b_j f(x_{n-j}, y_{n-j}) \right]$$

$$+ h b_{-1} f(x_{n+1}, y_{n+1}^{(l)}) \qquad l \geqslant 0 \tag{6.128}$$

The iteration converges if $hb_{-1}K < 1$, where K is the Lipschitz constant for $f(x,y)$, contained in (6.29). The linear rate of convergence will be bounded by $hb_{-1}K$.

We are looking for pairs of formulas, a corrector and a predictor. Suppose that the corrector formula has order m, that is, the truncation error is $O(h^{m+1})$ at each step. Usually only one iterate is computed in (6.128), and this means that the predictor must have order at least $m-1$ in order that the truncation error in $y_{n+1}^{(1)}$ is also $O(h^{m+1})$. See the discussion of the trapezoidal method iteration error in Section 6.5, using the Euler and midpoint predictors. The essential ideas transfer to (6.127) and (6.128) without any significant change.

The method of undetermined coefficients   If formula (6.127) is to have order $m \geq 1$, then from Theorem 6.5 it is necessary and sufficient that

$$\sum_{j=0}^{p} a_j = 1$$

$$\sum_{j=0}^{p} a_j(-j)^i + i \sum_{j=-1}^{p} b_j(-j)^{i-1} = 1 \qquad i = 1, 2, \cdots, m \qquad (6.129)$$

For an explicit method, there is the additional condition that $b_{-1} = 0$.

For a general implicit method, there are $2p+3$ parameters $\{a_j, b_j\}$ to be determined, and there are $m+1$ equations. It might be thought that we could take $m+1 = 2p+3$, but this would be extremely unwise from the viewpoint of the stability and convergence of (6.127). This will be illustrated in the next section and in the problems. Generally, it is best to let $m \leq p+2$ for an implicit method. For an explicit method, stability considerations are not as important because the method is usually a predictor for an implicit formula.

Example   Find all second-order, two-step methods. The formula (6.127) is

$$y_{n+1} = a_0 y_n + a_1 y_{n-1} + h\big[b_{-1} f(x_{n+1}, y_{n+1}) + b_0 f(x_n, y_n)$$

$$+ b_1 f(x_{n-1}, y_{n-1})\big] \qquad n \geq 1 \qquad (6.130)$$

The coefficients must satisfy (6.129) with $m = 2$:

$$a_0 + a_1 = 1 \qquad -a_1 + b_{-1} + b_0 + b_1 = 1 \qquad a_1 + 2b_{-1} - 2b_1 = 1$$

Solving

$$a_1 = 1 - a_0 \qquad b_{-1} = 1 - \tfrac{1}{4}a_0 - \tfrac{1}{2}b_0 \qquad b_1 = 1 - \tfrac{3}{4}a_0 - \tfrac{1}{2}b_0 \qquad (6.131)$$

with a_0, b_0 variable. The midpoint method is a special case, in which $a_0 = 0$, $b_0 = 2$. The coefficients a_0, b_0 can be chosen to improve the stability, give a small truncation error, give an explicit formula, or some combination of these. The conditions to ensure stability and convergence (other than $0 \leqslant a_0 \leqslant 1$ using Theorem 6.6) cannot be given until the general theory for (6.127) has been given in the next section.

With (6.129) satisfied, Theorem 6.5 implies that the truncation error satisfies

$$T_{n+1}(Y) = O(h^{m+1})$$

for all $Y(x)$ that are $m+1$ times continuously differentiable on $[x_0, b]$. But for practical purposes in constructing predictor-corrector algorithms, it is sometimes preferable to have an error of the form

$$T_{n+1}(Y) = d_m h^{m+1} Y^{(m+1)}(\xi_n) \qquad x_{n-p} \leqslant \xi_n \leqslant x_{n+1} \qquad (6.132)$$

for some constant d_m, independent of n. Examples are Euler's method, the midpoint method, and the trapezoidal method.

To obtain a formula (6.132) assuming (6.129) has been satisfied, we first express the truncation error $T_{n+1}(Y)$ as an integral, using the concept of the Peano kernel introduced in the last part of Section 5.1 of Chapter 5. Expand $Y(x)$ about x_{n-p},

$$Y(x) = \sum_{i=0}^{m} \frac{(x - x_{n-p})^i}{i!} Y^{(i)}(x_{n-p}) + R_{m+1}(x)$$

$$R_{m+1}(x) = \frac{1}{m!} \int_{x_{n-p}}^{x_{n+1}} (x - t)_+^m Y^{(m+1)}(t) dt$$

$$(x - t)_+^m = \begin{cases} (x - t)^m, & x \geqslant t \\ 0, & x \leqslant t \end{cases}$$

Substitute the expansion into the formula (6.64) for $T_{n+1}(Y)$. Use the assumption that the method is of order m to obtain

$$T_{n+1}(Y) = T_{n+1}(R_{m+1})$$

$$= R_{m+1}(x_{n+1}) - \left[\sum_{j=0}^{p} a_j R_{m+1}(x_{n-j}) + h \sum_{j=-1}^{p} b_j R'_{m+1}(x_{n-j}) \right]$$

$$T_{n+1}(Y) = \int_{x_{n-p}}^{x_{n+1}} G(t - x_{n-p}) Y^{(m+1)}(t) dt \qquad (6.133)$$

with the Peano kernel

$$G(s) = \frac{1}{m!} \left\{ (x_{p+1} - s)_+^m - \left[\sum_{j=0}^{p} a_j (x_{p-j} - s)_+^m + hm \sum_{j=-1}^{p} b_j (x_{p-j} - s)_+^{m-1} \right] \right\}$$

$$= T_{p+1}\left((x - s)_+^m \right) \qquad 0 \leqslant s \leqslant x_{p+1} \qquad (6.134)$$

The function $G(s)$ is also often called an *influence function*.

The $m+1$ conditions (6.129) on the coefficients in (6.127) mean there are $r=(2p+3)-(m+1)$ free parameters ($r=(2p+3)-(m+2)$ for an explicit formula). Thus there are r free parameters in determining $G(s)$. We determine those values of the parameters for which $G(s)$ is of one sign on $[0, x_{p+1}]$. And then by the integral mean-value theorem, we will have

$$T_{n+1}(Y) = Y^{(m+1)}(\xi_n) \int_{x_{n-p}}^{x_{n+1}} G(t - x_{n-p})\, dt$$

for some $x_{n-p} \leqslant \xi_n \leqslant x_{n+1}$. By further manipulation, we obtain (6.132) with

$$d_m = \frac{1}{m!} \int_0^{p+1} \left[v^m - \sum_{j=0}^{p} a_j (v-j-1)_+^m - m \sum_{j=-1}^{p} b_j (v-j-1)_+^{m-1} \right] dv$$

Again, this is dependent on $G(s)$ being of one sign for $0 \leqslant s \leqslant x_{p+1}$.

Example   Consider formula (6.130) in which we assume the formula is explicit $(b_{-1}=0)$. Then

$$y_{n+1} = a_0 y_n + a_1 y_{n-1} + h \left[b_0 f(x_n, y_n) + b_1 f(x_{n-1}, y_{n-1}) \right] \qquad n \geqslant 1 \quad (6.135)$$

with

$$a_1 = 1 - a_0 \qquad b_0 = 2 - \tfrac{1}{2} a_0 \qquad b_1 = -\tfrac{1}{2} a_0$$

$$G(s) = \frac{1}{2} \left[(x_2 - s)^2 - a_0 (x_1 - s)_+^2 - a_1 (x_0 - s)_+^2 - 2h b_0 (x_1 - s)_+ - 2h b_1 (x_0 - s)_+ \right]$$

$$= \begin{cases} \tfrac{1}{2} s \left[s(1 - a_0) + a_0 h \right] & 0 \leqslant s \leqslant h \\ \tfrac{1}{2} (x_2 - s)^2 & h \leqslant s \leqslant 2h \end{cases}$$

The condition that $G(s)$ be of one sign on $[0, 2h]$ is satisfied if and only if $a_0 \geqslant 0$. Then

$$T_{n+1}(Y) = \left(\tfrac{1}{12} a_0 + \tfrac{1}{3} \right) h^3 Y'''(\xi_n) \qquad x_{n-1} \leqslant \xi_n \leqslant x_{n+1} \qquad (6.136)$$

Note that the truncation error is a minimum when $a_0 = 0$; and thus the midpoint method is optimal in having a small truncation error, among all such explicit two-step, second-order methods. For a further discussion of this example, see Problem 24.

The following example is taken from the development in Ralston [R1, p. 186], and it leads to Hamming's method.

Example   Find an implicit fourth-order formula of the form

$$y_{n+1} = a_0 y_n + a_1 y_{n-1} + a_2 y_{n-2}$$

$$+ h\left[b_{-1} y'_{n+1} + b_0 y'_n + b_1 y'_{n-1} \right] \qquad n \geqslant 2 \tag{6.137}$$

in which $y'_k \equiv f(x_k, y_k)$. With $p = 2$ and $m = 4$, the coefficients must satisfy the following system of five equations,

$$a_0 + a_1 + a_2 = 1$$

$$-a_1 - 2a_2 + b_{-1} + b_0 + b_1 = 1$$

$$a_1 + 4a_2 + 2b_{-1} - 2b_1 = 1$$

$$-a_1 - 8a_2 + 3b_{-1} + 3b_1 = 1$$

$$a_1 + 16a_2 + 4b_{-1} - 4b_1 = 1$$

Note that arbitrarily $b_2 = 0$ in order to reduce the number of free parameters to be dealt with. The solution of the system is

$$a_0 = \tfrac{9}{8}(1 - a_1) \qquad a_2 = \tfrac{1}{8}(-1 + a_1) \qquad b_{-1} = \tfrac{1}{24}(9 - a_1)$$

$$b_0 = \tfrac{1}{12}(9 + 7a_1) \qquad b_1 = \tfrac{1}{24}(-9 + 17a_1)$$

The Peano kernel for the truncation error is of constant sign on $[0, 3h]$ for $-.6 \leqslant a_1 \leqslant 1.0$; and consequently, a formula of the form (6.132) is valid for that interval. We return to this example after completing the convergence and stability theory in the next section. Also, to see the possible danger in arbitrarily setting a coefficient equal to zero (as with $b_2 = 0$ above), see Problem 25.

Methods based on numerical integration   The general idea is the following. Reformulate the differential equation by integrating it over some interval $[x_{n-r}, x_{n+1}]$ to obtain

$$Y(x_{n+1}) = Y(x_{n-r}) + \int_{x_{n-r}}^{x_{n+1}} f(t, Y(t)) dt \tag{6.138}$$

for some $r \geqslant 0$, all $n \geqslant r$. Produce a polynomial $P(t)$ that interpolates to the integrand $Y'(t) = f(t, Y(t))$ at some set of node points $\{x_i\}$, and then integrate $P(t)$ over $[x_{n-r}, x_{n+1}]$ to approximate (6.138). The three previous methods, Euler, midpoint, and trapezoidal, can all be obtained in this way.

Example 1.   Simpson's rule. Use Simpson's integration rule, (5.20) and (5.21), on the equation

$$Y(x_{n+1}) = Y(x_{n-1}) + \int_{x_{n-1}}^{x_{n+1}} f(t, Y(t)) dt$$

This results in

$$Y_{n+1} = Y_{n-1} + \frac{h}{3} \left[Y'_{n-1} + 4Y'_n + Y'_{n+1} \right] - \frac{h^5}{90} Y^{(5)}(\xi_n)$$

for some $x_{n-1} \leqslant \xi_n \leqslant x_{n+1}$. This approximation is based on integrating the quadratic polynomial which interpolates to $Y'(t) = f(t, Y(t))$ on the nodes x_{n-1}, x_n, x_{n+1}. For simplicity, we will use the notation $Y_j = Y(x_j)$, and $Y'_j = Y'(x_j)$.

Dropping the error term results in the implicit fourth-order formula

$$y_{n+1} = y_{n-1} + \frac{h}{3} \left[f(x_{n-1}, y_{n-1}) + 4f(x_n, y_n) \right.$$

$$\left. + f(x_{n+1}, y_{n+1}) \right] \qquad n \geqslant 1 \qquad (6.139)$$

This is a well-known formula, and it is the corrector of a predictor-corrector algorithm known as Milne's method. From Theorem 6.6, the method converges and

$$\underset{x_0 \leqslant x_n \leqslant b}{\text{Max}} |Y(x_n) - y_n| = O(h^4)$$

But the method is only weakly stable, in the same manner as was true of the midpoint method in Section 6.4.

Example 2. A fourth-order explicit formula. Use

$$Y_{n+1} = Y_{n-3} + \int_{x_{n-3}}^{x_{n+1}} f(t, Y(t)) dt$$

with quadratic interpolation to the integrand $Y'(t)$ at x_{n-2}, x_{n-1}, x_n. This results in

$$Y_{n+1} = Y_{n-3} + \frac{4h}{3} (2Y'_{n-2} - Y'_{n-1} + 2Y'_n) + \frac{28}{90} h^3 Y^{(5)}(\xi_n)$$

for some $x_{n-3} \leqslant \xi_n \leqslant x_{n+1}$. The resulting formula is

$$y_{n+1} = y_{n-3} + \frac{4h}{3} \left[2f(x_{n-2}, y_{n-2}) - f(x_{n-1}, y_{n-1}) + 2f(x_n, y_n) \right] \qquad n \geqslant 3 \qquad (6.140)$$

This is the predictor formula for Milne's method, used with the corrector in (6.139). It too is only weakly stable.

The Adam's methods These are the most widely used multistep methods. They are used to produce predictor-corrector algorithms in which the error is controlled by varying both the stepsize h and the order of the method. This will be discussed in greater detail later in the section, and a numerical example will be given.

To derive the methods, use the integral equation reformulation

$$Y_{n+1} = Y_n + \int_{x_n}^{x_{n+1}} f(t, Y(t)) dt \qquad (6.141)$$

Polynomials that interpolate to $Y'(t) = f(t, Y(t))$ are constructed, and then they are integrated over $[x_n, x_{n+1}]$ to obtain an approximation to Y_{n+1}. We begin with the explicit or predictor formulas.

Case 1   The Adams-Bashforth methods. Let $P_p(t)$ denote the polynomial of degree $\leqslant p$, which interpolates to $Y'(t)$ at $x_{n-p}, \cdots, x_n$. The most convenient form for $P_p(t)$ will be the Newton backward difference formula (3.37), expanded about x_n,

$$P_p(t) = Y_n' + \frac{(t-x_n)}{h} \nabla Y_n' + \frac{(t-x_n)(t-x_{n-1})}{2!h^2} \nabla^2 Y_n' + \cdots$$

$$+ \frac{(t-x_n) \cdots (t-x_{n-p+1})}{p!h^p} \nabla^p Y_n' \qquad (6.142)$$

As an illustration of the notation, $\nabla Y_n' = Y'(x_n) - Y'(x_{n-1})$; and the higher-order backward differences are defined accordingly. The interpolation error is given by

$$E_p(t) = (t-x_{n-p}) \cdots (t-x_n) Y'[x_{n-p}, \cdots, x_n, t]$$

$$= \frac{(t-x_{n-p}) \cdots (t-x_n)}{(p+1)!} Y^{(p+2)}(\zeta_t) \qquad x_{n-p} \leqslant \zeta_t \leqslant x_{n+1} \qquad (6.143)$$

provided $x_{n-p} \leqslant t \leqslant x_{n+1}$ and $Y(x)$ is $p+2$ times continuously differentiable. See (3.8) and (3.22) of Chapter 3 for the justification of (6.143).

The integral of $P_p(t)$ is given by

$$\int_{x_n}^{x_{n+1}} P_p(t) = \sum_{j=0}^{p} \frac{1}{j!h^j} \nabla^j Y_n' \int_{x_n}^{x_{n+1}} (t-x_n) \cdots (t-x_{n+1-j}) dt$$

$$= h \sum_{j=0}^{p} \gamma_j \nabla^j Y_n' \qquad (6.144)$$

The coefficients γ_j are obtained by introducing the change of variable $s = (t-x_n)/h$, which leads to

$$\gamma_j = \frac{1}{j!} \int_0^1 s(s+1) \cdots (s+j-1) ds \qquad j \geqslant 1 \qquad (6.145)$$

Table 6.10 Adams-Bashforth coefficients

γ_0	γ_1	γ_2	γ_3	γ_4	γ_5
1	1/2	5/12	3/8	251/720	95/288

with $\gamma_0 = 1$. Table 6.10 contains the first few values of γ_j; and Gear [G1, pp. 104–111] contains additional information, including a generating function for the coefficients.

The truncation error in using the interpolating polynomial $P_p(t)$ in (6.141) is given by

$$T_{n+1}(Y) = \int_{x_n}^{x_{n+1}} E_p(t) dt$$

$$= \int_{x_n}^{x_{n+1}} (t - x_n) \cdots (t - x_{n-p}) Y' \left[x_{n-p}, \cdots, x_n, t \right] dt$$

Assuming $Y(x)$ is $p+2$ times continuously differentiable for $x_{n-p} \leqslant x \leqslant x_{n+1}$, Theorem 3.3 of Chapter 3 implies that the divided difference in the last integral is a continuous function of t. Since the polynomial $(t - x_n) \cdots (t - x_{n-p})$ is non-negative on $[x_n, x_{n+1}]$, use the integral mean-value theorem (1.3) to obtain

$$T_{n+1}(Y) = Y' \left[x_{n-p}, \cdots, x_n, \zeta \right] \int_{x_n}^{x_{n+1}} (t - x_n) \cdots (t - x_{n-p}) dt$$

for some $x_n \leqslant \zeta \leqslant x_{n+1}$. Use (6.145) to calculate the integral, and use (3.23) to convert the divided difference to a derivative. Then

$$T_{n+1}(Y) = \gamma_{p+1} h^{p+2} Y^{(p+2)}(\zeta_n) \qquad x_{n-p} \leqslant \zeta_n \leqslant x_{n+1} \qquad (6.146)$$

There is an alternative form which is very useful for estimating the truncation error. Use the mean-value theorem (1.2) and the analogue of Lemma 2 of Chapter 3 for backward differences to show

$$T_{n+1}(Y) = h\gamma_{p+1} \nabla^{p+1} Y_n' + O(h^{p+3}) \qquad (6.147)$$

The principal part of this error, $h\gamma_{p+1} \nabla^{p+1} Y_n'$, is the final term that would be obtained by using the interpolating polynomial $P_{p+1}(t)$ rather than $P_p(t)$. This form of the error is the basic tool used in algorithms that vary the order of the method to help control the truncation error.

Using (6.144) and (6.146), equation (6.141) becomes

$$Y_{n+1} = Y_n + h \sum_{j=0}^{p} \gamma_j \nabla^j Y_n' + \gamma_{p+1} h^{p+2} Y^{(p+2)}(\zeta_n) \qquad (6.148)$$

for some $x_{n-p} \leqslant \xi_n \leqslant x_{n+1}$. The corresponding numerical method is

$$y_{n+1} = y_n + h \sum_{j=0}^{p} \gamma_j \nabla^j y_n' \qquad n \geqslant p \tag{6.149}$$

In the formula, $y_j' \equiv f(x_j, y_j)$; and as an example of the backward differences, $\nabla y_j' = y_j' - y_{j-1}'$. Table 6.11 contains the formulas for $p = 0, 1, 2, 3$. They are written in the more usual form of (6.127), in which the dependence on the values $f(x_{n-j}, y_{n-j})$ is shown explicitly. Note that the $p = 0$ case is just Euler's method.

Table 6.11   Adams-Bashforth formulas

$p = 0$	$Y_{n+1} = Y_n + h Y_n' + \dfrac{1}{2} h^2 Y''(\xi_n)$
$p = 1$	$Y_{n+1} = Y_n + \dfrac{h}{2}(3 Y_n' - Y_{n-1}') + \dfrac{5}{12} h^3 Y'''(\xi_n)$
$p = 2$	$Y_{n+1} = Y_n + \dfrac{h}{12}[23 Y_n' - 16 Y_{n-1}' + 5 Y_{n-2}'] + \dfrac{3}{8} h^4 Y^{(4)}(\xi_n)$
$p = 3$	$Y_{n+1} = Y_n + \dfrac{h}{24}(55 Y_n' - 59 Y_{n-1}' + 37 Y_{n-2}' - 9 Y_{n-3}') + \dfrac{251}{720} h^5 Y^{(5)}(\xi_n)$

The Adams-Bashforth formulas satisfy the hypotheses of Theorem 6.6, and therefore they are convergent and stable methods. In addition, they do not have the instability of the type associated with the midpoint method (6.84) and Simpson's rule (6.139). The proof of this is given in the next section; and further discussion is postponed until near the end of that section, when the concept of relative stability is introduced.

Case 2 Adams-Moulton methods. Again use the integral formula (6.141), but interpolate to $Y'(t) = f(t, Y(t))$ at the $p + 1$ points $x_{n+1}, \cdots, x_{n-p+1}$ for $p \geqslant 0$. The derivation is exactly the same as for the Adams-Bashforth methods, and we give only the final results. Equation (6.141) is transformed to

$$Y_{n+1} = Y_n + h \sum_{j=0}^{p} \delta_j \nabla^j Y_{n+1}' + \delta_{p+1} h^{p+2} Y^{(p+2)}(\xi_n) \tag{6.150}$$

with $x_{n-p+1} \leqslant \xi_n \leqslant x_{n+1}$. The coefficients δ_j are defined by

$$\delta_j = \frac{1}{j!} \int_0^1 (s-1)s(s+1) \cdots (s+j-2) ds \qquad j \geqslant 1$$

with $\delta_0 = 1$; and a few values are given in Table 6.12. The truncation error can be put in the form

$$T_{n+1}(Y) = h \delta_{p+1} \nabla^{p+1} Y_{n+1}' + O(h^{p+3}) \tag{6.151}$$

just as with (6.147).

Table 6.12   Adams-Moulton coefficients

δ_0	δ_1	δ_2	δ_3	δ_4	δ_5
1	$-1/2$	$-1/12$	$-1/24$	$-19/720$	$-3/160$

The numerical method associated with (6.150) is

$$y_{n+1}=y_n+h\sum_{j=0}^{p}\delta_j\nabla^j y'_{n+1}\qquad n\geqslant p-1\qquad(6.152)$$

with $y'_j\equiv f(x_j,y_j)$ as before. Table 6.13 contains the low-order formulas for $p=0,1,2,3$. Note that the $p=1$ case is the trapezoidal method.

The formula (6.152) is an implicit method, and therefore a predictor is necessary for solving (6.152) by iteration. The basic ideas involved are exactly the same as those in Section 6.5 for the iterative solution of the trapezoidal method. If a fixed order predictor-corrector algorithm is desired, then an Adams-Moulton formula of order $m\geqslant2$ can use a predictor of order m or $m-1$ if only one iterate is to be computed. The advantage of using an order $m-1$ predictor is that the predictor and corrector would both use derivative values at the same nodes, namely, $x_n,x_{n-1},\cdots,x_{n-m+2}$. For example, the second order Adams-Moulton formula with the first order Adams-Bashforth formula as predictor is just the trapezoidal method with the Euler predictor. This was discussed in Section 6.5 and shown to be adequate; and both methods use the single past value of the derivative, $f(x_n,y_n)$.

Table 6.13   Adams-Moulton formulas

$p=0$	$Y_{n+1}=Y_n+hY'_{n+1}-\dfrac{1}{2}h^2Y''(\zeta_n)$
$p=1$	$Y_{n+1}=Y_n+\dfrac{h}{2}(Y'_{n+1}+Y'_n)-\dfrac{1}{12}h^3Y'''(\zeta_n)$
$p=2$	$Y_{n+1}=Y_n+\dfrac{h}{12}(5Y'_{n+1}+8Y'_n-Y'_{n-1})-\dfrac{1}{24}h^4Y^{(4)}(\zeta_n)$
$p=3$	$Y_{n+1}=Y_n+\dfrac{h}{24}[9Y'_{n+1}+19Y'_n-5Y'_{n-1}+Y'_{n-2}]-\dfrac{19}{720}h^5Y^{(5)}(\zeta_n)$

A less trivial example is the following fourth-order method:

$$y_{n+1}^{(0)}=y_n+\frac{h}{12}\left[23f(x_n,y_n)-16f(x_{n-1},y_{n-1})+5f(x_{n-2},y_{n-2})\right]$$

$$y_{n+1}^{(j+1)}=y_n+\frac{h}{24}\left[9f(x_{n+1},y_{n+1}^{(j)})+19f(x_n,y_n)\right.\qquad(6.153)$$

$$\left.-5f(x_{n-1},y_{n-1})+f(x_{n-2},y_{n-2})\right]$$

Generally only one iterate is calculated, although this will alter the form of the truncation error in (6.151). Let $u_n(x)$ denote the solution of $y'=f(x,y)$ passing through (x_n,y_n). Then for the truncation error in using the approximation $y_{n+1}^{(1)}$,

$$u_n(x_{n+1})-y_{n+1}^{(1)}=\left[u_n(x_{n+1})-y_{n+1}\right]+\left[y_{n+1}-y_{n+1}^{(1)}\right]$$

Using (6.151) and an expansion of the iteration error,

$$u_n(x_{n+1})-y_{n+1}^{(1)}=\delta_4h^5u_n^{(5)}(x_n)+\frac{3h}{8}f_y(x_n,y_n)(y_{n+1}-y_{n+1}^{(0)})+O(h^6) \tag{6.154}$$

The first two terms following the equality sign are of order h^5. If either (1) more iterates are calculated, or (2) a higher-order predictor is used, then the principal part of the truncation error will be simply $\delta_4h^5u_n^{(5)}(x_n)$. And this is a more desirable situation from the view point of estimating the error.

The Adams-Moulton formulas have a significantly smaller truncation error than the Adams-Bashforth formulas, for comparable order methods. For example, the fourth-order Adams-Moulton formula has a truncation error .076 times that of the fourth order Adams-Bashforth formula. This is the principal reason for using the implicit formulas, although there are other considerations. Note also that the fourth-order Adams-Moulton formula has over twice the truncation error of the Simpson method (6.139). The reason for using the Adams-Moulton formula is that it has much better stability properties than the Simpson rule (6.139).

Example The method (6.153) was used to solve

$$y'=\frac{1}{1+x^2}-2y^2 \qquad y(0)=0 \tag{6.155}$$

which has the solution $Y(x)=x/(1+x^2)$. The initial values y_1,y_2,y_3 were taken to be the true values to simplify the example. The solution values were computed with two values of h, and the resulting errors at a few node points are given in Table 6.14. The column labeled "Ratio" is the ratio of the error with $h=.125$ to that with $h=.0625$. Note that the values of the ratios are near 16, which is the theoretical ratio that would be expected since the method (6.153) is fourth order.

Table 6.14 Numerical example of Adams method (6.153)

x	Error for $h=.125$	Error for $h=.0625$	Ratio
2.0	$2.07E\text{-}5$	$1.21E\text{-}6$	17.1
4.0	$2.21E\text{-}6$	$1.20E\text{-}7$	18.3
6.0	$3.74E\text{-}7$	$2.00E\text{-}8$	18.7
8.0	$1.00E\text{-}7$	$5.24E\text{-}9$	19.1
10.0	$3.58E\text{-}8$	$1.83E\text{-}9$	19.6

Variable order methods At present, the most popular predictor-corrector algorithms control the truncation error by varying both the stepsize and the order of
the method; and all of these algorithms use the Adams family of formulas. The
first such computer program that was widely used was given by Gear in the text
[G1, pp. 158–166] in 1971; and other related programs are given in Krogh [K5]
and Shampine and Gordon [S2]. In all cases, the error estimation is based on the
formulas (6.150) or (6.151), although they (and the formulas (6.152) and (6.149))
must be modified to deal with a varying stepsize.

By allowing the order to vary, there is no difficulty in obtaining starting
values for the higher-order methods. The programs begin with the second order
trapezoidal formula with a Euler predictor; and the order is then generally
increased as extra starting values become available. If the solution is changing
rapidly, then the program will generally choose a low order formula; and for a
smoother and more slowly varying solution, the order will usually be larger.

In the program DE of Shampine and Gordon [S2, pp. 186–209], the
truncation error at x_{n+1} (call it *trunc*) is required to satisfy

$$|\text{trunc}_j| \leqslant \text{ABSERR} + \text{RELERR}*|y_{n,j}| \qquad (6.156)$$

This to hold for each component trunc_j of the truncation error and corresponding component $y_{n,j}$ of the solution y_n of the given system of differential
equations. The values ABSERR and RELERR are supplied by the user. The
value of trunc is given, roughly speaking, by

$$\text{trunc} \approx h\delta_{p+1} \nabla^{p+1} y'_{n+1}$$

assuming the spacing is uniform. This is the truncation error for the p-step
formula

$$y^{(p)}_{n+1} = y_n + h \sum_{j=0}^{p} \delta_j \nabla^j y'_{n+1}$$

Once the test (6.156) is satisfied, the value of y_{n+1} is

$$y_{n+1} \equiv y^{(p+1)}_{n+1} = y^{(p)}_{n+1} + \text{trunc}$$

Thus the actual truncation error is $O(h^{p+3})$; and combined with (6.156), it can
be shown that the truncation error in y_{n+1} satisfies an error per unit step criteria,
similar to that of (6.112) for the Detrap algorithm of Section 6.6. For a detailed
discussion, see Shampine and Gordon [S2, p. 100].

The program DE is very sophisticated in its error control, including the
choosing of the order and the stepsize. It cannot be discussed adequately in the
limited space available in this text; and the best reference is the text [S2] that is
devoted to variable-order Adams algorithms. The DE program has been well
designed from both the viewpoint of error control and user convenience. It is

also written in a portable form; and generally it is a well-recommended program for solving differential equations. A discussion of this and other programs is given in the Discussion of the Literature section at the end of the chapter.

Example   Consider the problem

$$y' = \frac{y}{4}\left(1 - \frac{y}{20}\right) \qquad y(0) = 1 \tag{6.157}$$

which has the solution

$$Y(x) = 20/[1 + 19e^{-x/4}]$$

DE was used to solve this problem with values output at $x = 2, 4, 6, \cdots, 20$. Three values of ABSERR were used, and RELERR $= 0$ in all cases. The true global errors are shown in Table 6.15. The column labeled "NFE" gives the number of evaluations of $f(x, y)$ necessary to obtain the value $y_h(x)$, beginning from x_0.

Table 6.15   Example of automatic program DE

x	ABSERR $= 10^{-3}$		ABSERR $= 10^{-6}$		ABSERR $= 10^{-9}$	
	Error	NFE	Error	NFE	Error	NFE
4.0	$4.8E\text{-}5$	17	$3.8E\text{-}7$	35	$4.4E\text{-}10$	65
8.0	$9.0E\text{-}4$	23	$9.0E\text{-}7$	50	$1.4E\text{-}9$	83
12.0	$2.7E\text{-}3$	27	$1.2E\text{-}6$	66	$3.0E\text{-}9$	103
16.0	$1.6E\text{-}3$	31	$1.5E\text{-}6$	76	$1.9E\text{-}9$	123
20.0	$-7.5E\text{-}5$	37	$2.3E\text{-}7$	88	$1.2E\text{-}9$	139

6.8 Convergence and Stability Theory for Multistep Methods

In this section, a complete theory of convergence and stability is presented for the multistep method

$$y_{n+1} = \sum_{j=0}^{p} a_j y_{n-j} + h \sum_{j=-1}^{p} b_j f(x_{n-j}, y_{n-j}) \qquad x_{p+1} \leqslant x_{n+1} \leqslant b \tag{6.158}$$

This will generalize the work of Section 6.3; and it will create the mathematical tools necessary for analyzing whether a method (6.158) is only weakly stable, due to instability of the type associated with the midpoint method.

We begin with a few definitions. The concept of stability was introduced with Euler's method [see (6.41) and (6.42)], and it will now be generalized. Let $\{y_n | 0 \leqslant n \leqslant N(h)\}$ be the solution of (6.158) for some differential equation

$y'=f(x,y)$, for all sufficiently small values of h, say $h \leqslant h_0$. Recall that $N(h)$ denotes the largest subscript N for which $x_N \leqslant b$. For each $h \leqslant h_0$, perturb the initial values $y_0, \cdots, y_p$ to new values $z_0, \cdots, z_p$ with

$$\underset{0 \leqslant n \leqslant p}{\text{Max}} |y_n - z_n| \leqslant \epsilon \qquad 0 < h \leqslant h_0 \tag{6.159}$$

We say the solution $\{y_n\}$ is stable if there is a constant c, independent of $h \leqslant h_0$ and all small ϵ, for which

$$\underset{0 \leqslant n \leqslant N(h)}{\text{Max}} |y_n - z_n| \leqslant c\epsilon \qquad 0 < h \leqslant h_0 \tag{6.160}$$

Consider all differential equation problems

$$y' = f(x,y) \qquad y(x_0) = Y_0 \tag{6.161}$$

with the derivative $f(x,y)$ continuous and satisfying the Lipschitz condition (6.29); and suppose the approximating solutions $\{y_n\}$ are all stable. Then we say that (6.158) is a stable numerical method.

To define convergence for a given problem (6.161), suppose the initial values $y_0, \cdots, y_p$ satisfy

$$\eta(h) \equiv \underset{0 \leqslant n \leqslant p}{\text{Max}} |Y(x_n) - y_n| \to 0 \qquad \text{as} \quad h \to 0 \tag{6.162}$$

Then the solution $\{y_n\}$ is said to converge to $Y(x)$ if

$$\underset{x_0 \leqslant x_n \leqslant b}{\text{Max}} |Y(x_n) - y_n| \to 0 \qquad \text{as} \quad h \to 0 \tag{6.163}$$

If (6.158) is convergent for all problems (6.161), then it is called a convergent numerical method.

Recall the definition of consistency, given in Section 6.3. Method (6.158) is consistent if

$$\frac{1}{h} \underset{x_p \leqslant x_n \leqslant b}{\text{Max}} |T_{n+1}(Y)| \to 0 \qquad \text{as} \quad h \to 0$$

for all functions $Y(x)$ continuously differentiable on $[x_0, b]$. Or equivalently from Theorem 6.5, the coefficients $\{a_j\}$ and $\{b_j\}$ must satisfy

$$\sum_{j=0}^{p} a_j = 1 \qquad -\sum_{j=0}^{p} j a_j + \sum_{j=-1}^{p} b_j = 1 \tag{6.164}$$

Convergence can be shown to imply consistency; and consequently we consider only methods satisfying (6.164). As an example of the proof of the necessity of (6.164), the assumption of convergence of (6.158) for the problem

$$y' \equiv 0 \qquad y(0) = 1$$

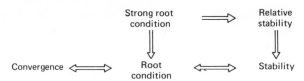

Figure 6.7   Schematic of the theory for consistent multistep methods.

implies the first condition in (6.164). Just take $y_0 = \cdots = y_p = 1$, and observe the consequences of the convergence of y_{p+1} to $Y(x) \equiv 1$.

The convergence and stability of (6.158) are linked to the roots of the polynomial

$$\rho(r) = r^{p+1} - \sum_{j=0}^{p} a_j r^{p-j} \qquad (6.165)$$

Note that $\rho(1) = 0$ from the consistency condition (6.164). Let $r_0, \cdots, r_p$ denote the roots of $\rho(r)$, repeated according to their multiplicity, and let $r_0 = 1$. The method (6.158) satisfies the *root condition* if

$$(1) \quad |r_j| \leqslant 1 \qquad j = 0, 1, \cdots, p \qquad (6.166)$$

$$(2) \quad |r_j| = 1 \Rightarrow \rho'(r_j) \neq 0 \qquad (6.167)$$

The first condition requires all roots of $\rho(r)$ to lie in the unit circle $\{z : |z| \leqslant 1\}$ in the complex plane. Condition (2) states that all roots on the boundary of the circle are to be simple roots of $\rho(r)$.

The main results of this section are pictured in Figure 6.7, although some of them will not be proven. The *strong root condition* and the concept of *relative stability* will be introduced later in the section.

Stability theory   All of the numerical methods presented in the preceding sections have been stable, but we now give an example of an unstable method. This is to motivate the need to develop a general theory of stability.

Example   Recall the general formula (6.135) for an explicit two-step second-order method, and choose $a_0 = 3$. Then we obtain the method

$$y_{n+1} = 3y_n - 2y_{n-1} + \frac{h}{2} \left[f(x_n, y_n) - 3f(x_{n-1}, y_{n-1}) \right] \qquad n \geqslant 1 \qquad (6.168)$$

with the truncation error

$$T_{n+1}(Y) = \tfrac{7}{12} h^3 Y'''(\xi_n) \qquad x_{n-1} \leqslant \xi_n \leqslant x_{n+1}$$

Consider solving the problem $y' \equiv 0, y(0) = 0$, which has the solution $Y(x) \equiv 0$. Using $y_0 = y_1 = 0$, the numerical solution is clearly $y_n = 0, n \geqslant 0$. Perturb the initial data to $z_0 = \epsilon, z_1 = 2\epsilon$, for some $\epsilon \neq 0$. Then the corresponding numerical solution can be shown to be

$$z_n = \epsilon \cdot 2^n \qquad n \geqslant 0$$

The reasoning used in deriving this solution will be given later in a more general context. To see the effect of the perturbation on the original solution,

$$\underset{x_0 \leqslant x_n \leqslant b}{\text{Max}} |y_n - z_n| = \underset{0 \leqslant n \leqslant N(h)}{\text{Max}} |\epsilon| 2^n = |\epsilon| 2^{N(h)}$$

Since $N(h) \to \infty$ as $h \to 0$, the deviation of $\{z_n\}$ from $\{y_n\}$ becomes increasingly greater as $h \to 0$. The method (6.168) is unstable, and it should never be used. Also note that the root condition is violated, since $\rho(r) = r^2 - 3r + 2$ has the roots $r_0 = 1, r_1 = 2$.

To investigate the stability of (6.158), we consider only the special equation

$$y' = \lambda y \qquad y(0) = 1 \tag{6.169}$$

with the solution $Y(x) = e^{\lambda x}$. The results obtained will transfer to the study of stability for a general differential equation problem. An intuitive reason for this is easily derived. Expand $Y'(x) = f(x, Y(x))$ to obtain

$$Y'(x) \approx f(x, Y_0) + f_y(x, Y_0)(Y(x) - Y_0)$$

$$\approx f(x, Y_0) + \lambda(Y(x) - Y_0) \qquad x_0 \leqslant x \leqslant b \tag{6.170}$$

with $\lambda = f_y(x_0, Y_0)$. This is a valid approximation if $b - x_0$ is sufficiently small. Introducing $V(x) = Y(x) - Y_0$,

$$V'(x) \approx \lambda V(x) + f(x, Y_0) \tag{6.171}$$

The inhomogeneous term $f(x, Y_0)$ will drop out of all derivations concerning numerical stability, because we are concerned with differences of solutions of the equation. Dropping the last term in (6.171) because it doesn't affect the results on stability, we obtain the model equation (6.169). As further motivation, refer back to the stability results (6.17) and (6.18) for studying the stability of solutions of a differential equation.

Applying (6.158) to the equation, we obtain

$$y_{n+1} = \sum_{j=0}^{p} a_j y_{n-j} + h\lambda \sum_{j=-1}^{p} b_j y_{n-j}$$

$$(1 - h\lambda b_{-1}) y_{n+1} - \sum_{j=0}^{p} (a_j + h\lambda b_j) y_{n-j} = 0 \qquad n \geqslant p \tag{6.172}$$

This is a homogeneous linear difference equation of order $p+1$, and the theory for its solvability is completely analogous to that of $(p+1)$st order homogeneous linear differential equations. As a general reference, see Henrici [H1, pp. 210–215] or Isaacson and Keller [I1, pp. 405–417].

We attempt to find a general solution by first looking for solutions of the special form

$$y_n = r^n \qquad n \geqslant 0 \tag{6.173}$$

If we can find $p+1$ linearly independent solutions, then an arbitrary linear combination will give the general solution of (6.172).

Substituting $y_n = r^n$ into (6.172) and cancelling r^{n-p}, we obtain

$$(1 - h\lambda b_{-1})r^{p+1} - \sum_{j=0}^{p} (a_j + h\lambda b_j)r^{p-j} = 0 \tag{6.174}$$

This is called the *characteristic equation*, and the lefthand side is the *characteristic polynomial*. The roots are called *characteristic roots*. Define

$$\sigma(r) = b_{-1}r^{p+1} + \sum_{j=0}^{p} b_j r^{p-j}$$

and recall the definition (6.165) of $\rho(r)$. Then (6.174) becomes

$$\rho(r) - h\lambda\sigma(r) = 0 \tag{6.175}$$

Denote the characteristic roots by

$$r_0(h\lambda), \cdots, r_p(h\lambda)$$

which can be shown to depend continuously on the value of $h\lambda$. When $h\lambda = 0$, the equation (6.175) becomes simply $\rho(r) = 0$; and we have $r_j(0) = r_j, j = 0, 1, \cdots, p$, for the earlier roots r_j of (6.165). Since $r_0 = 1$ is a root of $\rho(r)$, we let $r_0(h\lambda)$ be the root of (6.175) for which $r_0(0) = 1$. The root $r_0(h\lambda)$ is called the *principal root*, for reasons that will appear later. If the roots $r_j(h\lambda)$ are all distinct, then the general solution of (6.172) is

$$y_n = \sum_{j=0}^{p} \gamma_j [r_j(h\lambda)]^n \qquad n \geqslant 0 \tag{6.176}$$

But if $r_j(h\lambda)$ is a root of multiplicity $\nu > 1$, then the following are ν linearly independent solutions of (6.172):

$$\left\{ [r_j(h\lambda)]^n \right\}, \left\{ n[r_j(h\lambda)]^n \right\}, \cdots, \left\{ n^{\nu-1}[r_j(h\lambda)]^n \right\} \tag{6.177}$$

These can be used with the solutions arising from the other roots to generate a general solution for (6.172), comparable to (6.176).

Theorem 6.7 Assume the consistency condition (6.164). Then the multistep method (6.158) is stable if and only if the root condition (6.166)–(6.167) is satisfied.

Proof 1. We begin by showing the necessity of the root condition for stability. To do so, assume the opposite by letting

$$|r_j(0)| > 1$$

for some j. Consider the differential equation problem $y' \equiv 0, y(0) = 0$, with the solution $Y(x) \equiv 0$. Then (6.158) becomes

$$y_{n+1} = \sum_{j=0}^{p} a_j y_{n-j} \qquad n \geqslant p \tag{6.178}$$

If we take $y_0 = y_1 = \cdots = y_p = 0$, then the numerical solution is clearly $y_n = 0$, all $n \geqslant 0$. For the perturbed initial values, take

$$z_0 = \epsilon, z_1 = \epsilon r_j(0), \cdots, z_p = \epsilon r_j(0)^p \tag{6.179}$$

For these initial values,

$$\underset{0 < n < p}{\text{Max}} |y_n - z_n| = \epsilon |r_j(0)|^p$$

which is a uniform bound for all small values of h, since the right side is independent of h. As $\epsilon \to 0$, the bound also tends to zero.

The solution of (6.178) with the initial conditions (6.179) is simply

$$z_n = \epsilon r_j(0)^n \qquad n \geqslant 0$$

For the deviation from $\{y_n\}$,

$$\underset{0 < x_n < b}{\text{Max}} |y_n - z_n| = \epsilon |r_j(0)|^{N(h)}$$

As $h \to 0, N(h) \to \infty$ and the bound becomes infinite. This proves the method is unstable when some $|r_j(0)| > 1$. If the root condition is violated instead by assuming (6.167) is false, then a similar proof can be given. This is left as problem 22 for the reader.

2. Assume that the root condition is satisfied. The proof of stability will be restricted to the exponential equation (6.169). A proof can be given for the general equation $y' = f(x,y)$, but it is a fairly involved modification of the following proof. The general proof involves the solution of nonhomogeneous linear difference equations; see Isaacson and Keller [I1, pp. 405–417] for a complete development. To further simplify the proof, we assume that the roots $r_j(0), j=0, 1, \cdots, p$ are all distinct. The same will be true of $r_j(h\lambda)$, provided the value of h is kept sufficiently small, say $0 \leqslant h \leqslant h_0$.

Let $\{y_n\}$ and $\{z_n\}$ be two solutions of (6.172) on $[x_0, b]$, and assume

$$\underset{0 \leqslant n \leqslant p}{\text{Max}} \ |y_n - z_n| \leqslant \epsilon \qquad 0 < h \leqslant h_0 \qquad (6.180)$$

Introduce the error $e_n = y_n - z_n$. Subtracting using (6.172) for each solution,

$$(1 - h\lambda b_{-1})e_{n+1} - \sum_{j=0}^{p} (a_j + h\lambda b_j)e_{n-j} = 0 \qquad x_{p+1} \leqslant x_{n+1} \leqslant b \quad (6.181)$$

The general solution is

$$e_n = \sum_{j=0}^{p} \gamma_j [r_j(h\lambda)]^n \qquad n \geqslant 0 \qquad (6.182)$$

The coefficients $\gamma_0, \cdots, \gamma_p$ must be chosen so that

$$\gamma_0 + \gamma_1 + \cdots + \gamma_p = e_0$$

$$\gamma_0 r_0(h\lambda) + \cdots + \gamma_p r_p(h\lambda) = e_1$$

$$\vdots$$

$$\gamma_0 [r_0(h\lambda)]^p + \cdots + \gamma_p [r_p(h\lambda)]^p = e_p$$

The solution (6.182) will then agree with the given initial perturbations $e_0, \cdots, e_p$ and it will satisfy the difference equation (6.181). Using the bound (6.180) and the theory of linear systems of equations, it is fairly straightforward to show that

$$\underset{0 \leqslant i \leqslant p}{\text{Max}} \ |\gamma_i| \leqslant c_1 \epsilon \qquad 0 < h \leqslant h_0 \qquad (6.183)$$

for some constant $c_1 > 0$.

To bound the solution e_n on $[x_0, b]$, we must bound each term $[r_j(h\lambda)]^n$. To do so, consider the expansion

$$r_j(u) = r_j(0) + u r_j'(\zeta) \qquad (6.184)$$

for some ζ between 0 and u. To compute $r_j'(u)$, differentiate the identity

$$\rho(r_j(u)) - u\sigma(r_j(u)) = 0$$

Then

$$r_j'(u) = \frac{\sigma(r_j(u))}{\rho'(r_j(u)) - u\sigma'(r_j(u))} \qquad (6.185)$$

By the assumption that $r_j(0)$ is a simple root of $\rho(r) = 0, 0 \leqslant j \leqslant p$, it follows that $\rho'(r_j(0)) \neq 0$; and by continuity, $\rho'(r_j(u)) \neq 0$ for all sufficiently small values of u. The denominator in (6.185) is nonzero, and we can bound $r_j'(u)$,

$$|r_j'(u)| \leqslant c_2 \qquad \text{all} \quad |u| \leqslant u_0$$

for some $u_0 > 0$.

Using this with (6.184) and the root condition (6.166), we have

$$|r_j(h\lambda)| \leqslant |r_j(0)| + c_2|h\lambda| \leqslant 1 + c_2|h\lambda|$$

$$|[r_j(h\lambda)]^n| \leqslant [1 + c_2|h\lambda|]^n \leqslant e^{c_2 n|h\lambda|} \leqslant e^{c_2(b - x_0)|\lambda|}$$

for all $0 < h \leqslant h_0$. Combined with (6.182) and (6.183),

$$\operatorname*{Max}_{x_0 \leqslant x_n \leqslant b} |e_n| \leqslant c_3|\epsilon|e^{c_2(b - x_0)|\lambda|} \qquad 0 < h \leqslant h_0$$

for an appropriate constant c_3. This completes the proof. ∎

Convergence theory The following result generalizes Theorem 6.6 of Section 6.3, with necessary and sufficient conditions being given for the convergence of multistep methods.

Theorem 6.8 Assume the consistency condition (6.164). Then the multistep method (6.158) is convergent if and only if the root condition (6.166) to (6.167) is satisfied.

Proof 1. We begin by showing the necessity of the root condition for convergence, and again we use the problem $y' \equiv 0, y(0) = 0$, with the solution $Y(x) \equiv 0$. The multistep method (6.158) becomes

$$y_{n+1} = \sum_{j=0}^{p} a_j y_{n-j} \qquad n \geqslant p \qquad (6.186)$$

with $y_0, \cdots, y_p$ satisfying

$$\eta(h) \equiv \operatorname*{Max}_{0 \leqslant n \leqslant p} |y_n| \to 0 \qquad \text{as} \quad h \to 0 \tag{6.187}$$

Suppose that the root condition is violated. We will show that (6.186) is not convergent to $Y(x) \equiv 0$.

Assume that some $|r_j(0)| > 1$. Then a satisfactory solution of (6.186) is

$$y_n = h[r_j(0)]^n \qquad x_0 \leqslant x_n \leqslant b \tag{6.188}$$

Condition (6.187) is satisfied since

$$\eta(h) = h|r_j(0)|^p \to 0 \qquad \text{as} \quad h \to 0$$

But the solution (6.188) does not converge. First,

$$\operatorname*{Max}_{0 \leqslant x_n \leqslant b} |Y(x_n) - y_n| = h|r_j(0)|^{N(h)}$$

Consider those values of $h = b/N(h)$. Then L'Hospital's rule can be used to show that

$$\operatorname*{Limit}_{N \to \infty} \frac{b}{N} |r_j(0)|^N = \infty$$

showing (6.188) does not converge.

Assume (6.166) of the root condition is satisfied, but that some $r_j(0)$ is a multiple root of $\rho(r)$ and $|r_j(0)| = 1$. Then the above form of proof is still satisfactory, but we must use the solution

$$y_n = hn[r_j(0)]^n \qquad 0 \leqslant n \leqslant N(h)$$

This completes the proof of the necessity of the root condition.

2. Assume the root condition is satisfied. As with the previous theorem, it is too difficult to give a general proof of convergence for an arbitrary differential equation. For such, see the development in Isaacson and Keller [I1, pp. 405–417]. The present proof is restricted to the exponential equation (6.169); and again we assume the roots $r_j(0)$ are distinct, in order to simplify the proof. We will show that the term $\gamma_0[r_0(h\lambda)]^n$ in the solution

$$y_n = \sum_{j=0}^{p} \gamma_j[r_j(h\lambda)]^n$$

will converge to the solution $Y(x) = e^{\lambda x}$ on $[0,b]$. The remaining terms $\gamma_j [r_j(h\lambda)]^n, j = 1, 2, \cdots, p$, are parasitic solutions, and they can be shown to converge to zero as $h \to 0$; see Problem 23.

Expand $r_0(h\lambda)$ using Taylor's theorem,

$$r_0(h\lambda) = r_0(0) + h\lambda r_0'(0) + O(h^2)$$

From (6.185),

$$r_0'(0) = \sigma(1)/\rho'(1)$$

and using the consistency condition (6.164), this leads to $r'(0) = 1$. Then

$$r_0(h\lambda) = 1 + h\lambda + O(h^2) = e^{\lambda h} + O(h^2)$$

$$[r_0(h\lambda)]^n = e^{\lambda n h}[1 + O(h^2)]^n = e^{\lambda x_n}[1 + O(h)]$$

Thus

$$\underset{0 \leqslant x_n \leqslant b}{\text{Max}} |[r_0(h\lambda)]^n - e^{\lambda x_n}| \to 0 \qquad \text{as} \quad h \to 0 \qquad (6.189)$$

We must now show that the coefficient $\gamma_0 \to 1$ as $h \to 0$.

The coefficients $\gamma_0, \cdots, \gamma_p$ satisfy the linear system

$$\gamma_0 + \gamma_1 + \cdots + \gamma_p = y_0$$

$$\gamma_0 [r_0(h\lambda)] + \cdots + \gamma_p [r_p(h\lambda)] = y_1 \qquad (6.190)$$

$$\vdots$$

$$\gamma_0 [r_0(h\lambda)]^p + \cdots + \gamma_p [r_p(h\lambda)]^p = y_p$$

The initial values $y_0, \cdots, y_p$ are assumed to satisfy

$$\eta(h) \equiv \underset{0 \leqslant n \leqslant p}{\text{Max}} |e^{\lambda x_n} - y_n| \to 0 \qquad \text{as} \quad h \to 0$$

But this implies

$$\underset{h \to 0}{\text{Limit}}\, y_n = 1 \qquad 0 \leqslant n \leqslant p \qquad (6.191)$$

The coefficient γ_0 can be obtained by using Cramer's rule to solve (6.190). Then

$$\gamma_0 = \frac{\begin{vmatrix} y_0 & 1 & \cdots & 1 \\ y_1 & r_1 & \cdots & r_p \\ \vdots & & & \\ y_p & r_1^p & \cdots & r_p^p \end{vmatrix}}{\begin{vmatrix} 1 & 1 & \cdots & 1 \\ r_0 & r_1 & \cdots & r_p \\ \vdots & & & \\ r_0^p & r_1^p & \cdots & r_p^p \end{vmatrix}}$$

The denominator converges to the Vandermonde determinant for $r_0(0) = 1, r_1(0), \cdots, r_p(0)$; and this is nonzero since the roots are distinct (see Problem 1 of Chapter 3). By using (6.191), the numerator converges to the same quantity as $h \to 0$. Therefore, $\gamma_0 \to 1$ as $h \to 0$. Using this, along with (6.188) and Problem 23, the solution $\{y_n\}$ converges to $Y(x) = e^{\lambda x}$ on $[0, b]$. This completes the proof.   ∎

The following is a well-known result; it is a trivial consequence of Theorems 6.7 and 6.8.

Corollary   Let (6.158) be a consistent multistep method. Then it is convergent if and only if it is stable.

Relative stability and weak stability   Consider again the exponential equation (6.169) and its numerical solution (6.176). The past theorem stated that the parasitic solutions $\gamma_j[r_j(h\lambda)]^n$ will converge to zero as $h \to 0$. But for a fixed h with increasing x_n, we also would like them to remain small relative to the principal part of the solution $\gamma_0[r_0(h\lambda)]^n$. This will be true if the characteristic roots satisfy

$$|r_j(h\lambda)| \leqslant r_0(h\lambda) \qquad j = 1, 2, \cdots, p \tag{6.192}$$

for all sufficiently small values of h. This leads us to the definition of relative stability.

We say that the method (6.158) is *relatively stable* if the characteristic roots $r_j(h\lambda)$ satisfy (6.192) for all sufficiently small nonzero values of $|h\lambda|$. And the method is said to satisfy the *strong root condition* if

$$|r_j(0)| < 1 \qquad j = 1, 2, \cdots, p \tag{6.193}$$

This is an easy condition to check, and it implies relative stability. Just use the continuity of the roots $r_j(h\lambda)$ with respect to $h\lambda$ to have (6.193) imply (6.192). Relative stability does not imply the strong root condition, although they are equivalent for most methods. If a multistep method is stable but not relatively stable, then it will be called *weakly stable*.

Examples 1. For the midpoint method,

$$r_0(h\lambda) = 1 + h\lambda + O(h^2), \quad r_1(h\lambda) = -1 + h\lambda + O(h^2).$$

It is weakly stable according to (6.192) when $\lambda < 0$, which agrees with what was shown earlier in Section 6.4.

2. Adams-Bashforth and Adams-Moulton methods, (6.149) and (6.152), have the same characteristic polynomial when $h = 0$,

$$\rho(r) = r^{p+1} - r^p$$

The roots are $r_0 = 1, r_j = 0, j = 1, 2, \cdots, p$; and thus the strong root condition is satisfied and the Adams methods are relatively stable.

Example Return to the example (6.137) of the last section, a family of methods determined by the method of undetermined coefficients. The value of a_1 was restricted to lie in $[-.6, 1.0]$, and now it is chosen to make (6.192) true for as large a value of $|h\lambda|$ as possible. A convenient and nearly optimal choice is $a_1 = 0$, and it leads to the fourth-order method

$$y_{n+1} = \frac{1}{8}(9y_n - y_{n-2}) + \frac{3h}{8}\big[f(x_{n+1}, y_{n+1})$$

$$+ 2f(x_n, y_n) - f(x_{n-1}, y_{n-1})\big], \qquad n \geqslant 2$$

The combination of this corrector formula with the predictor (6.140) is known as Hamming's method. For a further discussion, see Ralston [R1, pp. 186-189].

Stability regions In the preceding discussions of stability, the values of h were required to be sufficiently small in order to carry through the derivations. But very little indication was given as to how small h should be. It is clear that if h is required to be extremely small, then the method is impractical; and thus we need to examine the permissible values of h. Since the stability depends on the characteristic roots, and since they depend on $h\lambda$, we are interested in determining the values of $h\lambda$ for which the method (6.158) is stable in some sense. To cover situations arising when solving systems of differential equations, it is necessary that the value of λ be allowed to be complex.

One natural region to consider is the *region of relative stability*. It is defined as all values $h\lambda$ for which (6.192) is satisfied. Comparison of the regions of relative stability for various multistep methods (6.158) is a useful way of comparing the methods to determine which is a better method. Generally, it is best to have as large a stability region as possible; although the size of the truncation error is also obviously quite important in choosing a method.

Another stability region has been of more use in the research literature, and we will concentrate on it rather than on the region of relative stability. The values of λ will be restricted to having a negative real part; and thus the solution $Y(x) = e^{\lambda x} \to 0$ as $x \to \infty$. The *region of absolute stability* is defined as the set of all $h\lambda$ for which the numerical solution $y_n \to 0$ as $x_n \to \infty$. Using the general formula

$$y_n = \sum_{j=0}^{p} \gamma_j [r_j(h\lambda)]^n$$

it is equivalent to require that $h\lambda$ satisfy

$$|r_j(h\lambda)| < 1 \qquad 0 \leqslant j \leqslant p \tag{6.194}$$

The larger the region of absolute stability, the less the restriction on h in order to have the numerical method give a numerical solution that is qualitatively the same as the true solution. For such values of $h\lambda$, we say the multistep method is absolutely stable.

Examples 1. Euler's method has the characteristic equation $r - 1 - h\lambda = 0$. The region of absolute stability is the set of all $h\lambda$ in a circle of radius 1 in the complex plane, centered at $(-1, 0)$.

2. Consider the second-order Adams-Bashforth method

$$y_{n+1} = y_n + \frac{h}{2}(3y_n' - y_{n-1}') \qquad n \geqslant 1$$

The characteristic equation is

$$r^2 - \left(1 + \tfrac{3}{2}h\lambda\right)r + \tfrac{1}{2}h\lambda = 0$$

and its roots are

$$r_0 = \tfrac{1}{2}\left(1 + \tfrac{3}{2}h\lambda + \sqrt{1 + h\lambda + \tfrac{9}{4}h^2\lambda^2}\,\right)$$

$$r_1 = \tfrac{1}{2}\left(1 + \tfrac{3}{2}h\lambda - \sqrt{1 + h\lambda + \tfrac{9}{4}h^2\lambda^2}\,\right)$$

The region of absolute stability is the set of $h\lambda$ for which

$$|r_0(h\lambda)| < 1 \qquad |r_1(h\lambda)| < 1$$

For λ real, the acceptable values of $h\lambda$ are $-1 < h\lambda < 0$. Determining the permissible complex values of $h\lambda$ is fairly complicated.

The regions of absolute stability for the Adams-Bashforth and Adams-Moulton methods are given in Figures 6.8 and 6.9, respectively. These are taken from Gear [G1, p. 131]. For Adams-Moulton formulas with one iteration of an Adams-Bashforth predictor, the regions of absolute stability are given in Shampine and Gordon [S2, pp. 135–140].

From these diagrams, it is clear that the region of absolute stability becomes smaller as the order of the method increases. And for formulas of the same order, the Adams-Moulton formula has a significantly larger region of absolute stability than the Adams-Bashforth formula. The size of these regions is usually quite acceptable from a practical point of view. For example, the real

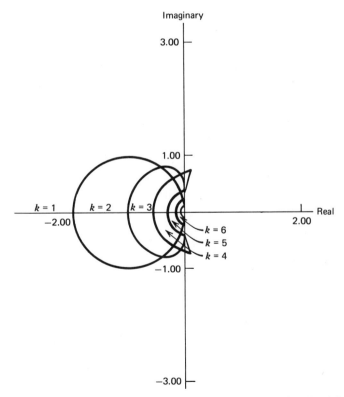

Figure 6.8 Stability regions for Adams-Bashforth methods. Method of order k is stable inside region indicated left of origin. (From C. William Gear, *Numerical Initial Value Problems in Ordinary Differential Equations*, ©1971, p. 131. Reprinted by permission of Prentice-Hall, Inc., Englewood Cliffs, N.J.)

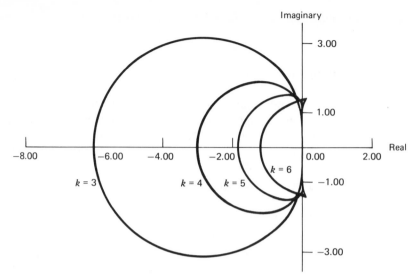

Figure 6.9 Stability regions for Adams-Moulton method. Method of order k is stable inside region indicated. (From C. William Gear, *Numerical Initial Value Problems in Ordinary Differential Equations*, ©1971, p. 131. Reprinted by permission of Prentice-Hall, Inc., Englewood Cliffs, N.J.)

values of $h\lambda$ in the region of absolute stability for the fourth-order Adams-Moulton formula are given by $-3 < h\lambda < 0$. This is not a serious restriction on h in most cases.

The Adams family of formulas is very convenient for creating a variable-order algorithm, and their stability regions are quite acceptable for most problems. They will have difficulty with problems for which λ is negative and quite large in magnitude, those that are called stiff equations; but these problems are best treated by special techniques. At present, there is no other family of formulas that is as well suited for a variable-order predictor-corrector algorithm as is the Adams family.

Stiff differential equations   The equation

$$y' = \lambda y + g(x), \qquad x_0 \leqslant x \leqslant b$$

is called stiff if λ is negative and is of large size (if λ is complex, we require its real part to be negative and have a large size). From (6.170) and (6.171) earlier in this section, we know that any general nonlinear equation can be approximated by such a linear equation, provided $[x_0, b]$ is not too large; and $\lambda = f_y(x_0, Y_0)$. For systems of equations, we consider the eigenvalues λ of the Jacobian matrix $\mathbf{f}_y(x, \mathbf{y})$.

From (6.2) of the introduction, the above linear non-homogeneous equation has the general solution

$$Y(x) = Y_0 e^{\lambda x} + \int_{x_0}^{x} e^{\lambda(x-t)} g(t) \, dt$$

Since λ is negative and of large magnitude, the term $Y_0 e^{\lambda x}$ quickly decreases to zero. Thus only the integral term is of any interest for large values of x. If we were concerned with only having a small truncation error, then h would be chosen on the basis of the smoothness properties of the integral term in $Y(x)$; and generally this would not require a very small value of h.

The difficulty with this analysis is that it ignores the stability of the numerical method, which is dependent on the size of $h\lambda$. With almost all the methods of this chapter, if $|\lambda|$ is large, then h will have to be correspondingly smaller in order that $h\lambda$ will still lie in the region of absolute stability of the numerical method. If $h\lambda$ is not in this region, then the numerical approximation to the term $Y_0 e^{\lambda x}$ will not decrease to zero, as it should. The choice of h when solving a very stiff system of equations must be very small to ensure stability of the numerical method. But the small value of h is unnecessary for accuracy, and it results in a much smaller truncation error than should be necessary. As a result of this source of inefficiency, much work has been done on developing special methods for stiff equations. A general introduction to some of these methods is given in Gear [G1, Chapter 11]; and we do not develop any such methods in this text.

Stiff equations occur in a number of contexts. Physically they are often associated with situations in which there is a transient solution which damps out rapidly. Another source that is becoming more important is in the use of the *method of lines* to solve time-dependent partial differential equations. For such applications, magnitudes for λ of 10^3 to 10^5 are not unusual; and these result in extremely small values for h when an ordinary multistep method is used.

One desirable feature for a numerical method to have for treating stiff differential equations is A-stability. This was defined earlier in Section 6.5, where it was shown that the trapezoidal method was A-stable. To be A-stable, the region of absolute stability of the numerical method must include any complex $h\lambda$ for which the real part is negative. It has been shown that there are no multistep methods (6.158) of order greater than two that are also A-stable. Thus the trapezoidal rule is the best that can be attained within the context of multistep methods, and it is an important tool for treating stiff differential equations. When it is used in connection with the method of lines for time-dependent partial differential equations, it results in the well-known Crank-Nicholson method.

6.9 Single-Step and Runge-Kutta Methods

Single-step methods for solving $y' = f(x,y)$ require only a knowledge of the numerical solution y_n in order to compute the next value y_{n+1}. This has obvious advantages over the p-step multistep methods that use several past values $\{y_n, \cdots, y_{n-p}\}$, and that require initial values $\{y_1, \cdots, y_p\}$ that have to be calculated by another method.

The best known one-step methods are the Runge-Kutta methods; and they are the usual means for calculating the initial values $\{y_1, \cdots, y_p\}$ for a $(p+1)$-step multistep method. They are fairly simple to program, and their truncation error can be controlled in a more straightforward manner than for the multistep methods. The major disadvantage of the Runge-Kutta methods is that they use many more evaluations of the derivative $f(x,y)$ to attain the same accuracy, compared with the multistep methods. And at present, there are no variable order Runge-Kutta methods comparable to the Adams-Bashforth and Adams-Moulton methods.

The most simple one-step method is based on using Taylor series. Assume $Y(x)$ is $r+1$ times continuously differentiable, where $Y(x)$ is the solution of the initial value problem

$$y' = f(x,y) \qquad y(x_0) = Y_0 \qquad (6.195)$$

Expand $Y(x_1)$ about x_0 using Taylor's theorem,

$$Y(x_1) = Y(x_0) + hY'(x_0) + \ldots + \frac{h^r}{r!} Y^{(r)}(x_0) + \frac{h^{r+1}}{(r+1)!} Y^{(r+1)}(\xi) \qquad (6.196)$$

for some $x_0 \leqslant \xi_0 \leqslant x_1$. By dropping the remainder term, we have an approximation for $Y(x_1)$, provided we can calculate $Y''(x_0), \cdots, Y^{(r)}(x_0)$. Differentiate $Y'(x) = f(x, Y(x))$ to obtain

$$Y''(x) = f_x(x, Y(x)) + f_y(x, Y(x)) Y'(x)$$

$$Y'' = f_x + f_y f$$

and proceed similarly to obtain the higher-order derivatives of $Y(x)$.

Example Consider the problem

$$y' = -y^2, \qquad y(0) = 1$$

with the solution $Y(x) = 1/(1+x)$. Then $Y'' = -2YY' = 2Y^3$; and (6.196) with $r = 2$ yields

$$Y(x_1) = Y_0 - hY_0^2 + h^2 Y_0^3 + \frac{h^3}{6} Y'''(\xi_0) \qquad x_0 \leqslant \xi_0 \leqslant x_1$$

We drop the remainder to obtain an approximation of $Y(x_1)$. This can then be used in the same manner to obtain an approximation for $Y(x_2)$, and so on. The numerical method is

$$y_{n+1} = y_n - hy_n^2 + h^2 y_n^3 \qquad n \geqslant 0 \qquad (6.197)$$

Table 6.16 contains the errors in this numerical solution at a selected set of node points. The grid sizes used are $h = .125$ and $h = .0625$; and the ratio of the resulting errors is also given. Note that when h is halved, the ratio is almost 4. This can be justified theoretically since the rate of convergence can be shown to be $O(h^2)$, with a proof similar to that given in Theorems 6.1 or 6.9.

Table 6.16 Example of Taylor series method (6.197)

x	$h = .0625$		$h = .125$	Ratio
	$y_h(x)$	$Y(x) - y_h(x)$	$Y(x) - y_h(x)$	
2.0	.333649	$-3.2E-4$	$-1.4E-3$	4.4
4.0	.200135	$-1.4E-4$	$-5.9E-4$	4.3
6.0	.142931	$-7.4E-5$	$-3.2E-4$	4.3
8.0	.111157	$-4.6E-5$	$-2.0E-4$	4.3
10.0	.090941	$-3.1E-5$	$-1.4E-4$	4.3

The Taylor series method can give excellent results. But it is bothersome to use because of the need to differentiate equations. The derivatives can be very difficult to calculate and very time-consuming to evaluate. To avoid the differentiation of $f(x,y)$, we turn to the Runge-Kutta formulas.

Runge-Kutta methods These are closely related to the Taylor series expansion of $Y(x)$ in (6.196), but no differentiations of f are necessary in the use of the method. For simplicity, we will abbreviate Runge-Kutta to RK. All RK methods will be written in the form

$$y_{n+1} = y_n + hF(x_n, y_n, h; f) \qquad n \geqslant 0 \qquad (6.198)$$

We begin with examples of F, and will later discuss hypotheses for it. But at this point, it should be intuitive that we want

$$F(x, Y(x), h; f) \approx Y'(x) = f(x, Y(x))$$

for all small values of h. Define the truncation error for (6.198) by

$$T_{n+1}(Y) = Y(x_{n+1}) - Y(x_n) - hF(x_n, Y(x_n), h; f) \qquad n \geqslant 0 \qquad (6.199)$$

and define $\tau_{n+1}(Y)$ implicitly by

$$T_{n+1}(Y)=h\tau_{n+1}(Y)$$

Rearranging (6.199), we obtain

$$Y(x_{n+1})=Y(x_n)+hF(x_n,Y(x_n),h;f)+h\tau_{n+1}(Y) \qquad n\geqslant 0 \qquad (6.200)$$

This will be compared with (6.198) to prove convergence in Theorem 6.9.

Examples 1.   Consider the trapezoidal method, solved with one iteration using Euler's method as the predictor,

$$y_{n+1}=y_n+\frac{h}{2}\left[f(x_n,y_n)+f(x_{n+1},y_n+hf(x_n,y_n))\right] \qquad n\geqslant 0 \qquad (6.201)$$

In the notation of (6.198),

$$F(x,y,h;f)=\tfrac{1}{2}\left[f(x,y)+f(x+h,y+hf(x,y))\right]$$

As can be seen in Figure 6.10, F is an average slope of $Y(x)$ on $[x,x+h]$.

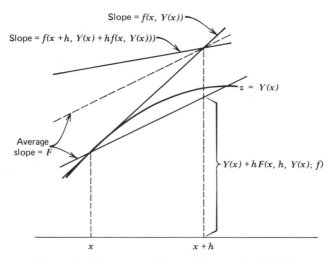

Slope = $f(x, Y(x))$

Slope = $f(x+h, Y(x)+hf(x, Y(x)))$

$z = Y(x)$

Average slope = F

$Y(x)+hF(x, h, Y(x); f)$

x

$x+h$

Figure 6.10   Illustration of Runge–Kutta method (6.201).

2.   The following method is also based on obtaining an average slope for the solution on $[x_n,x_{n+1}]$,

$$y_{n+1}=y_n+hf\left(x_n+\tfrac{1}{2}h, y_n+\tfrac{1}{2}hf(x_n,y_n)\right) \qquad n\geqslant 0 \qquad (6.202)$$

For this case,

$$F(x, y, h; f) = f\left(x + \tfrac{1}{2}h, y + \tfrac{1}{2}hf(x, y)\right)$$

The derivation of a formula for the truncation error is linked to the derivation of these methods, and this will also be true when considering RK methods of a higher order. The derivation of RK methods is illustrated by deriving a family of second-order formulas, which will include (6.201) and (6.202). We suppose F has the general form

$$F(x, y, h; f) = \gamma_1 f(x, y) + \gamma_2 f(x + \alpha h, y + \beta h f(x, y)) \qquad (6.203)$$

in which the four constants γ_1, γ_2, α, and β are to be determined.

We will use Taylor's theorem (1.5) for functions of two variables to expand the second term on the right side of (6.203), through the second derivative terms. This gives

$$F(x, y, h; f) = \gamma_1 f(x, y) + \gamma_2 \big[f(x, y) + h(\alpha f_x + \beta f f_y)$$
$$+ h^2 \left(\tfrac{1}{2} \alpha^2 f_{xx} + \alpha \beta f_{xy} f + \tfrac{1}{2} \beta^2 f^2 f_{yy} \right) \big] + O(h^3) \qquad (6.204)$$

Also we will need some derivatives of $Y'(x) = f(x, Y(x))$, namely

$$Y'' = f_x + f_y f$$

$$Y''' = f_{xx} + 2 f_{xy} f + f_{yy} f^2 + f_y f_x + f_y^2 f. \qquad (6.205)$$

For the truncation error,

$$T_{n+1}(Y) = Y(x_{n+1}) - Y(x_n) - hF(x_n, Y(x_n), h; f)$$

$$= hY_n' + \frac{h^2}{2} Y_n'' + \frac{h^3}{6} Y_n''' + O(h^4) - hF(x_n, Y_n, h; f)$$

Substituting from (6.204) and (6.205), and collecting together common powers of h, we obtain

$$T_{n+1}(Y) = h[1 - \gamma_1 - \gamma_2]f + h^2 \left[\left(\tfrac{1}{2} - \gamma_2 \alpha \right) f_x + \left(\tfrac{1}{2} - \gamma_2 \beta \right) f_y f \right]$$
$$+ h^3 \left[\left(\tfrac{1}{6} - \tfrac{1}{2} \gamma_2 \alpha^2 \right) f_{xx} + \left(\tfrac{1}{3} - \gamma_2 \alpha \beta \right) f_{xy} f + \left(\tfrac{1}{6} - \tfrac{1}{2} \gamma_2 \beta^2 \right) f_{yy} f^2 \right.$$
$$\left. + \tfrac{1}{6} f_y f_x + \tfrac{1}{6} f_y^2 f \right] + O(h^4) \qquad (6.206)$$

All derivatives are evaluated at (x_n, Y_n).

We wish to make the truncation error converge to zero as rapidly as possible. The coefficient of h^3 cannot be zero in general, if f is allowed to vary arbitrarily. The requirement that the coefficients of h and h^2 be zero leads to

$$\gamma_1 + \gamma_2 = 1 \qquad \gamma_2\alpha = 1/2 \qquad \gamma_2\beta = 1/2 \tag{6.207}$$

and this yields

$$T_{n+1}(Y) = O(h^3)$$

The system (6.207) is underdetermined, and its general solution is

$$\gamma_1 = 1 - \gamma_2 \qquad \alpha = \beta = \frac{1}{2\gamma_2} \tag{6.208}$$

with γ_2 arbitrary. Both (6.201) (with $\gamma_2 = \frac{1}{2}$) and (6.202) (with $\gamma_2 = 1$) are special cases of this solution.

By substituting into (6.206), we can obtain the leading term in the truncation error, dependent on only γ_2. In some cases, the value of γ_2 has been chosen to make the coefficient of h^3 as small as possible, while allowing f to vary arbitrarily. For example, if we write (6.206) as

$$T_{n+1}(Y) = c(f, \gamma_2)h^3 + O(h^4)$$

then the Cauchy-Schwartz inequality [from (7.4), Chapter 7] can be used to show

$$|c(f, \gamma_2)| \leq c_1(f)c_2(\gamma_2)$$

where

$$c_1(f) = \left[f_{xx}^2 + f_{xy}^2 f^2 + f_{yy}^2 f^4 + f_y^2 f_x^2 + f_y^4 f^2 \right]^{1/2}$$

$$c_2(\gamma_2) = \left[\left(\tfrac{1}{6} - \tfrac{1}{2}\gamma_2\alpha^2 \right)^2 + \left(\tfrac{1}{3} - \gamma_2\alpha\beta \right)^2 + \left(\tfrac{1}{6} - \tfrac{1}{2}\gamma_2\beta^2 \right)^2 + \tfrac{1}{18} \right]^{1/2}$$

with α, β given by (6.208). The minimum value $c_2(\gamma_2)$ is attained with $\gamma_2 = .75$, and $c_2(.75) = 1/\sqrt{18}$. The resulting second-order numerical method is

$$y_{n+1} = y_n + \frac{h}{4}\left[f(x_n, y_n) + 3f\left(x_n + \tfrac{2}{3}h, y_n + \tfrac{2}{3}hf(x_n, y_n)\right) \right] \qquad n \geq 0 \tag{6.209}$$

It is optimal in the sense of minimizing the coefficient $c_2(\gamma_2)$ of the term $c_1(f)h^3$ in the truncation error. For an extensive discussion of this means of analyzing the truncation error in RK methods, see Shampine [S1] or Shampine and Watts [S4].

Higher-order formulas can be created and analyzed in an analogous manner, although the algebra becomes very complicated. Assume a formula for

$F(x, y, h; f)$ of the form

$$F(x, y, h; f) = \sum_{j=1}^{p} \gamma_j V_j$$

$$V_1 = f(x, y)$$

$$V_j = f\left(x + \alpha_j h, y + h \sum_{i=1}^{j-1} \beta_{ji} V_i\right) \qquad j = 2, \cdots, p \tag{6.210}$$

These coefficients can be chosen to make the leading terms in the truncation error equal to zero, just as was done with (6.206) and (6.207). There is obviously a connection between the number of evaluations of $f(x, y)$, call it p, and the maximum possible order that can be attained for the truncation error. These are given in Table 6.17, which is due in part to Butcher [B2].

Table 6.17 Maximum order of Runge-Kutta methods

Number of function evaluations	1	2	3	4	5	6	7	8
Maximum order of method	1	2	3	4	4	5	6	6

Until recently, the most popular RK method has been the original classical formula that is a generalization of Simpson's rule. The method is

$$y_{n+1} = y_n + \frac{h}{6}(V_1 + 2V_2 + 2V_3 + V_4)$$

$$V_1 = f(x_n, y_n) \qquad V_2 = f\left(x_n + \tfrac{1}{2}h, y_n + \tfrac{1}{2}hV_1\right) \tag{6.211}$$

$$V_3 = f\left(x_n + \tfrac{1}{2}h, y_n + \tfrac{1}{2}hV_2\right) \qquad V_4 = f(x_n + h, y_n + hV_3)$$

It can be proven that (6.211) is a fourth-order formula with $T_{n+1}(Y) = O(h^5)$. If $f(x, y)$ does not depend on y, then this formula reduces to Simpson's integration rule.

Example Consider the problem

$$y' = \frac{1}{1 + x^2} - 2y^2 \qquad y(0) = 0 \tag{6.212}$$

with the solution $Y = x/(1 + x^2)$. The method (6.211) was used with a fixed stepsize, and the results are shown in Table 6.18. The stepsizes are $h = .25$ and $2h = .5$. The Ratio column gives the ratio of the errors for corresponding node

Table 6.18   Example of Runge-Kutta method (6.211)

x	$y_h(x)$	$Y(x) - y_h(x)$	$Y(x) - y_{2h}(x)$	Ratio	$\frac{1}{15}[y_h(x) - y_{2h}(x)]$
2.0	.39995699	4.3E−5	1.0E−3	24	6.7E−5
4.0	.23529159	2.5E−6	7.0E−5	28	4.5E−6
6.0	.16216179	3.7E−7	1.2E−5	32	7.7E−7
8.0	.12307683	9.2E−8	3.4E−6	36	2.2E−7
10.0	.09900987	3.1E−8	1.3E−6	41	8.2E−8

points as h is halved. The last column is an example of formula (6.214) from the following material on Richardson extrapolation. Because $T_{n+1}(Y) = O(h^5)$ for method (6.211), Theorem 6.9 implies that the rate of convergence of $y_h(x)$ to $Y(x)$ is $O(h^4)$. The theoretical value of Ratio is 16; and as h decreases further, this value will be realized more closely.

The RK methods have asymptotic error formulas, the same as the multistep methods. For (6.211),

$$Y(x) - y_h(x) = D(x)h^4 + O(h^5) \tag{6.213}$$

where $D(x)$ satisfies a certain initial value problem. The proof of this result is an extension of the proof of Theorem 6.9, and it is similar to the derivation used with Euler's method in Section 6.2. The result (6.213) can be used to produce an error estimate, just as was done with the trapezoidal method in formula (6.111) of Section 6.5. For a stepsize of $2h$,

$$Y(x) - y_{2h}(x) = 16D(x)h^4 + O(h^5)$$

Proceeding as for (6.111), we obtain

$$Y(x) - y_h(x) = \tfrac{1}{15}\left[y_h(x) - y_{2h}(x) \right] + O(h^5) \tag{6.214}$$

and the first term on the right side is an estimate of the left-hand error. This is illustrated in the last column of Table 6.18.

Convergence analysis   In order to obtain convergence of the general schema (6.198), we will need to have $\tau_{n+1}(Y) \to 0$ as $h \to 0$. Since

$$\tau_{n+1}(Y) = \frac{Y(x_{n+1}) - Y(x_n)}{h} - F(x_n, Y(x_n), h; f)$$

We require that

$$F(x, Y(x), h; f) \to Y'(x) = f(x, Y(x)) \text{ as } h \to 0$$

More precisely, define

$$\delta(h) = \underset{\substack{x_0 \leqslant x \leqslant b \\ -\infty < y < \infty}}{\text{Maximum}} |f(x, y) - F(x, y, h; f)|$$

and assume

$$\delta(h) \to 0 \text{ as } h \to 0 \qquad (6.215)$$

This is occasionally called the consistency condition for the method (6.198). We will also need a Lipschitz condition on F:

$$|F(x, y, h; f) - F(x, z, h; f)| \leqslant L|y - z| \qquad (6.216)$$

for all $x_0 \leqslant x \leqslant b$, $-\infty < y$, $z < \infty$, and all small $h > 0$. This condition is usually proven by using the Lipschitz condition (6.29) on $f(x, y)$. For example, with method (6.202),

$$|F(x, y, h; f) - F(x, z, h; f)| = \left| f\left(x + \frac{h}{2}, y + \frac{h}{2} f(x, y)\right) - f\left(x + \frac{h}{2}, z + \frac{h}{2} f(x, z)\right) \right|$$

$$\leqslant K\left| y - z + \frac{h}{2}[f(x, y) - f(x, z)] \right|$$

$$\leqslant K\left(1 + \frac{h}{2}K\right)|y - z|$$

Choose $L = K(1 + \frac{1}{2}K)$ for $h \leqslant 1$.

Theorem 6.9   Assume that the Runge-Kutta method (6.198) satisfies the Lipschitz condition (6.216). Then for the initial value problem (6.195), the solution $\{y_n\}$ satisfies

$$\underset{x_0 \leqslant x_n \leqslant b}{\text{Maximum}} |Y(x_n) - y_n|$$

$$\leqslant e^{(b - x_0)L}|Y_0 - y_0| + \left[\frac{e^{(b - x_0)L} - 1}{L} \right] \tau(h) \qquad (6.217)$$

where

$$\tau(h) \equiv \underset{x_0 \leqslant x_n \leqslant b}{\text{Max}} |\tau_{n+1}(Y)| \qquad (6.218)$$

If the consistency condition (6.215) is also satisfied, then the numerical solution $\{y_n\}$ converges to $Y(x)$.

Proof Subtract (6.198) from (6.200) to obtain

$$e_{n+1} = e_n + h\left[F(x_n, Y_n, h; f) - F(x_n, y_n, h; f)\right] + h\tau_{n+1}(Y) \quad (6.219)$$

in which $e_n = Y(x_n) - y_n$. Apply the Lipschitz condition (6.216) and use (6.218) to obtain

$$|e_{n+1}| \leq (1 + hL)|e_n| + h\tau(h) \qquad x_0 \leq x_n \leq b$$

As with the proof of the Euler method, this leads easily to the result (6.217); see (6.37)–(6.39).

In most cases, it is known by direct computation that $\tau(h) \to 0$ as $h \to 0$; and in that case, convergence of $\{y_n\}$ to $Y(x)$ is immediately proven. But all that we need to know is that (6.215) is satisfied. To see this, write

$$h\tau_{n+1}(Y) = Y(x_{n+1}) - Y(x_n) - hF(x_n, Y(x_n), h; f)$$

$$= hY'(x_n) + \frac{h^2}{2} Y''(\xi_n) - hF(x_n, Y(x_n), h; f)$$

$$h|\tau_{n+1}(Y)| \leq h\delta(h) + \frac{h^2}{2}\|Y''\|_\infty$$

$$\tau(h) \leq \delta(h) + \tfrac{1}{2}h\|Y''\|_\infty$$

Thus $\tau(h) \to 0$ as $h \to 0$, completing the proof. ∎

The following result is an immediate consequence of (6.217).

Corollary If the Runge-Kutta method (6.198) has a truncation error $T_{n+1}(Y) = O(h^{m+1})$, then the rate of convergence of $\{y_n\}$ to $Y(x)$ is $O(h^m)$.

It is not too difficult to derive an asymptotic error formula for the Runge-Kutta method (6.198), provided one is known for the truncation error. Assume

$$T_{n+1}(Y) = \phi(x_n)h^{m+1} + O(h^{m+2}) \qquad (6.220)$$

with $\phi(x)$ determined by $Y(x)$ and $f(x, Y(x))$. As an example see the result (6.206) to obtain this expansion for second-order RK methods. Strengthened forms of (6.215) and (6.216) are also necessary. Assume

$$F(x, y, h; f) - F(x, z, h; f) = \frac{\partial F(x, y, h; f)}{\partial y}(y - z) + O\left((y - z)^2\right) \quad (6.221)$$

and also that

$$\delta_1(h) \equiv \max_{\substack{x_0 \leqslant x \leqslant b \\ -\infty < y < \infty}} \left| \frac{\partial f(x,y)}{\partial y} - \frac{\partial F(x,y,h;f)}{\partial y} \right| \to 0 \text{ as } h \to 0 \qquad (6.222)$$

In practice, both of these results are straightforward to confirm. With these assumptions, we can derive the formula

$$Y(x) - y_h(x) = D(x)h^m + O(h^{m+1}) \qquad (6.223)$$

with $D(x)$ satisfying the linear initial value problem

$$D' = f_y(x, Y(x))D(x) + \phi(x) \qquad D(x_0) = 0$$

Stability results can be given for RK methods, in analogy with those for multistep methods. The basic type of stability defined near the beginning of Section 6.8 is easily proven. The proof is a simple modification of the proof of Theorem 6.9, and we leave it to Problem 34. An essential difference with the multistep theory is that there are no parasitic solutions created by the RK methods; and thus the concept of relative stability does not apply to RK methods. The regions of absolute stability can be studied as with the multistep theory, but we will omit it.

Estimation of the truncation error   In order to control the size of the truncation error, we must first be able to estimate it. Let $u_n(x)$ denote the solution of $y' = f(x, y)$ passing through (x_n, y_n). We wish to estimate the error in the numerical solution $y_h(x_n + 2h)$ relative to $u_n(x_n + 2h)$.

Using (6.219) for the error in $y_h(x)$ compared to $u_n(x)$, and using the asymptotic formula (6.220),

$$u_n(x_{j+1}) - y_h(x_{j+1}) = u_n(x_j) - y_h(x_j)$$

$$+ h\left[F(x_n, u_n(x_j), h; f) - F(x_n, y_h(x_j), h; f) \right]$$

$$+ \phi_n(x_j)h^{m+1} + O(h^{m+2}) \qquad j \geqslant n$$

From this, it is straightforward to prove

$$u_n(x_n + h) - y_h(x_n + h) = \phi_n(x_n)h^{m+1} + O(h^{m+2})$$

$$u_n(x_n + 2h) - y_h(x_n + 2h) = 2\phi_n(x_n)h^{m+1} + O(h^{m+2})$$

And applying the same procedure to $y_{2h}(x_j)$,

$$u_n(x_n + 2h) - y_{2h}(x_n + 2h) = 2^{m+1}\phi_n(x_n)h^{m+1} + O(h^{m+2})$$

From the last two equations,

$$u_n(x_n+2h)-y_h(x_n+2h)=\frac{1}{2^m-1}\left[y_h(x_n+2h)-y_{2h}(x_n+2h)\right]+O\left(h^{m+2}\right)$$

and the first term on the right is an asymptotic estimate of the error on the left side.

Consider the computation of $y_h(x_n+2h)$ from $y_h(x_n)\equiv y_n$ as a single step in an algorithm. Suppose that a user has given an error tolerance ϵ and that the value of

$$\text{trunc}\equiv u_n(x_n+2h)-y_h(x_n+2h)$$

$$\approx\frac{1}{2^m-1}\left[y_h(x_n+2h)-y_{2h}(x_n+2h)\right] \tag{6.224}$$

is to satisfy

$$.5\epsilon h\leqslant|\text{trunc}|\leqslant 2\epsilon h \tag{6.225}$$

This controls the error per unit step, and it requires that the error be neither too large or too small. Recall the concept of error per unit stepsize in (6.112) of Section 6.6.

If the test is satisfied, then computation continues with the same h. But if it is not satisfied, then a new value $\hat{h}$ must be chosen. Let it be chosen by

$$2|\phi_n(x_n)|\hat{h}^{m+1}\doteq\epsilon\hat{h}$$

where $\phi_n(x_n)$ is determined from

$$\phi_n(x_n)\doteq\frac{\text{trunc}}{2h^{m+1}}$$

With the new value of $\hat{h}$, the new truncation error should lie near the midpoint of (6.225). This form of algorithm has been implemented with a number of methods. For example, see Gear [G1, pp. 83–84] for a similar algorithm for a fourth-order RK method.

To better understand the expense of RK methods, consider only fourth-order RK methods with four evaluations of $f(x, y)$ per RK step. In going from x_n to x_n+2h, 8 evaluations are required to obtain $y_h(x_n+2h)$, and 3 additional evaluations to obtain $y_{2h}(x_n+2h)$. Thus a single step of the variable step algorithm will require 11 evaluations of f. Although fairly expensive to use when compared with a multistep method, a variable stepsize RK method is very stable, reliable, and is comparatively easy to program for a computer.

Runge-Kutta-Fehlberg methods These are RK methods in which the truncation error is computed by comparing the computed answer y_{n+1} with the result of an associated higher-order RK formula. The most popular of such methods are due to E. Fehlberg (for example, see [F1]); and these are currently the most popular RK methods. To clarify the presentation, we consider only the pair of Runge-Kutta-Fehlberg (RKF) formulas of order 4 and 5. These are computed simultaneously, and their difference is taken as an estimate of the truncation error in the fourth-order method.

Note from Table 6.17 that a fifth-order RK method requires six evaluations of f per step. Consequently, Fehlberg chose to use five evaluations of f for the fourth-order formula, rather than the usual four. This extra degree of freedom in choosing the fourth-order formula allowed it to be chosen with a smaller truncation error, and this is illustrated later.

As before, define

$$V_1 = f(x_n, y_n)$$

$$V_j = f\left(x_n + \alpha_j h, y_n + h \sum_{i=1}^{j-1} \beta_{ji} V_i\right) \qquad j = 2, \cdots, 6$$

The fourth- and fifth-order formulas are, respectively,

$$y_{n+1} = y_n + h \sum_{i=1}^{5} \gamma_j V_j \tag{6.226}$$

$$\hat{y}_{n+1} = y_n + h \sum_{i=1}^{6} \hat{\gamma}_j V_j \tag{6.227}$$

The truncation error in y_{n+1} is approximately

$$\text{trunc} = \hat{y}_{n+1} - y_{n+1} = h \sum_{j=1}^{6} c_j V_j \tag{6.228}$$

The coefficients are given in Tables 6.19 and 6.20.

Table 6.19 Coefficients α_j and β_{ji} for RKF method

| j | α_j | β_{ji} | | | | |
		$i=1$	$i=2$	$i=3$	$i=4$	$i=5$
2	1/4	1/4				
3	3/8	3/32	9/32			
4	12/13	1932/2197	−7200/2197	7296/2197		
5	1	439/216	−8	3680/513	−845/4104	
6	1/2	−8/27	2	−3544/2565	1859/4104	−11/40

Table 6.20 Coefficients $\gamma_j, \hat{\gamma}_j, c_j$ for RKF method

j	1	2	3	4	5	6
γ_j	25/216	0	1408/2565	2197/4104	$-1/5$	
$\hat{\gamma}_j$	16/135	0	6656/12825	28561/56430	$-9/50$	2/55
c_j	1/360	0	$-128/4275$	$-2197/75240$	1/50	2/55

To compare the truncation errors in a simple case, consider solving

$$y' = x^4 \qquad y(0) = 0$$

Then the truncation errors for (6.211) and (6.226) are, respectively,

$$\text{RKF (6.226)}: T_{n+1}(Y) \doteq .00048h^5$$

$$\text{RK (6.211)}: T_{n+1}(Y) \doteq -.0083h^5 \qquad n \geqslant 0$$

This suggests that the RKF method should generally have a smaller truncation error; although the difference will generally not be this great. Note that the classical method (6.211) with a stepsize h and using the error estimate (6.224), will require 11 evaluations of f to go from $y_h(x_n)$ to $y_h(x_n+2h)$. And the RKF method (6.226) requires 12 evaluations to go from $y_h(x_n)$ to $y_h(x_n+2h)$. Consequently, the computational effort in going from x_n to x_n+2h is comparable, and it is fair to compare their errors by using the same value of h.

Example Use the RKF method (6.226) to solve the problem (6.212). It was previously an example for the classical RK method (6.211). As before, $h=.25$; and the results are given in Table 6.21. The theoretical value for the ratios is again 16, and clearly it has not yet settled down to that value. As h decreases, it will approach 16 more closely. The use of the Richardson extrapolation formula (6.214) is given in the last column, and it clearly overestimates the error. Nonetheless, this is still a useful error estimate in that it gives some idea of the size of the global error.

Automatic Runge-Kutta-Fehlberg programs A variable stepsize RKF program can be written by using (6.228) to estimate and control the truncation error in the

Table 6.21 Example of RKF method (6.226)

x	$y_h(x)$	$Y(x)-y_h(x)$	$Y(x)-y_{2h}(x)$	Ratio	$\frac{1}{15}[y_h(x)-y_{2h}(x)]$
2.0	.40000881	$-8.8\text{E}-6$	$-5.0\text{E}-4$	57	$-3.3\text{E}-5$
4.0	.23529469	$-5.8\text{E}-7$	$-4.0\text{E}-5$	69	$-2.6\text{E}-6$
6.0	.16216226	$-9.5\text{E}-8$	$-7.9\text{E}-6$	83	$-5.2\text{E}-7$
8.0	.12307695	$-2.6\text{E}-8$	$-2.5\text{E}-6$	95	$-1.6\text{E}-7$
10.0	.09900991	$-9.4\text{E}-9$	$-1.0\text{E}-6$	106	$-6.6\text{E}-8$

fourth-order formula (6.226). Such a method has been written by L. Shampine and H. Watts, and it is described in [S4]. Its general features are as follows. The program is named RKF45; and a user of the program must specify two error parameters ABSERR and RELERR. The truncation error in (6.228) for y_{n+1} is forced to satisfy

$$|\text{trunc}_j| \leqslant \text{ABSERR} + \text{RELERR} * |y_{n,j}| \tag{6.229}$$

for each component $y_{n,j}$ of the computed solution y_n of the system of differential equations being solved. But then the final result of the computation, to be used in further calculations, is taken to be the fifth order formula $\hat{y}_{n+1}$ from (6.227) rather than the fourth-order formula y_{n+1}. The terms y_n on the right side in (6.226) and (6.227) should therefore be replaced by $\hat{y}_n$. Thus RKF45 is really a fifth-order method, which chooses the stepsize h by controlling the truncation error in the fourth-order formula (6.226). It can be shown that the result (6.229) for y_{n+1} implies that $\hat{y}_{n+1}$ satisfies an error per unit stepsize control on its truncation error. The argument is similar to that given in Shampine [S1, p. 100] for variable-order Adams methods. To review the concept of error per unit stepsize, recall (6.112) and the material near the end of Section 6.6.

The tests of Shampine, Watts, and Davenport [S5] show that RKF45 is a superior RK program, one that is an excellent candidate for inclusion in a library of programs for solving ordinary differential equations. And, in general, the comparisons given in Enright and Hull [E1] have shown RKF methods to be superior to other RK methods. Comparisons with multistep methods are more difficult. Multistep methods require fewer evaluations of the derivative $f(x, y)$ than RK methods; but the overhead costs per step are much greater with multistep than with RK methods. A judgment as to which kind of method to use depends on how costly it is to evaluate $f(x, y)$ as compared with the overhead costs in the multistep methods. There are other considerations, for example, the size of the system of differential equations being solved, and a general discussion of their influence is given in Hull et al. [H2], Enright and Hull [E1], and Shampine et al. [S5].

Example Solve the problem

$$y' = \frac{y}{4}\left(1 - \frac{y}{20}\right) \qquad y(0) = 1$$

which has the solution

$$Y(x) = 20 / [1 + 19e^{-(x/4)}]$$

The problem was solved using RKF45, and values were output at $x = 2, 4, 6, \cdots, 20$. Three values of ABSERR were used, and RELERR $= 10^{-12}$ in all cases. The resulting global errors are given in Table 6.22. The column headed by

Table 6.22   Example of Runge-Kutta-Fehlberg program RKF45

x	ABSERR = 10^{-3}		ABSERR = 10^{-6}		ABSERR = 10^{-9}	
	Error	NFE	Error	NFE	Error	NFE
4.0	−1.4E−5	19	−2.2E−7	43	−5.5E−10	121
8.0	−3.4E−5	31	−5.6E−7	79	−1.3E−9	229
12.0	−3.7E−5	43	−5.5E−7	103	−8.6E−10	312
16.0	−3.3E−5	55	−5.4E−7	127	−1.2E−9	395
20.0	−1.8E−6	67	−1.6E−7	163	−4.3E−10	503

"NFE" gives the number of evaluations of $f(x, y)$ needed to obtain the given answer. Compare the results with those given in Table 6.15, obtained using the variable order multistep program DE.

Discussion of the Literature

Ordinary differential equations are the principal form of mathematical model occurring in the sciences and engineering, and, consequently, the numerical solution of differential equations is a very large area of study. Two classic books, which reflect the state of knowledge before the widespread use of digital computers, are Collatz [C3] and Milne [M1]. Some important texts since 1960 are, in chronological order, Henrici [H1], Ceschino and Kuntzmann [C1], Lapidus and Seinfeld [L3], Gear [G1], Lambert [L1], Stetter [S6], and Shampine and Gordon [S2]. Most of these contain large bibliographies.

The modern theory of convergence and stability of multistep methods, introduced in Section 6.8, dates from Dahlquist [D1]. The text by Henrici [H1] has become a classic account of that theory, including extensions and applications of it. Gear [G1] is a more modern account of all methods, including variable-order methods. Stetter [S6] has given a very general and complete abstract analysis of the numerical theory for solving initial value problems. A complete account up to 1970 of Runge-Kutta methods, their development and error analysis, is given in Lapidus and Seinfeld [L3].

The first significant use of the concept of a variable-order method is due to Gear [G1] and Krogh [K5]. Such methods are superior to a fixed-order multistep method in efficiency, and they do not require any additional method for starting the integration or for changing the stepsize. A very good account of the variable-order Adams method is given in Shampine and Gordon [S2], and the excellent program DE is included.

A third class of methods has been ignored, those based on extrapolation. Current work in this area began with Gragg [G2] and Bulirsch and Stoer [B1], with the latter paper containing an Algol program. This has been translated into Fortran and improved by Fox [F2]. The main idea is to perform repeated extrapolation on some simple method, to obtain methods of a higher order. These methods have performed fairly well in the tests of Enright and Hull [E1] and Shampine et al. [S5]; but they were judged to not be as advanced in their practical and theoretical development as are the multistep and Runge-Kutta methods.

Global error estimation is an area in which little has been published. The principal papers are those of Stetter [S7], [S8] and Shampine and Watts [S3]. The latter paper also contains a program that estimates the global error in the Runge-Kutta-Fehlberg method based on the fourth- and fifth-order pair (6.226), (6.227). The truncation is controlled in the usual way, described near the end of Section 6.9; and Richardson extrapolation is used to estimate the global error. Additional work is needed in this area, especially for estimating the global error in multistep methods.

Because of the creation of a number of automatic programs for solving differential equations, several empirical studies have been made to assess their performance and to make comparisons between programs. Two major projects have been the ongoing testing program of Hull et al. at the University of Toronto [H2], [E1], [E2], and the review of Shampine, Watts, and Davenport [S5]. It is clear from their work that programs must be compared, as well as methods. Different program implementations of the same method can vary widely in their performance.

Stiff differential equations have become a major research topic within the 1970s. An important introductory account is given in Gear [G1, Chapter 11], along with the basis of the most widely used numerical methods for stiff equations. Examples of the developing theory of the area of stiff equations are given in the papers of Alexander [A1], Kreiss [K4], the contributions in the symposium proceedings of Willoughby [W1] and Lapidus and Schiesser [L2] and the book of Van der Houwen [V1]. A numerical comparison of some current algorithms is given in Enright et al. [E2].

Boundary value problems for ordinary differential equations are another important topic; but both their general theory and the numerical analysis of them are more sophisticated than for the initial value problem. Consequently, we just reference the texts of Keller for the two-point boundary value problem, [K1] and [K3], and his general survey of the entire field [K2]. All of these contain large bibliographies. Automatic computer programs for the two-point boundary value problem are now under development.

Finally, the development of all methods in this chapter was dependent on the solution $Y(x)$ being sufficiently differentiable over the interval of interest $[x_0, b]$. If this is not true, especially if it fails at x_0, then the results of this chapter are no longer valid. The program DE in Shampine and Gordon [S2] has been designed to recognize interior points at which the solution lacks sufficient differentiability, and to then take appropriate action. But it is expensive to try to integrate through such a singular point. It is best to recognize it analytically, and to then split the interval $[x_0, b]$ into intervals $[x_0, a]$, $[a, b]$ with a the singular point. The behavior of the numerical method on $[x_0, a]$ and on $[a, b]$ is better than on $[x_0, b]$ alone.

Bibliography

[A1] Alexander, R., Diagonally implicit Runge-Kutta methods for stiff O.D.E.'s, *SIAM J. Numer. Anal.*, **14**, 1977, pp. 1006–1021.

[B1] Boyce, W. and R. Diprima, Elementary Differential Equations, 3rd ed., John Wiley, New York, 1977.

[B2] Bulirsch, R. and J. Stoer, Numerical treatment of ordinary differential equations by extrapolation, *Numer. Math.*, **8**, 1966, pp. 1–13.

[B3] Butcher, J., On the attainable order of Runge-Kutta methods, *Math. Comp.*, **19**, 1965, pp. 408–417.

[C1] Ceschino, F. and J. Kuntzmann, *Numerical Solution of Initial Value Problems*, Prentice-Hall, Englewood Cliffs, N.J., 1966.

[C2] Coddington, E., and N. Levinson, *Theory of Ordinary Differential Equations*, McGraw-Hill, New York, 1955.

[C3] Collatz, L., *The Numerical Treatment of Differential Equations*, 3rd ed., Springer-Verlag, New York, 1966.

[D1] Dahlquist, G., Numerical integration of ordinary differential equations, *Math. Scandinavica*, **4**, 1956, pp. 33–50.

[E1] Enright, W., and T. Hull, Test results on initial value methods for non-stiff ordinary differential equations, *SIAM J. Numer. Anal.*, **13**, 1976, pp. 944–961.

[E2] Enright, W., T. Hull, and B. Lindberg, Comparing numerical methods for stiff systems of O.D.E.'s, *BIT*, **15**, 1975, pp. 10–48.

[F1] Fehlberg, E., Klassische Runge-Kutta-Formeln vierter und niedrigerer Ordnung mit Schrittweiten-Kontrolle und ihre Anwendung auf Wärmeleitungsprobleme, *Computing*, **6**, 1970, pp. 61–71.

[F2] Fox, P., DESUB: integration of a first-order system of ordinary differential equations, in *Mathematical Software*, J. Rice, ed., Academic Press, New York, 1971, pp. 477–507.

[G1] Gear, C. W., *Numerical Initial Value Problems in Ordinary Differential Equations*, Prentice-Hall, Englewood Cliffs, N.J., 1971.

[G2]   Gragg, W., On extrapolation algorithms for ordinary initial value problems, *SIAM J. Numer. Anal.*, **2**, 1965, pp. 384–403.

[H1]   Henrici, P., *Discrete Variable Methods in Ordinary Differential Equations*, John Wiley, New York, 1962.

[H2]   Hull, T., W. Enright, B. Fellen, and A. Sedgwick, Comparing numerical methods for ordinary differential equations, *SIAM J. Numer. Anal.*, **9**, 1972, pp. 603–637.

[I1]   Isaacson, E., and H. Keller, *Analysis of Numerical Methods*, John Wiley, New York, 1966.

[K1]   Keller, H., *Numerical Methods for Two-Point Boundary Value Problems*, Ginn-Blaisdell, Waltham, Mass., 1968.

[K2]   Keller, H., Numerical solution of boundary value problems for ordinary differential equations: Survey and some recent results on difference methods; in *Numerical Solutions of Boundary Value Problems for Ordinary Differential Equations*, A. Aziz, ed., Academic Press, New York, 1975, pp. 27–88.

[K3]   Keller, H., Numerical Solution of Two-Point Boundary Value Problems, Regional Conf. Series in Appl. Math. #24, SIAM, Philadelphia, 1976.

[K4]   Kreiss, H., Difference methods for stiff differential equations, Math. Research Center Tech. Rep. #1699, University of Wisconsin, Madison, Wis., 1976.

[K5]   Krogh, F., VODQ/SVDQ/DVDQ-variable order integrators for the numerical solution of ordinary differential equations, Section 314 subroutine writeup, Jet Propulsion Laboratory, Pasadena, Calif., 1969.

[L1]   Lambert, J., *Computational Methods in Ordinary Differential Equations*, John Wiley, New York, 1973.

[L2]   Lapidus, L. and W. Schiesser, eds., *Numerical Methods for Differential Equations: Recent Developments in Algorithms, Software, New Applications*, Academic Press, New York, 1976.

[L3]   Lapidus, L. and J. Seinfeld, *Numerical Solution of Ordinary Differential Equations*, Academic Press, New York, 1971.

[M1]   Milne, W. E., *Numerical Solution of Differential Equations*, John Wiley, New York, 1953.

[R1]   Ralston, A., *A First Course in Numerical Analysis*, McGraw-Hill, New York, 1965.

[S1] Shampine, L., The quality of Runge-Kutta Formulas, Sandia Labs Tech. Rep. SAND 76-0370, Albuquerque, New Mexico, 1976.

[S2] Shampine, L. and M. Gordon, *Computer Solution of Ordinary Differential Equations*, W. H. Freeman and Co., San Francisco, 1975.

[S3] Shampine, L. and H. Watts, Global error estimation for ordinary differential equations, *ACM Trans. Math. Soft.* **2**, 1976, pp. 172–186.

[S4] Shampine, L. and H. Watts, Practical solution of ordinary differential equations by Runge-Kutta methods, Sandia Labs Tech. Rep. SAND 76-0585, Albuquerque, New Mexico, 1976.

[S5] Shampine, L., H. Watts, and S. Davenport, Solving nonstiff ordinary differential equations—The state of the art, *SIAM Rev.*, **18**, 1976, pp. 376–411.

[S6] Stetter, H., *Analysis of Discretization Methods for Ordinary Differential Equations*, Springer-Verlag, New York, 1973.

[S7] Stetter, H., Local estimation of the global discretization error, *SIAM J. Numer. Anal.*, **8**, 1971, pp. 512–523.

[S8] Stetter, H., Economical global error estimation, in reference [W1].

[V1] Van der Houwen, P. J., *Construction of Integration Formulas for Initial Value Problems*, North-Holland Pub., Amsterdam, 1977.

[W1] Willoughby, R., ed., *Stiff Differential Systems*, Plenum Press, New York, 1974.

Problems

1. Draw the vector field for $y' = x - y^2$, including some sample solution curves. Attempt to guess the behavior of the solutions $Y(x)$ as $x \to \infty$.

2. Determine Lipschitz constants for the following functions, as in (6.4).

   (a) $f(x,y) = 2y/x, \ x \geqslant 1$

   (b) $f(x,y) = \tan^{-1} y$

   (c) $f(x,y) = (x^3 - 2)^{27}/(17x^2 + 4)$

   (d) $f(x,y) = x - y^2, \ |y| \leqslant 10$

3. Calculate the first three Picard iterates (6.7) for the problem $y' = y + 2\cos(x), \ y(0) = 1$. Compare these iterates with the true solution, and draw graphs of each of them.

4. Convert the following problems to first order systems.

   (a) $y'' - 3y' + 2y = 0$, $y(0) = 1$, $y'(0) = 1$

   (b) $y'' - 0.1(1 - y^2)y' + y = 0$, $y(0) = 1$, $y'(0) = 0$

   (Van der Pol's equation)

   (c) $x''(t) = -x/r^3$, $y''(t) = -y/r^3$, $r = \sqrt{x^2 + y^2}$
   (orbital equation)
   $x(0) = .4$, $x'(0) = 0$, $y(0) = 0$, $y'(0) = 2$

5. Write a computer program to solve $y' = f(x,y)$, $y(x_0) = y_0$, using Euler's method. Write it to be used with an arbitrary f, stepsize h, and interval $[x_0, b]$. Using the program, solve $y' = x^2 - y$, $y(0) = 1$, for $0 \le x \le 4$, with stepsizes of $h = .25$, .125, .0625, in succession. For each value of h, print the true solution, approximate solution, error, and relative error at the nodes $x = 0$, .25, .50, .75, $\cdots$, 4.00.

   Analyze your output and supply written comments on it (analysis of output is as important as obtaining it). Assuming the true answer was not known, what would you estimate the accuracy to be in $y_h(4.0)$, for each value of h. Hint: use the existence of the asymptotic error formula (6.46), even though $D(x)$ is unknown.

6. For the problem $y' = -y^2$, $y(0) = 1$, compare (a) the bound for (6.30), and (b) the asymptotic estimate from (6.46). For the Lipschitz constant K in (6.30), use $K = 2$.

7. Show that Euler's method fails to approximate the solution $Y(x) = (\frac{2}{3}x)^{3/2}$, $x \ge 0$, of the problem $y' = y^{1/3}$, $y(0) = 0$. Explain why.

8. For the equations in Problem 4, write out the approximating difference equations obtained by using Euler's method.

9. (a) Derive the Lipschitz condition (6.59) for a system of two differential equations.

   (b) Prove that the method (6.58) will converge to the solution of (6.57).

10. Consider the two-step method

$$y_{n+1} = \tfrac{1}{2}(y_n + y_{n-1}) + \frac{h}{4}(4y'_{n+1} - y'_n + 3y'_{n-1}) \qquad n \ge 1$$

   with $y'_n \equiv f(x_n, y_n)$. Show that it is a second-order method; and find the leading term in the truncation error, written as in (6.76).

11. Assume that the multistep method (6.63) is consistent and that is satisfies $a_j \ge 0$, $j = 0, 1, \cdots, p$, the same as in Theorem 6.6. Prove stability of (6.63), in complete analogy with (6.42) for Euler's method.

12. Write a program to solve $y' = f(x,y)$, $y(x_0) = y_0$, using the midpoint method (6.84). Use a fixed stepsize h; and for the initial value y_1, use the Euler method, $y_1 = y_0 + hf(x_0, y_0)$. With the program, solve the following problems.

(a) $y' = -y^2$, $y(0) = 1$ (c) $y' = y - 2\sin(x)$, $y(0) = 1$

(b) $y' = 1/[1 + (\tan y)^2]$, $y(0) = 0$

Solve on the interval $[x_0, b] = [0, 10]$, with $h = .5$ and $h = .25$; and print the approximate answer and true error at each node point. Discuss your results.

13. Write a program to solve $y' = f(x,y)$, $y(x_0) = y_0$, using the trapezoidal method (6.99), with a fixed stepsize h. Use the midpoint predictor (6.106) in the iteration (6.100); and allow the number J of iterates to be calculated to be an input variable. Solve the equations in Problem 12 with $h = .5$ and $.25$, and with $J = 1, 2, 3$. Discuss the results. Caution: for each J, solve the equation on $[x_0, b]$ without referring to the results obtained using any other value of J.

14. Derive the asymptotic error formula (6.109) for the trapezoidal method.

15. Write a program to implement the Detrap algorithm of Section 6.6. Using it, solve the problems given in Problem 12, with $\epsilon = .001$.

16. Find all explicit fourth-order formulas of the form

$$y_{n+1} = a_0 y_n + a_1 y_{n-1} + a_2 y_{n-2} + h(b_0 y_n' + b_1 y_{n-1}' + b_2 y_{n-2}') \qquad n \geqslant 2$$

17. Show that the Simpson method (6.139) is only weakly stable, in the same sense as was true of the midpoint method in Section 6.4.

18. Write a program to solve $y' = f(x,y)$, $y(x_0) = y_0$, $x_0 \leqslant x \leqslant b$, using the fourth-order Adams-Moulton formula and a fixed stepsize h. Use the fourth-order Adams-Bashforth formula as the predictor. Generate the initial values y_1, y_2, y_3 using the classical fourth-order Runge-Kutta method (6.211). Solve the equations of Problem 12 with $h = .5$ and $h = .25$. Print the calculated answers and the true errors. For comparison with method (6.153), also solve the example used in Table 6.14. Check the accuracy of the Richardson extrapolation error estimate (6.214), which is based on the global error being of the fourth order. Discuss all your results.

19. (a) For the coefficients γ_i and δ_i of the Adams-Bashforth and Adams-Moulton formulas, show that

$$\delta_i = \gamma_i - \gamma_{i-1} \qquad i \geqslant 1$$

(b) For the p-step Adams-Moulton formula (6.152), prove

$$y_{n+1} = y_{n+1}^{(0)} + h\gamma_p \nabla^{p+1} y_{n+1}' \tag{1}$$

with $y_{n+1}^{(0)}$ the $p+1$-step Adams-Bashforth formula from (6.146),

$$y_{n+1}^{(0)} = y_n + h \sum_{j=0}^{p} \gamma_j \nabla^j y_n' \tag{2}$$

These formulas are both of order $p+1$. The result (1) is of use in calculating the corrector from the predictor, and is based on carrying certain backward differences from one step to the next. There is a closely related result when a p-step predictor (order p) is used to solve the p-step corrector (order $p+1$); see [S2, p. 51].

20. Find the general solution of the linear difference equation

$$u_{n+1} = u_n + \tfrac{1}{4}(u_{n-1} - u_{n-2}) \qquad n \geq 2$$

What is the solution when $u_0 = 4, u_1 = \tfrac{3}{2}, u_2 = \tfrac{7}{4}$, and what is u_{1000}?

21. Consider the numerical method

$$y_{n+1} = 4y_n - 3y_{n-1} - 2hf(x_{n-1}, y_{n-1}) \qquad n \geq 1$$

Determine its order. Illustrate with an example that the method is unstable.

22. Complete part (1) of the proof of Theorem 6.7, in which the root condition is violated by assuming $|r_j| = 1$, $\rho'(r_j) = 0$.

23. For part (2) of the proof of Theorem 6.8, show that

$$\gamma_j [r_j(h\lambda)]^n \to 0 \quad \text{as} \quad h \to 0 \qquad 1 \leq j \leq p.$$

24. **(a)** Determine the values of a_0 in the explicit second-order method (6.135) for which the method is stable.

   (b) If only the truncation error (6.136) is considered, subject to the stability restriction in (a), how should a_0 be chosen?

   (c) To ensure a large region of stability, subject to (a), how should a_0 be chosen?

25. Show that every explicit three-step method of order 4 is unstable. As an aid, recall Problem 16.

26. Derive an implicit fourth-order multistep method, other than those given in the text. Make it be relatively stable.

27. For the polynomial $\rho(r) = r^{p+1} - \sum_0^p a_j r^{p-j}$, assume that $a_j \geq 0$, $0 \leq j \leq p$, and $\sum_0^p a_j = 1$. Show that the roots of $\rho(r)$ satisfies the root condition (6.166) to (6.167). This shows directly that Theorem 6.6 is a corollary of Theorem 6.8.

28. Derive the real values of $h\lambda$ that are in the region of relative stability of the second-order Adams-Bashforth method.

29. Consider methods of the form

$$y_{n+1} = y_{n-q} + h \sum_{j=-1}^{p} b_j f(x_{n-j}, y_{n-j})$$

with $q \geq 1$. Show they do not satisfy the strong root condition. Find an example with $q = 1$ that is relatively stable.

30. Show the region of absolute stability for the trapezoidal method is the set of all complex $h\lambda$ with $\text{Real}(\lambda) < 0$.

31. Using the Taylor series method of Section 6.9, produce a fourth-order method to solve $y' = x - y^2$, $y(0) = 0$. Use fixed step-sizes, $h = .5$, and .125 in succession; and solve for $0 \leq x \leq 10$. Estimate the global error using the error estimate (6.214) based on Richardson extrapolation.

32. Write a program to solve $y' = f(x,y)$, $y(x_0) = y_0$, using the classical Runge-Kutta method (6.211), and let the stepsize h be fixed. Using the program, solve $y' = x - y^2$, $y(0) = 0$, for $h = .5$, .25, and .125; and estimate the global error using (6.214). Compare the results with those of Problem 31.

33. Consider the three-stage Runge-Kutta formula

$$y_{n+1} = y_n + h(\gamma_1 V_1 + \gamma_2 V_2 + \gamma_3 V_3)$$

$$V_1 = f(x_n, y_n) \qquad V_2 = f(x_n + \alpha_2 h, y_n + h\beta_{21} V_1)$$

$$V_3 = f[x_n + \alpha_3 h, y_n + h(\beta_{31} V_1 + \beta_{32} V_2)]$$

Determine the set of equations that the coefficients $\{\gamma_j, \alpha_j, \beta_{ji}\}$ must satisfy if the formula is to be of order 3. Find a particular solution of these equations.

34. Prove that if the Runge-Kutta method (6.198) satisfies (6.216), then it is stable.

35. Investigate the differential equation programs provided by your computer center. Note those that automatically control the truncation error by varying the stepsize, and possibly the order. Classify the programs as multistep (fixed order or variable order), Runge-Kutta, or extrapolation.

Compare one of these with the programs DE (of Section 6.7 and [S2]) and RKF45 (of Section 6.9 and [S4]) by solving the problem

$$y' = \frac{y}{4}\left(1 - \frac{y}{20}\right) \qquad y(0) = 1$$

with desired absolute errors of 10^{-3}, 10^{-6}, and 10^{-9}. Compare the results with those given in Tables 6.15 and 6.22.

36. Consider the problem

$$y' = \frac{1}{t+1} + c\tan^{-1}(y(t)) - \tfrac{1}{2} \qquad y(0) = 0$$

with c a given constant. Since $y'(0) = \tfrac{1}{2}$, the solution $y(t)$ is initially increasing as t increases, regardless of the value of c. As best you can, show that there is a value of c, call it c^*, for which (a) if $c > c^*$, the solution $y(t)$ increases indefinitely, and (b) if $c < c^*$, then $y(t)$ increases initially, but then peaks and decreases. Determine c^* to within .00005, and then calculate the associated solution $y(t)$ for $0 \leqslant t \leqslant 50$.

37. Consider the system

$$x'(t) = Ax - Bxy \qquad y'(t) = Cxy - Dy$$

This is known as the Lotka-Volterra predator-prey model for two populations, with $x(t)$ being the number of prey and $y(t)$ the number of predators at time t.

(a) Let $A = 4, B = 2, C = 1, D = 3$, and solve the model to at least three significant digits for $0 \leqslant t \leqslant 5$. The initial values are $x(0) = 3, y(0) = 5$. Plot x and y as functions of t, and plot x versus y.

(b) Solve the same model with $x(0) = 3$ and, in succession, $y(0) = 1, 1.5, 2$. Plot x versus y in each case. What do you observe? Why would the point $(3, 2)$ be called an equilibrium point?

SEVEN

LINEAR
ALGEBRA

The solution of systems of simultaneous linear equations and the calculation of
the eigenvalues and eigenvectors of a matrix are two very important problems
that arise in a variety of contexts. As a preliminary to the discussion of these
problems in the following chapters, we present some results from linear algebra.
The first section contains a review of material on vector spaces, matrices, and
linear systems, which is taught in most undergraduate linear algebra courses.
These results are summarized only, and no derivations are included. The
remaining sections discuss eigenvalues, canonical forms for matrices, vector and
matrix norms, and perturbation theorems for matrix inverses. If necessary, this
chapter can be skipped; and the results can be referred back to as they are
needed in Chapters 8 and 9. For notation, Section 7.1 and the norm notation of
Section 7.3 should be skimmed.

7.1 Vector Spaces, Matrices, and Linear
Systems

Roughly speaking a *vector space* V is a set of objects, called vectors, for which
operations of *vector addition* and *scalar multiplication* have been defined. V has a
set of scalars associated with it; and in this text, this set can be either the real
numbers R or complex numbers C. The vector operations must satisfy certain
standard associative, commutative, and distributive rules, that we will not list. A
subset W of a vector space V is called a subspace of V if W is a vector space,
using the vector operations inherited from V. For a complete development of the
theory of vector spaces, see any undergraduate text on linear algebra, for
example Bradley [B1, Chapter 2], Halmos [H1, Chapter 1], or Noble [N1,
Chapters 4 and 14].

Examples 1. $V = R^n$, the set of all n-tuples $(x_1, \ldots, x_n)$ with real entries x_i, and R
is the associated set of scalars.

2. $V = C^n$, the set of all n-tuples with complex entries, and C is the set of scalars.

3. $V =$ the set of all polynomials of degree $\leq n$, for some given n, is a vector space. The scalars can be R or C, as desired for the application.

4. $V = C[a,b]$, the set of all continuous real-valued (or complex-valued) functions on the interval $[a,b]$, is a vector space with scalar set equal to R (or C). The example in (3) is a subspace of $C[a,b]$.

Definition Let V be a vector space and let $v_1, v_2, \ldots, v_m \in V$.

1. We say that $v_1, \ldots, v_m$ are *linearly dependent* if there is a set of scalars $\alpha_1, \ldots, \alpha_m$ with at least one nonzero scalar for which

$$\alpha_1 v_1 + \ldots + \alpha_m v_m = 0$$

Since at least one scalar is nonzero, say $\alpha_i \neq 0$, we can solve for v_i,

$$v_i = -\frac{\alpha_1}{\alpha_i} v_1 - \ldots - \frac{\alpha_{i-1}}{\alpha_i} v_{i-1} - \frac{\alpha_{i+1}}{\alpha_i} v_{i+1} - \ldots - \frac{\alpha_m}{\alpha_i} v_m$$

We say that v_i is a *linear combination* of the vectors $v_1, \ldots, v_{i-1}, v_{i+1}, \ldots, v_m$. For a set of vectors to be linearly dependent, one of them must be a linear combination of the remaining ones.

2. We say that $v_1, \ldots, v_m$ are *linearly independent* if they are not dependent. Equivalently, the only choice of scalars $\alpha_1, \ldots, \alpha_m$ for which

$$\alpha_1 v_1 + \ldots + \alpha_m v_m = 0$$

is the trivial choice $\alpha_1 = \ldots = \alpha_m = 0$

3. $\{v_1, \ldots, v_m\}$ is a *basis* for V if for every $v \in V$, there is a unique choice of scalars $\alpha_1, \ldots, \alpha_m$ for which

$$v = \alpha_1 v_1 + \ldots + \alpha_m v_m$$

Note that this implies $v_1, \ldots, v_m$ are independent. If such a finite basis exists, we say V is *finite dimensional*. Otherwise, it is called *infinite dimensional*.

Theorem 7.1 If V is a vector space with a basis $\{v_1, \ldots, v_m\}$, then every basis for V will contain exactly m vectors. The number m is called the dimension of V.

Examples 1. $\{1, x, x^2, \ldots, x^n\}$ is a basis for the space V of polynomials of degree $\leq n$. Thus dimension $V = n + 1$.

2. R^n and C^n have the basis $\{e_1, \ldots, e_n\}$, in which

$$e_i = (0, 0, \ldots, 0, 1, 0, \ldots, 0) \tag{7.1}$$

with the 1 in position i. Dimension $R^n, C^n = n$. This is called the standard basis for R^n and C^n; and the vectors in it are called unit vectors.

3. $C[a, b]$ is infinite dimensional.

Matrices and linear systems   Matrices are rectangular arrays of real or complex numbers, and the general matrix of *order* $m \times n$ has the form

$$A = \begin{bmatrix} a_{11} & a_{12} & \cdots & a_{1n} \\ \vdots & & & \vdots \\ a_{m1} & a_{m2} & \cdots & a_{mn} \end{bmatrix} \tag{7.1}$$

A matrix of order n is shorthand for a square matrix of order $n \times n$. Matrices are denoted by capital letters; and their entries are usually denoted by small letters, usually corresponding to the name of the matrix, as above. The following definitions give the common operations on matrices.

Definition **1.**   Let A and B have order $m \times n$. The sum of A and B is the matrix $C = A + B$, of order $m \times n$, given by

$$c_{ij} = a_{ij} + b_{ij}$$

2.   Let A have order $m \times n$, let α be a scalar. Then the scalar multiple $C = \alpha A$ is of order $m \times n$ and is given by

$$c_{ij} = \alpha a_{ij}$$

3.   Let A have order $m \times n$ and B have order $n \times p$. Then the product $C = AB$ is of order $m \times p$, and it is given by

$$c_{ij} = \sum_{k=1}^{n} a_{ik} b_{kj}$$

4.   Let A have order $m \times n$. The *transpose* $C = A^T$ has order $n \times m$, and is given by

$$c_{ij} = a_{ji}$$

The *conjugate transpose* $C = A^*$ also has order $n \times m$, and

$$c_{ij} = \bar{a}_{ji}$$

The notation $\bar{z}$ denotes the complex conjugate of the complex number z; and z is real if and only if $\bar{z} = z$. The conjugate transpose A^* is also called the *adjoint* of A.

The following arithmetic properties of matrices can be shown without much difficulty, and they are left to the reader.

(a) $A + B = B + A$ (b) $(A + B) + C = A + (B + C)$
(c) $A(B + C) = AB + AC$ (d) $A(BC) = (AB)C$
(e) $(A + B)^T = A^T + B^T$ (f) $(AB)^T = B^T A^T$

It is important for many applications to note that the matrices need not be square for the above properties to hold.

The vector spaces R^n and C^n are usually identified with the set of column vectors of order $n \times 1$, with real and complex entries, respectively. The linear system

$$a_{11} x_1 + \ldots + a_{1n} x_n = b_1$$
$$\vdots \qquad\qquad (7.2)$$
$$a_{m1} x_1 + \ldots + a_{mn} x_n = b_m$$

can be written as $Ax = b$, with A as in (7.1), and

$$x = [x_1, \ldots, x_n]^T \qquad b = [b_1, \ldots, b_m]^T$$

The vector b is a given vector in R^m, and the solution x is an unknown vector in R^n: The use of matrix multiplication reduces the linear system (7.2) to the simpler and more intuitive form $Ax = b$.

We now introduce a few additional definitions for matrices, including some special matrices.

Definition **1.** The *zero matrix* of order $m \times n$ has all entries equal to zero. It is denoted by $0_{m \times n}$, or more simply, by 0. For any matrix A of order $m \times n$,

$$A + 0 = 0 + A = A$$

2. The *identity matrix* of order n is defined by $I = [I_{ij}]$

$$I_{ij} = \begin{cases} 1, & i = j \\ 0, & i \neq j \end{cases}$$

for all $1 \leq i, j \leq n$. For all matrices A or order $m \times n$ and B of order $n \times p$,

$$AI = A \qquad IB = B$$

3. Let A be a square matrix of order n. If there is a square matrix B of order n for which $AB = BA = I$, then we say A is invertible, with *inverse* B. The matrix B can be shown to be unique, and we denote the inverse of A by A^{-1}.

4. A matrix A is called *symmetric* if $A^T = A$; and it is called *Hermitian* if $A^* = A$. The term symmetric is generally used only with real matrices. The matrix A is *skew symmetric* if $A^T = -A$. Of necessity, all matrices which are symmetric, Hermitian, or skew symmetric must also be square.

5. Let A be an $m \times n$ matrix. The row rank of A is the number of linearly independent rows in A, regarded as elements of R^n or C^n; and the column rank is the number of linearly independent columns. It can be shown (Problem 5) that these two numbers are always equal, and this is called the rank of A.

For the definition and properties of the determinant of a square matrix A, see any linear algebra text, for example Bradley [B1, Chapter 5], Noble [N1, Chapter 7], and Strang [S2, Chapter 4]. We now summarize many of the results on matrix inverses and the solvability of linear systems, in the following theorem.

Theorem 7.2 Let A be a square matrix with elements from R (or C), and let the vector space be $V = R^n$ (or C^n). Then the following are equivalent statements.

    1. $Ax = b$ has a unique solution $x \in V$ for every $b \in V$.

    2. $Ax = b$ has a solution $x \in V$ for every $b \in V$.

    3. $Ax = 0$ implies $x = 0$.

    4. A^{-1} exists.

    5. determinant $(A) \neq 0$.

    6. Rank $(A) = n$.

Although no proof is given here, it is an excellent exercise to prove the equivalence of some of these statements. Use the concepts of linear independence and basis, along with Theorem 7.1. Also, use the decomposition

$$Ax = x_1 A_{\bullet 1} + \ldots + x_n A_{\bullet n} \qquad x \in R^n \text{ or } C^n \qquad (7.3)$$

with $A_{\bullet j}$ denoting column j in A. This says that the space of all vectors of the form Ax is spanned by the columns of A, although they may be linearly dependent.

Inner-product vector spaces   One of the important reasons for reformulating problems as equivalent linear algebra problems is to introduce some geometric insight. Important to this process are the concepts of inner product and orthogonality.

Definition   1.   The *inner product* of two vectors $x, y \in R^n$ is defined by

$$(x,y) = \sum_{i=1}^{n} x_i y_i = x^T y = y^T x$$

and for vectors $x, y \in C^n$, define the inner product by

$$(x,y) = \sum_{i=1}^{n} x_i \bar{y}_i = y^* x$$

2.   The *Euclidean norm* of x in C^n or R^n is defined by

$$\|x\|_2 = \sqrt{(x,x)} = \sqrt{|x_1|^2 + \ldots + |x_n|^2}$$

The following results are fairly straightforward to prove, and they are left to the reader. Let V denote C^n or R^n.

1.   For all $x, y, z \in V$,

$$(x, y+z) = (x,y) + (x,z) \qquad (x+y, z) = (x,z) + (y,z)$$

2.   For all $x, y \in V$,

$$(\alpha x, y) = \alpha(x,y)$$

and for $V = C^n$, $\alpha \in C$,

$$(x, \alpha y) = \bar{\alpha}(x,y)$$

3.   In C^n, $(x,y) = \overline{(y,x)}$; and in R^n, $(x,y) = (y,x)$

4.   For all $x \in V$,

$$(x,x) \geq 0$$

and $(x,x) = 0$ if and only if $x = 0$

5.   For all $x, y \in V$,

$$|(x,y)|^2 \leq (x,x)(y,y) \tag{7.4}$$

This is called the Cauchy-Schwartz inequality; and it is proven in exactly the same manner as (4.34) in Chapter 4. Using the Euclidean norm, we can write it as

$$|(x,y)| \leq \|x\|_2 \|y\|_2 \tag{7.5}$$

6. For all $x, y \in V$,

$$\|x+y\|_2 \leq \|x\|_2 + \|y\|_2 \tag{7.6}$$

This is the triangle inequality. For a geometric interpretation, see the earlier comments in Section 4.1 of Chapter 4 for the norm $\|f\|_\infty$ on $C[a,b]$. For a proof of (7.6), see the derivation of (4.35) in Chapter 4.

7. For any square matrix A of order n, and for any $x, y \in C^n$,

$$(Ax, y) = (x, A^*y) \tag{7.7}$$

The inner product was used to introduce the Euclidean length, but it is also used to define a sense of angle, at least in spaces in which the scalar set is R.

Definition **1.** For x, y in R^n, the angle between x and y is defined by

$$\mathcal{C}(x,y) = \cos^{-1}\left[\frac{(x,y)}{\|x\|_2 \|y\|_2}\right]$$

Note that the argument is between -1 and 1, due to the Cauchy-Schwartz inequality (7.5). The above definition can be written implicitly as

$$(x,y) = \|x\|_2 \|y\|_2 \ \cos(\mathcal{C}) \tag{7.8}$$

a familiar formula from the use of the dot product in R^2 and R^3.

2. Two vectors x and y are *orthogonal* if and only if $(x,y) = 0$. This is motivated by (7.8). If $\{x^{(1)}, \ldots, x^{(n)}\}$ is a basis for C^n or R^n, and if $(x^{(i)}, x^{(j)}) = 0$ for all $i \neq j$, $1 \leq i, j \leq n$, then we say $\{x^{(1)}, \ldots, x^{(n)}\}$ is an *orthogonal basis*. If all basis vectors have Euclidean length 1, the basis is called *orthonormal*.

3. A square matrix U is called *unitary* if

$$U^*U = UU^* = I$$

If the matrix U is real, it is usually called *orthogonal*, rather than unitary. The rows (or columns) of an order n unitary matrix form an orthonormal basis for C^n, and similarly for orthogonal matrices and R^n.

Examples 1. The angle between

$$x = (1,2,3) \qquad y = (3,2,1)$$

is given by

$$\mathcal{Q} = \cos^{-1}\left[\frac{10}{14}\right] \doteq .775 \text{ radians}$$

2. The matrices

$$U_1 = \begin{bmatrix} \cos\theta & \sin\theta \\ -\sin\theta & \cos\theta \end{bmatrix} \qquad U_2 = \begin{bmatrix} 1/\sqrt{2} & i/\sqrt{2} \\ i/\sqrt{2} & 1/\sqrt{2} \end{bmatrix}$$

are unitary, with the first being orthogonal.

An orthonormal basis for a vector space $V = R^n$ or C^n is desirable, since it is then easy to decompose an arbitrary vector into its components in the direction of the basis vectors. More precisely, let $\{u^{(1)}, \ldots, u^{(n)}\}$ be an orthonormal basis for V, and let $x \in V$. Using the basis,

$$x = \alpha_1 u^{(1)} + \ldots + \alpha_n u^{(n)}$$

for some unique choice of coefficients $\alpha_1, \ldots, \alpha_n$. To find α_j, form the inner product of x with $u^{(j)}$; and then

$$(x, u^{(j)}) = x_1(u^{(1)}, u^{(j)}) + \cdots + x_n(u^{(n)}, u^{(j)})$$

$$= \alpha_j$$

using the orthonormality properties of the basis. Thus

$$x = \sum_{j=1}^{n} (x, u^{(j)})u^{(j)} \tag{7.9}$$

This can be given a geometric interpretation, which is shown in Figure 7.1. Using (7.8),

$$\alpha_j = (x, u^{(j)}) = \|x\|_2 \|u^{(j)}\|_2 \cos(\mathcal{Q}(x, u^{(j)}))$$

$$= \|x\|_2 \cos(\mathcal{Q}(x, u^{(j)}))$$

Thus the coefficient α_j is just the length of the orthogonal projection of x onto the axis determined by $u^{(j)}$. The formula (7.9) is a generalization of the decomposition of a vector x using the standard basis $\{e^{(1)}, \ldots, e^{(n)}\}$, defined earlier.

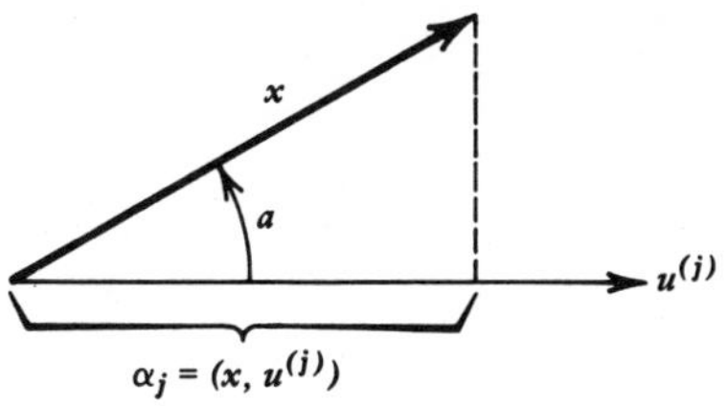

Figure 7.1 Illustration of (7.9).

Example Let $V = R^2$, and consider the orthonormal basis

$$u^{(1)} = \left(\frac{1}{2}, \frac{\sqrt{3}}{2} \right) \qquad u^{(2)} = \left(-\frac{\sqrt{3}}{2}, \frac{1}{2} \right)$$

Then for a given vector $x = (x_1, x_2)$, it can be written as

$$x = \alpha_1 u^{(1)} + \alpha_2 u^{(2)}$$

$$\alpha_1 = (x, u^{(1)}) = \frac{x_1 + x_2 \sqrt{3}}{2} \qquad \alpha_2 = (x, u^{(2)}) = \frac{x_2 - x_1 \sqrt{3}}{2}$$

For example,

$$(1,0) = \frac{1}{2} u^{(1)} - \frac{\sqrt{3}}{2} u^{(2)}$$

7.2 Eigenvalues and Canonical Forms for Matrices

The number λ, complex or real, is an *eigenvalue* of the square matrix A if there is a vector $x \in C^n, x \neq 0$, such that

$$Ax = \lambda x \tag{7.10}$$

The vector x is called an *eigenvector* corresponding to the eigenvalue λ. From Theorem 7.2, statements (3) and (5), λ is an eigenvalue of A if and only if

$$\det(A - \lambda I) = 0 \tag{7.11}$$

This is called the *characteristic equation* for A; and to analyze it we introduce the function

$$f_A(\lambda) \equiv \det(A - \lambda I)$$

If A has order n, then $f_A(\lambda)$ is a polynomial of degree exactly n, called the *characteristic polynomial* of A. To prove it is a polynomial, expand the determinant by minors repeatedly to get

$$f_A(\lambda) = \det(A - \lambda I)$$

$$= \det \begin{bmatrix} a_{11} - \lambda & a_{12} & \cdots & & \cdots & a_{1n} \\ a_{21} & a_{22} - \lambda & a_{23} & & \cdots & a_{2n} \\ \vdots & & \ddots & & & \vdots \\ \vdots & & & \ddots & & \\ a_{n1} & \cdots & & & \cdots & a_{nn} - \lambda \end{bmatrix}$$

$$= (a_{11} - \lambda)(a_{22} - \lambda) \ldots (a_{nn} - \lambda)$$

$$+ (\text{terms of degree} \leq n - 2)$$

$$f_A(\lambda) = (-1)^n \lambda^n + (-1)^{n-1}(a_{11} + \ldots + a_{nn})\lambda^{n-1}$$

$$+ (\text{terms of degree} \leq n - 2) \tag{7.12}$$

Also note that the constant term is

$$f_A(0) = \det(A) \tag{7.13}$$

From the coefficient of λ^{n-1}, define

$$\text{trace}(A) = a_{11} + a_{22} + \ldots + a_{nn} \tag{7.14}$$

often a quantity of interest in the study of A.

Since $f_A(\lambda)$ is of degree n, there are exactly n eigenvalues for A, if we count multiple roots according to their multiplicity. Every matrix has at least one eigenvalue-eigenvector pair, and the $n \times n$ matrix A has, at most, n distinct eigenvalues.

Examples 1. The characteristic polynomial for

$$A = \begin{bmatrix} 2 & 1 & 0 \\ 1 & 3 & 1 \\ 0 & 1 & 2 \end{bmatrix}$$

is

$$f_A(\lambda) = -\lambda^3 + 7\lambda^2 - 14\lambda + 8$$

The eigenvalues are $\lambda_1 = 1$, $\lambda_2 = 2$, and $\lambda_3 = 4$, and the corresponding eigenvectors are

$$u^{(1)} = \begin{bmatrix} 1 \\ -1 \\ 1 \end{bmatrix} \qquad u^{(2)} = \begin{bmatrix} 1 \\ 0 \\ -1 \end{bmatrix} \qquad u^{(3)} = \begin{bmatrix} 1 \\ 2 \\ 1 \end{bmatrix}$$

Note that these eigenvectors are orthogonal to each other, and therefore they are linearly independent. Since the dimension of R^3 (and C^3) is 3, these eigenvectors form an orthogonal basis for R^3 (and C^3). This illustrates Theorem 7.4, which is presented later in the section.

2. For the matrix

$$A = \begin{bmatrix} 1 & 0 & 0 \\ 0 & 1 & 0 \\ 0 & 0 & 1 \end{bmatrix}, \qquad f_A(\lambda) = (1-\lambda)^3$$

and there are three linearly independent eigenvectors for the eigenvalue $\lambda = 1$; for example,

$$[1,0,0]^T \qquad [0,1,0]^T \qquad [0,0,1]^T$$

All other eigenvectors are linear combinations of these three vectors.

3. For the matrix

$$A = \begin{bmatrix} 1 & 1 & 0 \\ 0 & 1 & 1 \\ 0 & 0 & 1 \end{bmatrix}, \qquad f_A(\lambda) = (1-\lambda)^3$$

but the matrix A has only one linearly independent eigenvector for the eigenvalue $\lambda = 1$, that is,

$$x = [1,0,0]^T$$

All other eigenvectors are multiples of this x.

The *algebraic multiplicity* of an eigenvalue of a matrix A is its multiplicity as a root of $f_A(\lambda)$; and its *geometric multiplicity* is the maximum number of linearly independent eigenvectors associated with the eigenvalue. The sum of the algebraic multiplicities of the eigenvalues of an $n \times n$ matrix A is constant with respect to small perturbations in A, that is, n. But the sum of the geometric multiplicities can vary greatly with small perturbations, and this causes the numerical calculation of eigenvectors to often be a very difficult problem. Also, the algebraic and geometric multiplicities need not be equal, as the last example shows.

Definition   Let A and B be square matrices of the same order. Then A is *similar* to B if there is a nonsingular matrix P for which

$$B = P^{-1}AP \tag{7.15}$$

Note that this is a symmetric relation since

$$A = Q^{-1}BQ \qquad Q = P^{-1}$$

The relation (7.15) can be interpreted to say that A and B are matrix representations of the same linear transformation T from V to V ($V = R^n$ or C^n), but with respect to different bases for V. The matrix P is called the *change of basis* matrix; and it relates the two representations of a vector $x \in V$ with respect to the two bases being used. See Noble [N1, Section 14.5] for greater detail.

We now present a few simple properties about similar matrices and their eigenvalues.

1. If A and B are similar, then $f_A(\lambda) = f_B(\lambda)$. To prove this, use (7.15) to show

$$f_B(\lambda) = \det(B - \lambda I) = \det\left[P^{-1}(A - \lambda I)P \right]$$

$$= \det(P^{-1})\det(A - \lambda I)\det(P) = f_A(\lambda)$$

since

$$\det(P)\det(P^{-1}) = \det(PP^{-1}) = \det(I) = 1$$

2. The eigenvalues of similar matrices A and B are exactly the same, and there is a one-to-one correspondence of the eigenvectors. If $Ax = \lambda x$, then using (7.15),

$$P^{-1}APP^{-1}x = \lambda P^{-1}x$$

$$Bz = \lambda z \qquad z = P^{-1}x$$

Trivially, $z \neq 0$, since otherwise x would be zero. And given any eigenvector z of B, this argument can be reversed to produce a corresponding eigenvector $x = Pz$ for A.

3. Since $f_A(\lambda)$ is invariant under similarity transformations of A, the coefficients of $f_A(\lambda)$ are also invariant under such similarity transformations. In particular, for A similar to B,

$$\text{trace}(A) = \text{trace}(B) \qquad \det(A) = \det(B) \tag{7.16}$$

Canonical forms   We now present several important canonical forms for matrices. These forms relate the structure of a matrix to its eigenvalues and eigenvectors, and they are used in a variety of applications in other areas of mathematics and science.

Theorem 7.3 (Schur normal form)   Let A have order n with elements from C. Then there exists a unitary matrix U such that

$$T \equiv U^*AU \tag{7.17}$$

is upper triangular.

Since T is triangular, and since $U^* = U^{-1}$,

$$f_A(\lambda) = f_T(\lambda) = (\lambda - t_{11})\ldots(\lambda - t_{nn}) \tag{7.18}$$

and thus the eigenvalues of A are the diagonal elements of T.

Proof   The proof is by induction on the order n of A. The result is trivially true for $n = 1$, using $U = [1]$. We assume that the result is true for all matrices of order $n \leq k - 1$, and we then prove it has to be true for all matrices of order $n = k$.

Let λ_1 be an eigenvalue of A, and let $u^{(1)}$ be an associated eigenvector with $\|u^{(1)}\|_2 = 1$. Beginning with $u^{(1)}$, pick an orthonormal basis for C^k, calling it $\{u^{(1)}, \ldots, u^{(k)}\}$. Define the matrix P_1 by

$$P_1 = [u^{(1)}, u^{(2)}, \ldots, u^{(k)}]$$

This is written in partitioned form, with columns $u^{(1)}, \ldots, u^{(k)}$ that are orthonormal. Then $P_1^* P_1 = I$, and thus $P_1^{-1} = P_1^*$. Define

$$B_1 = P_1^* A P_1$$

Claim:

$$B_1 = \begin{bmatrix} \lambda_1 & \alpha_2 \ldots \alpha_k \\ 0 & \\ \vdots & A_2 \\ \vdots & \\ 0 & \end{bmatrix}$$

with A_2 of order $k - 1$ and $\alpha_2, \ldots, \alpha_k$ some numbers. To prove this, multiply using partitioned matrices.

$$AP_1 = A[u^{(1)}, \ldots, u^{(k)}] = [Au^{(1)}, \ldots, Au^{(k)}]$$

$$= [\lambda_1 u^{(1)}, v^{(2)}, \ldots, v^{(k)}] \qquad v^{(j)} = Au^{(j)}$$

$$B_1 = P_1^* A P_1 = [\lambda_1 P_1^* u^{(1)}, P_1^* v^{(2)}, \ldots, P_1^* v^{(k)}]$$

Since $P_1^* P_1 = I$, it follows that $P_1^* u^{(1)} = e^{(1)} = [1, 0, \ldots, 0]^T$. Thus

$$B_1 = \left[\lambda_1 e^{(1)}, w^{(2)}, \ldots, w^{(k)}\right] \qquad w^{(j)} = P_1^* v^{(j)}$$

which has the desired form.

By the induction hypothesis, there exists an orthogonal matrix $\hat{P}_2$ of order $k - 1$ for which

$$\hat{T} = \hat{P}_2^* A_2 \hat{P}_2$$

is an upper triangular matrix of order $k - 1$. Define

$$P_2 = \begin{bmatrix} 1 & 0 & \cdots & 0 \\ 0 & & & \\ \vdots & & \hat{P}_2 & \\ 0 & & & \end{bmatrix}$$

Then P_2 is unitary, and

$$P_2^* B_1 P_2 = \begin{bmatrix} \lambda_1 & \gamma_2 & \cdots & \gamma_k \\ 0 & & & \\ \vdots & & \hat{P}_2^* A_2 \hat{P}_2 & \\ 0 & & & \end{bmatrix}$$

$$= \begin{bmatrix} \lambda_1 & \gamma_2 & \cdots & \gamma_k \\ 0 & & & \\ \vdots & & \hat{T} & \\ 0 & & & \end{bmatrix} \equiv T$$

an upper triangular matrix. Thus

$$T = P_2^* B_1 P_2 = P_2^* P_1^* A P_1 P_2 = (P_1 P_2)^* A (P_1 P_2)$$

$$T = U^* A U \qquad U = P_1 P_2$$

and U is easily unitary. This completes the induction and the proof. ∎

Example For the matrix

$$A = \begin{bmatrix} .2 & .6 & 0 \\ 1.6 & -.2 & 0 \\ -1.6 & 1.2 & 3.0 \end{bmatrix}$$

the matrices of the theorem and (7.17) are

$$T = \begin{bmatrix} 1 & 0 & -1 \\ 0 & 3 & 2 \\ 0 & 0 & -1 \end{bmatrix} \qquad U = \begin{bmatrix} .6 & 0 & -.8 \\ .8 & 0 & .6 \\ 0 & 1.0 & 0 \end{bmatrix}$$

This is not the usual way in which eigenvalues are calculated, but should be considered only as an illustration of the theorem. The theorem is used generally as a theoretical tool, rather than as a computational tool.

Using (7.16) and (7.17),

$$\text{trace}(A) = \lambda_1 + \lambda_2 + \dots + \lambda_n \qquad \det(A) = \lambda_1 \lambda_2 \dots \lambda_n \qquad (7.19)$$

where $\lambda_1, \dots, \lambda_n$ are the eigenvalues of A, which must form the diagonal elements of T. As a much more important application, we have the following well-known theorem.

Theorem 7.4 (Principal axes theorem)   Let A be a Hermitian matrix of order n, that is, $A^* = A$. Then A has n real eigenvalues $\lambda_1, \dots, \lambda_n$, not necessarily distinct, and n corresponding eigenvectors $u^{(1)}, \dots, u^{(n)}$ which form an orthonormal basis for C^n. If A is also real, the eigenvectors $u^{(1)}, \dots, u^{(n)}$ can be taken as real; and they form an orthonormal basis of R^n. Finally there is a unitary matrix U for which

$$U^* A U = D = \text{diag}[\lambda_1, \dots, \lambda_n] \qquad (7.20)$$

is a diagonal matrix with diagonal elements $\lambda_1, \dots, \lambda_n$. If A is also real, then U can be taken as orthogonal.

Proof   From Theorem (7.3), there is a unitary matrix U with

$$U^* A U = T$$

with T upper triangular. Form the conjugate transpose of both sides to obtain

$$T^* = (U^* A U)^* = U^* A^* (U^*)^* = U^* A U = T$$

Since T^* is lower triangular and T is upper triangular, we must have

$$T = \text{diagonal}[\lambda_1, \dots, \lambda_n]$$

Also, $T^* = T$ involves complex conjugation of all elements of T, and thus all diagonal elements of T must be real.

Write U as

$$U = [u^{(1)}, \ldots, u^{(n)}]$$

Then $T = U^*AU$ implies $AU = UT$,

$$A[u^{(1)}, \ldots, u^{(n)}] = [u^{(1)}, \ldots, u^{(n)}] \begin{bmatrix} \lambda_1 & & 0 \\ & \ddots & \\ 0 & & \lambda_n \end{bmatrix}$$

$$[Au^{(1)}, \ldots, Au^{(n)}] = [\lambda_1 u^{(1)}, \ldots, \lambda_n u^{(n)}]$$

and

$$Au^{(j)} = \lambda_j u^{(j)} \qquad j = 1, \ldots, n$$

Since the columns of U are orthonormal, and since the dimension of C^n is n, these must form an orthonormal basis for C^n. We skip the results that follow from A being real. This completes the proof.   ∎

Example   From an earlier example in this section, the matrix

$$A = \begin{bmatrix} 2 & 1 & 0 \\ 1 & 3 & 1 \\ 0 & 1 & 2 \end{bmatrix}$$

has the eigenvalues $\lambda_1 = 1$, $\lambda_2 = 2$, $\lambda_3 = 4$ and corresponding orthonormal eigenvectors

$$u^{(1)} = \frac{1}{\sqrt{3}} \begin{bmatrix} 1 \\ -1 \\ 1 \end{bmatrix} \qquad u^{(2)} = \frac{1}{\sqrt{2}} \begin{bmatrix} 1 \\ 0 \\ -1 \end{bmatrix} \qquad u^{(3)} = \frac{1}{\sqrt{6}} \begin{bmatrix} 1 \\ 2 \\ 1 \end{bmatrix}$$

These form an orthonormal basis for R^3 or C^3.

There is a second canonical form that has recently become important for problems in numerical linear algebra, especially for solving overdetermined systems of linear equations. These systems arise from the fitting of empirical data using the linear least squares procedure; see Lawson and Hanson [L1].

Theorem 7.5   (Singular value decomposition)   Let A be order $n \times m$. Then there are unitary matrices U and V, of orders m and n, respectively, such that

$$V^*AU = F \tag{7.21}$$

is a diagonal rectangular matrix of order $n \times m$,

$$F = \begin{bmatrix} \mu_1 & & & & & 0 \\ & \mu_2 & & & & \\ & & \ddots & & & \\ & & & \mu_r & & \\ 0 & & & & 0 & \\ & & & & & \ddots \end{bmatrix} \qquad (7.22)$$

with $F_{ii} = \mu_i, i = 1, \ldots, r$. The numbers $\mu_1, \ldots, \mu_r$ are called the singular values of A. They are all real and positive, and they can be arranged so that

$$\mu_1 \geq \mu_2 \geq \ldots \geq \mu_r > 0 \qquad (7.23)$$

where r is the rank of the matrix A.

Proof Consider the square matrix A^*A of order m. It is a Hermitian matrix, and consequently Theorem 7.4 can be applied to it. The eigenvalues of A^*A are all real; but moreover, they are all nonnegative. To see this, assume

$$A^*Ax = \lambda x \qquad x \neq 0$$

Then

$$(x, A^*Ax) = (x, \lambda x) = \lambda \|x\|_2^2$$

$$(x, A^*Ax) = (Ax, Ax) = \|Ax\|_2^2$$

$$\lambda = (\|Ax\|_2 / \|x\|_2)^2 \geq 0$$

This result also proves that

$$A^*Ax = 0 \qquad \text{if and only if} \qquad Ax = 0 \qquad x \in C^n \qquad (7.24)$$

From Theorem 7.3, there is an $m \times m$ unitary matrix U such that

$$U^*A^*AU = \text{diagonal}[\lambda_1, \ldots, \lambda_r, 0, \ldots, 0] \qquad (7.25)$$

where all $\lambda_i \neq 0$, $1 \leq i \leq r$, and all are positive, for some $r \geq 1$. Because A^*A has order m, the index $r \leq m$. Introduce the *singular values*

$$\mu_i = \sqrt{\lambda_i} \qquad i = 1, \ldots, r$$

The U can be chosen so that the ordering (7.23) is obtained. Using the diagonal matrix

$$D = \text{diagonal}\left[\mu_1, \ldots, \mu_r, 0, \ldots, 0 \right]$$

of order m, we can write (7.25) as

$$(AU)^*(AU) = D^2 \tag{7.26}$$

Let $W = AU$. Then (7.26) says $W^*W = D^2$. Writing W as

$$W = \left[W^{(1)}, \ldots, W^{(m)} \right], \quad W^{(j)} \in C^n$$

we have

$$(W^{(j)}, W^{(j)}) = \begin{cases} \mu_j^2, & 1 \le j \le r \\ 0, & j > r \end{cases} \tag{7.27}$$

and

$$(W^{(i)}, W^{(j)}) = 0 \quad \text{if} \quad i \ne j \tag{7.28}$$

From (7.27), $W^{(j)} = 0$ if $j > r$. And from (7.28), the first r columns of W are orthogonal elements in C^n. Thus the first r columns are linearly independent, and this implies $r \le n$.

Define

$$V^{(j)} = \frac{1}{\mu_j} W^{(j)} \quad j = 1, \ldots, r$$

This is an orthonormal set in C^n. If $r < n$, then choose $V^{(r+1)}, \ldots, V^{(n)}$ so that $\{ V^{(1)}, \ldots, V^{(n)} \}$ is an orthonormal basis for C^n. Define

$$V = \left[V^{(1)}, \ldots, V^{(n)} \right]$$

Easily V is an $n \times n$ unitary matrix, and it can be verified directly that $VF = W$, with F as in (7.22). Thus

$$VF = AU$$

which proves (7.21). The proof that $r = \text{rank}(A)$ and the derivation of other properties of the singular value decomposition are left to Problems 14 and 15. ∎

To give the most important canonical form, introduce the following nota-tion. Define the $n \times n$ matrix

$$J_n(\lambda) = \begin{bmatrix} \lambda & 1 & 0 & \cdots & & 0 \\ & \lambda & 1 & & & \\ & & \ddots & \ddots & & \\ & & & \ddots & \ddots & \\ & & & & \ddots & 1 \\ 0 & & & & & \lambda \end{bmatrix} \qquad n \geq 1$$

$J_n(\lambda)$ has the single eigenvalue λ, of algebraic multiplicity n and geometric multiplicity 1. It is called a *Jordan block*.

Theorem 7.6 (Jordan canonical form)   Let A have order n. Then there is a nonsingular matrix P for which

$$P^{-1}AP = \begin{bmatrix} J_{n_1}(\lambda_1) & & & 0 \\ & J_{n_2}(\lambda_2) & & \\ & & \ddots & \\ 0 & & & J_{n_r}(\lambda_r) \end{bmatrix} \qquad (7.29)$$

The eigenvalues $\lambda_1, \lambda_2, \ldots, \lambda_r$ need not be distinct. For A Hermitian, Theorem 7.4 implies we must have $n_1 = n_2 = \ldots = n_r = 1$; for in that case the sum of the geometric multiplicies must be n, the order of the matrix A.

It is often convenient to write (7.29) as

$$P^{-1}AP = D + N$$
$$D = \operatorname{diag}[\lambda_1, \ldots, \lambda_r] \qquad (7.30)$$

with each λ_i appearing n_i times on the diagonal of D. N is a matrix with all zero entries, except for possible ones on the superdiagonal. It is a nilpotent matrix, and more precisely, it satisfies

$$N^n = 0 \qquad (7.31)$$

The Jordan form is not an easy theorem to prove, and the reader is referred to any of the large number of linear algebra texts for a development of this rich topic. For example, see Halmos [H1, Section 58], Noble [N1, Chapter 11].

7.3 Vector and Matrix Norms

The Euclidean norm $\|x\|_2$ has already been introduced, and it is the way in which most people are used to measuring the size of a vector. But there are many situations in which it is more convenient to measure the size of a vector in other ways. Thus we introduce a general concept of the *norm* of a vector.

Definition Let V be a vector space, and let $N(x)$ be a real-valued function defined on V. Then $N(x)$ is a norm if

1. $N(x) \geq 0$ for all $x \in V$; and $N(x) = 0$ if and only if $x = 0$.

2. $N(\alpha x) = |\alpha| N(x)$, for all $x \in V$ and all scalars α.

3. $N(x + y) \leq N(x) + N(y)$, for all $x, y \in V$.

The usual notation is $\|x\| = N(x)$. The notation $N(x)$ is used to emphasize that the norm is a function. Define the distance from x to y as $\|x - y\|$. Simple consequences are the triangle inequality in its alternative form

$$\|x - z\| \leq \|x - y\| + \|y - z\|$$

and the *reverse triangle inequality*,

$$|\,\|x\| - \|y\|\,| \leq \|x - y\| \qquad x, y \in V$$

Examples 1. For $1 \leq p < \infty$, define the *p-norm*,

$$\|x\|_p = \left[\sum_1^n |x_j|^p \right]^{1/p} \qquad x \in C^n$$

2. The maximum norm is

$$\|x\|_\infty = \operatorname*{Max}_{1 \leq j \leq n} |x_j| \qquad x \in C^n$$

The use of the subscript ∞ on the norm is motivated by the result in Problem 18.

3. For the vector space $V = C[a,b]$, the function norms $\|f\|_2$ and $\|f\|_\infty$ were introduced in Chapters 4 and 1, respectively.

Example Consider the vector $x = (1, 0, -1, 2)$. Then

$$\|x\|_1 = 4 \qquad \|x\|_2 = \sqrt{6} \qquad \|x\|_\infty = 2$$

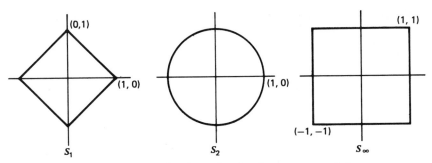

Figure 7.2 The unit spheres S_p using the vector norm $\|\cdot\|_p$.

To show that $\|\cdot\|_p$ is a norm for a general p is nontrivial. The cases $p=1$ and ∞ are straightforward, and $\|\cdot\|_2$ has been treated in Section 4.1. But for $1<p<\infty, p\neq2$, it is difficult to show that $\|\cdot\|_p$ satisfies the triangle inequality. This is not a significant problem for us since the main cases of interest are $p=1,2,\infty$. To give some geometrical intuition for these norms, the sets

$$S_p = \left\{ x\in R^2 \big| \; \|x\|_p = 1 \right\} \qquad p=1,2,\infty$$

are sketched in Figure 7.2.

We now prove some results relating different norms; and we begin with the following result on the continuity of $N(x)\equiv\|x\|$, as a function of x.

Lemma Let $N(x)$ be a norm on C^n (or R^n). Then $N(x)$ is a continuous function of the components $x_1, x_2, \ldots, x_n$ of x.

Proof We want to show that if $x_i\approx y_i$ for $i=1,\ldots,n$, then $N(x)\approx N(y)$. Using the reverse triangle inequality,

$$|N(x)-N(y)|\leq N(x-y) \qquad x,y\in C^n$$

Recall from Section 7.1 the definition of the standard basis $\{e^{(1)},\ldots,e^{(n)}\}$ for C^n. Then

$$x-y = \sum_{j=1}^{n} (x_j-y_j)e^{(j)}$$

$$N(x-y)\leq \sum_{j=1}^{n} |x_j-y_j|N(e^{(j)})\leq \|x-y\|_\infty \sum_{j=1}^{n} N(e^{(j)})$$

$$|N(x)-N(y)|\leq c\|x-y\|_\infty \qquad c=\sum_{j=1}^{n} N(e^{(j)}) \tag{7.32}$$

This completes the proof. ∎

Note that it also proves that for every vector norm N on C^n, there is a $c > 0$ with

$$N(x) \leq c\|x\|_\infty \qquad \text{all } x \in C^n \tag{7.33}$$

Just let $y = 0$ is (7.32). The following theorem proves the converse of this result.

Theorem 7.7   (Equivalence of norms)   Let N and M be norms on C^n. Then there are constants $c_1, c_2 > 0$ for which

$$c_1 M(x) \leq N(x) \leq c_2 M(x) \qquad \text{all } x \in V \tag{7.34}$$

Proof   It is sufficient to consider the case in which N is arbitrary and $M(x) = \|x\|_\infty$. Combining two such statements then leads to the general result. Thus we wish to show there are constants c_1, c_2 for which

$$c_1 \|x\|_\infty \leq N(x) \leq c_2 \|x\|_\infty$$

or equivalently,

$$c_1 \leq N(z) \leq c_2 \qquad \text{all } z \in S \tag{7.35}$$

in which S is the set of all points z in C^n for which $\|z\|_\infty = 1$. The upper inequality of (7.35) follows immediately from (7.33).

Note that S is a closed and bounded set in C^n, and N is a continuous function on S. It is then a standard result of advanced calculus that N attains its maximum and minimum on S at points of S, that is, there are constants c_1, c_2 and points z_1, z_2 in S for which

$$c_1 = N(z_1) \leq N(z) \leq N(z_2) = c_2 \qquad \text{all } z \in S$$

Clearly $c_1, c_2 \geq 0$. And if $c_1 = 0$, then $N(z_1) = 0$. But then $z_1 = 0$, contrary to the construction of S, which requires $\|z_1\|_\infty = 1$. This proves (7.35), completing the proof of the theorem. Note that this theorem does not generalize to infinite dimensional spaces. ∎

Many numerical methods for problems involving linear systems produce a sequence of vectors $\{x^{(m)} | m \geq 0\}$, and we want to speak of convergence of this sequence to a vector x.

Definition   A sequence of vectors $\{x^{(1)}, x^{(2)}, \ldots, x^{(m)}, \ldots\}$ in C^n or R^n is said to converge to a vector x if and only if

$$\|x - x^{(m)}\| \to 0 \qquad \text{as} \qquad m \to \infty$$

Note that the choice of norm is left unspecified. For finite dimensional spaces, it doesn't matter which norm is used. Let M and N be two norms on C^n. Then from (7.34),

$$c_1 M(x - x^{(m)}) \leq N(x - x^{(m)}) \leq c_2 M(x - x^{(m)}) \qquad m \geq 0$$

and $M(x - x^{(m)})$ converges to zero if and only if $N(x - x^{(m)})$ does the same. Thus $x^{(m)} \to x$ with the M norm if and only if it converges with the N norm. This is an important result, and it is not true for infinite dimensional spaces.

Matrix norms The set of all $n \times n$ matrices with complex entries can be considered as equivalent to the vector space C^{n^2}, with a special multiplicative operation added onto the vector space. Thus a matrix norm should satisfy the usual three requirements of a vector norm, listed at the beginning of the section. In addition, we also require two other conditions.

Definition **4.** $\|AB\| \leq \|A\| \|B\|$

5. Usually the vector space we will be working with, $V = C^n$ or R^n, will have some vector norm, call it $\|x\|_v, x \in V$. We require that the matrix and vector norms be compatible,

$$\|Ax\|_v \leq \|A\| \|x\|_v \qquad \text{all } x \in V, \text{ all } A$$

Example Let A be $n \times n$, $\|\cdot\|_v = \|\cdot\|_2$. Then for $x \in C^n$,

$$\|Ax\|_2 = \left[\sum_{i=1}^{n} \left| \sum_{j=1}^{n} a_{ij} x_j \right|^2 \right]^{1/2}$$

$$\leq \left[\sum_{i=1}^{n} \left\{ \sum_{j=1}^{n} |a_{ij}|^2 \right\} \left\{ \sum_{j=1}^{n} |x_j|^2 \right\} \right]^{1/2}$$

by using the Cauchy-Schwartz inequality (7.4). Then

$$\|Ax\|_2 \leq F(A) \|x\|_2 \qquad F(A) = \left[\sum_{i,j=1}^{n} |a_{ij}|^2 \right]^{1/2} \tag{7.36}$$

$F(A)$ is called the *Frobenius norm* of A. Property 5 is shown using (7.36) directly. Properties 1 through 3 are satisfied since $F(A)$ is just the Euclidean norm on

C^{n^2}. It remains to show 4 using the Cauchy-Schwartz inequality,

$$F(AB) = \left[\sum_{i,j=1}^{n} \left| \sum_{k=1}^{n} a_{ik} b_{kj} \right|^2 \right]^{1/2}$$

$$\leq \left[\sum_{i,j=1}^{n} \left\{ \sum_{k=1}^{n} |a_{ik}|^2 \right\} \left\{ \sum_{k=1}^{n} |b_{kj}|^2 \right\} \right]^{1/2}$$

$$= F(A)F(B)$$

Thus $F(A)$ is a matrix norm, compatible with the Euclidean vector norm.

Usually when given a vector space with a norm $\| \cdot \|_v$, an associated matrix norm is defined by

$$\|A\| = \underset{x \neq 0}{\text{Supremum}} \; \frac{\|Ax\|_v}{\|x\|_v} \tag{7.37}$$

It is often called the *operator norm*. By its definition, it satisfies property 5,

$$\|Ax\|_v \leq \|A\| \|x\|_v \qquad x \in C^n \tag{7.38}$$

For a matrix A, the operator norm induced by the vector norm $\|x\|_p$ is denoted by $\|A\|_p$. The most important cases are given in Table 7.1, and the derivations are given later. We need the following definition in order to define $\|A\|_2$.

Definition Let A be an arbitrary matrix. The *spectrum* of A is the set of all eigenvalues of A, and it is denoted by $\sigma(A)$. The *spectral radius* is the maximum size of these eigenvalues, and it is denoted by

$$r_\sigma(A) = \underset{\lambda \in \sigma(A)}{\text{Maximum}} |\lambda|$$

Table 7.1 Vector norms and associated operator matrix norms

Vector norm	Matrix norm
$\|x\|_1 = \sum_{i=1}^{n} |x_i|$	$\|A\|_1 = \underset{1 \leq j \leq n}{\text{Maximum}} \sum_{i=1}^{n} |a_{ij}|$
$\|x\|_2 = \left[\sum_{i=1}^{n} |x_i|^2 \right]^{1/2}$	$\|A\|_2 = \sqrt{r_\sigma(A^*A)}$
$\|x\|_\infty = \underset{1 \leq i \leq n}{\text{Maximum}} |x_i|$	$\|A\|_\infty = \underset{1 \leq i \leq n}{\text{Maximum}} \sum_{j=1}^{n} |a_{ij}|$

To show (7.37) is a norm in general, we begin by showing it is finite. Recall from Theorem 7.7 that there are constants $c_1, c_2 > 0$ with

$$c_1 \|x\|_2 \leqq \|x\|_v \leqq c_2 \|x\|_2 \qquad x \in C^n$$

Thus,

$$\frac{\|Ax\|_v}{\|x\|_v} \leqq \frac{c_2 \|Ax\|_2}{c_1 \|x\|_2} \leqq \frac{c_2}{c_1} F(A)$$

which proves $\|A\|$ if finite.

At this point it is interesting to note the geometric significance of $\|A\|$.

$$\|A\| = \operatorname*{Sup}_{x \neq 0} \frac{\|Ax\|_v}{\|x\|_v} = \operatorname*{Sup}_{x \neq 0} \left\| A\left(\frac{x}{\|x\|_v} \right) \right\|_v = \operatorname*{Sup}_{\|z\|_v = 1} \|Az\|_v$$

By noting that the supremum doesn't change if we let $\|z\|_v \leqq 1$,

$$\|A\| = \operatorname*{Sup}_{\|z\|_v \leqq 1} \|Az\|_v \tag{7.39}$$

Let

$$S = \{ z \in C^n \mid \|z\|_v \leqq 1 \}$$

the unit ball with respect to $\| \cdot \|_v$. Then

$$\|A\| = \operatorname*{Sup}_{z \in S} \|Az\|_v = \operatorname*{Sup}_{w \in A(S)} \|w\|_v$$

with $A(S)$ the image of S when A is applied to it. Thus $\|A\|$ measures the effect of A on the unit ball; and if $\|A\| > 1$, then $\|A\|$ denotes the maximum stretching of the ball S under the action of A.

The proof that $\|A\|$ is actually a norm is now given.

1. Clearly $\|A\| \geqq 0$; and if $A = 0$, then $\|A\| = 0$. Conversely, if $\|A\| = 0$, then $\|Ax\|_v = 0$ for all x. Thus $Ax = 0$ for all x, and this implies $A = 0$.

2. Let α be any scalar. Then using the fact that $\| \cdot \|_v$ is a vector norm,

$$\|\alpha A\| = \operatorname*{Sup}_{\|x\|_v \leqq 1} \|\alpha Ax\|_v = \operatorname*{Sup}_{\|x\|_v \leqq 1} |\alpha| \|Ax\|_v = |\alpha| \operatorname*{Sup}_{\|x\|_v \leqq 1} \|Ax\|_v$$

$$= |\alpha| \|A\|$$

3. For any $x \in C^n$,

$$\|(A + B)x\|_v = \|Ax + Bx\|_v \leqq \|Ax\|_v + \|Bx\|_v$$

since $\|\cdot\|_v$ is a norm. Using the property (7.38),

$$\|(A+B)x\|_v \leq \|A\|\,\|x\|_v + \|B\|\,\|x\|_v$$

$$\frac{\|(A+B)x\|_v}{\|x\|_v} \leq \|A\| + \|B\|$$

This implies

$$\|A+B\| \leq \|A\| + \|B\|$$

4.   For any $x \in C^n$, use (7.38) to get

$$\|(AB)x\|_v = \|A(Bx)\|_v \leq \|A\|\,\|Bx\|_v \leq \|A\|\,\|B\|\,\|x\|_v$$

$$\frac{\|ABx\|_v}{\|x\|_v} \leq \|A\|\,\|B\|$$

and this implies

$$\|AB\| \leq \|A\|\,\|B\|$$

We now comment more extensively on the results given in Table 7.1.

***Examples* 1.**   Use the vector norm

$$\|x\|_1 = \sum_{j=1}^{n} |x_j| \qquad x \in C^n$$

Then

$$\|Ax\|_1 = \sum_{i=1}^{n} \left| \sum_{j=1}^{n} a_{ij} x_j \right| \leq \sum_{i=1}^{n} \sum_{j=1}^{n} |a_{ij}|\,|x_j|$$

Changing the order of summation, we can separate the summands,

$$\|Ax\|_1 \leq \sum_{j=1}^{n} |x_j| \sum_{i=1}^{n} |a_{ij}|$$

Let

$$c = \operatorname*{Max}_{1 \leq j \leq n} \sum_{i=1}^{n} |a_{ij}| \tag{7.40}$$

Then

$$\|Ax\|_1 \leq c\|x\|_1$$

and thus

$$\|A\|_1 \leqq c$$

To show this is an equality, we demonstrate an x for which

$$\frac{\|Ax\|_1}{\|x\|_1} = c$$

Let k be the column index for which the maximum in (7.40) is attained. Let $x = e^{(k)}$, the kth unit vector. Then $\|x\|_1 = 1$ and

$$\|Ax\|_1 = \sum_{i=1}^{n} \left| \sum_{j=1}^{n} a_{ij} x_j \right| = \sum_{i=1}^{n} |a_{ik}| = c$$

This proves that for the vector norm $\|\cdot\|_1$, the operator norm is

$$\|A\|_1 = \operatorname*{Max}_{1 \leq j \leq n} \sum_{i=1}^{n} |a_{ij}| \tag{7.41}$$

This is often called the *column norm*.

2. For C^n with the norm $\|x\|_\infty$, the operator norm is

$$\|A\|_\infty = \operatorname*{Max}_{1 \leq i \leq n} \sum_{j=1}^{n} |a_{ij}| \tag{7.42}$$

This is called the *row norm* of A. The proof of the formula is left as problem 20, although it is similar to that for (7.41).

3. Use the norm $\|x\|_2$ on C^n. From (7.36) we conclude that

$$\|A\|_2 \leqq F(A) \tag{7.43}$$

In general, these are not equal. For example, with $A = I$, the identity matrix, use (7.36) and (7.37) to obtain

$$F(I) = \sqrt{n} \qquad \|I\|_2 = 1$$

We will prove

$$\|A\|_2 = \sqrt{r_\sigma(A^*A)} \tag{7.44}$$

as stated earlier in Table 7.1. The matrix A^*A is Hermitian and all of its eigenvalues are nonnegative, as shown in the proof of Theorem 7.5. Let it have the eigenvalues

$$\lambda_1 \geqq \lambda_2 \geqq \ldots \geqq \lambda_n \geqq 0$$

counted according to their multiplicity, and let $v^{(1)}, \ldots, v^{(n)}$ be the corresponding eigenvectors, arranged as an orthonormal basis for C^n.

For a general $x \in C^n$,

$$\|Ax\|_2^2 = (Ax, Ax) = (x, A^*Ax)$$

Write x as

$$x = \sum_{j=1}^{n} \alpha_j v^{(j)} \qquad \alpha_j \equiv (x, v^{(j)}) \tag{7.45}$$

Then

$$A^*Ax = \sum_{j=1}^{n} \alpha_j A^* Av^{(j)} = \sum_{j=1}^{n} \alpha_j \lambda_j v^{(j)}$$

and

$$\|Ax\|_2^2 = \left(\sum_{i=1}^{n} \alpha_i v^{(i)}, \sum_{j=1}^{n} \alpha_j \lambda_j v^{(j)} \right) = \sum_{j=1}^{n} \lambda_j |\alpha_j|^2$$

$$\leq \lambda_1 \sum_{j=1}^{n} |\alpha_j|^2 = \lambda_1 \|x\|_2^2$$

using (7.45) to calculate $\|x\|_2$. Thus

$$\|A\|_2^2 \leq \lambda_1$$

Equality follows by noting that if $x = v^{(1)}$, then $\|x\|_2 = 1$ and

$$\|Ax\|_2^2 = (x, A^*Ax) = (v^{(1)}, \lambda_1 v^{(1)}) = \lambda_1$$

This proves (7.44) since $\lambda_1 = r_\sigma(A^*A)$. Since AA^* and A^*A have the same nonzero eigenvalues (see Problem 14), we have $r_\sigma(AA^*) = r_\sigma(A^*A)$, an alternative formula for (7.44). And it also proves

$$\|A\|_2 = \|A^*\|_2 \tag{7.46}$$

This is not true for the previous matrix norms.

It can be shown fairly easily that if A is Hermitian, then

$$\|A\|_2 = r_\sigma(A) \tag{7.47}$$

This is left as Problem 21.

Example Consider the matrix

$$A = \begin{bmatrix} 1 & -2 \\ -3 & 4 \end{bmatrix}$$

Then

$$\|A\|_1 = 6 \qquad \|A\|_2 = \sqrt{15 + \sqrt{221}} \doteq 5.46 \qquad \|A\|_\infty = 7$$

And as an illustration of the inequality (7.48) of the following theorem,

$$r_\sigma(A) = \frac{5 + \sqrt{33}}{2} \doteq 5.37$$

Theorem 7.8 Let A be an arbitrary square matrix. Then for any operator matrix norm,

$$r_\sigma(A) \leq \|A\| \tag{7.48}$$

Moreover, if $\epsilon > 0$ is given, then there is an operator matrix norm, denoted here by $\|\cdot\|_\epsilon$, for which

$$\|A\|_\epsilon \leq r_\sigma(A) + \epsilon \tag{7.49}$$

Proof To prove (7.48), let $\|\cdot\|$ be any matrix norm with an associated compatible vector norm $\|\cdot\|_v$. Let λ be an eigenvalue in $\sigma(A)$ for which

$$|\lambda| = r_\sigma(A)$$

and let x be an associated eigenvector, $\|x\|_v = 1$. Then

$$r_\sigma(A) = |\lambda| = \|\lambda x\|_v = \|Ax\|_v \leq \|A\| \, \|x\|_v = \|A\|$$

which proves (7.48).

The proof of (7.49) is a nontrivial construction, and a proof is given in Isaacson and Keller [I1, p. 12]. ∎

The following corollary is an easy, but important, consequence of Theorem 7.8.

Corollary For a square matrix A, $r_\sigma(A) < 1$ if and only if $\|A\| < 1$ for some operator matrix norm.

This result can be used to prove Theorem 7.9 in the next section, but we prefer to use the Jordan canonical form, given in Theorem 7.6. The results (7.48) and (7.49) show that $r_\sigma(A)$ is almost a matrix norm; and this result is used in analyzing the rates of convergence for some of the iteration methods given in Chapter 8 for solving linear systems of equations.

7.4. Convergence and Perturbation Theorems

These results are the theoretical framework from which we later construct error analyses for numerical methods for linear systems of equations.

Theorem 7.9   Let A be a square matrix of order n. Then A^m converges to the zero matrix as $m \to \infty$ if and only if $r_\sigma(A) < 1$.

Proof   We use Theorem 7.6 as a fundamental tool. Let J be the Jordan canonical form for A,

$$P^{-1}AP = J$$

Then

$$A^m = (PJP^{-1})^m = PJ^mP^{-1} \tag{7.50}$$

and $A^m \to 0$ if and only if $J^m \to 0$. Recall from (7.30) and (7.31) that J can be written as

$$J = D + N$$

in which

$$D = \mathrm{diag}[\lambda_1, \ldots, \lambda_n]$$

contains the eigenvalues of J (and A), and N is a matrix for which

$$N^n = 0 \tag{7.51}$$

Thus

$$J^m = (D + N)^m = \sum_{j=0}^{m} \binom{m}{j} D^{m-j} N^j$$

and using (7.51) for $m \geq n$,

$$J^m = \sum_{j=0}^{n} \binom{m}{j} D^{m-j} N^j \tag{7.52}$$

Notice that the powers of D satisfy

$$m - j \geq m - n \to \infty \quad \text{as} \quad m \to \infty \tag{7.53}$$

independent of j.

We need the following limits: for any positive $c < 1$ and any $r \geq 0$,

$$\mathop{\mathrm{Limit}}_{m \to \infty} m^r c^m = 0 \tag{7.54}$$

This can be proven using L'Hospital's rule from elementary calculus.

In (7.52), there is a fixed number of terms, $n + 1$, regardless of the size of m, and we can consider the convergence of J^m by considering each of the individual terms. Assuming $r_\sigma(A)<1$, we know that all $|\lambda_i|<1, i=1,\ldots,n$. And for any matrix norm

$$\left\|\binom{m}{j}D^{m-j}N^j\right\| \leq \frac{m^j}{j!}\|N\|^j\|D^{m-j}\|$$

Using the row norm, we have the above is bounded by

$$\frac{1}{j!}\|N\|_\infty^j m^j \left[r_\sigma(A)\right]^{m-j}$$

which converges to zero as $m\to\infty$, using (7.53) and (7.54), for $0\leq j\leq n$. This proves half of the theorem, namely that if $r_\sigma(A)<1$, then J^m and A^m, from (7.50), converge to zero as $m\to\infty$.

Suppose that $r_\sigma(A)\geq 1$. Then let λ be an eigenvalue of A for which $|\lambda|\geq 1$, and let x be an associated eigenvector, $x\neq 0$. Then

$$A^m x=\lambda^m x$$

and clearly this does not converge to zero as $m\to\infty$. Thus it is not possible that $A^m\to0$, as that would imply $A^m x\to0$. This completes the proof.   ∎

Theorem 7.10 (Geometric series)   Let A be a square matrix. If $r_\sigma(A)<1$, then $(I-A)^{-1}$ exists, and it can be expressed as a convergent series,

$$(I-A)^{-1}=I+A+A^2+\ldots+A^m+\ldots \qquad (7.55)$$

Conversely, if the series in (7.55) is convergent, then $r_\sigma(A)<1$.

Proof Assume $r_\sigma(A)<1$. We show the existence of $(I-A)^{-1}$ by proving the equivalent statement (3) of Theorem 7.2. Assume

$$(I-A)x=0$$

Then $Ax=x$, and this implies 1 is an eigenvalue of A if $x\neq 0$. But we assumed $r_\sigma(A)<1$, and thus we must have $x=0$, concluding the proof of the existence of $(I-A)^{-1}$.

We need the following identity:

$$(I-A)(I+A+A^2+\ldots+A^m)=I-A^{m+1} \qquad (7.56)$$

which is true for any matrix A. Multiplying by $(I-A)^{-1}$,

$$I+A+A^2+\ldots+A^m=(I-A)^{-1}(I-A^{m+1})$$

The left-hand side has a limit if the right-hand side does. By the preceding Theorem 7.9, $r_\sigma(A) < 1$ implies $A^{m+1} \to 0$ as $m \to \infty$. Thus we have the result (7.55).

Conversely, assume that the series converges and denote it by

$$B = I + A + A^2 + \ldots + A^m + \ldots$$

Then $B - AB = B - BA = I$, and thus $I - A$ has an inverse, that is B. Taking limits on both sides of (7.56), the left-hand side has the limit $(I - A)B = I$, and thus the same must be true of the right-hand limit. But that implies

$$A^{m+1} \to 0 \quad \text{as} \quad m \to \infty$$

By Theorem 7.9, we must have $r_\sigma(A) < 1$. ∎

Theorem 7.11 Let A be a square matrix. If for some operator matrix norm, $\|A\| < 1$, then $(I - A)^{-1}$ exists and has the geometric series expansion (7.55). Moreover,

$$\|(I - A)^{-1}\| \leq \frac{1}{1 - \|A\|} \tag{7.57}$$

Proof Since $\|A\| < 1$, it follows from (7.48) of Theorem 7.8 that $r_\sigma(A) < 1$. Except for (7.57), the conclusions follow from Theorem 7.10. For (7.57), let

$$B_m = I + A + \ldots + A^m$$

From (7.56),

$$B_m = (I - A)^{-1}(I - A^{m+1})$$

$$(I - A)^{-1} - B_m = (I - A)^{-1}[I - (I - A^{m+1})] = (I - A)^{-1}A^{m+1} \tag{7.58}$$

Using the reverse triangle inequality,

$$\left| \|(I - A)^{-1}\| - \|B_m\| \right| \leq \|(I - A)^{-1} - B_m\|$$

$$\leq \|(I - A)^{-1}\| \|A\|^{m+1}$$

Since this converges to zero as $m \to \infty$, we have

$$\|B_m\| \to \|(I - A)^{-1}\| \quad \text{as} \quad m \to \infty \tag{7.59}$$

From the definition of B_m and the properties of a matrix norm,

$$\|B_m\| \leq \|I\| + \|A\| + \|A\|^2 + \ldots + \|A\|^m$$

$$= \frac{1 - \|A\|^{m+1}}{1 - \|A\|} \leq \frac{1}{1 - \|A\|}$$

Combined with (7.59), this concludes the proof of (7.57). ∎

Theorem 7.12 Let A and B be square matrices of the same order. Assume that A is nonsingular and suppose that

$$\|A - B\| < \frac{1}{\|A^{-1}\|} \tag{7.60}$$

Then B is also nonsingular,

$$\|B^{-1}\| \leq \frac{\|A^{-1}\|}{1 - \|A^{-1}\| \|A - B\|} \tag{7.61}$$

and

$$\|A^{-1} - B^{-1}\| \leq \frac{\|A^{-1}\|^2 \|A - B\|}{1 - \|A^{-1}\| \|A - B\|} \tag{7.62}$$

Proof Note the indentity

$$B = A - (A - B) = A[I - A^{-1}(A - B)] \tag{7.63}$$

The matrix $[I - A^{-1}(A - B)]$ is nonsingular using Theorem 7.11, based on the inequaltiy (7.60) which implies

$$\|A^{-1}(A - B)\| \leq \|A^{-1}\| \|A - B\| < 1$$

Since B is the product of nonsingular matrices, it too is nonsingular,

$$B^{-1} = [I - A^{-1}(A - B)]^{-1} A^{-1}$$

The bound (7.61) follows by taking norms and applying (7.57). To prove (7.62), use

$$A^{-1} - B^{-1} = A^{-1}(A - B)B^{-1}$$

Take norms again and apply (7.61). ∎

This theorem is important in a number of ways. But for the moment, it says that all sufficiently close perturbations of a nonsingular matrix are nonsingular.

Example We illustrate Theorem 7.11 by considering the invertibility of the matrix

$$A = \begin{bmatrix} 4 & 1 & 0 & 0 & \cdots & 0 \\ 1 & 4 & 1 & 0 & & \\ 0 & 1 & 4 & 1 & & \\ \vdots & & & \ddots & \ddots & \\ & & & 1 & 4 & 1 \\ 0 & \cdots & \cdots & 0 & 1 & 4 \end{bmatrix}$$

Rewrite A as

$$A = 4(I + B)$$

$$B = \begin{bmatrix} 0 & \frac{1}{4} & 0 & 0 & \cdots & 0 \\ \frac{1}{4} & 0 & \frac{1}{4} & 0 & \cdots & 0 \\ 0 & & \ddots & & & \vdots \\ \vdots & & & \ddots & & \frac{1}{4} \\ 0 & \cdots\cdots & & \frac{1}{4} & & 0 \end{bmatrix}$$

Using the row norm (7.42), $\|B\|_\infty = \frac{1}{2}$. Thus $(I+B)^{-1}$ exists from Theorem 7.11; and from (7.57),

$$\|(I+B)^{-1}\|_\infty \leq \frac{1}{1-\frac{1}{2}} = 2$$

Thus $A^{-1} = \frac{1}{4}(I+B)^{-1}$, and

$$\|A^{-1}\|_\infty \leq \frac{1}{2}$$

By use of the row norm and inequality (7.48),

$$r_\sigma(A) \leq 6 \qquad r_\sigma(A^{-1}) \leq \frac{1}{2}$$

Since the eigenvalues of A^{-1} are the reciprocals of those of A (see Problem 21), and since all eigenvalues of A are real because A is Hermitian, we have the bound

$$2 \leq |\lambda| \leq 6 \qquad \text{all} \quad \lambda \in \sigma(A)$$

Discussion of the Literature

The material of this chapter is linear algebra, especially selected for use in deriving and analyzing methods of numerical linear algebra. The books Bradley [B1] and Strang [S2] are introductory-level texts for undergraduate linear algebra. Franklin [F2] is a higher level introduction to matrix theory, Halmos [H1] is a well-known text on abstract linear algebra; Noble [N1] is a wide-ranging applied linear algebra text. Introductions to the foundations are also contained in Fadeeva [F1], Stewart [S1], and Wilkinson [W1], all of which are devoted entirely to numerical linear algebra. For additional theory at a more detailed and higher level, see the classical accounts of Gantmacher [G1] and Householder [H2]. Additional references are given in the bibliographies of Chapters 8 and 9.

Bibliography

[B1] Bradley, G., *A Primer of Linear Algebra*, Prentice-Hall, Englewood Cliffs, N. J., 1975.

[F1] Fadeeva, V., *Computational Methods of Linear Algebra*, Dover Press, New York, 1959.

[F2] Franklin, J., *Matrix Theory*, Prentice-Hall, Englewood Cliffs, N. J., 1968.

[G1] Gantmacher, F., *The Theory of Matrices*, Vol. I and II, Chelsea Pub., New York, 1960.

[H1] Halmos, P., *Finite-Dimensional Vector Spaces*, Van Nostrand, Princeton, N. J., 1958.

[H2] Householder, A., *The Theory of Matrices in Numerical Analysis*, Blaisdell Pub., New York, 1965.

[I1] Isaacson, E., and H. Keller, *Analysis of Numerical Methods*, John Wiley, New York, 1966.

[L1] Lawson, C., and R. Hanson, *Solving Least Squares Problems*, Prentice-Hall, Englewood Cliffs, N. J., 1974.

[N1] Noble, B., *Applied Linear Algebra*, Prentice-Hall, Englewood Cliffs, N. J., 1969.

[S1] Stewart, G., *Introduction to Matrix Computations*, Academic Press, New York, 1973.

[S2] Strang, G., *Linear Algebra and Its Applications*, Academic Press, New York, 1976.

[W1] Wilkinson, J., *The Algebraic Eigenvalue Problem*, Clarendon Press, Oxford, 1965.

Problems

1. Determine whether the following sets of vectors are dependent or independent.

    (a) $(1, 2, -1, 3)$, $(3, -1, 1, 1)$, $(1, 9, -5, 11)$

    (b) $(1, 1, 0)$, $(0, 1, 1)$, $(1, 0, 1)$

2.. Let A, B, and C be matrices of order $m \times n$, $n \times p$, and $p \times q$, respectively.

    (a) Prove the associative law $(AB)C = A(BC)$.

    (b) Prove $(AB)^T = B^T A^T$.

3. (a) Produce square matrices A and B for which $AB \neq BA$.

    (b) Produce square matrices A and B, with no zero entries, for which $AB = 0$, $BA \neq 0$.

4. Let A be a square matrix of order n with real entries. The function

$$q(x_1, \ldots, x_n) = (Ax, x) = \sum_{i=1}^{n} \sum_{j=1}^{n} a_{ij} x_i x_j \qquad x \in R^n$$

is called the *quadratic form* determined by A. It is a quadratic polynomial in the n variables $x_1, \ldots, x_n$, and it occurs when considering the maximization or minimization of a function of n variables.

    (a) Prove that if A is skew symmetric, then $q(x) \equiv 0$.

    (b) For a general square matrix A, define $A_1 = \frac{1}{2}(A + A^T)$, $A_2 = \frac{1}{2}(A - A^T)$. Then $A = A_1 + A_2$. Shown that A_1 is symmetric, A_2 is skew symmetric and $(Ax, x) = (A_1 x, x)$, all $x \in R^n$. This shows that the coefficient matrix A for a quadratic form can always be assumed to be symmetric, without any loss of generality.

5. Let A be a matrix of order $m \times n$; and let r and c denote the row and column rank of A, respectively. Prove that $r = c$. Hint: for convenience, assume that the first r rows of A are independent, with the remaining rows dependent on these first r rows. And assume the same for the first c columns of A. Let $\hat{A}$ denote the $r \times n$ matrix obtained by deleting the last $m - r$ rows of A; and let $\hat{r}$ and $\hat{c}$ denote the row and column rank of $\hat{A}$, respectively. Clearly $\hat{r} = r$. Also, the columns of $\hat{A}$ are elements of C^r, which has dimension r, and thus we must have $\hat{c} \leq r$. Show that $\hat{c} = c$, thus proving that $c \leq r$. The reverse inequality will follow by applying the same argument to A^T; and taken together, these two inequalities imply $r = c$.

6. Prove the equivalence of statements 1 to 4 and 6 of Theorem 7.2. Hint: use Theorem 7.1, the result in Problem 5, and the decomposition (7.3).

7. Given the orthogonal vectors

$$u^{(1)} = (1, 2, -1) \qquad u^{(2)} = (1, 1, 3)$$

produce a third vector $u^{(3)}$ such that $\{u^{(1)}, u^{(2)}, u^{(3)}\}$ is an orthogonal basis for R^3. Normalize these vectors to obtain an orthonormal basis.

8. For the column vector $w \in C^n$ with $\|w\|_2 = \sqrt{w^*w} = 1$, define the $n \times n$ matrix

$$A = I - 2ww^*$$

(a) For the special case $w = [\frac{1}{3}, \frac{2}{3}, \frac{2}{3}]^T$, produce the matrix A. Verify that it is symmetric and orthogonal.

(b) Show that, in general, all such matrices A are Hermitian and unitary.

9. Calculate the eigenvalues and eigenvectors of the following matrices.

(a) $\begin{bmatrix} 1 & 4 \\ 1 & 1 \end{bmatrix}$ **(b)** $\begin{bmatrix} 1 & 1 \\ -1 & 3 \end{bmatrix}$

10. Let U be an $n \times n$ unitary matrix.

(a) Show $\|Ux\|_2 = \|x\|_2$, all $x \in C^n$. Use this to prove that the distance between points x and y is the same as the distance between Ux and Uy, showing that unitary transformations of C^n preserve distances between all points.

(b) Let U be orthogonal, and show that

$$(Ux, Uy) = (x, y) \qquad x, y \in C^n$$

This shows that orthogonal transformations of R^n also preserve angles between lines, as defined in (7.8).

(c) Show that all eigenvalues of a unitary matrix have magnitude 1.

11. Let A be a Hermitian matrix of order n. It is called *positive definite* if and only if $(Ax,x)>0$ for all $x\neq 0$ in C^n. Show that A is positive definite if and only if all of its eigenvalues are real and positive. Hint: use Theorem 7.4, and expand (Ax,x) by using an eigenvector basis to express an arbitrary $x\in C^n$.

12. Let $y\neq 0$ in R^n, and define $A=yy^T$, an $n\times n$ matrix. Show that $\lambda=0$ is an eigenvalue of multiplicity exactly $n-1$. What is the single nonzero eigenvalue?

13. Demonstrate the relation (7.31).

14. Recall the notation used in Theorem 7.5 on the singular value decomposition of a matrix A.

    (a) Show that $\mu_1^2,...,\mu_r^2$ are the nonzero eigenvalues of A^*A and AA^*, with corresponding eigenvectors $U^{(1)},...,U^{(r)}$ and $V^{(1)},...,V^{(r)}$, respectively. The vector $U^{(j)}$ denotes column j of U, and similarly for $V^{(j)}$ and V.

    (b) Show that $AU^{(j)}=\mu_j V^{(j)}$, $A^*V^{(j)}=\mu_j U^{(j)}$, $1\leq j\leq r$.

    (c) Prove $r=\text{rank}(A)$.

15. Let A be a real $n\times m$ matrix, with $n>m$; and to further simplify the problem, assume rank $(A)=m$. Consider the linear system $Ax=b$. In general, it is overdetermined, and usually it does not have a solution. We define a *least squares solution* to be a vector $\hat{x}\in R^m$ that will minimize the least squares error

$$E(x)=\|Ax-b\|_2 \qquad x\in R^m$$

    The singular value decomposition of A can be used to calculate the least squares solution $\hat{x}$. From Theorem 7.5 and (7.21), $A=V^*FU$, the singular value decomposition. Define $y=U^*x$ (or $x=Uy$) and $c=V^*b$. Then show the least squares solution $\hat{x}$ exists and is unique, and is defined by $\hat{x}=U\hat{y}$,

$$\hat{y}_i=c_i/\mu_i \qquad i=1,...,m$$

    And for the resulting error, show that the least squares error is given by

$$E(\hat{x})=\|A\hat{x}-b\|_2=\sqrt{c_{m+1}^2+...+c_n^2}$$

    Hint: see Problem 10a. For an extensive discussion of the numerical solution of least-squares data-fitting problems, using the singular value decomposition, see Lawson and Hanson [L1].

16. For any polynomial $p(x)=b_0+b_1x+...+b_mx^m$, and for A any square matrix, define

$$p(A)=b_0I+b_1A+...+b_mA^m.$$

Let A be a matrix for which the Jordan canonical form is a diagonal matrix,

$$P^{-1}AD = D = \text{diag}[\lambda_1, \ldots, \lambda_n]$$

For the characteristic polynomial $f_A(\lambda)$ of A, prove $f_A(A) = 0$. (This result is known as the Cayley-Hamilton theorem. It is true for any square matrix A, not just those with a diagonal Jordan canonical form.)
Hint: use the result $A = PDP^{-1}$ to simplify $f_A(A)$.

17. Let A be a real nonsingular matrix of order n, and let $\|\cdot\|_v$ denote a vector norm on R^n. Define

$$\|x\|_* = \|Ax\|_v \qquad x \in R^n$$

Show that $\|\cdot\|_*$ is a vector norm on R^n.

18. Show

$$\underset{p \to \infty}{\text{Limit}} \left[\sum_{j=1}^{n} |x_i|^p \right]^{1/p} = \underset{1 \leq i \leq n}{\text{Max}} |x_i| \qquad x \in C^n$$

This justifies the use of the notation $\|x\|_\infty$ for the right side.

19. For any matrix norm, shown that (a) $\|I\| \geq 1$, and (b) $\|A^{-1}\| \geq 1/\|A\|$. For an operator norm, it is immediate from (7.37) that $\|I\| = 1$.
Hint: for (a), use $I^2 = I$, and take norms; and for (b), use $AA^{-1} = I$.

20. Derive formula (7.42) for the operator matrix norm $\|A\|_\infty$.

21. Let A be a square matrix of order $n \times n$.

(a) Given the eigenvalues and eigenvectors of A, determine those of (i) A^m for $m \geq 2$, (ii) A^{-1}, assuming that A is nonsingular, and (iii) $A + cI, c = $ constant.

(b) Prove $\|A\|_2 = r_\sigma(A)$ when A is Hermitian.

(c) For A arbitrary and U unitary of the same order, show

$$\|AU\|_2 = \|UA\|_2 = \|A\|_2$$

22. Let A be square of order $n \times n$.

(a) Show that $F(AU) = F(UA) = F(A)$, for any unitary matrix U.

(b) If A is Hermitian, then show that

$$(A) = \sqrt{\lambda_1^2 + \ldots + \lambda_n^2}$$

where $\lambda_1, \ldots, \lambda_n$ are the eigenvalues of A, repeated according to their multiplicity. Furthermore,

$$\frac{1}{\sqrt{n}}F(A) \leqq \|A\|_2 \leqq F(A)$$

23. Show that the infinite series

$$I + A + \frac{1}{2!}A^2 + \frac{1}{3!}A^3 + \ldots + \frac{1}{n!}A^n + \ldots$$

converges for any square matrix A, and denote the sum of the series by e^A.

(a) IF $A = P^{-1}BP$, show that $e^A = P^{-1}e^BP$.

(b) Let $\lambda_1, \ldots, \lambda_n$ denote the eigenvalues of A, repeated according to their multiplicity, and show that the eigenvalues of e^A are $e^{\lambda_1}, \ldots, e^{\lambda_n}$.

24. In producing cubic interpolating splines in Section 3.7 of Chapter 3, it was necessary to solve the linear system $AM = D$, with

$$A = \begin{bmatrix} \frac{h_1}{3} & \frac{h_1}{6} & 0 & \cdots & & 0 \\ \frac{h_1}{6} & \frac{h_1+h_2}{3} & \frac{h_2}{6} & & & \vdots \\ 0 & & \ddots & & & \\ \vdots & & & \frac{h_{m-1}}{6} & \frac{h_{m-1}+h_m}{3} & \frac{h_m}{6} \\ 0 & \cdots & & 0 & \frac{h_m}{6} & \frac{h_m}{3} \end{bmatrix}$$

All $h_i > 0, i = 1, \ldots, m$. Using one or more of the results of Section 7.4, show that A is nonsingular. In addition, derive the bounds

$$\tfrac{1}{6}\text{Min}(h_i) \leq |\lambda| \leq \text{Max}\, h_i \qquad \lambda \in \sigma(A)$$

for the eigenvalues of A.

EIGHT

NUMERICAL
SOLUTION
OF SYSTEMS
OF LINEAR
EQUATIONS

Systems of linear equations arise in a large number of areas, both directly in modeling physical situations and indirectly in the numerical solution of other mathematical models. These applications occur in virtually all areas of the physical, biological, and social sciences. And linear systems are involved in the numerical solution of optimization problems, systems of nonlinear equations, approximation of functions, boundary value problems for ordinary differential equations, partial differential equations, integral equations, statistical inference, and numerous other problems. Because of the widespread importance of linear systems, much research has been devoted to their numerical solution. Excellent algorithms have been developed for the most common types of problems for linear systems, and these are defined, analyzed, and illustrated in this chapter.

The most common type of problem is to solve a square linear system

$$Ax = b$$

of moderate order, with coefficients that are mostly nonzero. Such linear systems, of any order, are called *dense*. Since the coefficient matrix A must generally be stored in the main memory of the computer in order to efficiently solve the linear system, memory storage limitations in most larger computers will limit the order of the system to be less than 100 to 200, depending on the computer. Most algorithms for solving such systems are based on Gaussian elimination, and it will be defined in the Section 8.1. It is a direct method in the theoretical sense that if rounding errors are ignored, then the exact answer is found in a finite number of steps. Modifications for improved error behaviour with Gaussian elimination, variants for special classes of matrices, and error analyses are given in Sections 8.2 through 8.5.

A second important type of problem is to solve $Ax = b$ when A is square, sparse, and of large order. A *sparse* matrix is one in which most coefficients are zero. Such systems arise in a variety of ways, but we restrict our development to those for which there is a simple, known pattern for the nonzero coefficients. These systems arise commonly in the numerical solution of partial differential equations, and an example of this is given in Section 8.8. Because of the large order of most sparse systems of linear equations, sometimes as large as 100,000 or more, the linear system cannot usually be solved by a direct method such as Gaussian elimination. Iteration methods are the preferred method of solution, and these will be introduced in Sections 8.6 and 8.7.

For solving dense square systems of moderate order, most computer centers have a set of programs that can be used for a variety of problems. The student should become acquainted with those at his or her computer center, and use them to further illustrate the material of this chapter. For a general reference containing an excellent collection of algorithms (in ALGOL) for dense systems of moderate order, see Wilkinson and Reinsch [W3, Part I]. There are other types of problems for linear systems, and references to some of these are given at the end of the chapter.

8.1   Gaussian Elimination

This is the formal name given to the method of solving systems of linear equations by successively eliminating unknowns and reducing to systems of lower order. It is the method most people learn in high school algebra or in an undergraduate linear algebra course (in which it is often associated with producing the row-echelon form of a matrix). A precise definition will be given of Gaussian elimination, which is necessary when implementing it on a computer and when analyzing the effects of rounding errors which occur when computing with it.

To solve $Ax = b$, we will reduce it to an equivalent system $Ux = g$, in which U is upper triangular. This system can be easily solved by a process of *back substitution*. Denote the original linear system by $A^{(1)}x = b^{(1)}$,

$$A^{(1)} = \left[a_{ij}^{(1)} \right] \qquad b^{(1)} = \left[b_1^{(1)}, \ldots, b_n^{(1)} \right]^T \qquad 1 \le i,\ j \le n$$

in which n is the order of the system. We reduce the system to the triangular form $Ux = g$ by adding multiples of one equation to another equation, eliminating some unknown from the second equation. Additional row operations will be used in the modifications given in succeeding sections. To keep the presentation simple, we make some technical assumptions in defining the algorithm; they are removed in the next section.

Gaussian elimination algorithm   Step 1: Assume $a_{11}^{(1)} \neq 0$. Define the row multipliers by

$$m_{i1} = a_{i1}^{(1)} / a_{11}^{(1)} \qquad i = 2, 3, \ldots, n$$

These are used in eliminating the x_1 term from equations 2 through n. Define

$$a_{ij}^{(2)} = a_{ij}^{(1)} - m_{i1}a_{1j}^{(1)} \qquad i, j = 2,\ldots,n$$

$$b_i^{(2)} = b_i^{(1)} - m_{i1}b_1^{(1)} \qquad i = 2,\ldots,n$$

Also, the first rows of A and b are left undisturbed, and the first column of $A^{(1)}$, below the diagonal, is set to zero. The system $A^{(2)}x = b^{(2)}$ looks like

$$
\begin{bmatrix}
a_{11}^{(1)} & a_{12}^{(1)} & \cdots & a_{1n}^{(1)} \\
0 & a_{22}^{(2)} & \cdots & a_{2n}^{(2)} \\
\vdots & \vdots & & \vdots \\
0 & a_{n2}^{(2)} & \cdots & a_{nn}^{(2)}
\end{bmatrix}
\begin{bmatrix}
x_1 \\ x_2 \\ \vdots \\ x_n
\end{bmatrix}
=
\begin{bmatrix}
b_1^{(1)} \\ b_2^{(2)} \\ \vdots \\ b_n^{(2)}
\end{bmatrix}
$$

We continue to eliminate unknowns, going onto columns 2, 3, and so on; and this is expressed generally in the following.

Step k: Let $1 \le k \le n-1$. Assume that $A^{(k)}x = b^{(k)}$ has been constructed, with $x_1, x_2, \ldots, x_{k-1}$ eliminated at successive stages; and $A^{(k)}$ has the form

$$
A^{(k)} =
\begin{bmatrix}
a_{11}^{(1)} & a_{12}^{(1)} & & & & a_{1n}^{(1)} \\
0 & a_{22}^{(2)} & & & & a_{2n}^{(2)} \\
\vdots & & \ddots & & & \vdots \\
0 & \cdots & 0 & a_{kk}^{(k)} & \cdots & a_{kn}^{(k)} \\
\vdots & & \vdots & \vdots & & \vdots \\
0 & \cdots & 0 & a_{nk}^{(k)} & \cdots & a_{nn}^{(k)}
\end{bmatrix}
$$

Assume $a_{kk}^{(k)} \ne 0$. Define the multipliers

$$m_{ik} = a_{ik}^{(k)} / a_{kk}^{(k)} \qquad i = k+1,\ldots,n \tag{8.1}$$

Use these to remove the unknown x_k from equations $k+1$ through n. Define

$$a_{ij}^{(k+1)} = a_{ij}^{(k)} - m_{ik}a_{kj}^{(k)}$$
$$b_i^{(k+1)} = b_i^{(k)} - m_{ik}b_k^{(k)} \qquad i,j = k+1,\ldots,n \tag{8.2}$$

The earlier rows 1 through k are left undisturbed, and zeroes are introduced into column k below the diagonal element.

By continuing in this manner, after $n-1$ steps we obtain $A^{(n)}x=b^{(n)}$,

$$
\begin{bmatrix}
a_{11}^{(1)} & \cdots\cdots & a_{1n}^{(1)} \\
0 & \ddots & \vdots \\
\vdots & \ddots & \vdots \\
\vdots & & \vdots \\
0 & \cdots\cdots & a_{nn}^{(n)}
\end{bmatrix}
\begin{bmatrix}
x_1 \\ \vdots \\ \vdots \\ \vdots \\ x_n
\end{bmatrix}
=
\begin{bmatrix}
b_1^{(1)} \\ \vdots \\ \vdots \\ \vdots \\ b_n^{(n)}
\end{bmatrix}
$$

For notational convenience, let $U=A^{(n)}$ and $g=b^{(n)}$. The system $Ux=g$ is upper triangular, and it is quite easy to solve. First

$$x_n = g_n/u_{nn}$$

and then

$$x_k = \frac{1}{u_{kk}}\left(g_k - \sum_{j=k+1}^{n} u_{kj}x_j \right) \qquad k=n-1,n-2,\ldots,1 \qquad (8.3)$$

This completes the Gaussian elimination algorithm.

Example Solve the linear system

$$x_1+2x_2+x_3=0$$
$$2x_1+2x_2+3x_3=3 \qquad (8.4)$$
$$-x_1-3x_2=2$$

To simplify the notation, we note that the unknowns x_1,x_2,x_3 never enter into the algorithm until the final step. Thus we represent the above linear system with the *augmented matrix*

$$[A|b]=\begin{bmatrix} 1 & 2 & 1 & 0 \\ 2 & 2 & 3 & 3 \\ -1 & -3 & 0 & 2 \end{bmatrix}$$

The row operations are performed on this augmented matrix, and the unknowns are given in the final step. In the following diagram, the multipliers are given next to an arrow, corresponding to the changes they cause.

$$\begin{bmatrix} 1 & 2 & 1 & | & 0 \\ 2 & 2 & 3 & | & 3 \\ -1 & -3 & 0 & | & 2 \end{bmatrix} \xrightarrow[\substack{m_{21}=2 \\ m_{31}=-1}]{} \begin{bmatrix} 1 & 2 & 1 & | & 0 \\ 0 & -2 & 1 & | & 3 \\ 0 & -1 & 1 & | & 2 \end{bmatrix}$$

$$\downarrow m_{32}=\tfrac{1}{2}$$

$$[U|g] \equiv \begin{bmatrix} 1 & 2 & 1 & | & 0 \\ 0 & -2 & 1 & | & 3 \\ 0 & 0 & \tfrac{1}{2} & | & \tfrac{1}{2} \end{bmatrix}$$

Solving $Ux = g$,

$$x_3 = 1 \qquad x_2 = -1 \qquad x_1 = 1$$

Triangular factorization of a matrix   It is convenient to keep the multipliers m_{ij}, since we will often want to solve $Ax = b$ with the same A but a different vector b. In the computer the elements $a_{ij}^{(k+1)}, j \geq i$, are always stored into the storage for $a_{ij}^{(k)}$. The elements below the diagonal are being zeroed, and this provides a convenient storage for the elements m_{ij}. Store m_{ij} into the space originally used to store $a_{ij}, i > j$.

There is yet another reason for looking at the multipliers m_{ij} as the elements of a matrix. First, introduce the lower triangular matrix

$$L = \begin{bmatrix} 1 & 0 & 0 & \cdots & 0 \\ m_{21} & 1 & 0 & \cdots & 0 \\ \vdots & & \ddots & & \vdots \\ \vdots & & & \ddots & \\ m_{n1} & m_{n2} & \cdots & \cdots & 1 \end{bmatrix}$$

Theorem 8.1   If L and U are the lower and upper triangular matrices defined above using Gaussian elimination, then

$$A = LU \tag{8.5}$$

Proof   This proof is basically an algebraic manipulation, using the definitions (8.1) and (8.2). To visualize the matrix element $(LU)_{ij}$, use the vector multiplication

$$(LU)_{ij} = \left[m_{i1}, \ldots, m_{i,i-1}, 1, 0, \ldots, 0 \right] \begin{bmatrix} u_{1j} \\ \cdot \\ \cdot \\ \cdot \\ u_{jj} \\ 0 \\ \cdot \\ \cdot \\ \cdot \\ 0 \end{bmatrix}$$

For $i \leq j$,

$$(LU)_{ij} = m_{i1} u_{1j} + m_{i2} u_{2j} + \ldots + m_{i,i-1} u_{i-1,j} + u_{i,j}$$

$$= \sum_{k=1}^{i-1} m_{ik} a_{kj}^{(k)} + a_{ij}^{(i)}$$

$$= \sum_{k=1}^{i-1} \left[a_{ij}^{(k)} - a_{ij}^{(k+1)} \right] + a_{ij}^{(i)}$$

$$= a_{ij}^{(1)} = a_{ij}$$

For $i > j$,

$$(LU)_{ij} = m_{i1} u_{1j} + \ldots + m_{ij} u_{jj}$$

$$= \sum_{k=1}^{j-1} m_{ik} a_{kj}^{(k)} + m_{ij} a_{jj}^{(j)}$$

$$= \sum_{k=1}^{j-1} \left[a_{ij}^{(k)} - a_{ij}^{(k+1)} \right] + a_{ij}^{(j)}$$

$$= a_{ij}^{(1)} = a_{ij}.$$

This completes the proof. ■

The decomposition (8.5) is an important result, and extensive use is made of it in developing variants of Gaussian elimination for special classes of matrices. But for the moment, we give only the following corollary.

Corollary With the matrices A, L, and U as above in Theorem 8.1,

$$\text{Det}(A) = u_{11} u_{22} \ldots u_{nn}$$

$$= a_{11}^{(1)} a_{22}^{(2)} \ldots a_{nn}^{(n)}$$

Proof By the product rule for determinants,

$$\det(A) = \det(L)\det(U)$$

Since L and U are triangular, their determinants are the product of their diagonal elements. The desired result follows easily, since $\det(L) = 1$. ■

Example For the system (8.4) of the previous example,

$$L = \begin{bmatrix} 1 & 0 & 0 \\ 2 & 1 & 0 \\ -1 & \frac{1}{2} & 1 \end{bmatrix} \qquad U = \begin{bmatrix} 1 & 2 & 1 \\ 0 & -2 & 1 \\ 0 & 0 & \frac{1}{2} \end{bmatrix}$$

It is easily verified that $A = LU$. Also, $\det(A) = \det(U) = -1$.

Operation count To analyze the number of operations necessary to solve $Ax = b$ using Gaussian elimination, we consider separately the creation of L and U from A, the modification of b to g, and finally the solution of x.

1. Calculation of L and U. At step 1, $n-1$ divisions were used to calculate the multipliers m_i, $2 \le i \le n$. Then $(n-1)^2$ multiplications and $(n-1)^2$ additions were used to create the new elements $a_{ij}^{(2)}$. We can continue in this way for each step; and the results are summarized in Table 8.1. The total value for each column was obtained using the identities

$$\sum_{j=1}^{p} j = \frac{p(p+1)}{2} \qquad \sum_{j=1}^{p} j^2 = \frac{p(p+1)(2p+1)}{6} \qquad p \ge 1$$

Traditionally, it is the number of multiplications and divisions, counted together, that is used as the operation count for Gaussian elimination. This was based on the much greater speed of addition on early computers; but it is no longer true on most large modern computers, in which a multiplication is often only double the time of an addition. For a convenient notation, let $MD(\cdot)$ and $AS(\cdot)$ denote the number of multiplications and divisions, and the number of additions and subtractions, respectively, for the computation of the quantity in

Table 8.1 Operation count for LU decomposition of a matrix

Step k	Additions	Multiplications	Divisions
1	$(n-1)^2$	$(n-1)^2$	$n-1$
2	$(n-2)^2$	$(n-2)^2$	$n-2$
.	.	.	.
.	.	.	.
.	.	.	.
$n-1$	1	1	1
Total	$\dfrac{n(n-1)(2n-1)}{6}$	$\dfrac{n(n-1)(2n-1)}{6}$	$\dfrac{n(n-1)}{2}$

the parentheses. For the LU decomposition of A, we have

$$MD(LU) = \frac{n(n^2-1)}{3} \approx \frac{n^3}{3}$$

$$AS(LU) = \frac{n(n-1)(2n-1)}{6} \approx \frac{n^3}{3}$$

The final estimates are valid for larger values of n.

2. Modification of b to $g = b^{(n)}$.

$$MD(g) = (n-1) + (n-2) + \ldots + 1 = \frac{n(n-1)}{2}$$

$$AS(g) = \frac{n(n-1)}{2}$$

3. Solution of $Ux = g$

$$MD(x) = \frac{n(n+1)}{2} \qquad AS(x) = \frac{n(n-1)}{2}$$

4. Solution of $Ax = b$. Combine 1 through 3 to get

$$MD(LU,x) = \frac{n^3}{3} + n^2 - \frac{n}{3} \approx \tfrac{1}{3}n^3$$

$$AS(LU,x) = \frac{n(n-1)(2n+5)}{6} \approx \tfrac{1}{3}n^3 \tag{8.6}$$

The number of additions is always about the same as the number of multiplications and divisions, and thus from here on, we consider only the latter.

The first thing to note is that solving $Ax = b$ is comparatively cheap when compared to such a supposedly simple operation as multiplying two $n \times n$ matrices. The matrix multiplication requires n^3 operations, and the solution of $Ax = b$ requires only about $\frac{1}{3} n^3$ operations.

Second, the main cost of solving $Ax = b$ is in producing the decomposition $A = LU$. Once it has been found, only n^2 additional operations are necessary to solve $Ax = b$. After once solving $Ax = b$, it is comparatively cheap to solve additional systems with the same coefficient matrix, provided the LU decomposition has been saved.

Finally, Gaussian elimination is much cheaper than *Cramer's rule*, which uses determinants and is often taught in linear algebra courses (e.g., see Noble [N1, p. 209]). If the determinants in Cramer's rule are computed using *expansion by minors*, then the operation count is $(n+1)!$. For $n = 10$, Gaussian elimination uses 430 operations, and Cramer's rule uses 39,916,800 operations. This should emphasize the point that Cramer's rule is not a practical computational tool, and it should be considered as just a theoretical mathematics tool.

5. Inversion of A. The inverse A^{-1} is generally not needed, but it can be produced by using Gaussian elimination. Finding A^{-1} is equivalent to solving the equation $AX = I$, with X an $n \times n$ unknown matrix. If we write X and I in terms of their columns,

$$X = \left[x^{(1)}, \ldots, x^{(n)} \right] \qquad I = \left[e^{(1)}, \ldots, e^{(n)} \right]$$

then solving $AX = I$ is equivalent to solving the n systems

$$Ax^{(1)} = e^{(1)}, \ldots, Ax^{(n)} = e^{(n)}$$

all having the same coefficient matrix A. Using 1 to 3 from above,

$$MD(A^{-1}) = \tfrac{4}{3} n^3 - \frac{n}{3} \approx \tfrac{4}{3} n^3$$

Calculating A^{-1} is four times the expense of solving $Ax = b$ for a single vector b, not n times the work as one might first imagine. By careful attention to the details of the inversion process, taking advantage of the special form of the right-hand vectors $e^{(1)}, \ldots, e^{(n)}$, it is possible to further reduce the operation count to exactly

$$MD(A^{-1}) = n^3$$

However, it is still wasteful in most situations to produce A^{-1} to solve $Ax = b$. And there is no advantage in saving A^{-1} rather than the LU decomposition to solve future systems $Ax = b$. In both cases, the number of multiplications and divisions necessary to solve $Ax = b$ is exactly n^2.

8.2 Pivoting and Scaling in Gaussian Elimination

At each stage of the elimination process in the last section, we assumed the appropriate pivot element $a_{kk}^{(k)} \neq 0$. To remove this assumption, begin each step of the elimination process by switching rows to put a nonzero element in the pivot position. If none such exists, then the matrix must be singular, contrary to assumption.

But it is not enough to just ask that the pivot element be nonzero. Often an element would be zero except for rounding errors that have occurred in calculating it. Using such an element as the pivot element will result in gross errors in the further calculations in the matrix. To guard against this, and for other reasons involving the propagation of rounding errors, we introduce *partial pivoting* and *complete pivoting*.

Definition **1.** Partial pivoting. For $1 \le k \le n-1$, in the Gaussian elimination process at stage k, let

$$c_k = \operatorname*{Max}_{k \le i \le n} |a_{ik}^{(k)}| \tag{8.7}$$

Let i be the smallest row index, $i \ge k$, for which the maximum c_k is attained. If $i > k$, then switch rows k and i in A and b; and proceed with step k of the elimination process. All of the multipliers will now satisfy

$$|m_{ik}| \le 1 \qquad i = k+1, \ldots, n \tag{8.8}$$

This aids in preventing the growth of elements in $A^{(k)}$ of greatly varying size, and thus lessens the possibility for large loss of significance errors.

2. Complete pivoting. Define

$$c_k = \operatorname*{Max}_{k \le i,j \le n} |a_{ij}^{(k)}|$$

Switch rows of A and b and columns of A to bring to the pivot position an element giving the maximum c_k. Note that with a column switch, the order of the unknowns is changed. At the completion of the elimination process, this must be reversed.

Complete pivoting has been shown to cause the roundoff error in Gaussian elimination to propagate at a reasonably slow speed, compared with what can happen when no pivoting is used. The theoretical results on the use of partial pivoting are not quite as good; but in almost all practical problems, the error behavior is essentially the same as that for complete pivoting. Comparing operation times, complete pivoting is the more expensive strategy; and thus, partial pivoting is used in most practical algorithms. Henceforth we will always mean *partial pivoting* when we use the word *pivoting*. The entire question of roundoff error propagation in Gaussian elimination has been analyzed very thoroughly by J. H. Wilkinson (e.g., see [W2, pp. 209-220]), and some of his results are presented in Section 8.4.

Example We illustrate the effect of using pivoting by solving the system

$$.729x + .81y + .9z = .6867$$

$$x + y + z = .8338$$

$$1.331x + 1.21y + 1.1z = 1.000 \tag{8.9}$$

The exact solution, rounded to four significant digits, is

$$x = .2245 \qquad y = .2814 \qquad z = .3279 \tag{8.10}$$

Floating-point decimal arithmetic, with four digits in the mantissa, is used to solve the linear system. We use this arithmetic to show the effect of working with only a finite number of digits, while keeping the presentation manageable in size. The augmented matrix notation is used to represent the system (8.9), just as was done with the earlier example (8.4) in Section 8.1.

1. Solution without pivoting.

$$\begin{bmatrix} .7290 & .8100 & .9000 & | & .6867 \\ 1.000 & 1.000 & 1.000 & | & .8338 \\ 1.331 & 1.210 & 1.100 & | & 1.000 \end{bmatrix}$$

$$\downarrow \quad m_{21} = 1.372$$
$$\quad m_{31} = 1.826$$

$$\begin{bmatrix} .7290 & .8100 & .9000 & | & .6867 \\ 0.0 & -.1110 & -.2350 & | & -.1084 \\ 0.0 & -.2690 & -.5430 & | & -.2540 \end{bmatrix}$$

$$\downarrow m_{32} = 2.423$$

$$\begin{bmatrix} .7290 & .8100 & .9000 & | & .6867 \\ 0.0 & -.1110 & -.2350 & | & -.1084 \\ 0.0 & 0.0 & .02640 & | & .008700 \end{bmatrix}$$

The solution is

$$x = .2252 \qquad y = .2790 \qquad z = .3295 \tag{8.11}$$

2. Solution with pivoting. To indicate the interchange of rows i and j, we will use the notation $r_i \leftrightarrow r_j$.

$$\begin{bmatrix} .7290 & .8100 & .9000 & | & .6867 \\ 1.000 & 1.000 & 1.000 & | & .8338 \\ 1.331 & 1.210 & 1.100 & | & 1.000 \end{bmatrix}$$

$$r_1 \leftrightarrow r_3 \downarrow \begin{matrix} m_{21} = .7513 \\ m_{31} = .5477 \end{matrix}$$

$$\begin{bmatrix} 1.331 & 1.210 & 1.100 & | & 1.000 \\ 0.0 & .09090 & .1736 & | & .08250 \\ 0.0 & .1473 & .2975 & | & .1390 \end{bmatrix}$$

$$r_2 \leftrightarrow r_3 \downarrow m_{32} = .6171$$

$$\begin{bmatrix} 1.331 & 1.210 & 1.100 & | & 1.000 \\ 0.0 & .1473 & .2975 & | & .1390 \\ 0.0 & 0.0 & -.01000 & | & -.003280 \end{bmatrix}$$

The solution is

$$x = .2246 \qquad y = .2812 \qquad z = .3280 \tag{8.12}$$

The error in (8.11) is from 7 to 16 times larger than it is for (8.12), depending on the component of the solution being considered. The results in (8.12) have one more significant digit than do those of (8.11). This illustrates the positive effect that the use of pivoting can have on the error behavior for Gaussian elimination.

Pivoting changes the factorization result (8.5) given in Theorem 8.1. The result is still true, but in a modified form. If the row interchanges induced by pivoting are carried out on A before beginning elimination, then pivoting will be unnecessary. Row interchanges on A can be represented by premultiplication of A by an appropriate *permutation matrix* P, to get PA. Then Gaussian elimination on PA leads to

$$LU = PA \tag{8.13}$$

where U is the upper triangular matrix obtained in the elimination process with pivoting. The lower triangular matrix L can be constructed using the multipliers from Gaussian elimination with pivoting. We omit the details, as the actual construction is unimportant.

Example From the preceding example with pivoting, we form

$$L = \begin{bmatrix} 1.000 & 0.0 & 0.0 \\ .5477 & 1.000 & 0.0 \\ .7513 & .6171 & 1.000 \end{bmatrix} \qquad U = \begin{bmatrix} 1.331 & 1.210 & 1.100 \\ 0.0 & .1473 & .2975 \\ 0.0 & 0.0 & -.01000 \end{bmatrix}$$

When multiplied,

$$LU = \begin{bmatrix} 1.331 & 1.210 & 1.100 \\ .7290 & .8100 & .9000 \\ 1.000 & 1.000 & 1.000 \end{bmatrix} = PA$$

The result PA is the matrix A with first, rows 1 and 3 interchanged, and then rows 2 and 3 interchanged. This illustrates (8.13).

Scaling   It has been observed empirically that if the elements of the coefficient matrix A vary greatly in size, then it is likely that loss of significance errors will be introduced and that the rounding error propagation will be worse. To avoid this problem the matrix A should be *scaled* so that the elements vary less, usually by multiplying the rows and columns by suitable constants. The subject of scaling is not well understood currently, especially how to guarantee that loss of significance errors and rounding errors will be made smaller by scaling, and possibly even minimized. Computational experience has suggested that all rows should be scaled to make them approximately equal in magnitude. And usually, all components of the unknown x should be scaled to yield unknowns of about equal size. The latter is equivalent to column scaling of A; but generally, there is no known a priori strategy for picking these scaling factors, based solely on knowledge of A and b.

If we let B denote the result of row and column scaling in A, then

$$B = D_1 A D_2$$

with D_1 and D_2 diagonal matrices, with entries the scaling constants. To solve $Ax = b$, observe that

$$D_1 A D_2 (D_2^{-1} x) = D_1 b$$

Thus we solve for x by solving

$$Bz = D_1 b \qquad x = D_2 z$$

The remaining discussion will be restricted to row scaling, since the use of it is fairly widely agreed on.

Usually we attempt to choose the coefficients so as to have

$$\underset{1 \leq j \leq n}{\text{Max}} |b_{ij}| \approx 1 \qquad i = 1, \dots, n \tag{8.15}$$

where $B = [b_{ij}]$ is the result of scaling A. The most straightforward approach is to define

$$s_i = \underset{1 \leq j \leq n}{\text{Max}} |a_{ij}| \qquad i = 1, \dots, n$$

$$b_{ij} = a_{ij}/s_i \qquad i, j = 1, \dots, n \tag{8.16}$$

But because this introduces an additional rounding error into each element of the coefficient matrix, two other techniques are more widely used.

   1.   Scaling using computer number base. Let β denote the base used in the computer arithmetic, for example, $\beta = 2$ on binary machines. Let r_i be the smallest integer for which $\beta^{r_i} \geq s_i$. Define the scaled matrix B by

$$b_{ij} = a_{ij} / \beta^{r_i} \qquad i, j = 1, \ldots, n \tag{8.17}$$

No rounding is involved in defining b_{ij}, only a change in the exponent in the floating-point form for a_{ij}. The values of B satisfy

$$\beta^{-1} < \operatorname*{Max}_{1 \leq j \leq n} |b_{ij}| \leq 1$$

and thus (8.15) is satisfied fairly well.

   2.   Implicit scaling. The use of scaling generally changes the choice of pivot elements when pivoting is used with Gaussian elimination. And it is only with such a change of pivot elements that the results in Gaussian elimination will be changed. This is due to a result of F. Bauer (see [F1, p. 38]) who proved that if the scaling (8.17) is used, and if the choice of pivot elements is forced to remain the same as when solving $Ax = b$, then the solution of (8.14) yields exactly the same computed value for x. Thus the only significance of scaling is in the choice of the pivot elements.

   For implicit scaling, we continue to use the matrix A. But we choose the pivot element in step k of the Gaussian elimination algorithm by defining

$$c_k = \operatorname*{Max}_{k \leq i \leq n} \frac{|a_{ik}^{(k)}|}{s_i} \tag{8.18}$$

replacing the definition (8.7) used in defining partial pivoting. Choose the smallest index $i \geq k$ that yields c_k in (8.18); and if this $i \neq k$, then interchange rows i and k. Then proceed with the elimination algorithm of Section 8.1. This form of scaling seems to be the form most commonly used in current published algorithms.

An algorithm for Gaussian elimination   We first give an algorithm, called *Factor*, for the triangular factorization of a matrix A. It uses Gaussian elimination with partial pivoting, combined with the implicit scaling of (8.16) and (8.18). We then give a second algorithm, called *Solve*, for using the results of Factor to solve a linear system $Ax = b$. The reason for separating the elimination procedure into these two steps is that we will often want to solve several systems $Ax = b$, with the same A, but different values for b.

Algorithm   Factor $(A, n, \text{Pivot}, \det, \text{ier})$

1. Remarks: A is an $n \times n$ matrix, to be factored using the LU decomposition. Gaussian elimination is used, with partial pivoting and implicit scaling in the rows. Upon completion of the algorithm, the upper triangular matrix U will be stored in the upper triangular part of A; and the multipliers of (8.1), which make up L, below its diagonal, will be stored in the corresponding positions of A. The vector *Pivot* will contain a record of all row interchanges. If Pivot $(k) = k$, then no interchange was used in step k of the elimination process. But if Pivot $(k) = i \neq k$, then rows i and k were interchanged in step k of the elimination process. The variable *det* will contain $\det(A)$ on exit.

    The variable *ier* is an error indicator. If ier $= 0$, then the routine was completed satisfactorily. But for ier $= 1$, the matrix A was singular, in the sense that all possible pivot elements were zero at some step of the elimination process. In this case, all computation ceased, and the routine was exited. No attempt is made to check on the accuracy of the computed decomposition of A, and it can be nearly singular without being detected.

2. $\det := 1$

3. $s_i := \underset{1 \leq j \leq n}{\text{Max}} |a_{ij}|, \ i = 1, \ldots, n$

4. Do through step 16 for $k = 1, \ldots, n-1$.

5. $c_k := \underset{k \leq i \leq n}{\text{Max}} \left| \dfrac{a_{ik}}{s_i} \right|$

6. Let i_0 be the smallest index $i \geq k$ for which the maximum in step 5 is attained. Pivot $(k) := i_0$.

7. If $c_k = 0$, then ier $:= 1, \det := 0$, and exit from the algorithm.

8. If $i_0 = k$, then go to step 11.

9. $\det := -\det$

10. Interchange a_{kj} and $a_{i_0 j}, j = k, \ldots, n$.

11. Do through step 14 for $i = k+1, \ldots, n$.

12. $a_{ik} := m_i := a_{ik}/a_{kk}$

13. $a_{ij} := a_{ij} - m_i a_{kj}, j = k+1, \ldots, n$

14. End loop on i.

15. $\det := a_{kk} \det$

16.   End loop on k.

17.   ier: $=0$ and exit the algorithm.

Algorithm   Solve (A, n, b, Pivot)

1.   Remarks: This algorithm solves the linear system $Ax = b$. It is assumed that the original matrix A has been factored using the algorithm *Factor*, with the row interchanges recorded in Pivot. The solution will be stored in b on exit. The matrix A and vector Pivot are left unchanged.

2.   Do through step 5 for $k = 1, 2, \ldots, n-1$.

3.   If $j := \text{pivot}(k) \neq k$, then interchange b_j and b_k.

4. $b_i := b_i - a_{ik} b_k, i = k+1, \ldots, n$.

5.   End loop on k.

6. $b_n := b_n / a_{nn}$

7.   Do through step 9 for $i = n-1, \ldots, 1$.

8. $b_i := \dfrac{1}{a_{ii}} \left(b_i - \displaystyle\sum_{j=i+1}^{n} a_{ij} b_j \right)$.

9.   End loop on i.

10.   Exit from algorithm.

The earlier example (8.9) serves as an illustration. The use of implicit scaling in this case does not require a change in the choice of pivot elements with partial pivoting. Algorithms like Factor and Solve are given in a number of references. See Forsythe and Moler [F1, Chapter 16 and 17] for improved versions of these algorithms. In particular, they do not actually interchange rows in pivoting but do so implicitly; and this saves some operation time.

8.3   Variants of Gaussian Elimination

There are many variants of Gaussian elimination. Some are modifications or simplifications, based on the special properties of some class of matrices, for example, symmetric, positive definite matrices. Other variants are ways to rewrite Gaussian elimination in a more compact form, sometimes in order to use special techniques to reduce the error. We consider only a few such variants and later refer to others.

Gauss-Jordan method   This procedure is much the same as regular elimination, including the possible use of pivoting and scaling. It differs in eliminating the unknown in equations above the diagonal as well as below it. In step k of the elimination, choose the pivot element as before. Then define

$$a_{kj}^{(k+1)} = a_{kj}^{(k)} / a_{kk}^{(k)} \qquad j=k,\dots,n$$

$$b_k^{(k+1)} = b_k^{(k)} / a_{kk}^{(k)}$$

Eliminate the unknown x_k in equations both above and below equation k. Define

$$a_{ij}^{(k+1)} = a_{ij}^{(k)} - a_{ik}^{(k)} a_{kj}^{(k+1)}$$

$$b_i^{(k+1)} = b_i^{(k)} - a_{ik}^{(k)} b_k^{(k+1)}$$

for $j=k,\dots,n$, $i=1,\dots,n$, $i \neq k$.

This procedure converts $[A|b]$ to $[I|b^{(n)}]$, so that at the completion of the above elimination, $x=b^{(n)}$. To solve $Ax=b$ by this technique requires

$$\frac{n(n+1)^2}{2} \approx \frac{n^3}{2} \tag{8.19}$$

multiplications and divisions. This is 50 percent more than the regular elimination method; and consequently, the Gauss-Jordan method should not be used for solving linear systems. However, it can be used to produce a matrix inversion program that uses a minimum of storage. By taking special advantage of the special structure of the right side in $AX=I$, the Gauss-Jordan method can produce the solution $X=A^{-1}$ using only $n(n+1)$ storage locations, rather than the normal $2n^2$ storage locations. And pivoting and implicit scaling can still be used.

Compact methods   It is possible to move directly from a matrix A to its LU decomposition, and this can be combined with partial pivoting and scaling. If we disregard the possibility of pivoting for the moment, then the result

$$A = LU \tag{8.20}$$

leads directly to a set of recursive formulas for the elements of L and U.

There is some nonuniqueness in the choice of L and U, if we insist only that L and U be lower and upper triangular, respectively. If A is nonsingular, and if we have two decompositions

$$A = L_1 U_1 = L_2 U_2$$

then

$$L_2^{-1} L_1 = U_2 U_1^{-1} \tag{8.21}$$

The inverse and the products of lower triangular matrices are again lower triangular, and similarly for upper triangular matrices. The left and right sides of (8.21) are lower and upper triangular, respectively. Thus they must equal a diagonal matrix, call it D, and

$$L_1 = L_2 D \qquad U_1 = D^{-1} U_2$$

The choice of D is tied directly to the choice of the diagonal elements of either L or U; and once they have been chosen, D is uniquely determined.

If the diagonal elements of L are all required to equal 1, then the resulting decomposition (8.20) is that given by Gaussian elimination, as in Section 8.1. The associated compact method gives explicit formulas for l_{ij} and u_{ij}, and it is known as *Doolittle's method*. If we choose to have the diagonal elements of U all equal 1, the associated compact method for calculating (8.20) is called *Crout's method*. There is only a multiplying diagonal matrix to distinguish it from Doolittle's method. For an excellent algorithm using Crout's algorithm for the factorization (8.20), with partial pivoting and implicit scaling, see the program *unsymdet* in Wilkinson and Reinsch [W3, pp. 93–110].

The principal advantage of the compact formula is that the elements l_{ij} and u_{ij} all involve inner products, as illustrated below in formulas (8.23) to (8.24) for the factorization of a symmetric matrix. These inner products can be accumulated using double-precision arithmetic, possibly including a concluding division, and then be rounded to single precision. This limited use of double precision can greatly increase the accuracy of the factors L and U; and it is not possible to do this with the regular elimination method, unless all operations and storage are done in double precision. For a complete discussion of these compact methods, see Wilkinson [W2, pp. 221–228].

The Cholesky method   Let A be a symmetric and positive definite matrix of order n. The matrix A is positive definite if

$$(Ax, x) = \sum_{i=1}^{n} \sum_{j=1}^{n} a_{ij} x_i x_j > 0$$

for all $x \in R^n$, $x \neq 0$. Some of the properties of positive definite matrices are given in Problem 11 of Chapter 7 and Problems 7 and 8 of this chapter. Symmetric positive definite matrices occur in a wide variety of applications.

For such a matrix A, there is a very convenient factorization, and it can be carried through without any need for pivoting or scaling. This is called *Cholesky's method*, and it states that we can find a lower triangular matrix L such that

$$A = LL^T \tag{8.22}$$

The method requires only $\frac{1}{2} n(n+1)$ storage locations for L, rather than the usual n^2 locations; and the number of operations is about $\frac{1}{6}n^3$, rather than the number $\frac{1}{3}n^3$, required for the usual decomposition.

Let $L = [l_{ij}]$, with $l_{ij} = 0$ for $j > i$. Begin to construct L by multiplying the first row of L times the first column of L^T to get

$$l_{11}^2 = a_{11}$$

Because A is positive definite, $a_{11} > 0$, and $l_{11} = \sqrt{a_{11}}$. Multiply the second row of L times the first two columns of L^T to get

$$l_{21}l_{11} = a_{21} \qquad l_{21}^2 + l_{22}^2 = a_{22}$$

Again, we can solve for the unknowns l_{21} and l_{22}. In general for $i = 1, 2, \ldots, n$,

$$l_{ij} = \frac{a_{ij} - \sum\limits_{k=1}^{j-1} l_{ik} l_{jk}}{l_{jj}} \qquad j = 1, \ldots, i-1 \tag{8.23}$$

$$l_{ii} = \left[a_{ii} - \sum_{k=1}^{i-1} l_{ik}^2 \right]^{1/2} \tag{8.24}$$

The argument in the square root of (8.24) can be shown to be positive by using the positive definiteness of A. For a complete analysis of the method, see Wilkinson [W2, pp. 229–232]; and for programs, see Wilkinson and Reinsch [W3, pp. 10–30].

Note the inner products in (8.23) and (8.24). These can be accumulated in double precision, minimizing the number of rounding errors; and the elements l_{ij} will be in error by much less than if they had been calculated using only single-precision arithmetic. Also note that the elements of L remain bounded relative to A, since (8.24) yields a bound for the elements of row i, using

$$l_{i1}^2 + \ldots + l_{ii}^2 = a_{ii}$$

The square roots in (8.24) of Cholesky's method can be avoided by using a slight modification of (8.22). Find a diagonal matrix D and a lower triangular matrix $\tilde{L}$, with one's on the diagonal, such that

$$A = \tilde{L}D\tilde{L}^T \tag{8.25}$$

This factorization can be done with about the same number of operations as Cholesky's method, about $\frac{1}{6}n^3$, with no square roots. For further discussion and a program, see Wilkinson and Reinsch [W3, pp. 10–30].

Example Consider the Hilbert matrix of order 3,

$$A = \begin{bmatrix} 1 & \frac{1}{2} & \frac{1}{3} \\ \frac{1}{2} & \frac{1}{3} & \frac{1}{4} \\ \frac{1}{3} & \frac{1}{4} & \frac{1}{5} \end{bmatrix} \tag{8.26}$$

For the Cholesky decomposition,

$$L = \begin{bmatrix} 1 & 0 & 0 \\ \frac{1}{2} & \frac{1}{2\sqrt{3}} & 0 \\ \frac{1}{3} & \frac{1}{2\sqrt{3}} & \frac{1}{6\sqrt{5}} \end{bmatrix}$$

and for (8.25),

$$\tilde{L} = \begin{bmatrix} 1 & 0 & 0 \\ \frac{1}{2} & 1 & 0 \\ \frac{1}{3} & 1 & 1 \end{bmatrix} \qquad D = \begin{bmatrix} 1 & 0 & 0 \\ 0 & \frac{1}{12} & 0 \\ 0 & 0 & \frac{1}{180} \end{bmatrix}$$

Tridiagonal matrices The matrix $A = [a_{ij}]$ is tridiagonal if

$$a_{i,j} = 0 \qquad \text{for } |i-j| > 1$$

This gives the form

$$A = \begin{bmatrix} a_1 & c_1 & 0 & 0 & \cdots & \cdots & 0 \\ b_2 & a_2 & c_2 & 0 & & & \vdots \\ 0 & b_3 & a_3 & c_3 & & & \vdots \\ \vdots & & & \ddots & & & \vdots \\ \vdots & & & & \ddots & & \vdots \\ \vdots & & & & b_{n-1} & a_{n-1} & c_{n-1} \\ 0 & \cdots & \cdots & \cdots & 0 & b_n & a_n \end{bmatrix} \tag{8.27}$$

Tridiagonal matrices occur in a variety of applications. Recall the linear system (3.80) for spline functions in Section 3.7 of Chapter 3. And many numerical methods for solving boundary value problems for ordinary and partial differential equations involve the solution of tridiagonal systems. Virtually all of these applications yield tridiagonal matrices for which the LU factorization can be formed without pivoting, and for which there is no large increase in error as a consequence. The precise assumptions on A are given below in Theorem 8.2.

By considering the factorization $A = LU$ without pivoting, we find that most elements of L and U will be zero. And we are led to the following general formula for the decomposition,

$$A = LU = \begin{bmatrix} \alpha_1 & 0 & \cdots & & & \cdots & 0 \\ b_2 & \alpha_2 & 0 & & & & \vdots \\ 0 & b_3 & \alpha_3 & 0 & & & \vdots \\ \vdots & & & \ddots & & & \vdots \\ \vdots & & & & \ddots & & \vdots \\ \vdots & & & & & \ddots & \vdots \\ 0 & \cdots & & & \cdots & b_n & \alpha_n \end{bmatrix} \begin{bmatrix} 1 & \gamma_1 & 0 & \cdots & & \cdots & 0 \\ 0 & 1 & \gamma_2 & 0 & & \cdots & \vdots \\ \vdots & & \ddots & & & & \vdots \\ \vdots & & & \ddots & & & \vdots \\ \vdots & & & & \ddots & & \vdots \\ \vdots & & & & & 1 & \gamma_{n-1} \\ 0 & \cdots & & & \cdots & 0 & 1 \end{bmatrix}$$

We can multiply to obtain a way to compute $\{\alpha_i\}$ and $\{\gamma_i\}$ recursively.

$$a_1 = \alpha_1 \qquad \alpha_1 \gamma_1 = c_1$$

$$a_i = \alpha_i + b_i \gamma_{i-1} \qquad i = 2, \ldots, n$$

$$\alpha_i \gamma_i = c_i \qquad i = 2, 3, \ldots, n-1$$

These can be solved to give

$$\alpha_1 = a_1 \qquad \gamma_1 = c_1 / \alpha_1$$

$$\alpha_i = a_i - b_i \gamma_{i-1}, \quad \gamma_i = c_i / \alpha_i \qquad i = 2, 3, \ldots, n-1 \tag{8.28}$$

$$\alpha_n = a_n - b_n \gamma_{n-1}$$

To solve $LUx = f$, let $Ux = z$ and $Lz = f$. Then

$$z_1 = f_1 / \alpha_1 \qquad z_i = (f_i - b_i z_{i-1}) / \alpha_i \qquad i = 2, 3, \ldots, n$$

$$x_n = z_n \qquad x_i = z_i - \gamma_i x_{i+1} \qquad i = n-1, n-2, \ldots, 1 \tag{8.29}$$

The constants in (8.28) can be stored for later use in solving the linear system $Ax = f$, for as many right sides f as desired.

The number of operations to calculate L and U is $2n-2$, and to solve $Ax=f$ takes an additional $3n-2$ operations. Thus we need only $5n-4$ operations to solve $Ax=f$ the first time; and for each additional right side, with the same A, we need only $3n-2$ operations. This is extremely rapid. To illustrate this, note that A^{-1} generally has only nonzero entries; and thus the calculation of $x=A^{-1}f$ will require n^2 operations. In many applications, n may be larger than 1000, and thus there is a significant savings compared with other methods of solution.

To justify the above decomposition of A, especially to show that all the coefficients $\alpha_i \neq 0$, we have the following theorem.

Theorem 8.2 Assume the coefficients $\{a_i, b_i, c_i\}$ of (8.27) satisfy the following conditions:

1. $|a_1| > |c_1| > 0$

2. $|a_i| \geqslant |b_i| + |c_i| \qquad b_i, c_i \neq 0 \qquad i=2,\ldots,n-1$

3. $|a_n| > |b_n| > 0$.

Then A is nonsingular,

$$|\gamma_i| < 1 \qquad i=1,\ldots,n-1$$

$$|a_i| - |b_i| < |\alpha_i| < |a_i| + |b_i| \qquad i=2,\ldots,n$$

Proof See Isaacson and Keller [I1, p. 57] for the details. Note that the last bound shows $|\alpha_i| > |c_i|$, $i=2,\ldots,n-2$. Thus the coefficients of L and U remain bounded, and no divisions are used that are almost zero except for rounding error.

The condition that $b_i, c_i \neq 0$ is not essential. For example if some $b_i = 0$, then the linear system can be broken into two new systems, one of order $i-1$ and the other of order $n-i+1$. For example, if

$$A = \begin{bmatrix} a_1 & c_1 & 0 & 0 \\ b_2 & a_2 & c_2 & 0 \\ 0 & 0 & a_3 & c_3 \\ 0 & 0 & b_4 & a_4 \end{bmatrix}$$

then solve $Ax=f$ by reducing it to the following two linear systems,

$$\begin{bmatrix} a_3 & c_3 \\ b_4 & a_4 \end{bmatrix} \begin{bmatrix} x_3 \\ x_4 \end{bmatrix} = \begin{bmatrix} f_3 \\ f_4 \end{bmatrix} \qquad \begin{bmatrix} a_1 & c_1 \\ b_2 & a_2 \end{bmatrix} \begin{bmatrix} x_1 \\ x_2 \end{bmatrix} = \begin{bmatrix} f_1 \\ f_2 - c_2 x_3 \end{bmatrix}$$

This completes the proof. ∎

Example Consider the coefficient matrix for spline interpolation, following (3.80) in Chapter 3. Consider $h_i=$ constant in that matrix, and then factor $h/6$ from every row. Restricting our interest to the matrix of order 4, the resulting matrix is

$$A \equiv \begin{bmatrix} 2 & 1 & 0 & 0 \\ 1 & 4 & 1 & 0 \\ 0 & 1 & 4 & 1 \\ 0 & 0 & 1 & 2 \end{bmatrix}$$

Using the method (8.28) to (8.29), this has the LU factorization

$$L = \begin{bmatrix} 2 & 0 & 0 & 0 \\ 1 & \frac{7}{2} & 0 & 0 \\ 0 & 1 & \frac{26}{7} & 0 \\ 0 & 0 & 1 & \frac{45}{26} \end{bmatrix} \quad U = \begin{bmatrix} 1 & \frac{1}{2} & 0 & 0 \\ 0 & 1 & \frac{2}{9} & 0 \\ 0 & 0 & 1 & \frac{9}{38} \\ 0 & 0 & 0 & 1 \end{bmatrix}$$

This completes the example. And it should indicate that the solution of the cubic spline interpolation problem, described in Section 3.7 of Chapter 3, is not difficult to compute.

8.4   Error Analysis

We begin the error analysis of methods for solving $Ax=b$ by examining the stability of the solution x relative to small perturbations in the right side b. We will follow the general schemata of Section 1.5 of Chapter 1, and in particular, we study the condition number of (1.62).

Let $Ax=b$, of order n, be uniquely solvable; and consider the solution of the perturbed problem

$$A\tilde{x} = b + r$$

Let $e = \tilde{x} - x$, and subtract $Ax = b$ to get

$$Ae = r \qquad e = A^{-1}r \tag{8.30}$$

To examine the stability of $Ax=b$ as in (1.62), we want to bound the quantity

$$\frac{\|e\|}{\|x\|} \div \frac{\|r\|}{\|b\|} \tag{8.31}$$

as r ranges over all elements of R^n that are small relative to b.

From (8.30), take norms to obtain

$$\|r\| \le \|A\| \|e\| \qquad \|e\| \le \|A^{-1}\| \|r\|$$

Divide by $\|A\| \|x\|$ in the first inequality and by $\|x\|$ in the second one to obtain

$$\frac{\|r\|}{\|A\| \|x\|} \le \frac{\|e\|}{\|x\|} \le \frac{\|A^{-1}\| \|r\|}{\|x\|}$$

The matrix norm is the operator matrix norm, induced by the vector norm. Using the bounds

$$\|b\| \le \|A\| \|x\| \qquad \|x\| \le \|A^{-1}\| \|b\|$$

we obtain

$$\frac{1}{\|A\| \|A^{-1}\|} \cdot \frac{\|r\|}{\|b\|} \le \frac{\|e\|}{\|x\|} \le \|A\| \|A^{-1}\| \frac{\|r\|}{\|b\|} \tag{8.32}$$

Recalling (8.31), this result is justification for introducing the *condition number* of A,

$$\text{cond}(A) = \|A\| \|A^{-1}\| \tag{8.33}$$

The quantity (8.33) varies with the norm being used, but it is always bounded below by one, since

$$1 = \|I\| = \|AA^{-1}\| \le \|A\| \|A^{-1}\| = \text{cond}(A)$$

If the condition number is nearly 1, then we see from (8.32) that small relative perturbations in b will lead to similarly small relative perturbations in the solution x. But if cond(A) is large, then (8.33) suggests that there may be small relative perturbations of b that will lead to large relative perturbations in x.

Because (8.33) varies with the choice of norm, we sometimes use another condition number, independent of the norm. From Theorem 7.8 of Chapter 7,

$$\text{cond}(A) \ge r_\sigma(A) r_\sigma(A^{-1})$$

Since the eigenvalues of A^{-1} are the reciprocals of those of A, we have the result

$$\text{cond}(A) \ge \frac{\displaystyle\operatorname*{Max}_{\lambda \in \sigma(A)} |\lambda|}{\displaystyle\operatorname*{Min}_{\lambda \in \sigma(A)} |\lambda|} \equiv \text{cond}(A)_* \tag{8.34}$$

in which $\sigma(A)$ denotes the set of all eigenvalues of A.

Example   Consider the linear system

$$7x_1 + 10x_2 = b_1$$

$$5x_1 + 7x_2 = b_2$$

(8.35)

For the coefficient matrix,

$$A = \begin{bmatrix} 7 & 10 \\ 5 & 7 \end{bmatrix} \qquad A^{-1} = \begin{bmatrix} -7 & 10 \\ 5 & -7 \end{bmatrix}$$

Let the condition number in (8.33) be denoted by cond$(A)_p$ when it is generated using the matrix norm $\|\cdot\|_p$. For this example,

$$\text{cond}(A)_1 = \text{cond}(A)_\infty = (17)(17) = 289$$

$$\text{cond}(A)_2 \doteq 223, \quad \text{cond}(A)_* \doteq 198$$

These condition numbers all suggest that (8.35) may be sensitive to changes in the right side b. To illustrate this, consider the particular case

$$7x_1 + 10x_2 = 1$$

$$5x_1 + 7x_2 = .7$$

which has the solution

$$x_1 = 0 \qquad x_2 = .1$$

For the perturbed system, solve

$$7\tilde{x}_1 + 10\tilde{x}_2 = 1.01$$

$$5\tilde{x}_1 + 7\tilde{x}_2 = .69$$

It has the solution

$$\tilde{x}_1 = -.17 \qquad \tilde{x}_2 = .22$$

The relative changes in x are quite large when compared with the size of the relative changes in the right side b.

A linear system whose solution x is unstable with respect to small relative changes in the right side b is called ill-conditioned. The above system (8.35) is somewhat ill-conditioned, especially if only three or four decimal-digit floating-point arithmetic is used in solving it. The condition numbers cond(A) and cond$(A)_*$ are fairly good indicators of ill-conditioning. As they increase by a factor of 10, it is quite possible that one less digit of accuracy will be attainable in the solution.

In general if $\text{cond}(A)_*$ is large, then there will be values of b for which the system $Ax=b$ is quite sensitive to changes r in b. Let λ_l and λ_m denote eigenvalues of A for which

$$\lambda_l = \underset{\lambda\in\sigma(A)}{\text{Min}} |\lambda| \qquad \lambda_m = \underset{\lambda\in\sigma(A)}{\text{Max}} |\lambda|$$

and thus

$$\text{cond}(A)_* = \lambda_m/\lambda_l \qquad\qquad (8.36)$$

Let x_l and x_m be corresponding eigenvectors, with $\|x_l\|_\infty = \|x_m\|_\infty = 1$. Then

$$Ax = \lambda_m x_m$$

has the solution $x = x_m$. And the system

$$A\tilde{x} = \lambda_m x_m + \lambda_l x_l = \lambda_m\left[x_m + \frac{1}{\text{cond}(A)_*} x_l\right]$$

has the solution

$$\tilde{x} = x_m + x_l$$

If $\text{cond}(A)_*$ is large, then the right-hand side has only a small relative perturbation,

$$\frac{\|r\|_\infty}{\|b\|_\infty} = \frac{1}{\text{cond}(A)_*}$$

But for the solution, we have the much larger relative perturbation

$$\frac{\|\tilde{x}-x\|_\infty}{\|x\|_\infty} = \frac{\|x_l\|_\infty}{\|x_m\|_\infty} = 1$$

There are systems that are not ill-conditioned in actual practice, but for which the above condition numbers are quite large. For example,

$$A = \begin{bmatrix} 1 & 0 \\ 0 & 10^{-10} \end{bmatrix}$$

has all condition numbers $\text{cond}(A)_p$ and $\text{cond}(A)_*$ equal to 10^{10}. But the matrix is not ill-conditioned in any way. The difficulty is in using norms to measure changes in a vector, rather than looking at each component separately. If scaling has been carried out on the coefficient matrix and unknown vector, then this problem does not usually arise; and then the condition numbers are an accurate predictor of ill-conditioning.

As a final justification for the use of (8.33) as a condition number, we give the following theorem.

Theorem 8.3   (Gastinel)   Let A be any nonsingular matrix of order n, and let $\|\cdot\|$ denote an operator matrix norm. Then

$$\frac{1}{\text{cond}(A)} = \text{Minimum}\left\{\frac{\|A - B\|}{\|A\|} \,\middle|\, B \text{ is a singular matrix}\right\}$$

Proof   See Kahan [K1, p. 775].   ■

The theorem states that A can be well approximated by a singular matrix B if and only if cond(A) is quite large. And from our view, a singular matrix B is the ultimate in ill conditioning. There are nonzero perturbations of the solution, by the eigenvector for the eigenvalue $\lambda = 0$, which correspond to a zero perturbation in the right side b. And more importantly, there are values of b for which $Bx = b$ is no longer solvable.

The Hilbert matrix   The Hilbert matrix of order n is defined by

$$H_n = \begin{bmatrix} 1 & \frac{1}{2} & \frac{1}{3} & \cdots & \frac{1}{n} \\ \frac{1}{2} & \frac{1}{3} & \frac{1}{4} & & \frac{1}{n+1} \\ \vdots & & & & \vdots \\ \frac{1}{n} & \frac{1}{n+1} & & \cdots & \frac{1}{2n-1} \end{bmatrix}$$

This matrix occurs naturally in solving the continuous least squares approximation problem. Its derivation is given near the end of Section 4.3 of Chapter 4, with the resulting linear system given in (4.31). As was indicated in Section 4.3 and illustrated in Section 1.5 of Chapter 1, the Hilbert matrix is very ill conditioned, increasingly so as n increases. As such, it has been a favorite numerical example for checking programs for solving linear systems of equations, to determine the limits of effectiveness of the program when dealing with ill-conditioned problems. Table 8.2 gives the condition number cond(H_n)$_*$ for a few values of n. The inverse matrix $H_n^{-1} = [\alpha_{ij}^{(n)}]$ is known explicitly,

$$\alpha_{ij}^{(n)} = \frac{(-1)^{i+j}(n+i-1)!(n+j-1)!}{(i+j-1)[(i-1)!(j-1)!]^2(n-i)!(n-j)!} \qquad 1 \leq i, j \leq n \qquad (8.37)$$

Table 8.2   Condition numbers of Hilbert matrix

n	cond(H_n).	n	cond(H_n).
3	5.24E2	7	4.75E8
4	1.55E4	8	1.53E10
5	4.77E5	9	4.93E11
6	1.50E7	10	1.60E13

For additional information on H_n, including an asymptotic formula for cond(H_n)., see Gregory and Karney [G1, pp. 33–38, 66–73].

Although widely used as an example, some care must be taken as to what is the true answer. Let $\overline{H}_n$ denote the version of H_n after it is entered into the finite arithmetic of a computer. For a matrix inversion program, the results of the program should be compared with $\overline{H}_n^{-1}$, not with H_n^{-1}; and these two inverse matrices can be quite different. For example, if we use four-decimal digit, floating-point arithmetic with rounding, then

$$\overline{H}_3 = \begin{bmatrix} 1.000 & .5000 & .3333 \\ .5000 & .3333 & .2500 \\ .3333 & .2500 & .2000 \end{bmatrix} \tag{8.38}$$

Rounding has occurred only in expanding $\frac{1}{3}$ in decimal fraction form. Then

$$H_3^{-1} = \begin{bmatrix} 9.000 & -36.00 & 30.00 \\ -36.00 & 192.0 & -180.0 \\ 30.00 & -180.0 & 180.0 \end{bmatrix}$$

$$\overline{H}_3^{-1} = \begin{bmatrix} 9.062 & -36.32 & 30.30 \\ -36.32 & 193.7 & -181.6 \\ 30.30 & -181.6 & 181.5 \end{bmatrix} \tag{8.39}$$

Any program for matrix inversion, when applied to $\overline{H}_3$, should have its resulting solution compared with $\overline{H}_3^{-1}$ not H_3^{-1}. We return to this example in Section 8.5.

Error bounds   We consider the effects of rounding error on the solution $\hat{x}$ to $Ax = b$, obtained using Gaussian elimination. We begin by giving a result bounding the error when b and A are changed by small amounts. This is a useful result by itself, and it is necessary for the error analysis of Gaussian elimination that follows later.

Theorem 8.4   Consider the system $Ax = b$, with A nonsingular. Let δA and δb be perturbations of A and b, and assume

$$\| \delta A \| < \frac{1}{\| A^{-1} \|} \tag{8.40}$$

Then $A + \delta A$ is nonsingular. And if we define δx implicitly by

$$(A + \delta A)(x + \delta x) = b + \delta b \qquad (8.41)$$

then

$$\frac{\|\delta x\|}{\|x\|} \leqq \frac{\text{cond}(A)}{1 - \text{cond}(A)\dfrac{\|\delta A\|}{\|A\|}} \cdot \left\{ \frac{\|\delta A\|}{\|A\|} + \frac{\|\delta b\|}{\|b\|} \right\} \qquad (8.42)$$

Proof First note that δA represents any matrix satisfying (8.40), not a constant δ times the matrix A; and similarly for δb and δx. Using (8.40), the nonsingularity of $A + \delta A$ follows immediately from Theorem 7.12 of Chapter 7; and from (7.61),

$$\|(A + \delta A)^{-1}\| \leqq \frac{\|A^{-1}\|}{1 - \|A^{-1}\| \|\delta A\|} \qquad (8.43)$$

Solving for δx in (8.41), and using $Ax = b$,

$$(A + \delta A)\delta x + Ax + (\delta A)x = b + \delta b$$

$$\delta x = (A + \delta A)^{-1}\left[\delta b - (\delta A)x \right]$$

Using (8.43) and the definition (8.33) of cond(A),

$$\|\delta x\| \leqq \frac{\text{cond}(A)}{1 - \text{cond}(A)\dfrac{\|\delta A\|}{\|A\|}} \left\{ \frac{\|\delta b\|}{\|A\|} + \|x\| \frac{\|\delta A\|}{\|A\|} \right\}$$

Divide by $\|x\|$ on both sides, and use $\|b\| \leqq \|A\| \|x\|$ to obtain (8.42). ∎

The analysis of the effect of rounding errors on Gaussian elimination is due to J. H. Wilkinson, and can be found in Wilkinson [W1, pp. 94–99], [W2, pp. 209–215], and Forsythe and Moler [F1, Chap. 21]. Let $\hat{x}$ denote the computed solution of $Ax = b$. It is very difficult to compute directly the effects on x of rounding at each step, as a means of obtaining a bound on $\|x - \hat{x}\|$. Instead, it is easier, although nontrivial, to take $\hat{x}$ and the elimination algorithm and to work backward to show that $\hat{x}$ is the exact solution of a system

$$(A + \delta A)\hat{x} = b$$

in which bounds can be given for δA. This approach is known as *backward error analysis*. We can then use the preceding Theorem 8.4 to bound $\|x - \hat{x}\|$. In the following result, the matrix norm will be $\|A\|_\infty$, the row norm (7.42) induced by the vector norm $\|x\|_\infty$.

Theorem 8.5 Let A be of order n and nonsingular, and assume partial or complete pivoting is used in the Gaussian elimination process. Define

$$\rho = \frac{1}{\|A\|_\infty} \operatorname*{Maximum}_{1 \leq i,j,k \leq n} |a_{ij}^{(k)}| \qquad (8.44)$$

Let u denote the unit roundoff or truncation error on the computer being used, that is, the smallest positive number u on the computer for which $1 + u > 1$. (For more information on u, see Problem 7 in Chapter 1.) Then

1. The matrices L and U computed using Gaussian elimination satisfy

$$LU = A + E$$

$$\|E\|_\infty \leq n^2 \rho \|A\|_\infty u$$

2. The approximate solution $\hat{x}$ of $Ax = b$, computed using Gaussian elimination, satisfies

$$(A + \delta A)\hat{x} = b$$

with

$$\frac{\|\delta A\|_\infty}{\|A\|_\infty} \leq 1.01(n^3 + 3n^2)\rho u \qquad (8.45)$$

3. Using Theorem 8.4,

$$\frac{\|x - \hat{x}\|_\infty}{\|x\|_\infty} \leq \frac{\operatorname{cond}(A)_\infty}{1 - \operatorname{cond}(A)_\infty \frac{\|\delta A\|_\infty}{\|A\|_\infty}} \left[1.01(n^3 + 3n^2)\rho u \right] \qquad (8.46)$$

Proof See Forsythe and Moler [F1, Chapter 21] for a complete derivation, along with many additional results and comments. ■

Empirically, the bound (8.45) is too large, due to cancellation of rounding errors of varying magnitude and sign. According to Wilkinson [W1, p. 108], a better empirical bound for most cases is

$$\frac{\|\delta A\|_\infty}{\|A\|_\infty} \leq nu$$

The result (8.46) shows the importance of the size of cond(A).

The quantity ρ can be computed during the elimination process, and it can also be bounded a priori. For partial pivoting, an a priori bound is 2^{n-1}; and this is the best possible. For complete pivoting, an a priori bound is

$$\rho \leq 1.8n^{(\ln n)/4} \qquad n \geq 1$$

and it is conjectured that $\rho \leq cn$ for some c. Because of these differing bounds for ρ, complete pivoting is considered superior. But in actual practice with partial pivoting, the quantity ρ is proportional to n, at most; and consequently, partial pivoting is usually used rather than complete pivoting.

One of the most important consequences of the above analysis is to show that Gaussian elimination is a very stable process, provided only that the matrix A is not badly ill-conditioned. Historically, researchers in the early 1950s were uncertain as to the stability of Gaussian elimination for larger systems, for example $n \geq 10$, and that question has now been settled.

A posteriori error bounds We begin with error bounds for a computed inverse C of a given matrix A. Define the residual matrix by

$$R = I - CA$$

Theorem 8.6 If $\|R\| < 1$, then A and C are nonsingular, and

$$\frac{\|R\|}{\|A\|\,\|C\|} \leq \frac{\|A^{-1} - C\|}{\|C\|} \leq \frac{\|R\|}{1 - \|R\|} \tag{8.47}$$

Proof Since $\|R\| < 1$, $I - R$ is nonsingular by Theorem 7.10 of Chapter 7, and

$$\|(I-R)^{-1}\| \leq \frac{1}{1-\|R\|}$$

But

$$I - R = CA$$
$$0 \neq \det(I-R) = \det(CA) = \det(C)\det(A) \tag{8.48}$$

and thus both $\det(C)$ and $\det(A)$ are nonzero. This shows that both A and C are nonsingular.

For the lower bound in (8.47),

$$R = I - CA = (A^{-1} - C)A$$

$$\|R\| \le \|A^{-1} - C\| \|A\|$$

and by dividing $\|A\| \|C\|$ proves the result. For the upper bound, (8.48) implies

$$(I - R)^{-1} = A^{-1}C^{-1}$$

$$A^{-1} = (I - R)^{-1}C \tag{8.49}$$

For the error in C,

$$A^{-1} - C = (I - CA)A^{-1} = RA^{-1} = R(I - R)^{-1}C$$

$$\|A^{-1} - C\| \le \frac{\|R\| \|C\|}{1 - \|R\|}$$

This completes the proof. ∎

This result is generally of more theoretical than practical interest. Inverse matrices should not be produced for solving a linear system, as was pointed out earlier in Section 8.1. And consequently, there is seldom any real need for the above type of error bound. The main exception is when C has been produced as an approximation by means other than Gaussian elimination, often by some theoretical derivation. Such approximate inverses are then used to solve $Ax = b$ by the residual correction procedure (8.53) described in the next section. In this case, the above bound (8.47) can furnish some useful information on C.

Corollary Let A, C, and R be as given above in Theorem 8.6. Let $\hat{x}$ be an approximate solution to $Ax = b$, and define $r = b - A\hat{x}$. Then

$$\|x - \hat{x}\| \le \frac{\|Cr\|}{1 - \|R\|} \tag{8.50}$$

Proof Proceed as follows:

$$r = b - A\hat{x} = Ax - A\hat{x} = A(x - \hat{x})$$

$$x - \hat{x} = A^{-1}r = (I - R)^{-1}Cr \tag{8.51}$$

with (8.49) used in the last equality. Taking norms, we obtain (8.50). ∎

This bound (8.50) has been found to be quite accurate, especially when compared with a number of other bounds which are commonly used. For a complete discussion of computable error bounds, including a number of examples, see Aird and Lynch [A1].

The error bound (8.50) is relatively expensive to produce. If we suppose that $\hat{x}$ was obtained by Gaussian elimination, then about $n^3/3$ operations were used to calculate $\hat{x}$ and the LU decomposition of A. To produce $C \approx A^{-1}$ by elimination will take at least $\frac{2}{3}n^3$ additional operations, to produce CA requires n^3 operations, and Cr requires n^2 operations. Thus the error bound requires at least a fivefold increase in the number of operations. It is generally preferable to estimate the error by solving approximately the system

$$A(x-\hat{x})=r$$

using the LU decomposition stored earlier. This requires n^2 operations to evaluate r, and an additional n^2 to solve the linear system. Unless, the residual matrix $R = I - CA$ has norm nearly one, this approach will give a very reasonable error estimate. This is pursued and illustrated in the next section.

8.5   The Residual Correction Method

We assume that $Ax = b$ has been solved for an approximate solution $\hat{x} \equiv x^{(0)}$. Also, the LU decompostion along with a record of all row or column interchanges should have been stored. Calculate

$$r^{(0)} = b - Ax^{(0)}$$

Define $e^{(0)} = x - x^{(0)}$. Then as before in (8.51),

$$Ae^{(0)} = r^{(0)}$$

Solve this system using the stored LU decomposition, and call the resulting approximate solution $\hat{e}^{(0)}$. Define a new approximate solution to $Ax = b$ by

$$x^{(1)} = x^{(0)} + \hat{e}^{(0)} \tag{8.52}$$

The process can be repeated, calculating $x^{(2)}, \ldots,$ to continually decrease the error. To calculate $r^{(0)}$ takes n^2 operations, and the calculation of $\hat{e}^{(0)}$ takes an additional n^2 operations. Thus the calculation of the improved values $x^{(1)}, x^{(2)}, \ldots$ is inexpensive compared with the calculation of the original value $x^{(0)}$. This method is also known as *iterative improvement* or the *residual correction method*.

It is extremely important to obtain accurate values for $r^{(0)}$. Since $x^{(0)}$ approximately solves $Ax = b$, $r^{(0)}$ will generally involve loss of significance errors in its calculation. Thus to obtain accurate values for $r^{(0)}$, we must usually go to double-precision arithmetic. If only regular arithmetic is used to calculate $r^{(0)}$, the same arithmetic as used in calculating $x^{(0)}$ and LU, then the resulting inaccuracy in $r^{(0)}$ will usually lead to $\hat{e}^{(0)}$ being a poor approximation to $e^{(0)}$.

Example   Solve the system $Ax = b$, with $A = \overline{H}_3$ from (8.38). The arithmetic will be four-decimal digit floating point with rounding. For the right side, use

$$b = [1, 0, 0]^T$$

The true solution is the first column of $\overline{H}_3^{-1}$, which from (8.38) is

$$x = [9.062, -36.32, 30.30]^T$$

to four significant digits.
    Using elimination with partial pivoting,

$$x^{(0)} = [9.190, -37.04, 31.00]^T$$

The residual $r^{(0)}$ is calculated with double-precision arithmetic, and is then rounded to four significant digits. The value obtained is

$$r^{(0)} = [-.002300, .0004320, -.003027]^T$$

Solving $Ae^{(0)} = r^{(0)}$ with the stored LU decomposition,

$$\hat{e}^{(0)} = [-.1309, .7320, -.7122]^T$$

$$x^{(1)} = [9.059, -36.31, 30.29]^T$$

Repeating these operations,

$$r^{(1)} = [.0003430, .0001230, .0001353]^T$$

$$\hat{e}^{(1)} = [.002792, -.01349, .01289]^T$$

$$x^{(2)} = [9.062, -36.32, 30.30]^T$$

The vector $x^{(2)}$ is accurate to four significant digits. Also, note that $x^{(1)} - x^{(0)} = \hat{e}^{(0)}$ was an accurate predictor of the error $e^{(0)}$ in $x^{(0)}$.

    Formulas can be developed to estimate how many iterates should be calculated in order to get essentially full accuracy in the solution x. For a discussion of what is involved and for some algorithms implementing this method, see Forsythe and Moler [F1, Chapters 13, 16, and 17] or Wilkinson and Reinsch [W3, pp. 93–110].

Another residual correction method   There are situations in which we can calculate an approximate inverse C to the given matrix A. This is generally done by carefully considering the structure of A, and then using a variety of approximation techniques to estimate A^{-1}. Without considering the origin of C, we will show how to use it to iteratively solve $Ax = b$.

Let $x^{(0)}$ be an initial guess, and define $r^{(0)} = b - Ax^{(0)}$. As before, $A(x - x^{(0)}) = r^{(0)}$. Define $x^{(1)}$ implicitly by

$$x^{(1)} - x^{(0)} = Cr^{(0)}$$

In general, define

$$r^{(m)} = b - Ax^{(m)} \qquad x^{(m+1)} = x^{(m)} + Cr^{(m)} \qquad m = 0, 1, 2, \ldots \qquad (8.53)$$

If C is a good approximation to A^{-1}, the iteration will converge rapidly, as shown in the following analysis.

We first obtain a recursion formula for the error.

$$x - x^{(m+1)} = x - x^{(m)} - Cr^{(m)} = x - x^{(m)} - C\left[b - Ax^{(m)} \right]$$

$$= x - x^{(m)} - C\left[Ax - Ax^{(m)} \right]$$

$$x - x^{(m+1)} = (I - CA)(x - x^{(m)}) \qquad (8.54)$$

By induction

$$x - x^{(m)} = (I - CA)^m (x - x^{(0)}) \qquad m \geqslant 0 \qquad (8.55)$$

If

$$\|I - CA\| < 1 \qquad (8.56)$$

for some matrix norm, then using the associated vector norm,

$$\|x - x^{(m)}\| \leqslant \|I - CA\|^m \|x - x^{(0)}\| \qquad (8.57)$$

And this converges to zero as $m \to \infty$, for any choice of initial guess $x^{(0)}$. More generally, (8.55) implies that $x^{(m)}$ converges to x, for any choice of $x^{(0)}$, if and only if

$$(I - CA)^m \to 0 \qquad \text{as} \quad m \to \infty$$

And by Theorem 7.9 of Chapter 7, this is equivalent to

$$r_\sigma(I - CA) < 1 \qquad (8.58)$$

for the spectral radius of $I - CA$. This may be possible to show, even when (8.56) fails for the common matrix norms. Also note that

$$I - AC = A(I - CA)A^{-1}$$

and thus $I - AC$ and $I - CA$ are similar matrices and have the same eigenvalues. If

$$\|I - AC\| < 1$$

then (8.58) will be true, even if (8.56) is not true; and convergence will still occur.

Statement (8.54) shows that the rate of convergence of $x^{(m)}$ to x is linear,

$$\|x - x^{(m+1)}\| \le c\|x - x^{(m)}\| \qquad m \ge 0 \tag{8.59}$$

with $c < 1$ unknown. It is often estimated computationally with

$$c = \text{Max} \frac{\|x^{(m+2)} - x^{(m+1)}\|}{\|x^{(m+1)} - x^{(m)}\|} \tag{8.60}$$

with the maximum performed over all iterates that have been computed. This is not rigorous, but is motivated by the formula

$$x^{(m+2)} - x^{(m+1)} = (I - CA)(x^{(m+1)} - x^{(m)})$$

The proof is left to the reader.

If we assume (8.59) is valid for the iterates that we are calculating, and if we have an estimate for c, then we can produce an error bound.

$$\|x^{(m+1)} - x^{(m)}\| = \|[x - x^{(m)}] - [x - x^{(m+1)}]\|$$

$$\ge \|x - x^{(m)}\| - \|x - x^{(m+1)}\|$$

$$\ge \|x - x^{(m)}\| - c\|x - x^{(m)}\|$$

$$\|x - x^{(m)}\| \le \frac{1}{1 - c}\|x^{(m+1)} - x^{(m)}\|$$

$$\|x - x^{(m+1)}\| \le \frac{c}{1 - c}\|x^{(m+1)} - x^{(m)}\| \tag{8.61}$$

For slowly convergent iterates (with $c \approx 1$), this bound is important, since $\|x^{(m+1)} - x^{(m)}\|$ can then be much smaller than $\|x - x^{(m)}\|$. Also, recall the earlier derivation in Section 2.6 of Chapter 2. A similar bound, (2.55), was derived for the error in a linearly convergent method.

Example   Define $A(\epsilon) = A_0 + \epsilon B$, with

$$A_0 = \begin{bmatrix} 2 & 1 & 0 \\ 1 & 2 & 1 \\ 0 & 1 & 2 \end{bmatrix} \qquad B = \begin{bmatrix} 0 & 1 & 1 \\ -1 & 0 & 1 \\ -1 & -1 & 0 \end{bmatrix}$$

As an approximate inverse to $A(\epsilon)$, use

$$A(\epsilon)^{-1} \approx C = A_0^{-1} = \begin{bmatrix} \dfrac{3}{4} & -\dfrac{1}{2} & \dfrac{1}{4} \\[2mm] -\dfrac{1}{2} & 1 & -\dfrac{1}{2} \\[2mm] \dfrac{1}{4} & -\dfrac{1}{2} & \dfrac{3}{4} \end{bmatrix}$$

We can solve the system $A(\epsilon)x = b$ using the residual correction method (8.53). For the convergence analysis,

$$I - CA(\epsilon) = I - A_0^{-1}[A_0 + \epsilon B] = -\epsilon A_0^{-1} B$$

$$= -\epsilon \begin{bmatrix} \dfrac{1}{4} & \dfrac{1}{2} & \dfrac{1}{4} \\[2mm] -\dfrac{1}{2} & 0 & -\dfrac{1}{2} \\[2mm] -\dfrac{1}{4} & -\dfrac{1}{2} & -\dfrac{1}{4} \end{bmatrix}$$

Convergence is assured if

$$\|I - CA(\epsilon)\|_\infty = |\epsilon| < 1$$

and from (8.54),

$$\|x - x^{(m+1)}\|_\infty \leqslant |\epsilon| \, \|x - x^{(m)}\|_\infty \qquad m \geq 0$$

There are many situations of the kind in this example. We may have to solve linear systems of a general form $A(\epsilon)x = b$ for any ϵ near zero. To save time, we obtain either $A(0)^{-1}$ or the LU decomposition of $A(0)$. This is then used as an approximate inverse to $A(\epsilon)$, and we solve $A(\epsilon)x = b$ using the residual correction method.

8.6   Iteration Methods

As was mentioned in the introduction to this chapter, many linear systems are too large to be solved by direct methods based on Gaussian elimination. For these systems, iteration methods are often the only possible method of solution, as well as being faster than elimination in many cases. The largest area for the application of iteration methods is to the linear systems arising in the numerical solution of partial differential equations. Systems of orders 10,000 to 100,000 are not unusual, although almost all of the coefficients of the system will be zero. As an example of such problems, the numerical solution of Poisson's equation is studied in Section 8.8.

Besides being large, the linear systems to be solved, $Ax = b$, usually have several other important properties. They are usually sparse, which means that only a small percentage of the coefficients are nonzero. The nonzero coefficients generally have a special pattern in the way they occur in A; and there is usually a simple formula that can be used to generate the coefficients a_{ij} as they are needed, rather than having to store them. As one consequence of these properties, the storage space for the vectors x and b may be a more important consideration than is storage for A.

We begin by defining and analyzing two classical iteration methods; and following that, a general abstract framework is presented for studying iteration methods. The special properties of the linear system $Ax = b$ are very important when setting up an iteration method for its solution; and the results of this section are just a beginning to the design of a method for any particular area of applications.

The Gauss-Jacobi method   (Simultaneous displacements).   Rewrite $Ax = b$ as

$$x_i = \frac{1}{a_{ii}}\left[b_i - \sum_{\substack{j=1 \\ j \neq i}}^{n} a_{ij} x_j \right]   i = 1, 2, \ldots, n   \tag{8.62}$$

assuming all $a_{ii} \neq 0$. Define the iteration as

$$x_i^{(m+1)} = \frac{1}{a_{ii}}\left[b_i - \sum_{\substack{j=1 \\ j \neq i}}^{n} a_{ij} x_j^{(m)} \right]   i = 1, \ldots, n   m \geq 0   \tag{8.63}$$

and assume initial guesses $x_i^{(0)}, i = 1, \ldots, n$, are given. There are other forms to the method. For example, many problems are given naturally in the form

$$(I - B)x = b$$

and then we would usually first consider the iteration

$$x^{(m+1)} = b + Bx^{(m)}   m \geq 0   \tag{8.64}$$

Our initial error analysis will be restricted to (8.63), but the same ideas can be used for (8.64).

To analyze the convergence, let $e^{(m)} = x - x^{(m)}, m \geq 0$. Subtracting (8.63) from (8.62),

$$e_i^{(m+1)} = - \sum_{\substack{j=1 \\ j \neq i}}^{n} \frac{a_{ij}}{a_{ii}} e_j^{(m)}   i = 1, \ldots, n   m \geq 0   \tag{8.65}$$

$$|e_i^{(m+1)}| \leq \sum_{\substack{j=1 \\ j \neq i}}^{n} \left| \frac{a_{ij}}{a_{ii}} \right| \|e^{(m)}\|_\infty$$

Define

$$\mu = \operatorname*{Max}_{1 \leq i \leq n} \sum_{\substack{j=1 \\ j \neq i}}^{n} \left| \frac{a_{ij}}{a_{ii}} \right| \qquad (8.66)$$

Then

$$|e_i^{(m+1)}| \leq \mu \|e^{(m)}\|_\infty$$

and since the right side is independent of i,

$$\|e^{(m+1)}\|_\infty \leq \mu \|e^{(m)}\|_\infty \qquad (8.67)$$

If $\mu < 1$, then $e^{(m)} \to 0$ as $m \to \infty$ with a linear rate, and

$$\|e^{(m)}\|_\infty \leq \mu^m \|e^{(0)}\|_\infty$$

In order for $\mu < 1$ to be true, the matrix A must be *diagonally dominant*, that is, it must satisfy

$$\sum_{\substack{j=1 \\ j \neq i}}^{n} |a_{ij}| < |a_{ii}| \qquad i = 1, 2, \ldots, n \qquad (8.68)$$

Such matrices occur reasonably often in applications, and usually the associated matrix is sparse.

To have a more general result, write (8.65) as

$$e^{(m+1)} = Me^{(m)} \qquad m \geq 0 \qquad (8.69)$$

$$M = - \begin{bmatrix} 0 & \dfrac{a_{12}}{a_{11}} & \cdots & \cdots & \dfrac{a_{1n}}{a_{11}} \\ \dfrac{a_{21}}{a_{22}} & 0 & \dfrac{a_{23}}{a_{22}} & \cdots & \dfrac{a_{2n}}{a_{22}} \\ \vdots & & \ddots & & \vdots \\ \vdots & & & \ddots & \vdots \\ \dfrac{a_{n1}}{a_{nn}} & \cdots & \cdots & \cdots & 0 \end{bmatrix}$$

Inductively,

$$e^{(m)} = M^m e^{(0)}$$

If we want $e^{(m)} \to 0$ as $m \to \infty$, independent of the choice of $x^{(0)}$, and thus of $e^{(0)}$, it is necessary and sufficient that

$$M^m \to 0 \qquad \text{as} \quad m \to \infty$$

Or equivalently, from Theorem 7.9 of Chapter 7,

$$r_\sigma(M) < 1 \tag{8.70}$$

The condition $\mu < 1$ is merely the requirement that the row norm of M be less than 1, $\|M\|_\infty < 1$, and this implies (8.70). But now we see that $e^{(m)} \to 0$ if $\|M\| < 1$ for any operator matrix norm.

Example   Consider solving $Ax = b$ by the Gauss-Jacobi method, with

$$A = \begin{bmatrix} 10 & 3 & 1 \\ 2 & -10 & 3 \\ 1 & 3 & 10 \end{bmatrix} \qquad b = \begin{bmatrix} 14 \\ -5 \\ 14 \end{bmatrix} \qquad x^{(0)} = \begin{bmatrix} 0 \\ 0 \\ 0 \end{bmatrix} \tag{8.71}$$

Solving for unknown i in equation i, we have $x = g + Mx$,

$$M = \begin{bmatrix} 0 & -.3 & -.1 \\ .2 & 0 & .3 \\ -.1 & -.3 & 0 \end{bmatrix} \qquad g = \begin{bmatrix} 1.4 \\ .5 \\ 1.4 \end{bmatrix}$$

The true solution is $x = [1, 1, 1]^T$. To check for convergence, note that $\|M\|_\infty = .5, \|M\|_1 = .6$. Hence

$$\|e^{(m+1)}\|_\infty \le .5 \|e^{(m)}\|_\infty \qquad m \ge 0 \tag{8.72}$$

A similar statement holds for $\|e^{(m+1)}\|_1$. Thus convergence is guaranteed, and the errors will decrease by at least half with each iteration. Actual numerical results are given in Table 8.3, and they confirm the result (8.72).

Table 8.3   Numerical results for Gauss-Jacobi method

m	$x_1^{(m)}$	$x_2^{(m)}$	$x_3^{(m)}$	$\|e^{(m)}\|_\infty$	Ratio
0	0	0	0	1	
1	1.4	.5	1.4	.5	.5
2	1.11	1.20	1.11	.2	.4
3	.929	1.055	.929	.071	.36
4	.9906	.9645	.9906	.0355	.50
5	1.01159	.9953	1.01159	.01159	.33
6	1.000251	1.005795	1.000251	.005795	.50

The Gauss-Seidel method   (Successive displacements)   Using (8.62), define

$$x_i^{(m+1)} = \frac{1}{a_{ii}} \left[b_i - \sum_{j=1}^{i-1} a_{ij} x_j^{(m+1)} - \sum_{j=i+1}^{n} a_{ij} x_j^{(m)} \right] \qquad i = 1, 2, \ldots, n \tag{8.73}$$

Each new component $x_i^{(m+1)}$ is immediately used in the computation of the next component. This is convenient for computer calculations since the new value can be immediately stored in the location that held the old value; this minimizes the number of necessary storage locations. The storage requirements for x with the Gauss-Seidel method is only half what it would be with the Gauss-Jacobi method.

To analyze the error, subtract (8.73) from (8.62),

$$e_i^{(m+1)} = -\sum_{j=1}^{i-1} \frac{a_{ij}}{a_{ii}} e_j^{(m+1)} - \sum_{j=i+1}^{n} \frac{a_{ij}}{a_{ii}} e_j^{(m)} \qquad i=1,2,\ldots,n \qquad (8.74)$$

Define

$$\alpha_i = \sum_{1}^{i-1} \left| \frac{a_{ij}}{a_{ii}} \right| \qquad \beta_i = \sum_{j=i+1}^{n} \left| \frac{a_{ij}}{a_{ii}} \right| \qquad i=1,\ldots,n$$

with $\alpha_1 = \beta_n = 0$. Using the same value (8.66) for μ as with the Jacobi method,

$$\mu = \operatorname*{Max}_{1 \le i \le n} (\alpha_i + \beta_i)$$

We will assume that $\mu < 1$. Then define

$$\eta = \operatorname*{Max}_{1 \le i \le n} \frac{\beta_i}{1 - \alpha_i} \qquad (8.75)$$

From (8.74)

$$|e_i^{(m+1)}| \le \alpha_i \| e^{(m+1)} \|_\infty + \beta_i \| e^{(m)} \|_\infty \qquad i=1,\ldots,n \qquad (8.76)$$

Let k be the subscript for which

$$\| e^{(m+1)} \|_\infty = |e_k^{(m+1)}|$$

Then with $i=k$ in (8.76),

$$\| e^{(m+1)} \|_\infty \le \alpha_k \| e^{(m+1)} \|_\infty + \beta_k \| e^{(m)} \|_\infty$$

$$\| e^{(m+1)} \|_\infty \le \frac{\beta_k}{1 - \alpha_k} \| e^{(m)} \|_\infty$$

and thus

$$\| e^{(m+1)} \|_\infty \le \eta \| e^{(m)} \|_\infty \qquad (8.77)$$

Since for each i,

$$(\alpha_i + \beta_i) - \frac{\beta_i}{1 - \alpha_i} = \frac{\alpha_i[1 - (\alpha_i + \beta_i)]}{1 - \alpha_i} \geq \frac{\alpha_i}{1 - \alpha_i}(1 - \mu) \geq 0$$

we have

$$\eta \leq \mu < 1 \tag{8.78}$$

Combined with (8.77), this shows the convergence of $e^{(m)} \to 0$ as $m \to \infty$. Also, the rate of convergence will be linear, but with a faster rate than with the Jacobi method.

Example Use the system (8.71) of the previous example, and solve it with the Gauss-Seidel method. By a simple calculation from (8.75) and (8.71),

$$\eta = .4$$

The numerical results are given in Table 8.4. The speed of convergence is significantly better than for the previous example of the Gauss-Jacobi method, given in Table 8.3.

Table 8.4 Numerical results for Gauss-Seidel method

m	$x_1^{(m)}$	$x_2^{(m)}$	$x_3^{(m)}$	$\|e^{(m)}\|_\infty$	**Ratio**
0	0	0	0	1	
1	1.4	.78	1.026	.4	.4
2	.9234	.99248	1.1092	.1092	.27
3	.99134	1.0310	.99159	.031	.28
4	.99154	.99578	1.0021	.0085	.27

General framework for iteration methods To solve $Ax = b$, form a split of A,

$$A = N - P \tag{8.79}$$

and write $Ax = b$ as

$$Nx = b + Px \tag{8.80}$$

The matrix N is to be chosen in such a way that the linear system $Nz = f$ is easily solvable for any f. For example, N might be diagonal, triangular, or tridiagonal. Define the iteration method by

$$Nx^{(m+1)} = b + Px^{(m)} \qquad m \geq 0 \tag{8.81}$$

with $x^{(0)}$ given.

Examples 1.   The Jacobi method.

$$N = \text{diag}[a_{11}, a_{22}, \ldots, a_{nn}] \qquad P = N - A$$

2.   Gauss-Seidel method.

$$N = \begin{bmatrix} a_{11} & 0 & \cdot & \cdot & \cdot & 0 \\ a_{21} & a_{22} & 0 & & 0 \\ \cdot & & \cdot & & & \cdot \\ \cdot & & & \cdot & & \cdot \\ \cdot & & & & \cdot & \cdot \\ a_{n1} & \cdot & \cdot & \cdot & \cdot & a_{nn} \end{bmatrix} \qquad (8.82)$$

To analyze the error, subtract (8.81) from (8.80) to get

$$Ne^{(m+1)} = Pe^{(m)}$$

$$e^{(m+1)} = Me^{(m)} \qquad M = N^{-1}P \qquad (8.83)$$

By induction,

$$e^{(m)} = M^m e^{(0)} \qquad m \geq 0$$

In order that $e^{(m)} \to 0$ as $m \to \infty$, for arbitrary initial guesses $x^{(0)}$ (and thus arbitrary $e^{(0)}$), it is necessary and sufficient that

$$M^m \to 0 \qquad \text{as} \quad m \to \infty$$

Or equivalently, from Theorem 7.9,

$$r_\sigma(M) < 1 \qquad (8.84)$$

This general framework for iteration methods is adapted from Isaacson and Keller [I1, pp. 61–81].

The condition (8.84) was derived earlier in (8.70) for the Gauss-Jacobi method. For the Gauss-Seidel method the matrix $N^{-1}P$, with N given by (8.82), is more difficult to work with. We must examine the values of λ for which

$$\det(\lambda I - N^{-1}P) = 0$$

or equivalently,

$$\det(\lambda N - P) = 0 \qquad (8.85)$$

For applications to the numerical solution of partial differential equations, as in Section 8.8, the earlier convergence analysis of the Gauss-Seidel method is not adequate. The constants μ and η of (8.66) and (8.75) will both equal one, although empirically the method still converges. To deal with many of these systems, the following important theorem is often used.

Theorem 8.7 Let A be Hermitian with positive diagonal elements. Then the Gauss-Seidel method (8.73) for solving $Ax = b$ will converge, for any choice of $x^{(0)}$, if and only if A is positive definite.

Proof The proof is given in Isaacson and Keller [I1, pp. 70–71]. For the definition of a positive definite matrix, recall Problem 11 of Chapter 7. The theorem will be illustrated in Section 8.8. ∎

8.7 Error Prediction and Acceleration

From (8.83), we have the error relation

$$x - x^{(m+1)} = M(x - x^{(m)}) \qquad m \geq 0 \qquad (8.86)$$

The manner of convergence of $x^{(m)}$ to x can be quite complicated, depending on the eigenvalues and eigenvectors of M. But in most practical cases, the behavior of the errors is quite simple. The size of $\|x - x^{(m)}\|_\infty$ decreases by approximately a constant factor at each step,

$$\|x - x^{(m+1)}\|_\infty \leq c\|x - x^{(m)}\|_\infty \qquad m \geq 0 \qquad (8.87)$$

for some $c < 1$, closely related to $r_\sigma(M)$. To measure this constant c, note from (8.86) that

$$x^{(m+1)} - x^{(m)} = e^{(m)} - e^{(m+1)} = Me^{(m-1)} - Me^{(m)}$$

$$x^{(m+1)} - x^{(m)} = M(x^{(m)} - x^{(m-1)}) \qquad m \geq 0$$

This motivates the use of

$$c \approx \frac{\|x^{(m+1)} - x^{(m)}\|_\infty}{\|x^{(m)} - x^{(m-1)}\|_\infty} \qquad (8.88)$$

or for greater safety, the maximum of several successive ratios. In many applications, this ratio is about constant for large values of m.

Once this constant c has been obtained, and assuming (8.87), we can predict the error in $x^{(m+1)}$ by using (8.61),

$$\|x - x^{(m+1)}\|_\infty \leq \frac{c}{1-c}\|x^{(m+1)} - x^{(m)}\|_\infty \qquad (8.89)$$

This bound is important when $c \approx 1$ and the convergence is slow. In that case, the difference $\|x^{(m+1)} - x^{(m)}\|_\infty$ can be much smaller than the actual error $\|x - x^{(m+1)}\|_\infty$.

Example   The linear system (8.98) of Section 8.8 was solved using the Gauss-Seidel method. The linear system solved was of order 225, and the speed of convergence was typically slow. A selection of numerical results is shown in Table 8.5. The "Ratio" column is calculated from (8.88), and the "Error" column is the estimated error using the upper bound in (8.89).

Table 8.5   Example of Gauss-Seidel iteration

m	$\|x^{(m)} - x^{(m-1)}\|_\infty$	Ratio	Error
85	5.96 E-4	.9596	1.41 E-2
86	5.72 E-4	.9597	1.36 E-2
87	5.50 E-4	.9615	1.37 E-2
88	5.29 E-4	.9619	1.34 E-2
89	5.09 E-4	.9619	1.29 E-2
90	4.89 E-4	.9619	1.24 E-2

Speed of convergence   We want to discuss how many iterates to calculate in order to obtain a desired error. And when is iteration preferable to Gaussian elimination in solving $Ax = b$? We will find the value of m for which

$$\|x - x^{(m)}\|_\infty \leq \epsilon \|x - x^{(0)}\|_\infty \tag{8.90}$$

with ϵ a given factor by which the initial error is to be reduced. We base the analysis on the assumption (8.87). Generally the constant c is almost equal to $r_\sigma(M)$, with M as in (8.86).

The relation (8.87) implies

$$\|x - x^{(m)}\|_\infty \leq c^m \|x - x^{(0)}\|_\infty \qquad m \geq 1$$

Thus we find the smallest value of m for which

$$c^m \leq \epsilon$$

Solving this, we must have

$$m \geq \frac{-\ln \epsilon}{R(c)} \equiv m^* \qquad R(c) = -\ln c \tag{8.91}$$

Doubling $R(c)$ leads to halving the number of iterates that must be calculated.

To make this result more meaningful, we apply it to the solution of a dense linear system by iteration. Assume that the Gauss-Jacobi or Gauss-Seidel method is being used to solve $Ax = b$ to single-precision accuracy on an IBM 360 computer, that is, to about six significant digits. Assume that $x^{(0)} = 0$, and that we want to find m such that

$$\frac{\|x - x^{(m)}\|_\infty}{\|x\|_\infty} \leq 10^{-6} = \epsilon \tag{8.92}$$

Assuming that A has order n, the number of operations (multiplications and divisions) per iteration is n^2. To obtain the result (8.92), the necessary number of iterates is

$$m^* = \frac{6 \ln_e 10}{R(c)}$$

and the number of operations is

$$m^* n^2 = (6 \ln_e 10) \frac{n^2}{R(c)}$$

If Gaussian elimination is used to solve $Ax = b$ with the same accuracy, the number of operations is about $n^3/3$. The iteration method will be more efficient than the Gaussian elimination method if

$$m^* n^2 < n^3/3$$
$$m^* < n/3$$

Example Consider a matrix A of order $n = 51$. Then iteration is more efficient if $m^* < 17$. Table 8.6 gives the values of m^* for various values of c. For $c \leq .44$, the iteration method will be more efficient than Gaussian elimination. And if less than full precision accuracy is desired, then iteration will be more efficient with even larger values of c.

Table 8.6   Example of iteration count (8.91)

c	$R(c)$	m^*
.9	.105	131
.8	.223	62
.6	.511	27
.4	.916	15
.2	1.61	9

The main use of iteration methods is for the solution of large sparse systems, and then Gaussian elimination is usually not possible. We give some examples of such systems and of the values of c and m^* in Section 8.8. As the order n increases, the point at which iteration is superior to Gaussian elimination occurs with increasing values of c. When n is very large, then c can be very near to 1 with iteration still being superior.

Acceleration methods   Most iteration methods have a regular pattern in which the error decreases. This can often be used to accelerate the convergence, just as was done in earlier chapters with other numerical methods. Rather than giving a general theory for the acceleration of iteration methods for solving $Ax = b$, we merely describe an acceleration of the Gauss-Seidel method. This is the main case of interest in applications.

Recall the definition (8.73) of the Gauss-Seidel method. Introduce an *acceleration parameter* ω, and consider the following modification of (8.73).

$$z_i^{(m+1)} = \frac{1}{a_{ii}} \left[b_i - \sum_{j=1}^{i-1} a_{ij} x_j^{(m+1)} - \sum_{j=i+1}^{n} a_{ij} x_j^{(m)} \right]$$

$$x_i^{(m+1)} = \omega z_i^{(m+1)} + (1-\omega) x_i^{(m)} \qquad i = 1, \ldots, n \qquad (8.93)$$

for $m \geq 0$. The case $\omega = 1$ is the regular Gauss-Seidel method.

To understand how ω should be chosen, we rewrite (8.93) in matrix form. Decompose A as

$$A = D + L + U$$

with $D = \text{diag}[a_{11}, \ldots, a_{nn}]$, L lower triangular, and U upper triangular. Then (8.93) becomes

$$z^{(m+1)} = D^{-1} \left[b - Lx^{(m+1)} - Ux^{(m)} \right]$$

$$x^{(m+1)} = \omega z^{(m+1)} + (1-\omega) x^{(m)} \qquad m \geq 0$$

Eliminating $z^{(m+1)}$ and solving for $x^{(m+1)}$,

$$(I + \omega D^{-1}L) x^{(m+1)} = \omega D^{-1} b + \left[(1-\omega)I - \omega D^{-1}U \right] x^{(m)}$$

For the error,

$$e^{(m+1)} = M(\omega) e^{(m)} \qquad m \geq 0$$

$$M(\omega) = (I + \omega D^{-1}L)^{-1} \left[(1-\omega)I - \omega D^{-1}U \right] \qquad (8.94)$$

The parameter ω is to be chosen to minimize $r_\sigma(M(\omega))$, in order to make $x^{(m)}$ converge to x as rapidly as possible. Call the optimal value ω^*.

The calculation of ω^* is difficult except in the simplest cases. And usually it is obtained only approximately, based on trying several values of ω and observing the effect on the speed of convergence. In spite of the problem of calculating ω^*, the resulting increase in the speed of convergence of $x^{(m)}$ to x is very dramatic; and the calculation of ω^* is well worth the effort. This is illustrated in the next section.

8.8 The Numerical Solution of Poisson's Equation

The most important application of linear iteration methods is to the large linear systems arising from the numerical solution of partial differential equations by finite difference methods. To illustrate this, we will solve the Dirichlet problem for Poisson's equation on the unit square in the xy-plane,

$$\frac{\partial^2 u}{\partial x^2} + \frac{\partial^2 u}{\partial y^2} = g(x,y) \qquad 0 < x,y < 1$$

$$u(x,y) = f(x,y) \qquad (x,y) \text{ a boundary point}$$

(8.95)

The functions $g(x, y)$ and $f(x, y)$ are given, and we must find $u(x, y)$.
For $N > 1$, define $h = 1/N$, and

$$(x_j, y_k) = (jh, kh), \quad 0 \le j, k \le N$$

These are called the grid points or mesh points; see the intersection points in Figure 8.1. To approximate (8.95), we use approximations to the second derivatives. For a function $G(t)$ that is four times continuously differentiable for $x - h \le t \le x + h$, Taylor's theorem can be used to prove

$$G''(x) = \frac{G(x+h) - 2G(x) + G(x-h)}{h^2} - \frac{h^2}{12} G^{(4)}(\xi) \qquad x - h \le \xi \le x + h$$

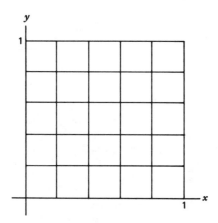

Figure 8.1 Finite difference mesh.

When applied to (8.95) at each interior grid point, we obtain

$$\frac{u(x_{j+1}, y_k) - 2u(x_j, y_k) + u(x_{j-1}, y_k)}{h^2} + \frac{u(x_j, y_{k+1}) - 2u(x_j, y_k) + u(x_j, y_{k-1})}{h^2}$$

$$= g(x_j, y_k) + \frac{h^2}{12} \left[\frac{\partial^4 u(\xi, y_k)}{\partial x^4} + \frac{\partial^4 u(x_j, \eta)}{\partial y^4} \right] \qquad (8.96)$$

for some $x_{j-1} \leq \xi \leq x_{j+1}$, $y_{k-1} \leq \eta \leq y_{k+1}$, $1 \leq j, k \leq N-1$.

For the numerical approximation $u_h(x, y)$ of (8.95), let

$$u_h(x_j, y_k) = f(x_j, y_k) \qquad (x_j, y_k) \text{ a boundary grid point} \qquad (8.97)$$

At all interior mesh points, drop the right-hand truncation errors in (8.96) and solve for the approximating solution $u_h(x_j, y_k)$,

$$u_h(x_j, y_k) = \frac{1}{4} \left[u_h(x_{j+1}, y_k) + u_h(x_j, y_{k+1}) + u_h(x_{j-1}, y_k) + u_h(x_j, y_{k-1}) \right]$$

$$- \frac{h^2}{4} g(x_j, y_k) \qquad 1 \leq j, k \leq N-1 \qquad (8.98)$$

The number of equations, (8.97) and (9.98), is equal to the number of unknowns, $(N+1)^2$.

Theorem 8.8 For each $N \geq 2$, the linear system (8.97) to (8.98) has a unique solution $\{u_h(x_j, y_k) | 0 \leq j, k \leq N\}$. If the solution $u(x, y)$ of (8.95) is four times continuously differentiable, then

$$\underset{0 \leq j, k \leq N}{\text{Max}} |u(x_j, y_k) - u_h(x_j, y_k)| \leq ch^2 \qquad (8.99)$$

$$c = \frac{1}{24} \left[\underset{0 \leq x, y \leq 1}{\text{Max}} \left| \frac{\partial^4 u(x, y)}{\partial x^4} \right| + \underset{0 \leq x, y \leq 1}{\text{Max}} \left| \frac{\partial^4 u(x, y)}{\partial y^4} \right| \right]$$

Proof *1.* We will prove the unique solvability of (8.97) to (8.98) by using Theorem 7.2 of Chapter 7. We consider the homogeneous system

$$v_h(x_j, y_k) = \frac{1}{4} \left[v_h(x_{j+1}, y_k) + v_h(x_j, y_{k+1}) \right.$$

$$\left. + v_h(x_{j-1}, y_k) + v_h(x_j, y_{k-1}) \right] \qquad 1 \leq j, k \leq N-1 \qquad (8.100)$$

$$v_h(x_j, y_k) = 0 \qquad (x_j, y_k) \text{ a boundary point} \qquad (8.101)$$

By showing that this system has only the trivial solution $v_h(x_j, y_k) \equiv 0$, it will follow from Theorem 7.2 that the nonhomogeneous system (8.97) to (8.98) will have a unique solution.

Let

$$\alpha = \operatorname*{Max}_{0 \le j, k \le N} v_h(x_j, y_k)$$

From (8.101), $\alpha \ge 0$. Assume that $\alpha > 0$. Then there must be an interior grid point $(\bar{x}_j, \bar{y}_k)$ for which this maximum is attained. But using (8.100), $v_h(\bar{x}_j, \bar{y}_k)$ is the average of the values of v_h at the four points neighboring $(\bar{x}_j, \bar{y}_k)$. The only way that this can be compatible with $(\bar{x}_j, \bar{y}_k)$ being a maximum point is if v_h also equals α at the four neighboring grid points. Continue the same argument to these neighboring points. Since there are only a finite number of grid points, we will eventually have $v_h(x_j, y_k) = \alpha$ for a boundary point (x_j, y_k). But then $\alpha > 0$ will contradict (8.101). Thus the maximum of $v_h(x_j, y_k)$ is zero. A similar argument will show that the minimum of $v_h(x_j, y_k)$ is also zero. Taken together, these results show that the only solution of (8.99) to (8.100) is $v_h(x_j, y_k) \equiv 0$.

2. To consider the convergence of $u_h(x_j, y_k)$ to $u(x_j, y_k)$, define

$$e_h(x_j, y_k) = u(x_j, y_k) - u_h(x_j, y_k)$$

Subtracting (8.98) from (8.96), we obtain

$$e_h(x_j, y_k) = \frac{1}{4} \left[e_h(x_{j+1}, y_k) + e_h(x_j, y_{k+1}) + e_h(x_{j-1}, y_k) + e_h(x_j, y_{k-1}) \right]$$

$$- \frac{h^2}{12} \left[\frac{\partial^4 u(\xi_j, y_k)}{\partial x^4} + \frac{\partial^4 u(x_j, \eta_k)}{\partial y^4} \right] \tag{8.102}$$

and from (8.97),

$$e_h(x_j, y_k) = 0 \qquad (x_j, y_k) \text{ a boundary grid point} \tag{8.103}$$

This system can be treated in a manner similar to that used in part 1, and the result (8.99) will follow. Because it is not central to the discussion of the linear systems, the argument is omitted. See Isaacson and Keller [I1, pp. 447–450]. ∎

Example Solve

$$\frac{\partial^2 u}{\partial x^2} + \frac{\partial^2 u}{\partial y^2} = 0, \quad 0 \le x, \ y \le 1$$

$$\begin{aligned} u(0, y) &= \cos(\pi y) & u(1, y) &= e^\pi \cos(\pi y) \\ u(x, 0) &= e^{\pi x} & u(x, 1) &= -e^{\pi x} \end{aligned} \tag{8.104}$$

The true solution is

$$u(x, y) = e^{\pi x} \cos(\pi y)$$

Numerical results for several values of N are given in Table 8.7. The error is the maximum over all grid points, and the "Ratio" column gives the factor by which the maximum error decreases when the grid size h is halved. Theoretically from (8.99), it should be 4.0; and the numerical results confirm this.

Table 8.7   Numerical solution of (8.104)

N	$\|u - u_h\|_\infty$	Ratio
4	.144	
		3.7
8	.0390	
		3.8
16	.0102	
		3.9
32	.00261	

Iterative solution   Because the Gauss-Seidel method is faster than the Gauss-Jacobi method for solving (8.97) to (8.98), we only consider the former. For $k = 1, 2, \ldots, N-1$, define

$$u_h^{(m+1)}(x_j, y_k) = \frac{1}{4}\left[u_h^{(m)}(x_{j+1}, y_k) + u_h^{(m)}(x_j, y_{k+1}) + u_h^{(m+1)}(x_{j-1}, y_k) \right.$$

$$\left. + u_h^{(m+1)}(x_j, y_{k-1}) \right] - \frac{h^2}{4} g(x_j, y_k) \qquad j = 1, 2, \ldots, N-1 \quad (8.105)$$

For boundary points, use

$$u_h^{(m)}(x_j, y_k) = f(x_j, y_k) \qquad \text{all } m \geq 0$$

The values of $u_h^{(m+1)}(x_j, y_k)$ are computed row by row, from the bottom row of grid points first to the top row of points last. And within each row, we solve from left to right.

For the iteration (8.105), the convergence analysis must be based on Theorem 8.7. It can be shown easily that the matrix is symmetric, and thus all eigenvalues are real. Moreover, the eigenvalues can all be shown to lie in the interval $0 < \lambda < 2$. From this and Problem 11 of Chapter 7, the matrix is positive definite. Since all diagonal coefficients of the matrix are positive, it then follows from Theorem 8.7 that the Gauss-Seidel method will converge. To show that all eigenvalues lie in $0 < \lambda < 2$, see Isaacson and Keller [I1, pp. 458–459] or Problem 3 of Chapter 9.

The calculation of the speed of convergence $r_\sigma(M)$ from (8.85) is nontrivial. The argument is quite sophisticated, and we will only refer to the very complete development in [I1, pp. 463–470], including the material on the acceleration of the Gauss-Seidel method. It can be shown that

$$r_o(M) = 1 - \pi^2 h^2 + O(h^4) \tag{8.106}$$

The Gauss-Seidel method converges, but the speed of convergence is quite slow for even moderately small values of h.

To accelerate the Gauss-Seidel method, use

$$v_h^{(m+1)}(x_j, y_k) = \frac{1}{4} \left[u_h^{(m)}(x_{j+1}, y_k) + u_h^{(m)}(x_j, y_{k+1}) + u_h^{(m+1)}(x_{j-1}, y_k) \right.$$

$$\left. + u_h^{(m+1)}(x_j, y_{k-1}) \right] - \frac{h^2}{4} g(x_j, y_k)$$

$$u_h^{(m+1)}(x_j, y_k) = \omega v_h^{(m+1)}(x_j, y_k) + (1-\omega) u_h^{(m)}(x_j, y_k) \qquad j = 1, \ldots, N-1 \tag{8.107}$$

for $k = 1, \ldots, N-1$. The optimal acceleration parameter is

$$\omega^* = \frac{2}{1 + \sqrt{1 - \xi^2}}$$

$$\xi = 1 - 2\sin^2\left(\frac{\pi}{2N}\right) \tag{8.108}$$

The corresponding rate of convergence is

$$r_o(M(\omega^*)) = \omega^* - 1 = 1 - 2h\pi + O(h^2) \tag{8.109}$$

This is a much better rate than that given by (8.106). The accelerated Gauss-Seidel method (8.107) with the optimal value ω^* of (8.108) is known as the SOR method. SOR is an abbreviation for *successive overrelaxation*, and is based on a physical interpretation of the method, first used in deriving it.

Example Recall the previous example (8.104). This was solved with both the Gauss-Seidel method and the SOR method. The initial guess for the iteration was taken to be

$$u_h^{(0)}(x, y) = \frac{1}{2} \left[(1-x)f(0, y) + xf(1, y) \right] + \frac{1}{2} \left[(1-y)f(x, 0) + yf(x, 1) \right]$$

at all interior grid points. The error test to stop the iteration was

$$\underset{1 \le j, k \le N-1}{\text{Max}} |u_h(x_j, y_k) - u_h^{(m)}(x_j, y_k)| \le \epsilon$$

with $\epsilon > 0$ given and using the right-hand side of (8.89) to predict the error in the iterate. The numerical results for the necessary number of iterates is given in

Table 8.8. The SOR method requires far fewer iterates for the smaller values of h than does the Gauss-Seidel method.

Table 8.8 Number of iterates necessary to solve (8.98)

N	ϵ	G-S	SOR
4	.01	7	6
8	.001	42	17
16	.001	155	33
32	.0001	794	73

Recall from the previous section that the number of iterates, called m^*, necessary to reduce the iteration error by a factor of ϵ is proportional to $1/\ln(c)$, where c is the ratio by which the iteration error decreases at each step. For the methods of this section, we take $c = r_\sigma(M)$. If we write $r_\sigma(M) = 1 - \delta$, then

$$\frac{1}{\ln r_\sigma(M)} = \frac{1}{\ln(1-\delta)} \approx \frac{1}{\delta}$$

When δ is halved, the number of iterates to be computed is doubled. For the Gauss-Seidel method,

$$\frac{1}{\ln r_\sigma(M)} \approx \frac{1}{\pi^2 h^2}$$

When h is halved, the number of iterates to be computed increases by a factor of four. For the SOR method,

$$\frac{1}{\ln r_\sigma(M(\omega^*))} \approx \frac{1}{2\pi h}$$

and when h is halved, the number of iterates is doubled. These two results are illustrated in Table 8.8 by the entries for $N=8$ and $N=16$.

In either case, note that doubling N will increase the number of equations to be solved by a factor of 4, and thus the work per iteration will increase by the same amount. The use of SOR greatly reduces the resulting work, although it still is large when N is large.

Discussion of the Literature

The references that have most influenced the presentation of Gaussian elimination in this chapter are the texts of Forsythe and Moler [F1], Isaacson and Keller [I1, Chapter 2], and Wilkinson [W1], [W2], along with the paper of Kahan [K1]. Other very good treatments are given in Conte and DeBoor [C1, Chapter 3], Noble [N1], Ortega [O1], and Stewart [S1].

Some excellent algorithms are given in [F1] and [W3]; and all computer centers should have a subroutine library for solving linear systems, similar to what is contained in these references. A new package of programs for solving linear systems is being developed by Bunch, Moler, and Stewart [B2]. It is called LINPACK and is being developed under the auspices of Argonne National Laboratory and the National Science Foundation. It is modeled after EISPACK, the successful package of computer programs for solving the matrix eigenvalue problem (see [S1] of Chapter 9). For a discussion of some of the problems involved in writing LINPACK, see Stewart [S2].

For general introductions to iteration methods for linear systems, see [I1, pp. 61–82, 463–479], Varga [V1], and Young [Y1]. Most of the theory in these texts is directed toward the linear systems arising in the numerical solution of partial differential equations. Other important references for the numerical solution of partial differential equations are the survey of Birkhoff [B1] and the classic account of Forsythe and Wasow [F2].

A very interesting iteration method for elliptic partial differential equations is given by Stone [S3]. Unlike the Gauss-Seidel and SOR methods of Section 8.8, the rate of convergence of Stone's method does not become worse as h decreases, and it converges quite rapidly. A second point to note in solving Poisson's equation is that often the system (8.98) can be solved directly. See Dorr [D1] for a general survey of such methods when the basic region is a rectangle.

Linear systems arise in solving many problems in addition to partial differential equations. For the general area of sparse linear systems, see the text of Tewarson [T1], the symposium proceedings of Rose and Willoughby [R1] and Bunch and Rose [B3], and the quite extensive survey of Duff [D2]. For linear integral equations, some iteration techniques for solving the resulting linear systems are given in Atkinson [A2, Part II, Chapter 4]. Finally, the least squares solution of over determined linear systems is described in Lawson and Hanson [L1], including some Fortran programs.

Bibliography

[A1] Aird, T. and R. Lynch, Computable accurate upper and lower error bounds for approximate solutions of linear algebraic systems, *ACM Trans. Math. Soft.*, **1**, 1975, pp. 217–231.

[A2] Atkinson, K., *A Survey of Numerical Methods for the Solution of Fredholm Integral Equations of the Second Kind*, SIAM Pub., Philadelphia, 1976.

[B1] Birkhoff, G., *The Numerical Solution of Elliptic Equations*, SIAM Pub., Philadelphia, 1972.

[B2] Bunch, J., C. Moler, and G. Stewart, LINPACK: A linear equations computer subroutine package, in preparation.

[B3]   Bunch, J. and D. Rose, eds., *Sparse Matrix Computations*, Academic Press, New York, 1976.

[C1]   Conte, S. and C. DeBoor, *Elementary Numerical Analysis*, 2nd ed., McGraw-Hill, New York, 1972.

[D1]   Dorr, F., The direct solution of the discrete Poisson equations on a rectangle, *SIAM Rev.*, **12** 1970, pp. 248–263.

[D2]   Duff, I. S., A survey of sparse matrix research, *Proc. IEEE*, 65 1977, pp. 500–535.

[F1]   Forsythe, G. and C. Moler, *Computer Solution of Linear Algebraic Systems*, Prentice-Hall Inc., Englewood Cliffs, N. J., 1967.

[F2]   Forsythe, G. and W. Wasow, *Finite Difference Methods for Partial Differential Equations*, John Wiley, New York, 1960.

[G1]   Gregory, R. and D. Karney, *A Collection of Matrices for Testing Computational Algorithms*, John Wiley, New York, 1969.

[I1]   Isaacson, E. and H. Keller, *Analysis of Numerical Methods*, John Wiley, New York, 1966.

[K1]   Kahan, W., Numerical linear algebra, *Canadian Math. Bull.*, **9** 1966, pp. 756–801.

[L1]   Lawson, C. and R. Hanson, *Solving Least Squares Problems*, Prentice-Hall, Inc., Englewood Cliffs, N. J., 1974.

[N1]   Noble, B., *Applied Linear Algebra*, Prentice-Hall, Inc., Englewood Cliffs, N. J., 1969.

[O1]   Ortega, J., *Numerical Analysis—A Second Course*, Academic Press, New York, 1972.

[R1]   Rose, D. and R. Willoughby, *Sparse Matrices and Their Applications*, Plenum Press, New York, 1972.

[S1]   Stewart, G., *Introduction to Matrix Computations*, Academic Press, New York, 1973.

[S2]   Stewart, G., Research, Development, and LINPACK, in *Mathematical Software III*, John Rice, ed., Academic Press, New York, 1977.

[S3]   Stone, H., Iterative solution of implicit approximations of multidimensional partial differential equations, *SIAM J. Num. Anal.*, **5**, 1968, pp. 530–558.

[T1]   Tewarson, P., *Sparse Matrices*, Academic Press, New York, 1973.

[V1]   Varga, R., *Matrix Iterative Analysis*, Prentice-Hall, Inc., Englewood Cliffs, N. J., 1962.

[W1] Wilkinson, J., *Rounding Errors in Algebraic Processes*, Prentice-Hall, Inc., Englewood Cliffs, N. J., 1963.

[W2] Wilkinson, J., *The Algebraic Eigenvalue Problem*, Oxford University Press, Oxford, 1965.

[W3] Wilkinson, J., and C. Reinsch, *Linear Algebra*, Handbook for Automatic Computation, Vol. 2, Springer-Verlag, New York, 1971.

[Y1] Young, D., *Iterative Solution of Large Linear Systems*, Academic Press, New York, 1971.

Problems

1. Solve the following systems $Ax = b$ by Gaussian elimination without pivoting. Check that $A = LU$, as in (8.5).

(a) $A = \begin{bmatrix} 1 & 1 & -1 \\ 1 & 2 & -2 \\ -2 & 1 & 1 \end{bmatrix}$ $b = \begin{bmatrix} 1 \\ 0 \\ 1 \end{bmatrix}$

(b) $A = \begin{bmatrix} 4 & 3 & 2 & 1 \\ 3 & 4 & 3 & 2 \\ 2 & 3 & 4 & 3 \\ 1 & 2 & 3 & 4 \end{bmatrix}$ $b = \begin{bmatrix} 1 \\ 1 \\ -1 \\ -1 \end{bmatrix}$

2. Consider the linear system

$$6x_1 + 2x_2 + 2x_3 = -2$$

$$2x_1 + \tfrac{2}{3}x_2 + \tfrac{1}{3}x_3 = 1$$

$$x_1 + 2x_2 - x_3 = 0$$

and verify its solution is

$$x_1 = 2.6 \qquad x_2 = -3.8 \qquad x_3 = -5.0$$

(a) Using four-digit, floating-point decimal arithmetic with rounding, solve the above system by Gaussian elimination without pivoting.

(b) Repeat part (a), but using partial pivoting.

In performing the arithmetic operations, remember to round to four significant digits after each operation, just as would be done on a computer.

3. **(a)** Implement the Factor and Solve algorithms of Section 8.2, or implement the analogous programs given in [F1, Chapters 16, 17].

   (b) To test the program, solve the system $Ax = b$ of order n, with $A = [a_{ij}]$ defined by

$$a_{ij} = \text{Max}(i,j)$$

   Also define $b = [1, 1, \ldots, 1]^T$. The true solution is $x = [0, 0, \ldots, 0, 1/n]^T$. This matrix is taken from [G1, p. 42].

4. **(a)** Show that the number of multiplications and divisions for the Gauss-Jordan method of Section 8.3 is about $n^3/2$.

   (b) Show how the Gauss-Jordan method, with partial pivoting, can be used to invert an $n \times n$ matrix within only $n(n+1)$ storage locations.

5. Use either the programs of Problem 3(a) or the Gauss-Jordan method to invert the matrices in problem 1.

6. Using the Cholesky method, calculate the decomposition $A = LL^T$ for

$$A = \begin{bmatrix} 2.25 & -3.0 & 4.5 \\ -3.0 & 5.0 & -10.0 \\ 4.5 & -10.0 & 34.0 \end{bmatrix}$$

7. Let A be real, symmetric, positive definite, and of order n. Consider solving $Ax = b$ using Gaussian elimination without pivoting. The purpose of this problem is to justify that the pivots will be nonzero.

   (a) Show that all of the diagonal elements satisfy $a_{ii} > 0$. This shows that a_{11} can be used as a pivot element.

   (b) After elimination of x_1 from equations 2 through n, let the resulting matrix $A^{(2)}$ be written as

$$A^{(2)} = \begin{bmatrix} a_{11} & a_{12} & \cdots & a_{1n} \\ 0 & & & \\ \vdots & & \hat{A}^{(2)} & \\ 0 & & & \end{bmatrix}$$

   Show that $\hat{A}^{(2)}$ is symmetric and positive definite.

   This procedure can be continued inductively to each stage of the elimination process, thus justifying the existence of nonzero pivots at every step.

Hint: to prove that $\hat{A}^{(2)}$ is positive definite, first prove the identity

$$\sum_{i,j=2}^{n} a_{ij}^{(2)} x_i x_j = \sum_{i,j=1}^{n} a_{ij} x_i x_j - a_{11}\left(x_1 + \sum_{j=2}^{n} \frac{a_{j1}}{a_{11}} x_j\right)^2$$

for any choice of $x_1, x_2, \ldots, x_n$.

8. Prove that if $A = LL^T$ with L real and nonsingular, then A is symmetric and positive definite.

9. Using the algorithm (8.28) to (8.29) for solving tridiagonal systems, solve $Ax = b$ with

$$A = \begin{bmatrix} 2 & -1 & 0 & 0 & 0 \\ -1 & 2 & -1 & 0 & 0 \\ 0 & -1 & 2 & -1 & 0 \\ 0 & 0 & -1 & 2 & -1 \\ 0 & 0 & 0 & -1 & 2 \end{bmatrix} \quad b = \begin{bmatrix} 1 \\ 0 \\ 0 \\ 0 \\ 0 \end{bmatrix}$$

Check that the hypotheses and conclusions of Theorem 8.2 are satisfied by this example.

10. Write a subroutine to solve tridiagonal systems using (8.28) to (8.29). Check it using the example in Problem 9. There are also a number of tridiagonal systems in [G1, Chapter 2], for which the true inverses are known.

11. (a) Consider solving $Ax = b$, with A and b complex and order$(A) = n$. Convert this problem to that of solving a real square system of order $2n$. Hint: write $A = A_1 + iA_2$, $b = b_1 + ib_2$, $x = x_1 + ix_2$, with $A_1, A_2, b_1, b_2, x_1, x_2$ all real. Determine equations to be satisfied by x_1 and x_2.

(b) Determine the storage requirements and the number of operations for this method of solving the complex system $Ax = b$. Compare these results with those based on directly solving $Ax = b$ using Gaussian elimination and complex arithmetic.

12. (a) Calculate the condition numbers cond$(A)_p$, $p = 1, 2, \infty$, for

$$A = \begin{bmatrix} 100 & 99 \\ 99 & 98 \end{bmatrix}$$

(b) Find the eigenvalues and eigenvectors of A, and use them to illustrate the remarks following (8.36) in Section 8.4.

13. Prove that if A is unitary, then cond$(A)_* = 1$.

14. As in Section 8.4, let $H_n = [1/(i+j-1)]$ denote the Hilbert matrix of order n; and let $\overline{H}_n$ denote the matrix obtained when H_n is entered into your computer in single precision arithmetic. To compare H_n^{-1} and $\overline{H}_n^{-1}$, convert $\overline{H}_n$ to a double precision matrix by appending additional zeroes to the mantissa of each entry. Then use a double-precision matrix inversion computer program to calculate $\overline{H}_n^{-1}$ numerically. This will give an accurate value of $\overline{H}_n^{-1}$ to single-precision accuracy, for lower values of n. After obtaining $\overline{H}_n^{-1}$, compare it with H_n^{-1}, given in (8.37) or [G1, pp. 34–37].

15. Using the programs of Problem 3, or using similar programs from your computer center, solve $H_n x = b$ for several values of n. Use

$$b = [1, -1, 1, -1, \ldots,]^T,$$

and calculate the true answer using $\overline{H}_n^{-1}$ from Problem 14.

16. Using the residual correction method, described at the beginning of Section 8.5, calculate accurate single precision answers to the linear systems of Problem 15. Print the residuals and corrections; and examine the rate of decrease in the correction terms as the order n is increased. Attempt to explain your results.

17. Consider solving the linear system $(\lambda I - K)x = b$ of order n, with $K = [k_{ij}]$,

$$k_{ij} = \frac{1}{n} \cos\left[\frac{(2i-1)(2j-1)\pi}{4n^2}\right] \qquad 1 \le i, j \le n$$

and $b = [1, 1, \ldots, 1]^T$. The value of λ is nonzero, and it is not to be an eigenvalue of the matrix K. Occasionally we want to solve such a system for several values of λ that are close together. Write a program to first solve the system for $\lambda_0 = 4$, and then save the LU decomposition of $\lambda_0 I - K$. To solve $(\lambda I - K)x = b$ with other values of λ nearby λ_0, use the residual correction method (8.53) with $C = [LU]^{-1}$. For example, solve the system when $\lambda = 4.1$, 4.5, 5, and 10. In each case, print the iterates and calculate the ratio in (8.60). Comment on the behavior of the iterates as λ increases.

18. The system $Ax = b$,

$$A = \begin{bmatrix} 4 & -1 & 0 & -1 & 0 & 0 \\ -1 & 4 & -1 & 0 & -1 & 0 \\ 0 & -1 & 4 & 0 & 0 & -1 \\ -1 & 0 & 0 & 4 & -1 & 0 \\ 0 & -1 & 0 & -1 & 4 & -1 \\ 0 & 0 & -1 & 0 & -1 & 4 \end{bmatrix} \qquad b = \begin{bmatrix} 2 \\ 1 \\ 2 \\ 2 \\ 1 \\ 2 \end{bmatrix}$$

has the solution $x = [1, 1, 1, 1, 1, 1]^T$. Solve the system using the Gauss-Jacobi iteration method, and then solve it again using the Gauss-Seidel method.

Use the initial guess $x^{(0)}=0$. Note the rate at which the iteration error decreases. Find the answers with an accuracy of $\epsilon=.001$.

19. For the error equation (8.83), show that $r_\sigma(M)<1$ if

$$\|P\|<\frac{1}{2\|A^{-1}\|}$$

for some matrix norm.

20. Let A and B have order n, with A nonsingular. Consider solving the linear system

$$Az_1+Bz_2=b_1 \qquad Bz_1+Az_2=b_2$$

with $z_1,z_2,b_1,b_2\in R^n$.

(a) Find necessary and sufficient conditions for convergence of the iteration method

$$Az_1^{(m+1)}=b_1-Bz_2^{(m)} \qquad Az_2^{(m+1)}=b_2-Bz_1^{(m)} \qquad m\geq 0$$

(b) Repeat part (a) for the iteration method

$$Az_1^{(m+1)}=b_1-Bz_2^{(m)} \qquad Az_2^{(m+1)}=b_2-Bz_1^{(m+1)} \qquad m\geq 0$$

Compare the convergence rates of the two methods.

21. (a) Let C_0 be an approximate inverse to A. Define $R_0=I-AC_0$, and assume $\|R_0\|<1$ for some matrix norm. Define the iteration method

$$C_{m+1}=C_m(I+R_m) \qquad R_{m+1}=I-AC_{m+1} \qquad m\geq 0$$

This is a well-known iteration method for calculating the inverse A^{-1}. Show the convergence of C_m to A^{-1} by first relating the error $A^{-1}-C_m$ to the residual R_m. And then examine the behavior of the residual R_m by showing that $R_{m+1}=R_m^2$, $m\geq 0$.

(b) Relate C_m to the expansion

$$A^{-1}=C_0(I-R_0)^{-1}=C_0\sum_{j=0}^{\infty} R_0^j$$

Observe the relation of this method for inverting A to the iteration method (2.5) of Chapter 2 for calculating $1/a$, for nonzero numbers a. Also, see Problem 1 of Chapter 2.

22. (a) Program the Gauss-Seidel method for solving the system (8.97) to

(8.98), as indicated in (8.105). Print the successive differences

$$d_m = \underset{1 \le j,k \le N-1}{\text{Maximum}} |u_h^{(m)}(x_j, y_k) - u_h^{(m-1)}(x_j, y_k)|$$

and the ratios d_{m+1}/d_m, $m \ge 1$. When h is halved, compare the resulting ratios with the convergence rate implied by (8.106).

(b) Solve the same system using the accelerated Gauss-Seidel method, given in (8.107) with the acceleration parameter ω^* in (8.108). Again, print the successive differences and ratios. Compare the results with those of part (a).

As a possible sample problem, consider solving

$$\frac{\partial^2 u}{\partial x^2} + \frac{\partial^2 u}{\partial y^2} = 0, \quad 0 < x, y < 1$$

with the true solution $u = x^3 - 3xy^2$ used to generate the boundary values $f(x, y)$. In this case, the discretization error in (8.96) is zero. And consequently, for the solution of the system (8.98),

$$u_h(x_j, y_k) = x_j^3 - 3x_j y_k^2 \qquad 1 \le j, k \le N-1$$

23. **(a)** Generalize the discretization of the Poisson equation in (8.95) to the equation

$$\frac{\partial^2 u}{\partial x^2} + \frac{\partial^2 u}{\partial y^2} - c(x, y)u = g(x, y) \qquad 0 < x, y < 1$$

with $u = f(x, y)$ on the boundary as before.

(b) Assume $c(x, y) \ge 0$ for $0 \le x, y \le 1$. Generalize part 1 of the proof of Theorem 8.8 to show that the linear system of part (a) will have a unique solution.

NINE

THE MATRIX EIGENVALUE PROBLEM

We will study the problem of calculating the eigenvalues and eigenvectors of a square matrix. This problem occurs in a number of contexts and the resulting matrices may take a variety of forms. These matrices may be sparse or dense, may have greatly varying order, and often are symmetric. In addition, what is to be calculated can vary enough as to affect the choice of method to be used. If only a few eigenvalues are to be calculated, then the numerical method will be different than if all eigenvalues are required.

The general problem of finding all eigenvalues and eigenvectors of an arbitrary nonsymmetric matrix is quite difficult. The eigenvalues and eigenvectors of a nonsymmetric matrix A can be quite unstable with respect to perturbations in the coefficients of A; and this makes more difficult the design of general methods and computer programs. The eigenvalues of a symmetric matrix A are quite stable with respect to perturbations in A. This is investigated in Section 9.1, along with the possible instability for nonsymmetric matrices. Because of the greater stability of the eigenvalue problem for symmetric matrices and because of its common occurrence, many methods have been developed especially for it. This will be a major emphasis of the development of this chapter, although methods for the nonsymmetric matrix eigenvalue problem are also discussed.

The eigenvalues of a matrix are usually calculated first; and they are used in calculating the eigenvectors, if these are desired. The main exception to this rule is the *power method* described in Section 9.2, which is useful in calculating a single dominant eigenvalue and associated eigenvector of a matrix. The usual procedure for calculating the eigenvalues of a matrix A is two-stage. First, similarity transformations are used to reduce A to a simpler form, which is usually tridiagonal for symmetric matrices. And second, this reduced matrix is used to calculate the eigenvalues, and also the eigenvectors if they are required. The main form of similarity transformations used are certain special unitary or orthogonal matrices; these are discussed in Section 9.3. For the calculation of the eigenvalues of a symmetric tridiagonal matrix, the theory of Sturm sequences is introduced in Section 9.4 and the QR algorithm is discussed in Section 9.5. Once the eigenvalues have been calculated, the most powerful technique for calculating the eigenvectors is the method of *inverse iteration*. It is discussed and illustrated in Section 9.6. It should be noted that we will use the words *symmetric* and *nonsymmetric* quite generally, where they ordinarily should be used only in connection with real matrices. For complex matrices, always substitute *Hermitian* and *non-Hermitian*, respectively.

Most numerical methods used at present have been developed since 1950. They are nontrivial to implement as computer programs, especially those that are to be used for nonsymmetric matrices. Beginning in the middle 1960s, algorithms for a variety of matrix eigenvalue problems were published, in ALGOL, in the journal *Numerische Mathematik*. These were tested extensively, and were subsequently revised based on the tests and on new theoretical results. These algorithms have been collected together in Wilkinson and Reinsch [W4, Part II]. A group within the Applied Mathematics Division of Argonne National Laboratory has translated these programs into FORTRAN, and further testing and improvement was carried out. This package of programs is called EISPACK and it is available from Argonne National Laboratory. A complete description of the package, including all programs, is given in Smith et al. [S1].

9.1   Eigenvalue   Location,   Error,   and Stability Results

We begin by giving some results for bounding the eigenvalues of a matrix A. For a crude upper bound, recall from Theorem 7.8 of Chapter 7 that

$$\underset{\lambda \in \sigma(A)}{\text{Max}} |\lambda| \le \|A\| \tag{9.1}$$

for any matrix norm. The notation $\sigma(A)$ denotes the set of all eigenvalues of A. The next result is a simple computational technique for giving better estimates for the location of the eigenvalues of A.

For $A = [a_{ij}]$ of order n, define

$$r_i = \sum_{\substack{j=1 \\ j \ne i}}^{n} |a_{ij}| \qquad i = 1, 2, \ldots, n \tag{9.2}$$

and let Z_i denote the circle in the complex plane with center a_{ii} and radius r_i,

$$Z_i = \{ z \in C \, | \, |z - a_{ii}| \le r_i \} \tag{9.3}$$

Theorem 9.1   (Gerschgorin) Let A have order n and let λ be an eigenvalue of A. Then λ belongs to one of the circles Z_i. Moreover if m of the circles form a connected set S, disjoint from the remaining $n-m$ circles, then S contains exactly m of the eigenvalues of A, counted according to their multiplicity as roots of the characteristic polynomial of A.

Since A and A^T have the same eigenvalues and characteristic polynomial, these results are also valid if summation within the column, rather than in the row, is used in defining the radii in (9.2).

Proof Figure 9.1 gives a picture in the complex plane of what the circles might look like for a complex matrix of order 3. The solid circles are the ones given by (9.3), and the dotted ones occur later in the proof. According to the theorem, there should be one eigenvalue in Z_3, and two eigenvalues in the union of Z_1 and Z_2.

Let λ be an eigenvalue of A, and let x be a corresponding eigenvector. Let k be the subscript of a component of x for which

$$|x_k| = \max_{1 \leq i \leq n} |x_i| = \|x\|_\infty$$

Then from $Ax = \lambda x$, the kth component yields

$$\sum_{j=1}^{n} a_{kj} x_j = \lambda x_k$$

$$(\lambda - a_{kk}) x_k = \sum_{\substack{j=1 \\ j \neq k}}^{n} a_{kj} x_j$$

$$|\lambda - a_{kk}| |x_k| \leq \sum_{\substack{j=1 \\ j \neq k}}^{n} |a_{kj}| |x_j| \leq r_k \|x\|_\infty$$

Canceling $\|x\|_\infty$ proves the first part of the theorem.
Define

$$D = \text{diag}[a_{11}, a_{22}, \ldots, a_{nn}] \qquad E = A - D$$

For $0 \leq \epsilon \leq 1$, define

$$A(\epsilon) = D + \epsilon E \tag{9.4}$$

and denote its eigenvalues by $\lambda_1(\epsilon), \ldots, \lambda_n(\epsilon)$. Note that $A(1) = A$, the original matrix. The eigenvalues are the roots of the characteristic polynomial

$$f_\epsilon(\lambda) \equiv \det[A(\epsilon) - \lambda I]$$

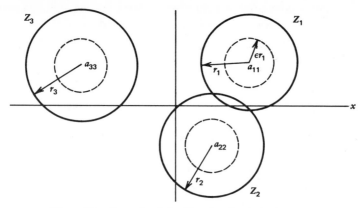

Figure 9.1   Example of the Gerschgorin circle theorem.

Since the coefficients of $f_\epsilon(\lambda)$ are continuous functions of ϵ, and since the roots of any polynomial are continuous functions of its coefficients (see Henrici [H1, p. 281]), we have that $\lambda_1(\epsilon),\dots,\lambda_n(\epsilon)$ are continuous functions of ϵ. As the parameter ϵ changes, each eigenvalue $\lambda_i(\epsilon)$ will vary in the complex plane, marking out a path from $\lambda_i(0)$ to $\lambda_i(1)$.

From the first part of the theorem, we know the eigenvalues $\lambda_i(\epsilon)$ are contained in the circles

$$Z_i(\epsilon) = \{z \in C \,||z - a_{ii}| \le \epsilon r_i\} \qquad i = 1,\dots,n \qquad (9.5)$$

with r_i defined as before in (9.2). Examples of these circles are given in Figure 9.1 by the dotted circles. These circles decrease as ϵ goes from 1 to 0, and the eigenvalues $\lambda_i(\epsilon)$ must remain within them. When $\epsilon = 0$, the eigenvalues are simply

$$\lambda_i(0) = a_{ii}$$

Considering the path $\lambda_i(\epsilon)$, $0 \le \epsilon \le 1$, it must remain in the circle $Z_i(1)$ in which it begins at $\epsilon = 0$. Thus if m of the circles $Z_i(1)$ form a connected set S, disjoint from the remaining $n - m$ circles, then S must contain exactly m eigenvalues $\lambda_i(1)$, as it contains the m eigenvalues $\lambda_i(0)$. This proves the second result.

Since

$$\det[A - \lambda I] = \det[A - \lambda I]^T = \det[A^T - \lambda I]$$

we have $\sigma(A) = \sigma(A^T)$. Thus apply the theorem to the rows of A^T in order to prove it for the columns of A. This completes the proof. ∎

This theorem can be used in a number of ways, but we will just provide two simple numerical examples.

Example Consider the matrix

$$A = \begin{bmatrix} 4 & 1 & 0 \\ 1 & 0 & -1 \\ 1 & 1 & -4 \end{bmatrix}$$

From the above theorem, the eigenvalues must be contained in the circles

$$|\lambda - 4| \leq 1 \qquad |\lambda| \leq 2 \qquad |\lambda + 4| \leq 2 \qquad (9.6)$$

Since the first circle is disjoint from the remaining ones, there must be a single root in the circle. Since the coefficients of

$$f(\lambda) = \det[A - \lambda I]$$

are real, the complex eigenvalues must occur in conjugate pairs, if they occur at all. This will easily imply, with (9.6), that there is a real eigenvalue in the interval $[3, 5]$. The last two circles touch at the single point $\lambda = -2$. Using the same reasoning as before, the eigenvalues in these two circles must be real. And by using the construction (9.4) of $A(\epsilon)$, $\epsilon < 1$, there is one eigenvalue in $[-6, -2]$ and one in $[-2, 2]$. Since it is easily checked that $\lambda = -2$ is not an eigenvalue, we can conclude that A has one real eigenvalue in each of the intervals

$$[-6, -2) \qquad (-2, 2] \qquad [3, 5]$$

The true eigenvalues are

$$-3.76010 \qquad -.442931 \qquad 4.20303$$

Example Consider

$$A = \begin{bmatrix} 4 & 1 & 0 & \cdots & & \cdots & 0 \\ 1 & 4 & 1 & 0 & & & 0 \\ 0 & 1 & 4 & 1 & & 0 & \vdots \\ \vdots & & & & & & \vdots \\ \vdots & & & & 1 & & 4 & 1 \\ 0 & & & & & 0 & 1 & 4 \end{bmatrix}$$

a matrix of order n. Since A is symmetric, all eigenvalues of A are real. The radii r_i of (9.2) are all 1 or 2, and all the centers of the circles are $a_{ii} = 4$. Thus from the above theorem, the eigenvalues must all lie in the interval $[2, 6]$. Since the eigenvalues of A^{-1} are the reciprocals of those of A, we must have

$$\frac{1}{6} \leq \mu \leq \frac{1}{2}$$

for all eigenvalues μ of A^{-1}. Using the matrix norm (7.47) induced by the Euclidean vector norm, we have

$$\|A^{-1}\|_2 = r_\sigma(A^{-1}) \leq \frac{1}{2}$$

independent of the size of n.

Bounds for perturbed eigenvalues   Given a matrix A, we wish to perturb it and to then observe the effect on the eigenvalues of A. Analytical bounds will be derived for the perturbations in the eigenvalues based on the perturbations in the matrix A. These bounds will also suggest a definition of *condition number* that can be used to indicate the degree of stability or instability present in the eigenvalues.

To simplify the arguments considerably we will assume that the Jordan canonical form of A is diagonal (see Theorem 7.6):

$$P^{-1}AP = \operatorname{diag}[\lambda_1, \ldots, \lambda_n] \equiv D \tag{9.7}$$

for some nonsingular matrix P. The columns of P will be the eigenvectors of A, corresponding to the eigenvalues $\lambda_1, \ldots, \lambda_n$. Matrices for which (9.7) holds are the most important case in practice. For a brief discussion of the case in which the Jordan canonical form is not diagonal, see the last topic of this section.

We also need to assume a special property for the matrix norms to be used. For any diagonal matrix

$$G = \operatorname{diag}[g_1, \ldots, g_n]$$

we must have that

$$\|G\| = \operatorname*{Max}_{1 \leq i \leq n} |g_i| \tag{9.8}$$

All of the operator matrix norms induced by the vector norms $\|x\|_p$, $1 \leq p \leq \infty$, have this property. We can now state the following result.

Theorem 9.2   (Bauer-Fike)   Let A be a matrix with a diagonal Jordan canonical form, as in (9.7). And assume the matrix norm satisfies (9.8). Let $A + E$ be a perturbation of A, and let λ be an eigenvalue of $A + E$. Then

$$\operatorname*{Min}_{1 \leq i \leq n} |\lambda - \lambda_i| \leq \|P\| \|P^{-1}\| \|E\| \tag{9.9}$$

Proof   If λ is also an eigenvalue of A, then (9.9) is trivially true. Thus assume $\lambda \neq \lambda_1, \lambda_2, \ldots, \lambda_n$. And let x be an eigenvector for $A + E$ corresponding to

λ. Then

$$(A + E)x = \lambda x$$

$$(\lambda I - A)x = Ex$$

Substitute from (9.7) and multiply by P^{-1} to obtain

$$(\lambda I - PDP^{-1})x = Ex$$

$$(\lambda I - D)(P^{-1}x) = (P^{-1}EP)(P^{-1}x)$$

Since $\lambda \neq \lambda_1, \ldots, \lambda_n$, $\lambda I - D$ is nonsingular,

$$(\lambda I - D)^{-1} = \operatorname{diag}\left[(\lambda - \lambda_1)^{-1}, \ldots, (\lambda - \lambda_n)^{-1}\right]$$

Then

$$P^{-1}x = (\lambda I - D)^{-1}(P^{-1}EP)(P^{-1}x)$$

$$\|P^{-1}x\| \leq \|(\lambda I - D)^{-1}\| \, \|P^{-1}EP\| \, \|P^{-1}x\|$$

Canceling $\|P^{-1}x\|$ and using (9.8),

$$1 \leq \left[\operatorname{Max}|\lambda - \lambda_i|^{-1}\right] \|P^{-1}\| \, \|P\| \, \|E\|.$$

This is equivalent to (9.9), completing the proof. ∎

Corollary If A is Hermitian, and if $A + E$ is any perturbation of A, then

$$\operatorname*{Min}_{1 \leq i \leq n} |\lambda - \lambda_i| \leq \|E\|_2 \tag{9.10}$$

for any eigenvalue λ of $A + E$.

Proof Since A is Hermitian, the matrix P can be chosen to be unitary. And using the operator norm (7.44) induced by the Euclidean vector norm, $\|P\|_2 = \|P^{-1}\|_2 = 1$ (see Problem 10 of Chapter 7). This completes the proof. ∎

The statement (9.10) proves that small perturbations of a Hermitian matrix lead to equally small perturbations in the eigenvalues, as was asserted in the introduction. Note that the relative error in some or all of the eigenvalues may still be large, and this occurs commonly when the eigenvalues of a matrix vary greatly in magnitude.

Example Consider the Hilbert matrix of order three,

$$H_3 = \begin{bmatrix} 1 & \frac{1}{2} & \frac{1}{3} \\ \frac{1}{2} & \frac{1}{3} & \frac{1}{4} \\ \frac{1}{3} & \frac{1}{4} & \frac{1}{5} \end{bmatrix}$$

Its eigenvalues to seven significant digits are

$$\lambda_1 = 1.408319 \qquad \lambda_2 = .1223271 \qquad \lambda_3 = .002687340 \qquad (9.11)$$

Now consider the perturbed matrix $\overline{H}_3$, representing H_3 to four significant digits,

$$\hat{H}_3 = \begin{bmatrix} 1.000 & .5000 & .3333 \\ .5000 & .3333 & .2500 \\ .3333 & .2500 & .2000 \end{bmatrix} \qquad (9.12)$$

Its eigenvalues to seven significant digits are

$$\hat{\lambda}_1 = 1.408294 \qquad \hat{\lambda}_2 = .1223415 \qquad \hat{\lambda}_3 = .002664489 \qquad (9.13)$$

To verify the validity of (9.10) for this case, it is straightforward to calculate

$$\|E\| = r_\sigma(E) = \tfrac{1}{3} \times 10^{-4} \doteq .000033$$

For the errors and relative errors in (9.13),

$$\lambda_1 - \hat{\lambda}_1 = .0000249 \qquad \text{Rel}(\hat{\lambda}_1) = .0000177$$

$$\lambda_2 - \hat{\lambda}_2 = -.0000144 \quad \text{Rel}(\hat{\lambda}_2) = -.000118$$

$$\lambda_3 - \hat{\lambda}_3 = .0000229 \qquad \text{Rel}(\hat{\lambda}_3) = .0085$$

All of the errors satisfy (9.10). But the relative error in $\hat{\lambda}_3$ is quite significant compared to the perturbation in $\hat{H}_3$.

For a nonsymmetric matrix A with P as in (9.7), the number

$$K(A) = \|P\| \|P^{-1}\| \qquad (9.14)$$

is called the condition number for the eigenvalue problem for A. This is based on the bound (9.9) for the perturbations in the eigenvalues of the matrix A when it is perturbed. Another choice, even more difficult to compute, would be to use

$$K(A) = \text{Infimum} \|P\| \|P^{-1}\|$$

with the infimum taken over all matrices P for which (9.7) holds and over all matrix norms satisfying (9.8). The reason for having condition numbers is that for nonsymmetric matrices A, small perturbations E can lead to relatively large perturbations in the eigenvalues of A.

Example To illustrate the pathological problems that can occur with nonsymmetric matrices, consider

$$A = \begin{bmatrix} 101 & -90 \\ 110 & -98 \end{bmatrix} \qquad A + E = \begin{bmatrix} 101 - \epsilon & -90 - \epsilon \\ 110 & -98 \end{bmatrix}$$

The eigenvalues of A are $\lambda = 1, 2$; and the eigenvalues of $A + E$ are

$$\lambda = \frac{3 - \epsilon \pm \sqrt{1 - 828\epsilon + \epsilon^2}}{2}$$

As a specific example to give better intuition, take $\epsilon = .001$. Then

$$A + E = \begin{bmatrix} 100.999 & -90.001 \\ 110 & -98 \end{bmatrix} \tag{9.15}$$

and its eigenvalues are

$$\lambda \doteq 1.298, 1.701 \tag{9.16}$$

This example should not be taken to imply that all nonsymmetric matrices are ill-conditioned. Most cases in practice are fairly well conditioned. But in writing a general algorithm, we always seek to cover as many cases as possible; and this example shows that this is likely to be difficult for the class of all nonsymmetric matrices.

For symmetric matrices, the result (9.10) can be improved upon in several ways. There is a *minimax characterization* for the eigenvalues of symmetric matrices. For a discussion of this theory and the resultant error bounds, see Wilkinson [W2, p. 101]. Instead, we give the following result, which will be more useful for error analyses of methods presented later.

Theorem 9.3 (Wielandt-Hoffman) Let A and E be real, symmetric matrices of order n, and define $\hat{A} = A + E$. Let λ_i and $\hat{\lambda}_i$, $i = 1, \ldots, n$, be the eigenvalues of A and $\hat{A}$, respectively, arranged in increasing order.

Then

$$\left[\sum_{j=1}^{n} \left(\lambda_j - \hat{\lambda}_j \right)^2 \right]^{1/2} \leq F(E) \qquad (9.17)$$

where $F(E)$ is the Frobenius norm of E, defined in (7.36).

Proof See [W2, pp. 104–108].   ∎

This result will be used later in bounding the effect of the rounding errors which occur in reducing a symmetric matrix to tridiagonal form.

A computable error bound for symmetric matrices Let A be a symmetric matrix for which an approximate eigenvalue λ and approximate eigenvector x has been computed. Assume that $\|x\|_2 = 1$, and define the residual

$$\eta = Ax - \lambda x$$

Since A is symmetric, there is a unitary matrix U for which

$$U^*AU = \text{diag}[\lambda_1, \dots, \lambda_n] \equiv D \qquad (9.18)$$

Then we will show that

$$\underset{1 \leq i \leq n}{\text{Min}} |\lambda - \lambda_i| \leq \|\eta\|_2 \qquad (9.19)$$

Using (9.18),

$$\eta = UDU^*x - \lambda x$$

$$U^*\eta = DU^*x - \lambda U^*x = (D - \lambda I)U^*x.$$

If λ is an eigenvalue of A, then (9.19) is trivially true. Thus there is no loss of generality is assuming $\lambda \neq \lambda_1, \dots, \lambda_n$. Thus $D - \lambda I$ is nonsingular, and

$$U^*x = (D - \lambda I)^{-1} U^*\eta$$

$$\| U^*x \|_2 \leq \| (D - \lambda I)^{-1} \|_2 \| U^*\eta \|_2$$

Recall Problem 10 of Chapter 7, which implies

$$\| U^*x \|_2 = \| x \|_2 = 1 \qquad \| U^*\eta \|_2 = \| \eta \|_2$$

Then using the definition of the matrix norm, we have

$$1 \leq \left[\underset{1 \leq i \leq n}{\text{Max}} |(\lambda - \lambda_i)^{-1}| \right] \| \eta \|_2$$

which is equivalent to (9.19).

The use of (9.19) will be illustrated later in (9.51) of Section 9.2, using an approximate eigenvalue-eigenvector pair produced by the power method.

Stability of eigenvalues for nonsymmetric matrices   In order to deal effectively with the potential for instability in the nonsymmetric matrix eigenvalue problem, it is necessary to have a better understanding of the nature of that instability. For example, one consequence of the analysis of instability will be that unitary similarity transformations will not make worse the conditioning of the problem. As before, assume that A has a diagonal Jordan canonical form,

$$P^{-1}AP = \text{diag}[\lambda_1, \ldots, \lambda_n] \equiv D \tag{9.20}$$

Then $\lambda_1, \ldots, \lambda_n$ are the eigenvalues of A; and the columns of P are the corresponding eigenvectors, call them $u_1, \ldots, u_n$. The matrix P is not unique. For example, if F is any nonsingular diagonal matrix, then

$$(PF)^{-1}A(PF) = F^{-1}DF = D$$

By choosing F appropriately, the columns of PF will have length one. Thus without loss of generality, assume the columns of P will have length one, that is,

$$Au_i = \lambda_i u_i \qquad u_i^* u_i = 1 \qquad i = 1, \ldots, n \tag{9.21}$$

with

$$P = [u_1, \ldots, u_n]$$

By taking the conjugate transpose in (9.20),

$$P^*A^*(P^*)^{-1} = D^* = \text{diag}[\bar{\lambda}_1, \ldots, \bar{\lambda}_n]$$

which shows that the eigenvalues of A^* are the complex conjugates of those of A. Writing

$$(P^*)^{-1} = [w_1, \ldots, w_n]$$

we have

$$A^* w_i = \bar{\lambda}_i w_i \qquad i = 1, \ldots, n \tag{9.22}$$

Since $P^{-1}P = I$, and since

$$P^{-1} = \begin{bmatrix} w_1^* \\ \vdots \\ w_n^* \end{bmatrix}$$

we have

$$w_i^* u_j = \begin{cases} 1, & i=j \\ 0, & i \neq j \end{cases} \qquad (9.23)$$

This says the eigenvectors $\{u_i\}$ of A and the eigenvectors $\{w_i\}$ of A^* form a *biorthogonal* set.

Normalize the eigenvectors w_i by

$$v_i = w_i / \|w_i\|_2 \qquad i = 1, \ldots, n$$

Define

$$s_i = v_i^* u_i = 1 / \|w_i\|_2 \qquad (9.24)$$

a positive real number. The matrix $(P^*)^{-1}$ can now be written

$$(P^*)^{-1} = \left[\frac{v_1}{s_1}, \ldots, \frac{v_n}{s_n} \right]$$

And

$$A^* v_i = \bar{\lambda}_i v_i \qquad \|v_i\|_2 = 1 \qquad i = 1, \ldots, n \qquad (9.25)$$

We will now examine the stability of a simple eigenvalue λ_k of A. Being simple means that λ_k has multiplicity one as a root of the characteristic polynomial of A. The results can be extended to eigenvalues of multiplicity greater than one, but we omit that case. Consider the perturbed matrix

$$A(\epsilon) = A + \epsilon B \qquad \epsilon > 0$$

for some matrix B, independent of ϵ. Denote the eigenvalues of $A(\epsilon)$ by $\lambda_1(\epsilon), \ldots, \lambda_n(\epsilon)$. Then

$$P^{-1}A(\epsilon)P = D + \epsilon C \qquad C = P^{-1}BP$$

$$c_{ij} = \frac{1}{s_i} v_i^* B u_j \qquad 1 \leq i, j \leq n \qquad (9.26)$$

We will prove that

$$\lambda_k(\epsilon) = \lambda_k + \frac{\epsilon}{s_k} v_k^* B u_k + 0(\epsilon^2) \qquad (9.27)$$

The derivation of this result uses the Gerschgorin Theorem 9.1. We also need to note that for any nonsingular diagonal matrix F,

$$FP^{-1}A(\epsilon)PF^{-1} = D + \epsilon FCF^{-1} \qquad (9.28)$$

and this leaves the eigenvalues of $A(\epsilon)$ unchanged. Pick F as follows:

$$F_{ii} = \begin{cases} \epsilon\alpha, & i = k \\ 1, & i \neq k \end{cases}$$

with α a positive constant to be determined later. Most of the coefficients of the matrix (9.28) are not changed, and only those in row k and column k need to be considered. They are

$$\left[D + \epsilon FCF^{-1}\right]_{kj} = \begin{cases} \epsilon^2 \alpha c_{kj}, & j \neq k \\ \lambda_k + \epsilon c_{kk}, & j = k \end{cases}$$

$$\left[D + \epsilon FCF^{-1}\right]_{ik} = \frac{1}{\alpha} c_{ik} \qquad i \neq k$$

Apply Theorem 9.1 to the matrix (9.28). The circle centers and radii are

$$\text{center} = \lambda_k + \epsilon c_{kk} \qquad r_k = \epsilon^2 \alpha \sum_{j \neq k} |c_{kj}|$$

$$\text{center} = \lambda_i + \epsilon c_{ii} \qquad r_i = \epsilon \sum_{j \neq i,k} |c_{ij}| + \frac{1}{\alpha} |c_{ik}| \qquad i \neq k \qquad (9.29)$$

We wish to pick α so large and ϵ sufficiently small so as to isolate the circle about $\lambda_k + \epsilon c_{kk}$ from the remaining circles; and in that way, know there is exactly one eigenvalue of (9.28) within circle k. The distance between the centers of circles k and $i \neq k$ is bounded from below by

$$|\lambda_i - \lambda_k| - \epsilon|c_{ii} - c_{kk}|$$

which is about $|\lambda_i - \lambda_k|$ for small values of ϵ. Pick α such that

$$\frac{1}{\alpha}|c_{ik}| \leq \tfrac{1}{2}|\lambda_i - \lambda_k| \qquad \text{all } i \neq k \qquad (9.30)$$

Then choose ϵ_0 such that for all $0 < \epsilon \leq \epsilon_0$, circle k does not intersect any of the remaining circles. This can be done because λ_k is distinct from the remaining eigenvalues λ_i, and because of the inequality (9.30).

Using this construction, Theorem 9.1 implies that circle k contains exactly one eigenvalue of $A(\epsilon)$, call it $\lambda_k(\epsilon)$. From (9.29),

$$|\lambda_k(\epsilon) - \lambda_k - \epsilon c_{kk}| \leq r_k = 0(\epsilon^2)$$

and using the formula for c_{kk} in (9.26), this proves (9.27). Taking bounds in (9.27), we obtain

$$|\lambda_k(\epsilon) - \lambda_k| \leq \frac{\epsilon}{s_k} \|v_k\|_2 \|B\|_2 \|u_k\|_2 + 0(\epsilon^2)$$

and using (9.21) and (9.25),

$$|\lambda_k(\epsilon) - \lambda_k| \leq \frac{\epsilon}{s_k} \|B\|_2 + 0(\epsilon^2) \tag{9.31}$$

The number s_k is intimately related to the stability of the eigenvalue of λ_k, when the matrix A is perturbed by small amounts $E = \epsilon B$. If A were symmetric, we would have $u_k = v_k$; and thus $s_k = 1$, giving the same qualitative result for symmetric matrices as derived previously. For nonsymmetric matrices, if s_k is quite small, then small perturbations $E = \epsilon B$ can lead to a large perturbation in the eigenvalue λ_k. Such problems are called ill-conditioned.

Example Recall the earlier example (9.15). Then $\epsilon = .001$,

$$P^{-1}AP = \begin{bmatrix} 1 & 0 \\ 0 & 2 \end{bmatrix} \qquad B = \begin{bmatrix} -1 & -1 \\ 0 & 0 \end{bmatrix}$$

$$P = \begin{bmatrix} 9 & -10 \\ 10 & -11 \end{bmatrix} \qquad P^{-1} = \begin{bmatrix} -11 & 10 \\ -10 & 9 \end{bmatrix}$$

If we use the row norm, then the condition number (9.14) is

$$K(A) = 441$$

To calculate (9.31) for $\lambda_1 = 1$, we have

$$s_1 = \frac{1}{\sqrt{40001}} \doteq .005 \qquad \|B\|_2 = \sqrt{2} \qquad \epsilon = .001$$

Then

$$|\lambda_1(\epsilon) - \lambda_1| \leq \frac{\sqrt{2}}{.005} \cdot \epsilon + 0(\epsilon^2) \doteq 283\epsilon + 0(\epsilon^2) \tag{9.32}$$

and for $\epsilon = .001$, this gives the previously observed perturbation in (9.16),

$$\lambda_1(\epsilon) \doteq 1.298 \qquad \lambda_1(\epsilon) - \lambda_1 \doteq .298$$

In the following sections some of the numerical methods will first convert the matrix A to a simpler form using similarity transformation. We wish to use transformations that will not make the numbers s_k even smaller, which would make an ill-conditioned problem even worse. From this viewpoint, unitary transformations are the best ones to use.

Let U be unitary, and let $\hat{A} = U^*AU$. For a simple eigenvalue λ_k, let s_k and $\hat{s}_k$ denote the numbers (9.24) for the two matrices. If $\{u_i\}$ and $\{v_i\}$ are the eigenvectors of A and A^*, then $\{U^*u_i\}$ and $\{U^*v_i\}$ are the corresponding eigenvectors for $\hat{A}$ and $\hat{A}^*$. For $\hat{s}_k$,

$$\hat{s}_k = (U^*v_k)^*(U^*u_k) = v_k^* UU^*u_k = v_k^* u_k$$

$$= s_k$$

Thus the stability of the eigenvalue λ_k is made neither better nor worse. Unitary transformations also preserve vector length and the angles between vectors; see Problem 10 of Chapter 7. In general unitary matrix operations on a given matrix A will not cause any deterioration in the conditioning of the eigenvalue problem; and for that reason they are the preferred form of similarity transformation in numerical analysis.

Matrices with nondiagonal Jordan canonical form   We have avoided discussing the eigenvalue problem for those matrices for which the Jordan canonical form is not diagonal. There are problems of instability in the eigenvalue problem, worse than that given in Theorem 7.2. And there are significant problems in the determination of a correct basis for the eigenvectors.

Rather than giving a general development, we illustrate the difficulties for this class of matrices by examining one simple case in detail. Let

$$A = \begin{bmatrix} 1 & 1 & 0 & \cdots & \cdots & 0 \\ 0 & 1 & 1 & 0 & \cdots & \\ \vdots & & \ddots & \ddots & \ddots & \vdots \\ \vdots & & & \ddots & 1 & 1 \\ 0 & \cdots & \cdots & \cdots & 0 & 1 \end{bmatrix} \tag{9.33}$$

be a matrix of order n. The characteristic polynomial is

$$f(\lambda) = (1-\lambda)^n$$

and $\lambda = 1$ is a root of multiplicity n. There is only a one-dimensional set of eigenvectors, spanned by

$$x = [1, 0, \ldots, 0]^T \tag{9.34}$$

For $\epsilon > 0$, perturb A to

$$A(\epsilon) = \begin{bmatrix} 1 & 1 & 0 & \cdots & \cdots & 0 \\ 0 & 1 & 1 & 0 & & \vdots \\ \vdots & & \ddots & \ddots & & \vdots \\ 0 & & & & 1 & 1 \\ \epsilon & 0 & \cdots & \cdots & 0 & 1 \end{bmatrix} \tag{9.35}$$

Its characteristic polynomial is

$$f_\epsilon(\lambda) = (1-\lambda)^n - (-1)^n \epsilon$$

There are n distinct roots,

$$\lambda_k(\epsilon) = 1 + \omega_k \sqrt[n]{\epsilon} \qquad k=1,\ldots,n \tag{9.36}$$

with $\{\omega_k\}$ the nth roots of unity,

$$\omega_k = e^{2\pi k i/n} \qquad k=1,\ldots,n$$

For the perturbations in the eigenvalue of A,

$$|\lambda_k(\epsilon) - \lambda_k| = \sqrt[n]{\epsilon} \tag{9.37}$$

For example, if $n=10$ and $\epsilon = 10^{-10}$, then

$$|\lambda_k(\epsilon) - \lambda_k| = .1 \tag{9.38}$$

The earlier result (9.9) gave a bound that was linear in ϵ; and (9.37) is much worse, as shown by (9.38).

Since $A(\epsilon)$ has n distinct eigenvalues, it also has a complete set of n linearly independent eigenvectors, call them $x_1(\epsilon),\ldots,x_n(\epsilon)$. First we must give the relationship of these eigenvectors to the single eigenvector x in (9.34). This is a difficult problem to deal with, and it always must be dealt with for matrices whose Jordan form is not diagonal.

The matrices A and $A(\epsilon)$ are in extremely simple form, and they merely hint at the difficulties that can occur when a matrix is not similar to a diagonal matrix. And this example is correct from the qualitative point of view in showing the difficulties that arise.

9.2   The Power Method

This is a classical method, used mainly in finding the dominant eigenvalue and associated eigenvector of a matrix. It is not a general method, but is useful in a number of situations. For example, it is sometimes a satisfactory method with large sparse matrices, where the methods of later sections cannot be used because of computer memory size limitations. In addition the method of inverse iteration, described in Section 9.6, is the power method applied to an appropriate inverse matrix. And the considerations of this section are an introduction to that later material. It is extremely difficult to implement the power method as a general purpose computer program, treating a large and quite varied class of matrices. But it is easy to implement for more special classes.

We assume that A is an $n \times n$ matrix for which the Jordan canonical form is diagonal. Let $\lambda_1,\ldots,\lambda_n$ denote the eigenvalues of A, and let $x_1,\ldots,x_n$ be the corresponding eigenvectors, which form a basis for C^n. We further assume that

$$|\lambda_1| > |\lambda_2| \geq |\lambda_3| \geq \cdots \geq |\lambda_n| \geq 0 \qquad (9.39)$$

Although quite special, this is the main case of interest for the application of the power method. And the development can be extended fairly easily to the case of a single dominant eigenvalue of geometric multiplicity $r > 1$; see Problem 8.

Let $z^{(0)}$ be an initial guess of some multiple of the eigenvector x_1. If there is no rational method for choosing $z^{(0)}$, then use a random number generator to choose each component. For the power method, define

$$w^{(m+1)} = Az^{(m)} \qquad z^{(m+1)} = w^{(m+1)}/\|w^{(m+1)}\|_\infty \qquad m \geq 0 \qquad (9.40)$$

We show that the vectors $z^{(m)}$ will converge to $\pm x_1/\|x_1\|_\infty$; and the analysis will also give us a sequence of approximate eigenvalues $\lambda_1^{(m)}$ converging to λ_1.

We begin by showing that

$$z^{(m)} = \frac{A^m z^{(0)}}{\|A^m z^{(0)}\|_\infty} \qquad m \geq 1 \qquad (9.41)$$

First,

$$w^{(1)} = Az^{(0)} \qquad z^{(1)} = w^{(1)}/\|w^{(1)}\|_\infty = Az^{(0)}/\|Az^{(0)}\|_\infty$$

For $m = 2$,

$$w^{(2)} = Az^{(1)} = A^2 z^{(0)}/\|Az^{(0)}\|_\infty$$

$$z^{(2)} = \frac{w^{(2)}}{\|w^{(2)}\|_\infty} = \frac{A^2 z^{(0)}}{\|Az^{(0)}\|_\infty} \div \frac{\|A^2 z^{(0)}\|_\infty}{\|Az^{(0)}\|_\infty} = \frac{A^2 z^{(0)}}{\|A^2 z^{(0)}\|_\infty}$$

For general $m > 2$, use an induction proof following the same way as the case $m = 2$.

To examine the convergence of (9.40), first expand $z^{(0)}$ using the eigenvector basis $\{x_j\}$,

$$z^{(0)} = \sum_{j=1}^{n} \alpha_j x_j \qquad (9.42)$$

We will assume $\alpha_1 \neq 0$, which a random choice of $z^{(0)}$ will generally ensure. From (9.42),

$$A^m z^{(0)} = \sum_{j=1}^{n} \alpha_j A^m x_j = \sum_{j=1}^{n} \alpha_j \lambda_j^m x_j$$

$$A^m z^{(0)} = \lambda_1^m \left[\alpha_1 x_1 + \sum_{j=2}^{n} \alpha_j \left(\frac{\lambda_j}{\lambda_1} \right)^m x_j \right] \qquad m \geq 1 \qquad (9.43)$$

From (9.39),

$$\left(\frac{\lambda_j}{\lambda_1}\right)^m \to 0 \quad \text{as} \quad m \to \infty \quad 2 \leq j \leq n \tag{9.44}$$

Using this in (9.43), and applying it to (9.41), we obtain

$$\sigma_m z^{(m)} \to \frac{x_1}{\|x_1\|_\infty} \quad \text{as} \quad m \to \infty \tag{9.45}$$

with $|\sigma_m| = 1$, depending on both α_1 and m. The rate of convergence will depend on the ratio $|\lambda_2/\lambda_1|$, since it converges more slowly than the other ratios in (9.44).

To obtain a sequence of approximate eigenvalues, let k be the index of a nonzero component of x_1. Generally we would pick k as the subscript of a maximal component of some iterate $z^{(m)}$. Define

$$\lambda_1^{(m)} = \frac{w_k^{(m)}}{z_k^{(m-1)}} \quad m \geq 1 \tag{9.46}$$

To examine the rate of convergence, using the kth components we have

$$\lambda_1^{(m)} = \left[\frac{A^m z^{(0)}}{\|A^{m-1} z^{(0)}\|_\infty}\right]_k \div \left[\frac{A^{m-1} z^{(0)}}{\|A^{m-1} z^{(0)}\|_\infty}\right]_k$$

$$= \frac{\lambda_1^m \left[\alpha_1 x_1 + \sum_{j=2}^{n} \alpha_j \left(\frac{\lambda_j}{\lambda_1}\right)^m x_j\right]_k}{\lambda_1^{m-1} \left[\alpha_1 x_1 + \sum_{j=2}^{m} \alpha_j \left(\frac{\lambda_j}{\lambda_1}\right)^{m-1} x_j\right]_k}$$

$$\lambda_1^{(m)} = \lambda_1 \left[1 + 0\left(\left(\frac{\lambda_2}{\lambda_1}\right)^m\right)\right] \quad m \geq 1 \tag{9.47}$$

The rate of convergence is linear, and it depends on the size of λ_2/λ_1. Before continuing with the theoretical development, we present an example.

Example   Let

$$A = \begin{bmatrix} 1 & 2 & 3 \\ 2 & 3 & 4 \\ 3 & 4 & 5 \end{bmatrix} \tag{9.48}$$

The true eigenvalues are

$$\lambda_1 = 9.623475383 \qquad \lambda_2 = -.6234753830 \qquad \lambda_3 = 0 \qquad (9.49)$$

An initial guess $z^{(0)}$ was generated with a random number generator. The first five iterates are shown in Table 9.1, along with the ratios

$$R_m = \frac{\lambda_1^{(m)} - \lambda_1^{(m-1)}}{\lambda_1^{(m-1)} - \lambda_1^{(m-2)}} \qquad (9.50)$$

According to a later discussion, these ratios should approximate the ratio λ_2/λ_1 as $m \to \infty$.

Table 9.1 Example of power method

m	$z_1^{(m)}$	$z_2^{(m)}$	$z_3^{(m)}$	$\lambda_1^{(m)}$	R_m
1	.50077	.75038	1.0000	11.7628133	
2	.52626	.76313	1.0000	9.5038496	
3	.52459	.76230	1.0000	9.6313231	−.05643
4	.52470	.76235	1.0000	9.6229674	−.06555
5	.52469	.76235	1.0000	9.6235083	−.06474

We will use the computable error bound (9.19) that was derived earlier in Section 9.1. Calculate

$$\eta = Ax - \lambda x$$

with

$$x = z^{(5)}/\|z^{(5)}\| \qquad \lambda = \lambda_1^{(5)}$$

Then from (9.19),

$$\operatorname*{Min}_i |\lambda_i - \lambda_1^{(5)}| \le \|\eta\|_2 = 3.30 \times 10^{-5} \qquad (9.51)$$

A direct comparison with the true answer λ, gives

$$\lambda_1 - \lambda_1^{(5)} = -.0000329$$

which shows that (9.51) is a very accurate estimate in this case.

Acceleration methods Since there is a known regular pattern in which the error decreases for both $\lambda_1^{(m)}$ and $z^{(m)}$, this can be used to obtain more rapidly convergent methods. We give three different approaches for accelerating the convergence.

Case 1 Translation of the eigenvalues. Choose a constant b, and replace the calculation of the eigenvalues of A by those of

$$B = A - bI \tag{9.52}$$

The eigenvalues of B are $\lambda_i - b$, $i = 1, \ldots, n$. Pick b so that $\lambda_1 - b$ is the dominant eigenvalue of B, and choose b to minimize the ratio of convergence.

As a particular case in order to be more explicit, suppose that all eigenvalues of A are real and they have been so arranged that

$$\lambda_1 > \lambda_2 \geq \lambda_3 \geq \cdots \geq \lambda_n \qquad \lambda_1 > |\lambda_n|$$

Then the dominant eigenvalue of B could be either $\lambda_1 - b$ or $\lambda_n - b$, depending on the size of b. We first require that b satisfy

$$|\lambda_1 - b| > |\lambda_n - b|$$

The rate of convergence will be

$$\text{Maximum} \left\{ \frac{|\lambda_2 - b|}{|\lambda_1 - b|}, \frac{|\lambda_n - b|}{|\lambda_1 - b|} \right\} \tag{9.53}$$

If we look carefully at the behavior of these two ratios as b varies, we see that the minimum of (9.53) occurs when

$$(\lambda_1 - b) - (\lambda_2 - b) = (\lambda_n - b) - \left[-(\lambda_1 - b) \right]$$

and

$$b^* = \tfrac{1}{2}(\lambda_2 + \lambda_n) \tag{9.54}$$

is the optimal choice of b. The resulting ratio of convergence is

$$\frac{\lambda_2 - b^*}{\lambda_1 - b^*} = -\frac{\lambda_n - b^*}{\lambda_1 - b^*} = \frac{\lambda_2 - \lambda_n}{2\lambda_1 - \lambda_2 - \lambda_n} \tag{9.55}$$

Experimental methods based on this formula and on (9.47) can be used to determine approximate values of b^*.

Transformation other than (9.52) can be used to transform the set of eigenvalues in such a way as to obtain even more rapid convergence. For a further discussion of these ideas, see Wilkinson [W2, pp. 570–584].

Example   In the previous example (9.48), the theoretical ratio of convergence was

$$\frac{\lambda_2}{\lambda_1} \doteq -.0648$$

Using the optimal value b given by (6.54), and using (9.49) in a rearranged order,

$$b = \tfrac{1}{2}(\lambda_2 + \lambda_n) \doteq -.31174$$

The eigenvalues of $A - bI$ are

$$9.93522 \qquad .31174 \qquad -.31174 \tag{9.56}$$

The ratio of convergence for the power method applied to $A - bI$ will be

$$\pm \frac{.31174}{9.93522} \doteq \pm .0314$$

which is less than half the magnitude of the original ratio.

Case 2 Aitken extrapolation. The form of convergence in (9.47) is completely analogous to the linearly convergent rootfinding methods of Section 2.4 of Chapter 2. Following the same development as in Section 2.5, we will consider the use of Aitken extrapolation to accelerate the convergence in both (9.45) and (9.47). To use the following development, we must assume that

$$|\lambda_2| \neq |\lambda_3| \tag{9.57}$$

this can be weakened to

$$|\lambda_2| = |\lambda_j| \qquad \text{implies} \quad \lambda_2 = \lambda_j$$

But we cannot allow two ratios of convergence of equal magnitude and opposite sign (see the preceding example and (9.56) for such a case).

With (9.57), and using (9.43) and (9.47),

$$\lambda_1 - \lambda_1^{(m)} \approx c r^m \qquad \hat{x}_1 - z^{(m)} \approx r^m \hat{x}_2 \tag{9.58}$$

c is some constant; and $\hat{x}_1$ and $\hat{x}_2$ of norm 1. The unknown rate of convergence is r, and theoretically $r = \lambda_2 / \lambda_1$. Proceeding exactly as in Section 2.5, (9.58) implies that

$$R_m = \frac{\lambda_1^{(m)} - \lambda_1^{(m-1)}}{\lambda_1^{(m-1)} - \lambda_1^{(m-2)}} \to r \qquad \text{as} \quad m \to \infty$$

And Aitken extrapolation gives the improved value $\hat{\lambda}_1$,

$$\hat{\lambda}_1 = \lambda_1^{(m)} - \frac{\left[\lambda_1^{(m)} - \lambda_1^{(m-1)}\right]^2}{\left[\lambda_1^{(m)} - \lambda_1^{(m-1)}\right] - \left[\lambda_1^{(m-1)} - \lambda_1^{(m-2)}\right]} \qquad m \geq 3 \tag{9.59}$$

Using the second part of (9.58), the same formula could also be applied to each component of the sequence $\{z^{(m)}\}$, although there are some difficulties if $\sigma_m \neq$ constant in (9.45).

Example   Consider again the example (9.48). In Table 9.1,

$$R_5 = -.06474 \approx \frac{\lambda_2}{\lambda_1} \doteq -.06479$$

As an example of (9.59), extrapolate with the values $\lambda_1^{(2)}$, $\lambda_1^{(3)}$, and $\lambda_1^{(4)}$ from that table. Then

$$\hat{\lambda}_1 = 9.6234814 \qquad \lambda_1 - \hat{\lambda}_1 = -6.03 \times 10^{-6}$$

In comparison using the more accurate table value $\lambda_1^{(5)}$, the error is $\lambda_1 - \lambda_1^{(5)} = -3.29 \times 10^{-5}$. This again shows the value of using extrapolation, whenever the theory justifies its use.

Case 3   The Rayleigh-Ritz quotient. Whenever A is symmetric, it is better to use the following eigenvalue approximations,

$$\lambda_1^{(m+1)} = \frac{(Az^{(m)}, z^{(m)})}{(z^{(m)}, z^{(m)})} = \frac{(w^{(m+1)}, z^{(m)})}{(z^{(m)}, z^{(m)})} \qquad m \geq 0 \qquad (9.60)$$

We are using standard inner product notation,

$$(w, z) = \sum_1^n w_i z_i \qquad w, z \in R^n$$

To analyze this sequence (9.60), note that all eigenvalues of A are real and that the eigenvectors $x_1, \ldots, x_n$ can be chosen to be orthonormal. Then (9.41) and (9.43), together with (9.60), imply

$$\lambda_1^{(m+1)} = \frac{\displaystyle\sum_{j=1}^n |\alpha_j|^2 \lambda_j^{2m+1}}{\displaystyle\sum_{j=1}^n |\alpha_j|^2 \lambda_j^{2m}}$$

$$= \lambda_1 \left[1 + 0 \left[\left(\frac{\lambda_2}{\lambda_1} \right)^{2m} \right] \right] \qquad (9.61)$$

The ratio of convergence of $\lambda_1^{(m)}$ to λ_1 is $(\lambda_2/\lambda_1)^2$, an improvement on the original ratio in (9.47) of λ_2/λ_1.

This is a well-known classical procedure, and it has many additional aspects that are useful in some problems. For additional discussion, see Wilkinson [W2, pp. 172 – 178].

Example   In the example (9.48), use the approximate eigenvector $z^{(2)}$ in (9.60). Then

$$\lambda_1^{(3)} = \frac{(Az^{(2)}, z^{(2)})}{(z^{(2)}, z^{(2)})} \doteq 9.623464$$

which is as accurate as the value $\lambda_1^{(5)}$ obtained earlier.

The power method can be used when there is not a single dominant eigenvalue, but the algorithm is more complicated. The power method can also be used to detemine eigenvalues other than the dominant one. This involves a process called *deflation* of A to remove λ as an eigenvalue. For a complete discussion of all aspects of the power method, see Wilkinson [W2, Chapter 9]. Although it is a useful method in some circumstances, it should be stressed that the methods of the following sections are usually more efficient for moderate-sized dense matrices.

9.3   Orthogonal Transformations Using Householder Matrices

As one step in finding the eigenvalues of a matrix, it is often reduced to a simpler form using similarity transformations. Orthogonal matrices will be the class of matrices we use for these transformations. It was shown in Section 9.1 that orthogonal transformations will not worsen the condition or stability of the eigenvalues of a nonsymmetric matrix. And orthogonal matrices have other desirable error propagation properties, an example of which is given later in the section. For these reasons, we restrict our transformations to those using orthogonal matrices.

We begin the section by looking at a special class of orthogonal matrices known as *Householder matrices*. Then we show how to construct a Householder matrix that will transform a given vector to a simpler form. With this construction as a tool, we look at two transformations of a given matrix A: (1) obtain its QR factorization, and (2) construct a similar tridiagonal matrix when A is a symmetric matrix. These forms will be used in the next two sections in the calculation of the eigenvalues of A. As a matter of notation, note that we should be restricting the use of the term *orthogonal* to real matrices. But it has become common usage in this area to use *orthogonal* rather than *unitary* for the general complex case, and we adopt the same convention. The reader should understand *unitary* when *orthogonal* is used for a complex matrix.

Let $w \in C^n$ with $\|w\|_2 = \sqrt{w^*w} = 1$. Define

$$U = I - 2ww^* \tag{9.62}$$

This is the general form of a Householder matrix.

Example   For $n=3$, we require

$$w = [w_1, w_2, w_3]^T \qquad |w_1|^2 + |w_2|^2 + |w_3|^2 = 1$$

The matrix $\acute{U}$ is given by

$$U = \begin{bmatrix} 1-2|w_1|^2 & -2w_1\bar{w}_2 & -2w_1\bar{w}_3 \\ -2\bar{w}_1 w_2 & 1-2|w_2|^2 & -2w_2\bar{w}_3 \\ -2\bar{w}_1 w_3 & -2\bar{w}_2 w_3 & 1-2|w_3|^2 \end{bmatrix}$$

For the particular case

$$w = \left[\frac{1}{3}, \frac{2}{3}, \frac{2}{3} \right]$$

$$U = \frac{1}{9} \begin{bmatrix} 7 & -4 & -4 \\ -4 & 1 & -8 \\ -4 & -8 & 1 \end{bmatrix}$$

We first prove that U is Hermitian and orthogonal. To show it is Hermitian,

$$U^* = (I - 2ww^*)^* = I^* - 2(ww^*)^*$$

$$= I - 2(w^*)^* w^* = I - 2ww^* = U$$

To show it is orthogonal,

$$U^* U = U^2 = (I - 2ww^*)^2$$

$$= I - 4ww^* + 4(ww^*)(ww^*)$$

$$= I,$$

since using the associative law and $w^*w = 1$ implies

$$(ww^*)(ww^*) = w(w^*w)w^* = ww^*$$

The above matrix U of the preceding example illustrates these properties.
We usually use vectors w with leading zero components,

$$w = [0, \ldots, 0, w_r, \ldots, w_n]^T = [0_{r-1}, \hat{w}^T]^T \tag{9.63}$$

with $\hat{w} \in C^{n-r+1}$. Then

$$U = \left[\begin{array}{c|c} I_{r-1} & 0 \\ \hline 0 & I_{n-r+1} - 2\hat{w}\hat{w}^* \end{array} \right] \tag{9.64}$$

Premultiplication of a matrix A by this U will leave the first $r-1$ rows of A unchanged, and postmultiplication of A will leave its first $r-1$ columns unchanged. For the remainder of this section we assume all matrices and vectors are real, in order to simplify the presentation. This is also the principal case of interest in applications.

The Householder matrices will be used to transform a nonzero vector into a new vector containing mainly zeroes. Let $b \neq 0$ be given, $b \in R^n$, and suppose we want to produce U of form (9.62) such that Ub contains zeroes in positions $r+1$ through n, for some given $r \geq 1$. Choose w as in (9.63). Then the first $r-1$ elements of b and Ub are the same.

To simplify the later work, write $m = n - r + 1$,

$$w = \left[\begin{array}{c} 0_{r-1} \\ v \end{array} \right] \qquad b = \left[\begin{array}{c} c \\ d \end{array} \right]$$

with $c \in R^{r-1}, v, d \in R^m$. Then our restriction on the form of Ub requires the first $r-1$ components of Ub to be c, and

$$(I - 2vv^T)d = [\alpha, 0, \ldots, 0]^T \qquad \|v\|_2 = 1 \tag{9.65}$$

for some α. Since $I - 2vv^T$ is orthogonal, the length of d is preserved (Problem 10 of Chapter 7); and thus

$$|\alpha| = \|d\|_2 \equiv S$$

$$\alpha = \pm S = \pm \sqrt{d_1^2 + \ldots + d_m^2} \tag{9.66}$$

Define

$$p = v^T d$$

From (9.65),

$$d - 2pv = [\alpha, 0, \ldots, 0]^T \tag{9.67}$$

Multiplication by v^T and use of $\|v\|_2 = 1$ implies

$$p = -\alpha v_1 \tag{9.68}$$

Substituting this into the first component of (9.67) gives

$$d_1 + 2\alpha v_1^2 = \alpha$$

$$v_1^2 = \frac{1}{2}\left(1 - \frac{d_1}{\alpha}\right) \tag{9.69}$$

Choose the sign of α in (9.66) by

$$\text{sign}(\alpha) = -\text{sign}(d_1) \tag{9.70}$$

This choice maximizes v_1^2, and it avoids any possible loss of significance errors in the calculation of v_1. The sign for v_1 is irrelevant. Having v_1, obtain p from (9.68). Return to (9.67), and using components 2 through m,

$$v_j = \frac{d_j}{2p} \qquad j = 2, 3, \ldots, m \tag{9.71}$$

The statements (9.66) and (9.68) to (9.71) completely define v, and thus w and U. The operation count is $2m + 2$ multiplications and divisions, and one square root. A sequence of such operations is used to systematically reduce matrices to simpler forms.

Example Consider the given vector

$$b = [2, 2, 1]^T$$

We calculate a matrix U for which Ub will have zeroes in its last two positions. To help in following the construction, some of the intermediate calculations are listed. Note that $w = v$ and $b = d$ for this case. Then

$$\alpha = -3 \qquad v_1 = \sqrt{\frac{5}{6}} \qquad p = \sqrt{\frac{15}{2}}$$

$$v_2 = \frac{2}{\sqrt{30}} \qquad v_3 = \frac{1}{\sqrt{30}}$$

The matrix U is given by

$$U = \begin{bmatrix} -\dfrac{2}{3} & -\dfrac{2}{3} & -\dfrac{1}{3} \\[2ex] -\dfrac{2}{3} & \dfrac{11}{15} & -\dfrac{2}{15} \\[2ex] -\dfrac{1}{3} & -\dfrac{2}{15} & \dfrac{14}{15} \end{bmatrix}$$

and

$$Ub = [-3,0,0]^T$$

The QR factorization of a matrix Given a real matrix A, we will show there is an orthogonal matrix Q and an upper triangular matrix R for which

$$A = QR \qquad (9.72)$$

Let

$$P_r = I - 2w^{(r)}w^{(r)T} \qquad r = 1,\dots,n-1 \qquad (9.73)$$

with $w^{(r)}$ as in (9.63) with $r-1$ leading zeroes. Writing A in terms of its columns $A_{\bullet 1},\dots,A_{\bullet n}$, we have

$$P_1 A = [P_1 A_{\bullet 1},\dots,P_1 A_{\bullet n}]^T$$

Pick P_1 and $w^{(1)}$ using the preceding construction (9.66) to (9.71) with $b = A_{\bullet 1}$. Then $P_1 A$ contains zeroes below the diagonal in its first column.

Choose P_2 similarly, so that $P_2 P_1 A$ will contain zeroes in its second column below the diagonal. First note that because $w^{(2)}$ contains a zero in position 1, and because $P_1 A$ is zero in the first column below position 1, the products $P_2 P_1 A$ and $P_1 A$ contain the same elements in row and column 1. Now choose P_2 and $w^{(2)}$ as before in (9.66) to (9.71), with b equal to the second column of $P_1 A$.

By carrying this out with each column of A, we obtain an upper triangular matrix

$$R \equiv P_{n-1}\dots P_1 A \qquad (9.74)$$

If at step r of the construction, all elements below the diagonal of column r are zero, then just choose $P_r = I$ and go onto the next step. To complete the construction, define

$$Q^T = P_{n-1}\dots P_1 \qquad (9.75)$$

which is orthogonal. Then $A = QR$, as desired.

Example Consider

$$A = \begin{bmatrix} 4 & 1 & 1 \\ 1 & 4 & 1 \\ 1 & 1 & 4 \end{bmatrix}$$

Then

$$w^{(1)} \doteq [.985599, .119573, .119573]^T$$

$$A_2 = P_1 A = \begin{bmatrix} -4.24264 & -2.12132 & -2.12132 \\ 0 & 3.62132 & .621321 \\ 0 & .621321 & 3.62132 \end{bmatrix}$$

$$w^{(2)} = [0, .996393, .0848572]^T,$$

$$R = P_2 A_2 = \begin{bmatrix} -4.24264 & -2.12132 & -2.12132 \\ 0 & -3.67423 & -1.22475 \\ 0 & 0 & 3.46410 \end{bmatrix}$$

For the factorization $A = QR$, evaluate $Q = P_1 P_2$. But in most applications, it would be inefficient to explicitly produce Q. We will comment further on this shortly.

Since Q orthogonal implies $\det(Q) = \pm 1$, we have

$$|\det(A)| = |\det(Q)\det(R)| = |\det(R)| = 53.9999$$

This number is consistent with the fact that the eigenvalues of A are $\lambda = 3, 3, 6$, and their product is $\det(A) = 54$.

Discussion of the QR factorization   It is useful to know to what extent the factorization $A = QR$ is unique. For A nonsingular, suppose

$$A = Q_1 R_1 = Q_2 R_2 \tag{9.76}$$

Then R_1 and R_2 must also be nonsingular, and

$$Q_2^T Q_1 = R_2 R_1^{-1}$$

The inverse of an upper triangular matrix is upper triangular, and the product of two upper triangular matrices is upper triangular. Thus $R_2 R_1^{-1}$ is upper triangular. Also, the product of two orthogonal matrices is orthogonal; and thus the product $Q_2^T Q_1$ is orthogonal. This says $R_2 R_1^{-1}$ is orthogonal. But it is not hard to show that the only upper triangular orthogonal matrices are the diagonal matrices. For some diagonal matrix D,

$$R_2 R_1^{-1} = D$$

Since $R_2 R_1^{-1}$ is orthogonal,

$$D^2 = I$$

Since we are only dealing with real matrices, D has diagonal elements equal to

$+1$ or -1. Combining these results,

$$Q_2 = Q_1 D \qquad R_2 = DR_1 \tag{9.77}$$

This says the signs of the diagonal elements of R in $A = QR$ can be chosen arbitrarily, but then the rest of the decomposition is uniquely determined. This result will be used later in the analysis of the QR method in Section 9.5.

Another practical matter is deciding how to evaluate the matrix R of (9.74). Let

$$A_r = P_r A_{r-1} = \left[I - 2w^{(r)} w^{(r)T} \right] A_{r-1} \qquad r = 1, 2, \dots, n-1 \tag{9.78}$$

with $A_0 = A, A_{n-1} = R$. If we calculate P_r and then multiply it times A_{r-1} to form A_r, the number of multiplications will be

$$(n - r + 1)^3 + \tfrac{1}{2}(n - r + 2)(n - r + 1)$$

There is a much more efficient method for calculating A_r. Rewrite (9.78) as

$$A_r = A_{r-1} - 2w^{(r)} \left(w^{(r)T} A_{r-1} \right) \tag{9.79}$$

First calculate $w^{(r)T} A_{r-1}$; and then calculate $w^{(r)} (w^{(r)T} A_{r-1})$ and A_r. This requires about

$$2(n - r + 1)^2 + (n - r + 1)$$

multiplications, which shows (9.79) is a preferable way to evaluate each A_r and finally $R = A_{n-1}$.

If it is necessary to store the matrices $P_1, \dots, P_{n-1}$ for later use, just store each column $w^{(r)}, r = 1, \dots, n-1$. Save the first nonzero element of $w^{(r)}$, the one in position r, in a special storage location; and save the remaining nonzero elements of $w^{(r)}$, those in positions $r+1$ through n, in column r of the matrix A_r and R, below the diagonal. The matrix Q of (9.75) could be produced explicitly. But as the construction (9.79) shows, we do not need Q explicitly in order to multiply it times some other matrix.

The main use of the QR factorization of A will be in defining the QR method for calculating the eigenvalues of A, which is presented in Section 9.5. The factorization can also be used to solve a linear system of equations $Ax = b$. The factorization leads directly to the equivalent system $Rx = Q^T b$; and very little error is introduced because Q is orthogonal. The system $Rx = Q^T b$ is upper triangular, and it can be solved in a stable manner using back substitution. For A an ill-conditioned matrix, this may be a superior way to solve the linear system $Ax = b$. For a discussion of the errors involved in obtaining and using the QR factorization and for a comparison of it and Gaussian elimination for solving $Ax = b$, see Wilkinson [W2, pp. 236, 244–249].

The transformation of a symmetric matrix to tridiagonal form   Let A be a real symmetric matrix. To find the eigenvalues of A, it is usually first reduced to tridiagonal form by orthogonal similarity transformations. The eigenvalues of the tridiagonal matrix are then calculated using the theory of Sturm sequences, presented in Section 9.4, or the QR method, presented in Section 9.5. For the orthogonal matrices, we will use the Householder matrices of (9.64).

Let

$$P_r = I - 2w^{(r+1)}w^{(r+1)T} \qquad r = 1,\ldots,n-2 \tag{9.80}$$

with $w^{(r+1)}$ defined as in (9.63),

$$w^{(r+1)} = [0,\ldots,0,w_{r+1},\ldots,w_n]$$

The matrix

$$A_2 = P_1^T A P_1 = P_1 A P_1$$

is similar to A, the element a_{11} is unchanged, and A_2 will be symmetric. Produce $w^{(2)}$ and P_1 to obtain

$$P_1 A_{*1} = [a_{11}, \hat{a}_{21}, 0, \ldots, 0]^T$$

for some $\hat{a}_{21}$. The vector A_{*1} is the first column of A. Use (9.66) to (9.71) with $m = n-1$ and

$$d = [a_{21}, a_{31}, \ldots, a_{n1}]^T$$

For example, from (9.69) with $r = 1$,

$$w_2^2 = v_1^2 = \tfrac{1}{2}\left(1 - \frac{a_{21}}{\alpha}\right)$$

$$\alpha = -\operatorname{sign}(a_{21})\sqrt{a_{21}^2 + \ldots + a_{n1}^2}$$

Having obtained P_1 and $P_1 A$, postmultiplication by P_1 will not change the first column of $P_1 A$. (This should be checked by the reader.) The symmetry of A_2 follows from

$$A_2^T = (P_1 A P_1)^T = P_1^T A^T P_1^T = P_1 A P_1 = A_2$$

Since A_2 is symmetric, the construction on the first column of A will imply that A_2 has zeroes in positions 3 through n of both the first row and column.

Continue this process, letting

$$A_{r+1} = P_r^T A_r P_r \qquad r = 1, 2, \ldots, n-2 \tag{9.81}$$

with $A_1 = A$. Pick P_r to introduce zeroes into positions $r+2$ through n of column r. Columns 1 through $r-1$ will remain undisturbed in calculating $P_r A_{r-1}$, due to the special form of P_r. Pick the vector $w^{(r+1)}$ in analogy with the above description for $w^{(2)}$.

The final matrix $T \equiv A_{n-1}$ is tridiagonal and symmetric.

$$
T = \begin{bmatrix}
\alpha_1 & \beta_1 & 0 & \cdots & \cdots & \cdots & 0 \\
\beta_1 & \alpha_2 & \beta_2 & & & & \vdots \\
0 & \beta_2 & \alpha_3 & & \beta_3 & & \vdots \\
\vdots & & & \ddots & & & \vdots \\
\vdots & & & \beta_{n-2} & \alpha_{n-1} & \beta_{n-1} & \\
0 & \cdots & \cdots & \cdots & \cdots & \beta_{n-1} & \alpha_n
\end{bmatrix}
\tag{9.82}
$$

This will be a much more convenient form for the calculation of the eigenvalues of A; and the eigenvectors of A can easily be obtained from those of T.

A is related to T by

$$
T = Q^T A Q \qquad Q = P_1, \ldots, P_{n-2}
\tag{9.83}
$$

As before with the QR factorization, we seldom produce Q explicitly, preferring to work with the individual matrices P_r in analogy with (9.79). For an eigenvector x of A, say $Ax = \lambda x$, we have

$$
Tz = \lambda z \qquad x = Qz
\tag{9.84}
$$

If we produce an orthonormal set of eigenvectors $\{z_i\}$ for T, then $\{Qz_i\}$ will be an orthonormal set of eigenvectors for A, since Q preserves length and angles (see Problem 10 of Chapter 7).

Example Let

$$
A = \begin{bmatrix} 1 & 3 & 4 \\ 3 & 1 & 2 \\ 4 & 2 & 1 \end{bmatrix}
$$

Then

$$
w^{(2)} = \left[0, 2/\sqrt{5}, 1/\sqrt{5} \right]^T
$$

$$
P_1 = \begin{bmatrix} 1 & 0 & 0 \\ 0 & -3/5 & -4/5 \\ 0 & -4/5 & 3/5 \end{bmatrix}
$$

$$
T = P_1^T A P_1 = \begin{bmatrix} 1 & -5 & 0 \\ -5 & 73/25 & 14/25 \\ 0 & 14/25 & -23/25 \end{bmatrix}
$$

For an error analysis of this reduction to tridiagonal form, we give some results from Wilkinson [W2]. Let the computer arithmetic be binary floating point with rounding, with t binary digits in the mantissa. Furthermore, assume that all inner products

$$\sum_{i=1}^{m} a_i b_i$$

that occur in the calculation are accumulated in double precision, with rounding to single precision at the completion of the summation. These inner products occur in a variety of places in the computation of T from A. Let $\hat{T}$ denote the actual symmetric tridiagonal matrix that is computed from A using the above computer arithmetic. Let $\hat{P}_r$ denote the actual matrix produced in converting A_{r-1} to A_r, let P_r be the theoretically exact version of this matrix if no rounding errors occurred, and let $Q = P_1 \ldots P_{r-1}$ be the exact product of these P_r's, an orthogonal matrix.

Theorem 9.4   Let A be a real symmetric matrix of order n. Let $\hat{T}$ be the real symmetric tridiagonal matrix resulting from applying the Householder similarity transformations (9.80) to A, as in (9.81). Assume the floating-point arithmetic used has the characteristics described in the preceding paragraph. Let $\{\lambda_i\}$ and $\{\tau_i\}$ be the eigenvalues of A and $\hat{T}$, respectively, arranged in increasing order. Then

$$\left[\frac{\sum_{i=1}^{n} (\tau_i - \lambda_i)^2}{\sum_{i=1}^{n} \lambda_i^2} \right]^{1/2} \le c_n 2^{-t} \tag{9.85}$$

with

$$c_n = 25(n-1)\left[1 + (12.36)2^{-t}\right]^{2n-4}$$

For small and moderate values of n, $c_n \doteq 25(n-1)$.

Proof   From Wilkinson [W2, p. 161] using the Frobenius matrix norm F,

$$F(\hat{T} - Q^T A Q) \le 2x(n-1)(1+x)^{2n-4} F(A) \tag{9.86}$$

with $x = (12.36)2^{-t}$. From the Wielandt-Hoffman result (9.17) of Theorem 9.3, we have

$$\left[\sum_{i=1}^{n} (\tau_i - \lambda_i)^2 \right]^{1/2} \le F(\hat{T} - Q^T A Q) \tag{9.87}$$

since A and Q^TAQ have the same eigenvalues. And from Problem 22(b) of Chapter 7,

$$F(A) = \left[\sum_{i=1}^{n} \lambda_i^2 \right]^{1/2}$$

Combining these results yields (9.85).   ∎

For a further discussion of the error, including the case in which inner products are not accumulated in double-precision, see Wilkinson [W2, pp. 297–299]. The result (9.85) shows that the reduction to tridiagonal form is an extremely stable operation, with little new error introduced for the eigenvalues.

Planar rotation orthogonal matrices   There are other classes of orthogonal matrices that can be used in place of the Householder matrices. The principal class is the set of plane rotations, which can be given the geometric interpretation of rotating a pair of coordinate axes through a given angle θ in the plane of the axes. For integers $k,l, 1 \le k < l \le n$, define the $n \times n$ orthogonal matrix $R^{(k,l)}$ by altering four elements of the identity matrix I_n. For any real number θ, define the elements of $R^{(k,l)}$ by

$$R_{i,j}^{(k,l)} = \begin{cases} \cos\theta, & (i,j) = (k,k) \text{ or } (l,l) \\ \sin\theta, & (i,j) = (k,l) \\ -\sin\theta, & (i,j) = (l,k) \\ (I_n)_{ij}, & \text{all other } (i,j) \end{cases} \tag{9.88}$$

for $1 \le i,j \le n$.

Example   For $n = 3$,

$$R^{(1,3)} = \begin{bmatrix} \cos\theta & 0 & \sin\theta \\ 0 & 1 & 0 \\ -\sin\theta & 0 & \cos\theta \end{bmatrix}$$

As a particular case, take $\theta = \pi/4$. Then

$$R^{(1,3)} = \begin{bmatrix} 1/\sqrt{2} & 0 & 1/\sqrt{2} \\ 0 & 1 & 0 \\ -1/\sqrt{2} & 0 & 1/\sqrt{2} \end{bmatrix}$$

The plane rotations $R^{(k,l)}$ can be used to accomplish the same reductions for which the Householder matrices are used. In most situations, the Householder matrices are more efficient; although the plane rotations are more efficient for part of the QR method. The idea of solving the symmetric matrix eigenvalue problem by first reducing it to tridiagonal form is due to W. Givens in 1954. He also proposed the use of the techniques of the next section for the calculation of the eigenvalues of the tridiagonal matrix. Givens used the plane rotations $R^{(k,l)}$, and the Householder matrices were introduced in 1958 by A. Householder. For additional discussion of these matrices and their properties and uses, see Wilkinson [W2] and Problems 12 and 13.

9.4   The Eigenvalues of a Symmetric Tridiagonal Matrix

Let T be a real symmetric tridiagonal matrix of order n, as in (9.82). We will compute the characteristic polynomial of T and use it to calculate the eigenvalues of T.

To compute

$$f_n(\lambda) \equiv \det(T - \lambda I)$$

introduce the sequence

$$f_k(\lambda) = \det \begin{bmatrix} \alpha_1 - \lambda & \beta_1 & 0 \cdots \cdots 0 \\ \beta_1 & \alpha_2 - \lambda & \beta_2 & \\ 0 & & \ddots & \vdots \\ \vdots & & & \ddots & \beta_{k-1} \\ 0 \cdots \cdots \cdots & & \beta_{k-1} & \alpha_k - \lambda \end{bmatrix}, \quad 1 \le k \le n \qquad (9.89)$$

and $f_0(\lambda) \equiv 1$. By direct evaluation,

$$f_1(\lambda) = \alpha_1 - \lambda$$

$$f_2(\lambda) = (\alpha_2 - \lambda)(\alpha_1 - \lambda) - \beta_1^2$$

$$= (\alpha_2 - \lambda)f_1(\lambda) - \beta_1^2 f_0(\lambda)$$

The formula for $f_2(\lambda)$ illustrates the general triple recursion relation which the sequence $\{f_k(\lambda)\}$ satisfies,

$$f_k(\lambda) = (\alpha_k - \lambda)f_{k-1}(\lambda) - \beta_{k-1}^2 f_{k-2}(\lambda) \qquad 2 \le k \le n \qquad (9.90)$$

To prove (9.90), expand the determinant (9.89) in its last row using minors, and the result will follow easily. This method for evaluating $f_n(\lambda)$ will require $2n - 3$ multiplications, once the coefficients $\{\beta_k^2\}$ have been evaluated.

Example Let

$$T = \begin{bmatrix} 2 & 1 & 0 & 0 & 0 & 0 \\ 1 & 2 & 1 & 0 & 0 & 0 \\ 0 & 1 & 2 & 1 & 0 & 0 \\ 0 & 0 & 1 & 2 & 1 & 0 \\ 0 & 0 & 0 & 1 & 2 & 1 \\ 0 & 0 & 0 & 0 & 1 & 2 \end{bmatrix} \tag{9.91}$$

Then

$$f_0(\lambda) = 1 \qquad f_1(\lambda) = 2 - \lambda$$

$$f_j(\lambda) = (2 - \lambda) f_{j-1}(\lambda) - f_{j-2}(\lambda) \qquad j = 2, 3, 4, 5, 6 \tag{9.92}$$

Without the triple recursion relation (9.90), the evaluation of $f_6(\lambda)$ would be much more complicated.

At this point, we might consider the problem as solved since $f_n(\lambda)$ is a polynomial and there are many polynomial rootfinding methods. Or we might use a more general method such as the secant method or Brent's method, both described in Chapter 2. But the sequence $\{f_k(\lambda) | 0 \le k \le n\}$ has special properties that make it a *Sturm sequence*; and these properties make it comparatively easy to isolate the eigenvalues of T. Once the eigenvalues have been isolated, a method such as Brent's method (see Section 2.8) can be used to rapidly calculate the roots. The theory of Sturm sequences is discussed in Henrici [H1, pp. 444], but we will only consider the special case of $\{f_k(\lambda)\}$.

Before beginning the derivation of the special properties of $\{f_k(\lambda)\}$, we consider what will happen when some $\beta_l = 0$. Then the eigenvalue problem can be broken apart into two smaller eigenvalue problems of orders l and $n - l$. For example, consider

$$T = \begin{bmatrix} \alpha_1 & \beta_1 & 0 & 0 & 0 \\ \beta_1 & \alpha_2 & 0 & 0 & 0 \\ 0 & 0 & \alpha_3 & \beta_3 & 0 \\ 0 & 0 & \beta_3 & \alpha_4 & \beta_4 \\ 0 & 0 & 0 & \beta_4 & \alpha_5 \end{bmatrix}$$

Define T_1 and T_2 as the two blocks along the diagonal, of orders 2 and 3, respectively; and then

$$T = \begin{bmatrix} T_1 & 0 \\ 0 & T_2 \end{bmatrix}$$

From this,

$$\det[T - \lambda I_5] = \det[T_1 - \lambda I_2] \det[T_2 - \lambda I_3]$$

and we can find the eigenvalues of T by finding those of T_1 and T_2. The eigenvector problem also can be solved in the same way. For example, if $T_1 \hat{x} = \lambda x$, with $\hat{x} \neq 0$ in R^2, define

$$x = [\hat{x}^T, 0, 0, 0]^T$$

Then $Tx = \lambda x$. This construction can be used to calculate a complete set of eigenvectors for T from those for T_1 and T_2. For the remainder of the section, we assume that all $\beta_l \neq 0$ in the matrix T.

The Sturm sequence property of $\{f_k(\lambda)\}$ The sequences $\{f_k(a)\}$ and $\{f_k(b)\}$ can be used to determine the number of roots of $f_n(\lambda)$ that are contained in $[a,b]$. To do this, introduce the integer valued function $s(\lambda)$. Let $s(\lambda)$ be the number of agreements in sign of consecutive members of the sequence $\{f_k(\lambda)\}$; and if the value of some member $f_j(\lambda) = 0$, let its sign be chosen opposite to that of $f_{j-1}(\lambda)$. We will show later that $f_j(\lambda) = 0$ implies $f_{j-1}(\lambda) \neq 0$.

Example   Consider the sequence $f_0(\lambda), \ldots, f_6(\lambda)$ given in (9.92) of the last example. For $\lambda = 3$,

$$(f_0(\lambda), \ldots, f_6(\lambda)) = (1, -1, 0, 1, -1, 0, 1)$$

The corresponding sequence of signs is

$$(+, -, +, +, -, +, +)$$

and $s(3) = 2$.

We now state the basic result used in computing the roots of $f_n(\lambda)$ and thus the eigenvalues of T.

Theorem 9.5   Let T be a real symmetric tridiagonal matrix of order n, as given in (9.82). Let the sequence $\{f_k(\lambda) | 0 \leq k \leq n\}$ be defined as in (9.89), and assume all $\beta_l \neq 0, l = 1, \ldots, n-1$. Then the number of roots of $f_n(\lambda)$ that are greater than $\lambda = a$ is given by $s(a)$, which is defined in the preceding paragraph. For $a < b$, the number of roots in the interval $a < \lambda \leq b$ is given by $s(a) - s(b)$.

Proof   The proof of this result is fairly lengthy; and those who wish to skip it may go directly to the example following the proof. This theorem is characteristic of all Sturm sequences, but we will consider only $\{f_k(\lambda)\}$.

The proof is divided into several parts, beginning with a derivation of properties of $\{f_k(\lambda)\}$.

1. No two consecutive polynomials have a common zero. For a proof by contradiction, suppose that

$$f_j(\lambda) = f_{j-1}(\lambda) = 0$$

for some $j \geq 2$. Then from (9.90) with $k = j$ and solving for $f_{j-2}(\lambda)$,

$$f_{j-2}(\lambda) = \frac{1}{\beta_{j-2}^2} \left[(\alpha_j - \lambda) f_{j-1}(\lambda) - f_j(\lambda) \right]$$

$$= 0$$

Continue the argument inductively using (9.90) to prove

$$f_k(\lambda) = 0 \qquad \text{all} \quad k \leq j$$

But this will contradict the definition $f_0(\lambda) \equiv 1$.

2. The zeroes of $f_{k-1}(\lambda)$ interlace those of $f_k(\lambda)$, for $k = 2, 3, \ldots, n$. For the simplest case, consider the zeroes of $f_1(\lambda)$ and $f_2(\lambda)$. The single zero of $f_1(\lambda)$ is $\lambda = \alpha_1$. Since $f_2(\pm\infty) = +\infty$ and $f_2(\alpha_1) = -\beta_1^2 < 0$, there must be zeroes of $f_2(\lambda)$ to the left and right of $\lambda = \alpha_1$.

Assume the result is true for the zeroes of $f_1(\lambda), \ldots, f_{k-1}(\lambda)$. We will prove it is true for the zeroes of $f_k(\lambda)$. Let the roots of $f_{k-1}(\lambda)$ and $f_{k-2}(\lambda)$ be $\{\lambda_1, \ldots, \lambda_{k-1}\}$ and $\{\mu_1, \ldots, \mu_{k-2}\}$, respectively. By assumption,

$$\lambda_{k-1} < \mu_{k-2} < \lambda_{k-2} < \cdots < \lambda_2 < \mu_1 < \lambda_1 \tag{9.93}$$

Note that since degree $f_{k-1}(\lambda) = k-1$ and since there are exactly $k-1$ roots $\lambda_1, \ldots, \lambda_{k-1}$, they must all be simple roots. The same is true of the roots $\mu_1, \ldots, \mu_{k-2}$ for $f_{k-2}(\lambda)$.

Examine the sign of $f_k(\lambda)$ at $\lambda = \lambda_{j-1}$ and $\lambda = \lambda_j$. From (9.90),

$$f_k(\lambda_j) = (\alpha_k - \lambda_j) f_{k-1}(\lambda_j) - \beta_{k-1}^2 f_{k-2}(\lambda_j)$$

$$= -\beta_{k-1}^2 f_{k-2}(\lambda_j)$$

$$f_k(\lambda_{j-1}) = -\beta_{k-1}^2 f_{k-2}(\lambda_{j-1}) \tag{9.94}$$

Because $f_{k-2}(\lambda)$ has a single simple root μ_{j-1} between λ_{j-1} and λ_j, we have

$$\operatorname{sign} f_{k-2}(\lambda_j) = -\operatorname{sign} f_{k-2}(\lambda_{j-1})$$

But from (9.94) the same must be true of $f_k(\lambda_j)$ and $f_k(\lambda_{j-1})$; and therefore, $f_k(\lambda)$ must have a root of odd multiplicity between λ_{j-1} and λ_j. This is true for each $j = 2, 3, \ldots, k-1$.

Since

$$f_l(\lambda) = (-1)^l \lambda^l + \text{lower-degree terms}, \quad \text{all} \quad l \geq 1$$

we have

$$\begin{aligned} \operatorname{sign} f_k(-\infty) &= \operatorname{sign} f_{k-2}(-\infty) \\ \operatorname{sign} f_k(+\infty) &= \operatorname{sign} f_{k-2}(+\infty) \end{aligned} \tag{9.95}$$

Also from (9.94),

$$\operatorname{sign} f_k(\lambda_1) = -\operatorname{sign} f_{k-2}(\lambda_1) = -\operatorname{sign} f_{k-2}(+\infty)$$

The last equality follows since f_{k-2} has no zeroes to the right of μ_1. Using (9.95), $f_k(\lambda)$ must have a zero to the right of λ_1. A similar argument shows $f_k(\lambda)$ has a root to the left of λ_{k-1}. Counting roots of $f_k(\lambda)$, there are k so far. But degree $f_k(\lambda) = k$ implies these must be all the roots, and they must all be simple. This completes the proof that the zeroes of $f_{k-1}(\lambda)$ interlace those of $f_k(\lambda)$.

3. We now prove the main conclusion of the theorem: the number of roots of $f_n(\lambda)$ that are greater than a is equal to $s(a)$. The proof will be an induction on n.

For the sequence with $n = 1$,

$$\{ f_0(a), f_1(a) \} = \{ 1, \alpha_1 - a \}$$

it is clear that the number of agreements in sign equal the number of roots of $f_1(\lambda)$ greater than a.

Assume the result is true for the sequence

$$f_0(a), \ldots, f_{k-1}(a)$$

and let the number of roots of $f_{k-1}(\lambda)$ that are greater than a be denoted by m. Denote the roots of $f_{k-1}(\lambda)$ by $\{\lambda_i\}$ and the roots of $f_k(\lambda)$ by $\{\nu_i\}$. Then the induction hypothesis and part 2 imply

$$\lambda_1 > \lambda_2 > \ldots > \lambda_m > a \geq \lambda_{m+1} > \ldots > \lambda_{k-1} \tag{9.96}$$

$$\nu_1 > \lambda_1 > \nu_2 > \ldots > \nu_m > \lambda_m > \nu_{m+1} > \lambda_{m+1} > \ldots > \lambda_{k-1} > \nu_k \tag{9.97}$$

We wish to prove that the number of roots of $f_k(\lambda)$ that are greater than a equals the number of agreements in sign in the sequence

$$\{ f_0(a), \ldots, f_k(a) \} \tag{9.98}$$

From (9.96) and (9.97), the number of roots of $f_k(\lambda)$ that are greater than a must be either m or $m+1$. The proof of our desired result breaks into three cases.

Case 1 $a \neq \lambda_{m+1}, \nu_{m+1}$. Write

$$f_{k-1}(\lambda) = (\lambda_1 - \lambda) \ldots (\lambda_{k-1} - \lambda)$$

$$f_k(\lambda) = (\nu_1 - \lambda) \ldots (\nu_k - \lambda) \tag{9.99}$$

There are two subcases. First, if $\lambda_{m+1} < a < \nu_{m+1}$, then the number of roots of $f_k(\lambda)$ greater than a is $m+1$. Also from (9.99),

$$\text{sign} f_{k-1}(a) = (-1)^{(k-1)-m}, \text{sign} f_k(a) = (-1)^{k-(m+1)}$$

which are equal. Since these agree, the number of agreements in sign in the sequence (9.50) is $m+1$, using the original number m from the sequence

$$\{ f_0(a), \ldots, f_{k-1}(a) \}$$

This is the desired result. For the second subcase, $\nu_{m+1} < a < \lambda_m$, the number of roots of $f_k(\lambda)$ that are greater than a is just m. Also,

$$\text{sign} f_{k-1}(a) = (-1)^{(k-1)-m}, \quad \text{sign} f_k(a) = (-1)^{k-m}$$

which are different; and the desired result follows as before.

Case 2 $a = \nu_{m+1}$. Thus $f_k(a) = 0$; and by agreement, the sign of $f_k(a)$ is opposite to that of $f_{k-1}(a)$. Thus the number of agreements in sign in (9.98) is still m, the same as the number of zeroes of $f_k(\lambda)$ that are greater than $a = \nu_{m+1}$.

Case 3 $a = \lambda_{m+1}$. Then $f_{k-1}(a) = 0$ and there are $m+1$ zeroes of $f_k(\lambda)$ which are greater than a. Using (9.90),

$$f_k(a) = -\beta_{k-1}^2 f_{k-2}(a)$$

and by construction in the hypothesis,

$$\text{sign} f_{k-1}(a) = -\text{sign} f_{k-2}(a)$$

Combining these gives

$$\text{sign} f_k(a) = \text{sign} f_{k-1}(a)$$

and thus the number of agreements in sign in (9.98) is $m+1$, the desired result.

This completes the proof. ∎

Calculation of the eigenvalues   The past theorem will be the basic tool in locating and separating the roots of $f_n(\lambda)$. To begin, calculate an interval which contains the roots. Using the Gerschgorin circle Theorem 9.1, all eigenvalues are contained in the interval $[a, b]$, with

$$a = \underset{1 \le i \le n}{\mathrm{Min}} \{\alpha_i - |\beta_i| - |\beta_{i-1}|\}$$

$$b = \underset{1 \le i \le n}{\mathrm{Max}} \{\alpha_i + |\beta_i| + |\beta_{i-1}|\}$$

where $\beta_0 = \beta_n = 0$.

We use the bisection method on $[a, b]$ to divide it into smaller subintervals. Theorem 9.5 is used to determine how many roots are contained in a subinterval, and we seek to obtain subintervals that will each contain one root. If some eigenvalues are nearly equal, then we continue subdividing until the root is found with sufficient accuracy. Once a subinterval is known to contain a single root, we can switch to a more rapidly convergent method.

Example   Consider further the example (9.91). By the Gerschgorin Theorem 9.1, all eigenvalues lie in $[0, 4]$. And it is easily checked that neither $\lambda = 0$ or $\lambda = 4$ is an eigenvalue. A systematic bisection process was carried out on $[0, 4]$ to separate the six roots of $f_6(\lambda)$ into six separate subintervals. The results are shown in Table 9.2 in the order they were calculated. The roots are labeled as follows:

$$0 \le \lambda_6 \le \lambda_5 \le \dots \le \lambda_1 \le 4$$

The roots can be found by continuing with the bisection method, although Theorem 9.5 is no longer needed. But it would be better to use some other rootfinding method.

Table 9.2   Example of use of Theorem 9.5

λ	$f_6(\lambda)$	$s(\lambda)$	Comment
0.0	7.0	6	$\lambda_6 > 0$
4.0	7.0	0	$\lambda_1 < 4$
2.0	-1.0	3	$\lambda_4 < 2 < \lambda_3$
1.0	1.0	4	$\lambda_5 < 1 < \lambda_4 < 2$
0.5	-1.421875	5	$0 < \lambda_6 < 0.5 < \lambda_5 < 1$
3.0	1.0	2	$2 < \lambda_3 < 3 < \lambda_2$
3.5	-1.421875	1	$3 < \lambda_2 < 3.5 < \lambda_1 < 4$

Although all roots of a tridiagonal matrix may be found by this technique, it is generally faster in that case to use the QR algorithm of the next section. With large matrices, we usually do not want all of the roots; and then the methods of this section are preferable. If we want only certain specific roots, for example, the five largest or all roots in a given interval, for example, all roots in $[1, 3]$, then it is easy to locate them using Theorem 9.5.

9.5 The *QR* Method

At the present time this is the most efficient and widely used general method for the calculation of all of the eigenvalues of a matrix. The method was first published in 1961 by J. G. F. Francis and it has since been the subject of intense investigation. The *QR* method is quite complex in both its theory and application, and we are able to give only an introduction to the theory of the method. For actual algorithms for both symmetric and nonsymmetric matrices, refer to those in Wilkinson and Reinsch [W4] and Smith et al. [S1].

Given a matrix A, there is a factorization

$$A = QR$$

with R upper triangular and Q orthogonal. With A real, both Q and R can be chosen real; and their construction is given in Section 9.3. We assume that A is real throughout this section. Let $A_1 = A$, and define a sequence of matrices A_m, Q_m, and R_m by

$$A_m = Q_m R_m \qquad A_{m+1} = R_m Q_m \qquad m = 1, 2, \ldots \qquad (9.100)$$

The sequence $\{A_m\}$ will converge to either a triangular matrix with the eigenvalues of A on its diagonal or to a near-triangular matrix from which the eigenvalues can be easily calculated. In this form the convergence is usually slowly convergent; and a technique known as *shifting* is used to accelerate the convergence. The technique of shifting is introduced and illustrated later in the section.

Before illustrating (9.100) with an example, a few properties of (9.100) will be derived. From (9.100) $R_m = Q_m^T A_m$, and thus

$$A_{m+1} = Q_m^T A_m Q_m \qquad (9.101)$$

A_{m+1} is orthogonally similar to A_m, and thus by induction to $A_1 = A$. From (9.101),

$$A_{m+1} = Q_m^T \cdots Q_1^T A_1 Q_1 \cdots Q_m \qquad (9.102)$$

Introduce the matrices P_m and U_m by

$$P_m = Q_1 \cdots Q_m \qquad U_m = R_m \cdots R_1 \qquad (9.103)$$

Then from (9.102)

$$A_{m+1} = P_m^T A_1 P_m \qquad m \geq 1 \qquad (9.104)$$

The matrix P_m is orthogonal, and U_m is upper triangular.

For later use, we derive a further relation involving P_m and U_m.

$$P_m U_m = Q_1 \cdots Q_m R_m \cdots R_1$$

$$= Q_1 \cdots Q_{m-1} A_m R_{m-1} \cdots R_1$$

From (9.102) with m replacing $m+1$,

$$Q_1 \cdots Q_{m-1} A_m = A_1 Q_1 \cdots Q_{m-1}$$

Using this,

$$P_m U_m = A_1 Q_1 \cdots Q_{m-1} R_{m-1} \cdots R_1 = A_1 P_{m-1} U_{m-1}$$

Since $P_1 U_1 = Q_1 R_1 = A_1$, induction on the last statement implies

$$P_m U_m = A_1^m \qquad m \geq 1 \tag{9.105}$$

Example Let

$$A_1 = \begin{bmatrix} 2 & 1 & 0 \\ 1 & 3 & 1 \\ 0 & 1 & 4 \end{bmatrix} \tag{9.106}$$

The eigenvalues are

$$\lambda_1 = 3 + \sqrt{3} \doteq 4.7321 \qquad \lambda_2 = 3.0 \qquad \lambda_3 = 3 - \sqrt{3} \doteq 1.2679$$

The iterates A_m do not converge rapidly, and only a few are given to indicate the qualitative behavior of the convergence.

$$A_2 = \begin{bmatrix} 3.0000 & 1.0954 & 0 \\ 1.0954 & 3.0000 & -1.3416 \\ 0 & -1.3416 & 3.0000 \end{bmatrix} \qquad A_3 = \begin{bmatrix} 3.7059 & .9558 & 0 \\ .9558 & 3.5214 & .9738 \\ 0 & .9738 & 1.7727 \end{bmatrix}$$

$$A_7 = \begin{bmatrix} 4.6792 & .2979 & 0 \\ .2979 & 3.0524 & .0274 \\ 0 & .0274 & 1.2684 \end{bmatrix} \qquad A_8 = \begin{bmatrix} 4.7104 & .1924 & 0 \\ .1924 & 3.0216 & -.0115 \\ 0 & -.0115 & 1.2680 \end{bmatrix}$$

$$A_9 = \begin{bmatrix} 4.7233 & .1229 & 0 \\ .1229 & 3.0087 & .0048 \\ 0 & .0048 & 1.2680 \end{bmatrix} \qquad A_{10} = \begin{bmatrix} 4.7285 & .0781 & 0 \\ .0781 & 3.0035 & -.0020 \\ 0 & -.0020 & 1.2680 \end{bmatrix}$$

The elements in the $(1,2)$ position decrease geometrically with a ratio of about .64 per iterate, and those in the $(2,3)$ position decrease with a ratio of about .42 per iterate. The value in the $(3,3)$ position of A_{15} will be 1.2679, which is correct to five places.

The preliminary reduction of A to simpler form   The QR method (9.100) can be relatively expensive because the QR factorization is time consuming when repeated many times. To decrease the expense the matrix is prepared for the QR method by reducing it to a simpler form, one for which the QR factorization is much less expensive.

If A is symmetric, it is reduced to a similar symmetric tridiagonal matrix exactly as described in Section 9.3. If A is nonsymmetric, it is reduced to a similar *Hessenberg matrix*. A matrix B is Hessenberg if

$$B_{ij}=0 \qquad \text{for all} \quad i>j+1$$

It is upper triangular except for a single nonzero subdiagonal. The matrix A is reduced to Hessenberg form using the same algorithm as was used for reducing symmetric matrices to tridiagonal form.

With A tridiagonal or Hessenberg, the Householder matrices of Section 9.3 take a simple form when calculating the QR factorization. But generally the plane rotations (9.88) are used in place of the Householder matrices because they are slightly efficient to compute and apply in this situation. Having produced $A_1 = Q_1 R_1$ and $A_2 = R_1 Q_1$, we have to know that the form of A_2 is the same as that of A_1 in order to continue using the less expensive form of QR factorization.

Suppose A_1 is in Hessenberg form. From Section 9.3 the factorization $A_1 = Q_1 R_1$ has the following value for Q_1:

$$Q_1 = H_1 \ldots H_{n-1} \tag{9.107}$$

with each H_k a Householder matrix (9.64),

$$H_k = I - 2w^{(k)}w^{(k)T} \qquad 1 \le k \le n-1 \tag{9.108}$$

Because the matrix A_1 is of Hessenberg form, the vectors $w^{(k)}$ can be shown to have the special form

$$w_i^{(k)} = 0 \qquad \text{for} \quad i<k \qquad \text{and} \quad i>k+1 \tag{9.109}$$

This can be shown from the equations for the components of $w^{(k)}$, and in particular (9.71). From (9.109), the matrix H_k will differ from the identity in only the four elements in positions (k,k), $(k,k+1)$, $(k+1,k)$, and $(k+1,k+1)$. And from this it is a fairly straightforward computation to show that Q_1 must be Hessenberg in form. Another necessary lemma is that the product of an triangular matrix and a Hessenberg matrix is again Hessenberg. Just multiply the two forms of matrices, observing the respective patterns of zeros, in order to prove this lemma. Combining these results, observing that R_1 is upper triangular, we have that $A_2 = R_1 Q_1$ must be in Hessenberg form.

If A_1 is symmetric and tridiagonal, then it is trivially Hessenberg. From the preceding result, A_2 must also be Hessenberg. But A_2 is symmetric since

$$A_2^T = \left(Q_1^T A_1 Q_1\right)^T = Q_1^T A_1^T Q_1 = Q_1^T A_1 Q_1 = A_2$$

Since any symmetric Hessenberg matrix is tridiagonal, we have shown that A_2 is tridiagonal. Note that the iterates in the example (9.106) illustrate this result.

A convergence result for the QR method   A convergence result is given for a large class of matrices, symmetric and nonsymmetric. Part of the proof will show the factor that determines the speed of convergence of the method. For a more general discussion of the method, see Wilkinson [W2, Chapter 8] and Parlett [P1].

Theorem 9.6   Let A be a real matrix of order n, and let its eigenvalues $\{\lambda_i\}$ satisfy

$$|\lambda_1| > |\lambda_2| > \ldots > |\lambda_n| > 0 \qquad (9.110)$$

Then the iterates A_m of the QR method, defined in (9.100), will converge to an upper triangular matrix that contains the eigenvalues $\{\lambda_i\}$ in the diagonal positions. If A is symmetric, the sequence $\{A_m\}$ converges to a diagonal matrix.

Proof   The proof is fairly lengthy and involved, and the reader may wish to skip it and go onto the discussion following the proof. The main factor in determining the speed of convergence is contained in (9.115), and it is illustrated following the proof. This proof follows closely that of Wilkinson [W2, pp. 517–519].

Since A has distinct eigenvalues, there is a nonsingular matrix X for which

$$X^{-1}AX = D = \operatorname{diag}[\lambda_1, \ldots, \lambda_n]$$

Then

$$A^m = XD^mX^{-1} \qquad (9.111)$$

Since A is real and all of its eigenvalues are of distinct magnitude, it cannot have any complex roots, as they would have to occur in conjugate pairs of equal magnitude.

We now derive some alternative forms for A^m, beginning with (9.111). Assume that X^{-1} has the decomposition

$$X^{-1} = LU \qquad (9.112)$$

For the appropriate modification to use when this is not possible and pivoting must be used, see Wilkinson [W2, p. 519]. Combining (9.111) and (9.112),

$$A^m = X(D^mLD^{-m})D^mU \qquad (9.113)$$

Recall that in the derivation of the decomposition (9.112) in Section 8.1 of Chapter 8, the diagonal elements of L could be chosen to be 1. Then the matrix D^mLD^{-m} is lower triangular with diagonal elements equal to

1, and

$$(D^mLD^{-m})_{ij} = \left(\frac{\lambda_i}{\lambda_j}\right)^m L_{ij} \qquad 1 \leqq j < i \leqq n \qquad (9.114)$$

Define E_m implicitly by

$$D^mLD^{-m} = I + E_m$$

E_m is a lower triangular matrix that converges to zero. Using (9.114) and (9.110),

$$\|E_m\|_\infty \leqq c \cdot \underset{1 \leqq j \leqq n-1}{\text{Max}} \left|\frac{\lambda_{j+1}}{\lambda_j}\right|^m \qquad m \geqq 1 \qquad (9.115)$$

for some constant $c > 0$.

The matrix X can be factored,

$$X = QR$$

for some orthogonal Q and nonsingular upper triangular R. Returning to (9.113), this leads to

$$A^m = QR(I + E_m)D^mU$$

$$= Q(I + RE_mR^{-1})RD^mU \qquad (9.116)$$

Using another QR factorization,

$$I + RE_mR^{-1} = \tilde{Q}_m\tilde{R}_m \qquad (9.117)$$

We require the diagonal elements of $\tilde{R}_m$ to be positive, which is possible from the construction for the factorization given in Section 8.3. Also see the discussion between (9.76) and (9.77), which shows that with this positivity assumption, the decomposition (9.117) is unique.

We can show that

$$\tilde{Q}_m, \tilde{R}_m \to I \qquad \text{as} \quad m \to \infty \qquad (9.118)$$

Using (9.117) it is straightforward to show that

$$\tilde{R}_m\tilde{R}_m^T - I \to 0 \qquad \text{as} \quad m \to \infty$$

A detailed examination of the coefficients of $\tilde{R}_m\tilde{R}_m^T$ will then show that $\tilde{R}_m \to I$, using the positivity of the diagonal elements. Using this result in (9.117) will then show that $\tilde{Q}_m \to I$.

Using (9.117) in (9.116),

$$A^m = (Q\tilde{Q}_m)(\tilde{R}_mRD^mU) \qquad (9.119)$$

Clearly $Q\tilde{Q}_m$ is orthogonal. And because $\tilde{R}_m, R, U$ are upper triangular and D^m is diagonal, we have their product is upper triangular. Thus (9.119) is a QR factorization of A^m. Returning to (9.105) we have the second QR factorization

$$A^m = P_m U_m$$

Comparing these results and using the uniqueness of the QR factorization expressed in (9.76) and (9.77), we have

$$P_m = (Q\tilde{Q}_m)\tilde{D}_m \qquad U_m = \tilde{D}_m(\tilde{R}_m RD^m U) \qquad (9.120)$$

for some diagonal matrix $\tilde{D}_m$ with

$$\tilde{D}_m^2 = I \qquad m \geq 1 \qquad (9.121)$$

We now examine the behavior of the sequence $\{A_m\}$ as $m \to \infty$. From (9.104) and (9.120),

$$A_{m+1} = P_m^T A P_m$$

$$= \tilde{D}_m \tilde{Q}_m^T Q^T A Q \tilde{Q}_m \tilde{D}_m$$

From earlier $X = QR$, and

$$Q = XR^{-1}$$

$$Q^T = Q^{-1} = RX^{-1}$$

Substituting above,

$$A_{m+1} = \tilde{D}_m^T \tilde{Q}_m^T RX^{-1} A X R^{-1} \tilde{Q}_m \tilde{D}_m$$

$$= \tilde{D}_m^T \tilde{Q}_m^T RDR^{-1} \tilde{Q}_m \tilde{D}_m \qquad (9.122)$$

Consider just the diagonal elements of A_{m+1} since they are the main point of interest. The matrix RDR^{-1} is upper triangular and its diagonal elements are just $\{\lambda_1, \ldots, \lambda_n\}$. Using (9.118) and (9.121), that $\tilde{Q}_m \to I$ and $\tilde{D}_m^2 = I$, we will then have that the diagonal elements of A_{m+1} will converge to the eigenvalues of A, ordered from largest to smallest in magnitude. In addition, since RDR^{-1} is upper triangular, the elements below the diagonal in A_{m+1} will converge to zero. The speed of convergence will depend completely on the speed of convergence of $\tilde{Q}_m$ to I; and this depends on the bound in (9.115).

If A is symmetric, then the iterates A_m are also symmetric. Since the lower triangular part of A_m converges to zero, the same is true of the part above the diagonal. This proves that for a symmetric matrix satisfying (9.110), A_m converges to a diagonal matrix. This completes the proof. ∎

As pointed out in the proof, the critical factor in determing the speed of convergence are the ratios $\lambda_{j+1}/\lambda_j, 1 \le j \le n-1$. Thus there is a geometric rate of convergence, and this can be very slow. In the example (9.106), the ratios of successive eigenvalues are

$$\frac{\lambda_2}{\lambda_1} \doteq .63 \qquad \frac{\lambda_3}{\lambda_2} \doteq .42$$

And if the off-diagonal elements are observed, we see that they decrease with about these ratios.

For matrices whose eigenvalues do not satisfy (9.110), the iterates A_m may not converge to a triangular matrix. For A symmetric, the sequence $\{A_m\}$ converges to a block diagonal matrix

$$A_m \to D = \begin{bmatrix} B_1 & & & 0 \\ & B_2 & & \\ & & \ddots & \\ 0 & & & B_r \end{bmatrix} \qquad (9.123)$$

in which all blocks B_i have order 1 or 2. Thus the eigenvalues of A can be easily computed from those of D. If A is real and nonsymmetric, the situation is more complicated, but acceptable. For a discussion, see Wilkinson [W2, Chapter 8] and Parlett [P1].

To see that $\{A_m\}$ does not always converge to a diagonal matrix, consider the simple symmetric example

$$A = \begin{bmatrix} 0 & 1 \\ 1 & 0 \end{bmatrix}$$

Its eigenvalues are $\lambda = \pm 1$. Since A is orthogonal, we have

$$A = Q_1 R_1 \qquad Q_1 = A \qquad R_1 = I$$

And thus

$$A_2 = R_1 Q_1 = A$$

and all iterates $A_m = A$. The sequence $\{A_m\}$ does not converge to a diagonal matrix.

The QR method with shift   The QR algorithm is generally applied with a shift of origin for the eigenvalues in order to increase the speed of convergence. For a sequence of constants $\{c_m\}$, define $A_1 = A$ and

$$A_m - c_m I = Q_m R_m$$

$$A_{m+1} = c_m I + R_m Q_m \qquad m = 1, 2, \dots \tag{9.124}$$

The matrices A_m are similar to A_1, since

$$R_m = Q_m^T (A_m - c_m I)$$

$$A_{m+1} = c_m I + Q_m^T (A_m - c_m I) Q_m$$

$$= c_m I + Q_m^T A_m Q_m - c_m I$$

$$A_{m+1} = Q_m^T A_m Q_m \qquad m \geq 1 \tag{9.125}$$

The eigenvalues of A_{m+1} are the same as those of A_m, and thence the same as those of A.

To be more specific on the choice of shifts $\{c_m\}$, we will consider only a symmetric tridiagonal matrix A. For A_m, let

$$A_m = \begin{bmatrix} \alpha_1^{(m)} & \beta_1^{(m)} & 0 & \cdots & & & 0 \\ \beta_1^{(m)} & \alpha_2^{(m)} & \beta_2^{(m)} & & & & \vdots \\ 0 & \ddots & \ddots & \ddots & & & \\ \vdots & & \ddots & \ddots & \ddots & & \beta_{n-1}^{(m)} \\ 0 & \cdots & & & & \beta_{n-1}^{(m)} & \alpha_n^{(m)} \end{bmatrix} \tag{9.126}$$

There are two methods by which $\{c_m\}$ is chosen. (1) Let $c_m = \alpha_n^{(m)}$; and (2) let c_m be the eigenvalue of

$$\begin{bmatrix} \alpha_{n-1}^{(m)} & \beta_{n-1}^{(m)} \\ \beta_{n-1}^{(m)} & \alpha_n^{(m)} \end{bmatrix} \tag{9.127}$$

which is closest to $\alpha_n^{(m)}$. The second strategy is preferred; but in either case the matrices A_m converge to a block diagonal matrix in which the blocks have order 1 or 2, as in (9.123). Either choice of $\{c_m\}$ ensures

$$\beta_{n-1}^{(m)} \beta_{n-2}^{(m)} \to 0 \qquad \text{as } n \to \infty \tag{9.128}$$

generally at a much more rapid rate than with the original QR method (9.100).
From (9.125),

$$\|A_{m+1}\|_2 = \|Q_m^T A_m Q_m\|_2 = \|A_m\|_2$$

using the operator matrix norm (7.44) and Problem 21(c) of Chapter 7. The matrices $\{A_m\}$ are uniformly bounded, and consequently the same is true of their elements. From (9.128) and the uniform boundedness of $\{\beta_{n-1}^{(m)}\}$ and $\{\beta_{n-2}^{(m)}\}$, we have either $\beta_{n-1}^{(m)} \to 0$ or $\beta_{n-2}^{(m)} \to 0$ as $m \to \infty$. In the former case, $\alpha_n^{(m)}$ converges to an eigenvalue of A. And in the latter case, two eigenvalues can easily be extracted from the submatrix (9.127).

Once one or two eigenvalues have been obtained due to $\beta_{n-1}^{(m)}$ or $\beta_{n-2}^{(m)}$ being essentially zero, the matrix A_m can be reduced in order by one or two rows, respectively. Following this, the QR method with shift can be applied to the reduced matrix. The choice of shifts is designed to make the convergence to zero be more rapid for $\beta_{n-1}^{(m)}, \beta_{n-2}^{(m)}$ than for the remaining off-diagonal elements of the matrix. In this way, the QR method becomes a rapid general purpose method, faster than any other method at the present time. For a proof of convergence of the QR method with shift, see Wilkinson [W3].

Example   Use the previous example (9.106), and use the first method of choosing the shift, $c_m = \alpha_n^{(m)}$. The iterates are

$$A_1 = \begin{bmatrix} 2 & 1 & 0 \\ 1 & 3 & 1 \\ 0 & 1 & 4 \end{bmatrix} \qquad A_2 = \begin{bmatrix} 1.4000 & .4899 & 0 \\ .4899 & 3.2667 & .7454 \\ 0 & .7454 & 4.3333 \end{bmatrix}$$

$$A_3 = \begin{bmatrix} 1.2915 & .2017 & 0 \\ .2017 & 3.0202 & .2724 \\ 0 & .2724 & 4.6884 \end{bmatrix} \qquad A_4 = \begin{bmatrix} 1.2737 & .0993 & 0 \\ .0993 & 2.9943 & .0072 \\ 0 & .0072 & 4.7320 \end{bmatrix}$$

$$A_5 = \begin{bmatrix} 1.2694 & .0498 & 0 \\ .0498 & 2.9986 & 0 \\ 0 & 0 & 4.7321 \end{bmatrix}$$

The element $\beta_2^{(m)}$ converges to zero extremely rapidly, but the element $\beta_1^{(m)}$ converges to zero geometrically with a ratio of only about .5.

We note the antecedent to the QR method, motivating much of it. In 1958, H. Rutishauser introduced an LR method based on the Gaussian elimination decomposition of a matrix into a lower triangular matrix times an upper triangular matrix. Define

$$A_m = L_m R_m \qquad A_{m+1} = R_m L_m = L_m^{-1} A_m L_m$$

with L_m lower triangular, R_m upper triangular. When applicable, this method will generally be more efficient than the QR method. But the nonorthogonal similarity transformations can cause a deterioration of the conditioning of the eigenvalues of some nonsymmetric matrices. And generally it is a more complicated algorithm to implement in an automatic program. A complete discussion is given in Wilkinson [W2, Chapter 8].

9.6 The Calculation of Eigenvectors and Inverse Iteration

The most powerful tool for the calculation of the eigenvectors of a matrix is inverse iteration, a method due to H. Wielandt in 1944. We first define and illustrate inverse iteration, and then comment more generally on the calculation of eigenvectors.

To simplify the analysis, let A be a matrix whose Jordan canonical form is diagonal,

$$P^{-1}AP = \text{diag}[\lambda_1, \ldots, \lambda_n]$$

Let the columns of P be denoted by $x_1, \ldots, x_n$. Then

$$Ax_i = \lambda_i x_i \qquad i = 1, \ldots, n \tag{9.129}$$

Without loss of generality, it can also be assumed that $\|x_i\|_\infty = 1$, for all i.

Let λ be an approximation to a simple eigenvalue λ_k of A. Given an initial $z^{(0)}$, define $\{w^{(m)}\}$ and $\{z^{(m)}\}$ by

$$(A - \lambda I)w^{(m+1)} = z^{(m)} \qquad z^{(m+1)} = w^{(m+1)}/\|w^{(m+1)}\|_\infty \qquad m \geq 0 \tag{9.130}$$

This is essentially the power method, with $(A - \lambda I)^{-1}$ replacing A in (9.40). The matrix $A - \lambda I$ is ill-conditioned from the viewpoint of the material in Section 8.4 of Chapter 8. But any resulting large perturbations in the solution will be rich in the eigenvector x_k of the eigenvalue $\lambda_k - \lambda$ for $A - \lambda I$; and this is the vector we desire. For a further discussion of this source of instability in solving the linear system, see the material following formula (8.36) in Section 8.4. For the method (9.130) to work, we do not want $A - \lambda I$ to be singular. Thus λ shouldn't be exactly λ_k, although it can be quite close, as a later example will demonstrate.

For a more precise analysis, let $z^{(0)}$ be expanded in terms of the eigenvector basis of (9.129),

$$z^{(0)} = \sum_{i=1}^{n} \alpha_i x_i \tag{9.131}$$

And assume $\alpha_k \neq 0$. In analogy with formula (9.41) for the power method, we have

$$z^{(m)} = \frac{(A - \lambda I)^{-m} z^{(0)}}{\|(A - \lambda I)^{-m} z^{(0)}\|_\infty} \tag{9.132}$$

Using (9.131),

$$(A - \lambda I)^{-m} z^{(0)} = \sum_{i=1}^{n} \alpha_i \left(\frac{1}{\lambda_i - \lambda}\right)^m x_i \tag{9.133}$$

Let $\lambda_k - \lambda = \epsilon$, and assume

$$|\lambda_i - \lambda| \geq c > 0 \qquad i = 1, \ldots, n \qquad i \neq k$$

From (9.133) and (9.132),

$$z^{(m)} = \frac{\sigma_m x_k + \epsilon^m \sum_{i \neq k} \left(\frac{\alpha_i}{\alpha_k}\right)\left(\frac{1}{\lambda_i - \lambda}\right)^m x_i}{\left\| x_k + \epsilon^m \sum_{i \neq k} \left(\frac{\alpha_i}{\alpha_k}\right)\left(\frac{1}{\lambda_i - \lambda}\right)^m x_i \right\|_\infty} \tag{9.134}$$

with some $|\sigma_m| = 1$. If $|\epsilon| < c$, then

$$\left\| \epsilon^m \sum_{i \neq k} \left(\frac{\alpha_i}{\alpha_k}\right)\left(\frac{1}{\lambda_i - \lambda}\right)^m x_i \right\|_\infty \leq \left(\frac{\epsilon}{c}\right)^m \sum_{i \neq k} \left|\frac{\alpha_i}{\alpha_k}\right| \tag{9.135}$$

This quantity goes to zero as $m \to \infty$. Combined with (9.134), this shows $z^{(m)}$ converges to x_k as $m \to \infty$. The convergence is linear, with a ratio of $|\epsilon/c|$ decrease in the error in each iterate. In practice $|\epsilon|$ is quite small and this will usually mean $|\epsilon/c|$ is also quite small, insuring rapid convergence.

In implementing (9.130), begin by factoring $A - \lambda I$ using the *LU* decomposition of Section 8.1 of Chapter 8. To simplify the notation, write

$$A - \lambda I = LU$$

in which pivoting is not involved. In practice, pivoting would be used. Solve for each iterate $z^{(m+1)}$ as follows:

$$Ly^{(m+1)} = z^{(m)} \qquad Uw^{(m+1)} = y^{(m+1)}$$
$$z^{(m+1)} = w^{(m+1)} / \|w^{(m+1)}\|_\infty \tag{9.136}$$

Since $A - \lambda I$ is nearly singular, the last diagonal element of U will be nearly zero. If it is exactly zero, then change it to some small number or else change λ very slightly and recalculate L and U.

For the initial guess $z^{(0)}$, Wilkinson [W1, p. 147] suggests using

$$z^{(0)} = Le \qquad e = [1, 1, \ldots, 1]^T$$

And thus in (9.136),

$$y^{(1)} = e \qquad Uw^{(m+1)} = e$$

This choice is intended to insure that α_k is neither nonzero nor small in (9.131). But even if it were small, the method would usually converge rapidly. For example, suppose that some or all of the values α_i / α_k in (9.135) are about 10^4.

And suppose $|\epsilon/c| = 10^{-5}$, a realistic value for many cases. Then the bound in (9.135) becomes

$$(10^{-5})^m \cdot n \cdot 10^4$$

and this will decrease very rapidly as m increases.

Example   Use the earlier matrix (9.106),

$$A = \begin{bmatrix} 2 & 1 & 0 \\ 1 & 3 & 1 \\ 0 & 1 & 4 \end{bmatrix} \tag{9.137}$$

Let $\lambda = 1.2679 \approx \lambda_3 = 3 - \sqrt{3}$, which is accurate to five places. This leads to

$$L = \begin{bmatrix} 1.0 & 0 & 0 \\ 1.3659 & 1.0 & 0 \\ 0 & 2.7310 & 1.0 \end{bmatrix} \qquad U = \begin{bmatrix} .7321 & 1.0 & 0 \\ 0 & .3662 & 1.0 \\ 0 & 0 & .0011 \end{bmatrix}$$

subject to the effects of rounding errors. Using $y^{(1)} = [1, 1, 1]^T$,

$$w^{(1)} = [3385.2, -2477.3, 908.20]^T$$

$$z^{(1)} = [1.0000, -.73180, .26828]$$

$$w^{(2)} = [20345, -14894, 5451.9]^T$$

$$z^{(2)} = [1.0000, -.73207, .26797]^T \tag{9.138}$$

and the vector of $z^{(3)} = z^{(2)}$. The true answer is

$$x_3 = [1, 1 - \sqrt{3}, 2 - \sqrt{3}]^T$$

$$\doteq [1.0000, -.73205, .26795]^T \tag{9.139}$$

and $z^{(2)}$ equals x_3 to within the limits of rounding error accumulations.

Eigenvectors for symmetric tridiagonal matrices   Let A be a real symmetric tridiagonal matrix of order n. As above we assume that some or all of its eigenvalues have been computed accurately. Inverse iteration is the preferred method of calculation for the eigenvectors, and it is quite easy to implement. For λ an approximate eigenvalue of A, the calculation of the LU decomposition is inexpensive in time and storage, even with pivoting. For example, see the material on tridiagonal systems in Section 8.3 of Chapter 8. The above numerical example also illustrates the method for tridiagonal matrices.

Some error results will be given as further justification for the use of inverse iteration. Suppose that the computer arithmetic is binary floating point with rounding and with t digits in the mantissa. In Wilkinson [W1, pp. 143–147] it is shown that the computed solution $\hat{w}$ of

$$(A - \lambda I)w^{(m+1)} = z^{(m)}$$

is the exact solution of

$$(A - \lambda I + E)\hat{w} = z^{(m)} \tag{9.140}$$

with

$$\|E\|_2 \leq K\sqrt{n} \cdot 2^{-t}$$

for some constant K of order unity. If the solution $\hat{w}$ of (9.140) is quite large, then it will be a good approximation to an eigenvector of A.

To show this, we begin by introducing

$$\hat{z} = \frac{\hat{w}}{\|\hat{w}\|_2}$$

Then

$$(A - \lambda I + E)\hat{z} = z^{(m)}/\|\hat{w}\|_2$$

$$\eta \equiv (A - \lambda I)\hat{z} = -E\hat{z} + \hat{z}/\|\hat{w}\|_2$$

For the residual η,

$$\|\eta\|_2 \leq \|E\|_2 + 1/\|\hat{w}\|_2$$

$$\leq K\sqrt{n} \cdot 2^{-t} + 1/\|\hat{w}\|_2 \tag{9.141}$$

which is small if $\|\hat{w}\|_2$ is large. To prove that this implies $\hat{z}$ is close to an eigenvector of A, we let $\{x_i | i = 1, \ldots, n\}$ be an orthonormal set of eigenvectors. And assume

$$\lambda \approx \lambda_k \qquad \hat{z} = \sum_{i=1}^{n} \alpha_i x_i \qquad \|\hat{z}\|_2^2 = \sum_{1}^{n} \alpha_i^2 = 1$$

with λ_k an isolated eigenvalue of A. Also, suppose

$$|\lambda_i - \lambda| \geq c > 0 \qquad \text{all } i \neq k$$

with

$$c \gg |\lambda_k - \lambda|$$

With these assumptions, we can now derive a bound for the error in $\hat{z}$. Expanding η using the eigenvector basis,

$$\eta = Az - \lambda z = \sum_1^n \alpha_i (\lambda_i - \lambda) x_i$$

$$\|\eta\|_2^2 = \sum_1^n \alpha_i^2 (\lambda_i - \lambda)^2$$

$$\geq \sum_{i \neq k} \alpha_i^2 (\lambda_i - \lambda)^2 \geq c^2 \sum_{i \neq k} \alpha_i^2$$

$$\sum_{i \neq k} \alpha_i^2 \leq \frac{1}{c^2} \|\eta\|_2^2$$

which is quite small using (9.141). Thus $|\alpha_k| \approx 1$ and

$$\|\hat{z} - \alpha_k x_k\|_2 = \sqrt{\sum_{i \neq k} \alpha_i^2} \leq \frac{1}{c} \|\eta\|_2 \tag{9.142}$$

showing the desired result. For a further discussion of the error, see Wilkinson [W1, pp. 142–146] and [W2, pp. 321–330].

Another method for calculating eigenvectors would appear to be the direct solution of

$$(A - \lambda I)x = 0$$

after deleting one equation and setting one of the unknown components to a nonzero constant, for example $x_1 = 1$. This is often the procedure used in undergraduate linear algebra courses. But as a general numerical method, it can be disastrous. A complete discussion of this problem is given in Wilkinson [W2, pp. 315–321], including an excellent example. We will just use the previous example to show that the results need not be as good as those obtained with inverse iteration.

Example Consider the preceding example (9.137) with $\lambda = 1.2679$. We evaluate $A - \lambda I$ and delete the last equation to obtain

$$.7321 x_1 + x_2 = 0$$

$$x_1 + 1.7321 x_2 + x_3 = 0$$

Taking $x_1 = 1.0$, we have the approximate eigenvector

$$x = [\, 1.0000, \, -.73210, .26807 \,]$$

Comparing with the true answer (9.139), this is a somewhat poorer result than (9.138) obtained by inverse iteration.

The inverse iteration method requires a great deal of care in its implementation. For dealing with a particular matrix, any difficulties can be dealt with on an ad hoc basis. But for a general computer program we have to deal with eigenvalues that are multiple or close together, which can cause some difficulty if not dealt with carefully. For nonsymmetric matrices whose Jordan canonical form is not diagonal, there are additional difficulties in selecting a correct basis of eigenvectors. The best reference for this topic is Wilkinson [W2], and for several excellent programs, see Wilkinson [W4] and Smith et al. [S1].

Discussion of the Literature

The main source of information for this chapter was the encyclopedic and well-known book [W2] of J. H. Wilkinson. Other sources were Gourlay and Watson [G1], Householder [H2], [Noble, Chapters 9–12], Stewart [S2], and Wilkinson [W1]. For matrices of moderate size, the numerical solution of the eigenvalue problem is fairly well understood; and excellent algorithms for most problems are given in the book [W4], edited by Wilkinson and Reinsch.

The computer library of eigenvalue-eigenvector programs called EISPACK was developed under the auspices of Argonne National Laboratory. It is a major event within the growth of the area of computer science known as *mathematical software*. The programs in EISPACK are based on those of [W4], with additional development based on new theoretical and experimental results. The programs are written in FORTRAN and are available on all major lines of computing equipment in the United States. For IBM equipment, a driver program EISPAC was also developed to make it much less complex to use the EISPACK programs for a variety of eigenvalue-eigenvector problems. For a complete description of EISPACK, see [S1]. It contains listings of the programs, timings of the programs for a variety of computers, and very complete descriptions of what routines to use for a particular task. The second version of EISPACK (1976) also contains a number of programs that were originally not given in [W4].

A number of numerical methods were not discussed in this chapter. For an account of some of these methods, see [H2, Chapters 6, 7] and [W2, Chapter 7]. The major one deleted is the Jacobi method, and it is an elegant method for calculating all of the eigenvalues of a symmetric matrix. It is somewhat less efficient than the QR method, but it is easier to program. For a description of the Jacobi method, see [G1, Chapter 7] or [W2, pp. 266–282]. An ALGOL program is given in [W4, pp. 202–211].

When the matrices have large order, (e.g., $n \geq 1000$), most of the methods of this chapter are more difficult to apply because of storage considerations. In addition, the methods often do not take special account of the sparseness of most matrices occurring in practice; and it is important from the viewpoint of both time and storage to take advantage of the sparseness. One common form of problem involves a symmetric matrix with all elements near the diagonal. Programs for this problem are given in [W4, pp. 266–283], based on both the *QR* algorithm and on reduction to a tridiagonal matrix. A discussion of other problems and techniques for sparse matrices is given in [B1], especially the two papers by G. Stewart and by W. Kahan and B. Parlett. Also see the general survey of methods for this problem in the survey of Duff [D1].

Bibliography

[**B1**]   Bunch, J. and D. Rose, eds., *Sparse Matrix Computations*, Academic Press, New York, 1976.

[**D1**]   Duff, I., A survey of sparse matrix research, *Proc. IEEE*, **65**, 1977, pp. 500–535.

[**G1**]   Gourlay, A. and G. Watson, *Computational Methods for Matrix Eigenproblems*, John Wiley, New York, 1976.

[**G2**]   Gregory, R. and D. Karney, *A Collection of Matrices for Testing Computational Algorithms*, John Wiley, New York, 1969.

[**H1**]   Henrici, P., *Applied and Computational Complex Analysis*, John Wiley, New York, 1974.

[**H2**]   Householder, A., *The Theory of Matrices in Numerical Analysis*, Blaisdell, New York, 1964.

[**N1**]   Noble, B., *Applied Linear Algebra*, Prentice-Hall, Englewood Cliffs, N.J., 1969.

[**P1**]   Parlett, B., Global convergence of the basic *QR* algorithm on Hessenberg matrices, *Math. Comp.*, **22**, 1968, pp. 803–817.

[**S1**]   Smith, B. T., J. Boyle, B. Garbow, Y. Ikebe, V. Klema, and C. Moler, *Matrix Eigensystem Routines—EISPACK Guide*, 2nd ed., Lecture Notes in Comp. Sci., Vol. 6, Springer-Verlag, New York, 1976.

[**S2**]   Stewart, G., *Introduction to Matrix Computations*, Academic Press, New York, 1973.

[**W1**]   Wilkinson, J., *Rounding Errors in Algebraic Processes*, Prentice-Hall, Englewood Cliffs, N.J., 1963.

[W2] Wilkinson, J., *The Algebraic Eigenvalue Problem*, Oxford University Press, Oxford, England, 1965.

[W3] Wilkinson, J., Global convergence of tridiagonal QR algorithm with origin shifts, *Linear Algebra and Its Applic.*, **1**, 1968, pp. 409–420.

[W4] Wilkinson, J. and C. Reinsch, ed., *Handbook for Automatic Computation*, *Vol. II, Linear Algebra*, Springer-Verlag, New York, 1971.

Problems

1. Use the Gerschgorin Theorem 9.1 to determine the approximate location of the eigenvalues of

$$A = \begin{bmatrix} 1 & -1 & 0 \\ 1 & 5 & 1 \\ -2 & -1 & 9 \end{bmatrix}$$

Use these results to infer whether the eigenvalues are real or complex. To check these results, compute the eigenvalues directly by finding the roots of the characteristic polynomial.

2. **(a)** Given a polynomial

$$p(\lambda) = \lambda^n + a_{n-1}\lambda^{n-1} + \ldots + a_0$$

show $p(\lambda) = \det[\lambda I - A]$ for the matrix

$$A = \begin{bmatrix} 0 & 1 & 0 & \cdots & \cdots & 0 \\ 0 & 0 & 1 & 0 & & \vdots \\ \vdots & & \ddots & \ddots & & \vdots \\ \vdots & & & \ddots & \ddots & \\ 0 & 0 & \cdots & \cdots & 0 & 1 \\ -a_0 & -a_1 & \cdots & \cdots & & -a_{n-1} \end{bmatrix}$$

The roots of $p(\lambda)$ are the eigenvalues of A. The matrix A is called the *companion matrix* for the polynomial $p(\lambda)$.

(b) Apply the Gerschgorin Theorem 9.1 to obtain the following bounds for the roots r of $p(\lambda)$: $|r| \le 1$, or $|r + a_{n-1}| \le |a_0| + \ldots + |a_{n-2}|$. If these bounds give disjoint regions in the complex plane, what can be said about the number of roots within each region?

(c) Use the Gerschgorin theorem on the columns of A to obtain additional bounds for the roots of $p(\lambda)$.

(d) Use the results of (b) and (c) to bound the roots of the following polynomials

$$\text{(i) } \lambda^{10}+8\lambda^9+1=0 \qquad \text{(ii) } \lambda^6-4\lambda^5+\lambda^4-\lambda^3+\lambda^2-\lambda+1=0$$

3. Recall the linear system (8.98) of Chapter 8, which arises when numerically solving Poisson's equation. If the equations are ordered in the manner described in (8.105) and following, then the linear system is symmetric with positive diagonal elements. For the Gauss-Seidel iteration method in (8.105) to converge, it is necessary and sufficient that A be positive definite, according to Theorem 8.7. Use the Gerschgorin Theorem 9.1 to prove A is positive definite. It will also be necessary to quote Theorem 8.8, that $\lambda=0$ is not an eigenvalue of A.

4. The values $\lambda=-8.02861$ and

$$x=[1.0, \ 2.50146, \ -.75773, \ -2.56421]$$

are an approximate eigenvalue and eigenvector for the matrix

$$A = \begin{bmatrix} 2 & 1 & 3 & 4 \\ 1 & -3 & 1 & 5 \\ 3 & 1 & 6 & -2 \\ 4 & 5 & -2 & -1 \end{bmatrix}$$

Use the result (9.19) to compute an error bound for λ. Note that $\|x\|_2 \neq 1$ and that x or (9.19) must be modified accordingly.

5. For the matrix example (9.15) with $\lambda=2$, compute the perturbation error bound (9.31). The same bound was given in (9.32) for the other eigenvalue $\lambda=1$.

6. For the following matrices $A(\epsilon)$, determine the eigenvalues and eigenvectors for both $\epsilon=0$ and $\epsilon>0$. Observe the behavior as $\epsilon \to 0$.

(a) $\begin{bmatrix} 1 & 1 \\ \epsilon & 1 \end{bmatrix}$ **(b)** $\begin{bmatrix} 1 & 1 \\ 0 & 1+\epsilon \end{bmatrix}$ **(c)** $\begin{bmatrix} 1 & \epsilon \\ 0 & 1 \end{bmatrix}$

7. Use the power method to calculate the dominant eigenvalue and associated eigenvector for the following matrices.

(a) $\begin{bmatrix} 6 & 4 & 4 & 1 \\ 4 & 6 & 1 & 4 \\ 4 & 1 & 6 & 4 \\ 1 & 4 & 4 & 6 \end{bmatrix}$ **(b)** $\begin{bmatrix} 2 & 1 & 3 & 4 \\ 1 & -3 & 1 & 5 \\ 3 & 1 & 6 & -2 \\ 4 & 5 & -2 & -1 \end{bmatrix}$

Check the speed of convergence, calculating the ratios R_m of (9.50). When the ratios R_m are fairly constant, use Aitken extrapolation to improve the speed of convergence of both the eigenvalue and eigenvector.

8. For a matrix A of order n, assume its Jordan canonical form is diagonal and denote the eigenvalues by $\lambda_1, \ldots, \lambda_n$. Assume that $\lambda_1 = \lambda_2 = \ldots = \lambda_r$ for some $r > 1$, and

$$|\lambda_r| > |\lambda_{r+1}| \geq \ldots \geq |\lambda_n| \geq 0$$

Show that the power method (9.40) will still converge to λ_1 and some associated eigenvector, for most choices of initial vector $z^{(0)}$.

9. Let A be a symmetric matrix of order n, with the eigenvalues ordered by

$$\lambda_1 \geq \lambda_2 \geq \ldots \geq \lambda_n$$

Define

$$R(x) = \frac{(Ax, x)}{(x, x)} \qquad x \neq 0, x \in R^n$$

using the standard inner product. Show

$$\operatorname{Max} R(x) = \lambda_1 \qquad \operatorname{Min} R(x) = \lambda_n$$

as $x \neq 0$ ranges over R^n. The function $R(x)$ is called the *Rayleigh quotient*; and it can be used to characterize the remaining eigenvalues of A, in addition to λ_1 and λ_n. Using these maximizations and minimizations for $R(x)$ forms the basis of some classical numerical methods for calculating the eigenvalues of A.

10. **(a)** Let A be a symmetric matrix, and let λ and x be an eigenvalue-eigenvector pair for A with $\|x\|_2 = 1$. Let P be an orthogonal matrix for which

$$Px = e_1 \equiv [1, 0, \ldots, 0]^T$$

Consider the similar matrix $B = PAP^T$; and show that the first row and column are zero except for the diagonal element which equals λ. Hint: calculate and use Be_1.

(b) For the matrix

$$A = \begin{bmatrix} 2 & 10 & 2 \\ 10 & 5 & -8 \\ 2 & -8 & 11 \end{bmatrix}$$

$\lambda = 9$ is an eigenvalue with associated eigenvector $x = [\frac{2}{3}, \frac{1}{3}, \frac{2}{3}]$. Produce a Householder matrix P for which $Px = e_1$, and then produce $B = PAP^T$. The matrix eigenvalue problem for B can then be reduced easily to a problem for a 2×2 matrix. Use this procedure to calculate the remaining eigenvalues and eigenvectors of A. The process of changing A to B and then solving a matrix eigenvalue problem of order one less than for A, is known as *deflation*. It can be used to extend the applicability of the power method to other than the dominant eigenvalue. For an extensive discussion, see Wilkinson [W2, pp. 584–598].

11. Use Householder matrices to produce the QR factorization of

$$A = \begin{bmatrix} 1 & 1 & 1 \\ 2 & -1 & -1 \\ 2 & -4 & 5 \end{bmatrix}$$

12. Consider the rotation matrix of order n,

$$R^{(k,l)} = \begin{bmatrix} 1 & 0 & 0 \cdots & & & & & & \cdots & 0 \\ 0 & 1 & 0 & & & & & & & \\ \vdots & & \ddots & & & & & & & \vdots \\ 0 & & & \alpha & 0 \cdots & & \cdots \beta & \cdot 0 & & 0 \\ \vdots & & & 0 & 1 & & & 0 & & \vdots \\ & & & \vdots & & \ddots & & \vdots & & \\ 0 & & & -\beta & 0 & & & \alpha & 0 \cdot \cdot 0 \\ & & & & & & & & 1 & \\ \vdots & & & & & & & & & \\ 0 & \cdots & & & & & & & \cdots & 1 \end{bmatrix}$$

with $\alpha^2 + \beta^2 = 1$. If we compute Rb for a given $b \in R^n$, then the only elements that will be changed are in positions k and l. By choosing α and β suitably, we can force Rb to have a zero in position l. Choose α, β so that

$$\begin{bmatrix} \alpha & \beta \\ -\beta & \alpha \end{bmatrix} \begin{bmatrix} b_k \\ b_l \end{bmatrix} = \begin{bmatrix} \gamma \\ 0 \end{bmatrix}$$

for some γ.

(a) Derive formulas for α, β, and show $\gamma = \sqrt{b_k^2 + b_l^2}$.

(b) Reduce $b = [1, 1, 1, 1]^T$ to the form $\hat{b} = [c, 0, 0, 0]$ by a sequence of multiplications by rotation matrices,

$$\hat{b} = R^{(1,2)} R^{(1,3)} R^{(1,4)} b$$

13. Show how the rotation matrices $R^{(k,l)}$ can be used to produce the QR factorization of a matrix.

14. **(a)**   Do an operation count for producing the QR factorization of a matrix using Householder matrices, as in Section 9.3. As usual, combine multiplications and divisions; and keep a separate count for the number of square roots.

   (b)   Repeat part (a), but use the rotation matrices $R^{(k,l)}$ for the reduction.

15.   Give the explicit formulas for the calculation of the QR factorization of a symmetric tridiagonal matrix. Do an operations count, and compare the result with those of Problem 14.

16.   Use Theorem 9.5 to separate the roots of

$$\begin{bmatrix} 0 & 1 & 0 & 0 & 0 \\ 1 & 1 & 1 & 0 & 0 \\ 0 & 1 & 1 & 1 & 0 \\ 0 & 0 & 1 & 1 & 1 \\ 0 & 0 & 0 & 1 & 2 \end{bmatrix}$$

   Then obtain accurate approximations using the bisection method or some other rootfinding technique.

17.   **(a)**   Write a program to reduce a symmetric matrix to tridiagonal form using Householder matrices for the similarity transformations. For efficiency in the matrix multiplications, use the analog of the form of multiplication shown in (9.79).

   (b)   Use the program to reduce the matrix

$$A = \begin{bmatrix} 5 & 4 & 1 & 1 \\ 4 & 5 & 1 & 1 \\ 1 & 1 & 4 & 2 \\ 1 & 1 & 2 & 4 \end{bmatrix}$$

   to tridiagonal form. The true eigenvalues of A are $\lambda = 10, 5, 2, 1$. Calculate the eigenvalues of your reduced tridiagonal matrix as accurately as possible, and compare them to the true eigenvalues of A.

18.   Let A be a symmetric tridiagonal matrix in which all subdiagonal elements $A_{i,i-1}$ are nonzero. Show that the eigenvalues of A are all distinct.

19.   Use the QR method (a) without shift, and (b) with shift, to calculate the eigenvalues of

$$A = \begin{bmatrix} 3 & 1 & 0 \\ 1 & 2 & 1 \\ 0 & 1 & 1 \end{bmatrix}$$

20. Let A be a Hessenberg matrix and consider the factorization $A = QR$, Q orthogonal and R upper triangular.

   (a) Recalling the discussion following (9.107), show that (9.109) is true.

   (b) Show that the result (9.109) implies a form for H_k in (9.108) such that Q will be a Hessenberg matrix.

   (c) Show the product of a Hessenberg matrix and an upper triangular matrix is again a Hessenberg matrix.

   When combined, these results show that RQ is again a Hessenberg matrix, as claimed in the paragraph following (9.109).

21. For the matrix A of Problem 4, two additional approximate eigenvalues are $\lambda = 7.9329$ and $\lambda = 5.6689$. Use inverse iteration to calculate the associated eigenvectors.

22. Investigate the programs available at your computer center for the calculation of the eigenvalues of a real symmetric matrix. Using such a program, compute the eigenvalues of the Hilbert matrices H_n for $n = 3, 4, 5, 6, 7$. To check your answers, see the very accurate values given in Gregory and Karney [G2, pp. 66–73].

23. Consider calculating the eigenvalues and associated eigenfunctions $x(t)$ for which

$$\int_0^1 \frac{x(t)dt}{1+(s-t)^2} = \lambda x(s) \qquad 0 \le s \le 1$$

One way to obtain approximate eigenvalues is to discretize the equation using numerical integration. Let $h = 1/n$ for some $n \ge 1$ and define $t_j = (j - \frac{1}{2})h$, $j = 1, \ldots, n$. Substitute t_i for s in the equation, and approximate the integral using the midpoint numerical integration method. This leads to the system

$$h \sum_{j=1}^n \frac{\hat{x}(t_j)}{1+(t_i - t_j)^2} = \lambda \hat{x}(t_i) \qquad i = 1, \ldots, n$$

in which $\hat{x}(s)$ denotes a function that hopefully approximates $x(s)$. This system is the eigenvalue problem for a symmetric matrix of order n. Find the two largest eigenvalues of this matrix for $n = 2, 4, 8, 16, 32$. Examine the convergence of these eigenvalues as n increases; and attempt to predict the error in the most accurate case, $n = 32$, as compared with the unknown true eigenvalues for the integral equation.

Answers to Selected Exercises

Chapter 1

2. **(a)** $p_{2n-2}(x) = \sum_{j=0}^{n-1} (-1)^j \dfrac{x^{2j}}{(2j+1)(j!)}$

   $|\text{Error}| \le \dfrac{|x|^{2n}}{(2n+1)(n!)}$

3. $\sqrt{1+x-y} \approx 1 + \frac{1}{2}x - \frac{1}{2}y$ for small x and y

6. **(a)** 4 **(b)** 2 **(c)** 3

7. For CDC 6000 series computers, $u = 2^{-47} \doteq 7.1 \times 10^{-15}$; and for IBM 360 series computers using single precision arithmetic, $u = 16^{-5} \doteq 9.5 \times 10^{-7}$.

8. **(a)** $[2.05265, 2.05375]$

   (d) $\left[\dfrac{8.4725}{.0645}, \dfrac{8.4735}{.0635} \right] \doteq [131.3566, 133.4409]$

11. For $x \approx 0$, the quadratic Taylor series approximation is

    $f(x) \approx \dfrac{-5}{24} - \dfrac{11}{48}x - \dfrac{379}{1920}x^2$.

    Also, $f(0) = \dfrac{-5}{24}$ exactly.

13. **(a)** $-.0030$ **(c)** $.00068$

15. **(a)** Error $\approx -1.46(x_T - x_A) - 1.71(y_T - y_A)$, with $x_A = 3.14, y_A = 2.685$, and x_T, y_T denoting the true unrounded numbers associated with x_A, y_A. Also, $|\text{Relative error}| \le .0098$

Chapter 2

1. $\Pi_{j=0}^{\infty}(1+r^{2^j})=1/(1-r)$, and the infinite product converges if and only if $|r|<1$.

2. root$=4.493409458$

4. **(a)** 1.8392868   **(b)**   1.1284251   **(c)**   0.42403104

5.

B	Root
1	$-.5884017765$
5	$-.4049115482$
10	$-.3264020101$
25	$-.2374362439$
50	$-.1832913333$

7. **(a)**   4.493409458   **(b)**   98.95006282

8. **(e)** $|\text{Rel}(x_4)|\leq 2^{-15}10^{-16}\doteq 3.05\times 10^{-21}$

9. The iterate x_4 will be sufficiently accurate.

10. **(a)**   does not converge   **(b)**   converges

11. **(b)** $2.470638970+4.640533162i$

13. root$=.4501836113$, convergence rate $\doteq -.218$

16.
$$\underset{n\rightarrow\infty}{\text{Limit}}\frac{\sqrt{a}-x_{n+1}}{\left(\sqrt{a}-x_n\right)^3}=\frac{1}{4a}$$

22. **(a)**   This polynomial is the degree 12 Legendre polynomial on the interval $-1\leq x\leq 1$.   Its roots are

$\pm.1252334085$	$\pm.7699026742$
$\pm.3678314990$	$\pm.9041172564$
$\pm.5873179543$	$\pm.9815606342$

26.

(x_0, y_0)	Method Converges To
$(1.2,2.5)$	$(1.336355377, 1.754235198)$
$(-2,2.5)$	$(-.9012661908, -2.086587595)$
$(-1.2,-2.5)$	$(-.9012661908, -2.086587595)$
$(2.0,-2.5)$	$(-3.001624887,.1481079950)$

In the last case, the method jumps around in a seemingly random fashion, and after 16 iterations it begins to converge to the root given above. The equations do have another root.

27. There are two roots, (a,b) and (b,a), with $a = .673007170$, $b = 1.94502682$

Chapter 3

3. $|\text{Interpolation error}| \le \dfrac{h^2}{8} e^2 \doteq 9.2 \times 10^{-5}$

   $|\text{Error due to rounding}| \le 5 \times 10^{-5}$
   $|\text{Total error}| \le 1.42 \times 10^{-4}$

5. If $10 \le x < 100$, the error due to interpolation is bounded by .000988 and the relative error is bounded by .000313. Choose the rounding error to be bounded by .0005, which means choosing the table entries to have four significant digits. If $100 \le x \le 999$, the error and relative errors are bounded by .0000313 and .00000313, respectively. Choosing the rounding error to be bounded by .00005 requires the table entries to have six significant digits; but for practical purposes, five is probably sufficient, leading to a total error comparable to that for the range of $10 \le x \le 100$.

7. Choose $h = .001$, and let the table entries have seven significant digits. Then the total error will be bounded by 8.4×10^{-7}.

10. Choose the table entries to have six significant digits. One possible partition of $1 \le x \le 10$ with the suggested grid size and resulting total interpolation error is given below.

Interval	h	Total error
$1 \le x \le 2$	.01	6.81×10^{-7}
$2 \le x \le 3$	.025	7.35×10^{-7}
$3 \le x \le 6$	.05	8.85×10^{-7}
$6 \le x \le 10$	.1	8.85×10^{-7}

16. degree 3

18. The sixth entry 419327 should be changed to 419237 or 419238. There are two other errors, and their higher-order differences have overlapping effects.

24. In order that there be a unique interpolating polynomial of degree ≤ 2, it is necessary and sufficient that $x_1 \ne \frac{1}{2}(x_0 + x_2)$.

Chapter 4

1. The Bernstein polynomial of degree 4 is

$$p_4(x) = 2\sqrt{2}\, x(1-x)^3 + 6x^2(1-x)^2 + 2\sqrt{2}\, x^3(1-x)$$

The Taylor polynomial of degree 4 is

$$f_4(x) = 1 - \frac{\pi^2}{2}\left(x - \frac{1}{2}\right)^2 + \frac{\pi^4}{24}\left(x - \frac{1}{2}\right)^4$$

The following table gives the values of $p_4(x)$ and $f_4(x)$ at several points, along with the errors when compared to $\sin(\pi x)$.

x	$p_4(x)$	Error	$f_4(x)$	Error
0.0	0.0	0.0	.0200	$-2.0E-2$
0.1	.2573	.0517	.3143	$-5.3E-3$
0.2	.4613	.1265	.5887	$-9.6E-4$
0.3	.6091	.1999	.8091	$-8.5E-5$
0.4	.6986	.2525	.9511	$-1.3E-6$
0.5	.7286	.2714	1.0000	0.0

4. **(b)** Let $S = \displaystyle\sum_1^\infty \frac{(-1)^j x^{2j}}{j^2}$. The series has the interval of convergence $|x| \le 1$. For the error,

$$\left| S - \sum_{j=1}^n (-1)^j \frac{x^{2j}}{j^2} \right| \le \frac{|x|^{2n+2}}{(n+1)^2}$$

If $n \ge 316$, the error is less than 10^{-5} for all $|x| \le 1$. But this is likely to be much too large for $|x| < 1$ since it ignores the factor of $|x|^{2n+2}$ in the error.

7. The Pade approximation is $R(x) = \dfrac{1 + \frac{1}{2}x}{1 - \frac{1}{2}x}$. For the error, $e^x - R(x) = -[\frac{1}{12}x^3 + \frac{1}{12}x^4 + \frac{13}{240}x^5 + \ldots]$; and for the Taylor series error, $e^x - p_2(x) = \frac{1}{6}x^3 + \frac{1}{24}x^4 + \ldots$. The error in $R(x)$ is about half that of $p_2(x)$ for small $|x|$. But for $x = 1$, the error is larger with $R(x)$.

8. **(a)** $p_1(x) = -.3799 + .5379x$; $\|\ln(x) - p_1\|_\infty = .16$
    (b) $q_1^*(x) = -.5203 + .5820x$; $\|\ln(x) - q_1^*\|_\infty = .062$

12. Minimizing value of α is $\alpha = \sqrt{e}$. Minimum value is $e + 1 - 2\sqrt{e} \doteq .42$.

14. $a_0 = \dfrac{1}{\pi} \displaystyle\int_0^\pi f(x)\,dx, \quad a_j = \dfrac{2}{\pi} \displaystyle\int_0^\pi \cos(jx) f(x)\,dx \ \text{for } j \ge 1$

15. $\varphi_0(x) \equiv 1, \quad \varphi_1(x) = \dfrac{12}{\sqrt{7}}\left(x - \dfrac{1}{4}\right)$

$\varphi_2(x) = \sqrt{\dfrac{226800}{647}}\left(x^2 - \dfrac{5}{7}x + \dfrac{17}{252}\right)$

23. $C_3(x) = .970563x - .189514x^3; \quad \|\tan^{-1}x - C_3\|_\infty = .00542$
$I_3(x) = .968706x - .187130x^3; \quad \|\tan^{-1}x - I_3\|_\infty = .00590$
$F_3(x) = .972420x - .191898x^3; \quad \|\tan^{-1}x - F_3\|_\infty = .00500$
For comparison, the cubic Taylor polynomial and its error are

$$p_3(x) = x - \tfrac{1}{3}x^3 \qquad \|\tan^{-1}x - p_3\|_\infty = .119$$

Also from examining the relative maxima and minima of $\tan^{-1}x - F_3(x)$, and using Theorem 4.10, we obtain

$$.00488 \le \rho_3(\tan^{-1}x) \le .00500$$

26. For the error in $I_n(x)$ approximating e^x on $[-1,1]$,

$$e^x - I_n(x) = \dfrac{1}{(n+1)!2^n} T_{n+1}(x) e^{\xi_x}$$

for some ξ_x in $[-1,1]$. Thus

$$\alpha_n = \dfrac{e^{-1}}{(n+1)!2^n} \qquad \beta_n = \dfrac{e}{(n+1)!2^n}$$

Chapter 5

1.   (a) $\displaystyle\int_0^1 e^{-x^2}\,dx$ (c) $\displaystyle\int_{-4}^4 \dfrac{dx}{1+x^2}$

n	I_n	R_n	n	I_n	R_n
2	.73137025		2	4.235294	
4	.74298410		4	2.917647	
8	.74586561	4.03	8	2.658824	5.09
16	.74658460	4.01	16	2.650507	31.1
32	.74676425	4.00	32	2.651347	-9.90
64	.74680916	4.00	64	2.651563	3.89
128	.74682039	4.00	128	2.651617	4.00
256	.74682320	4.00	256	2.651631	4.00

2.   (a) $\displaystyle\int_0^1 e^{-x^2}\,dx$ (c) $\displaystyle\int_{-4}^4 \frac{dx}{1+x^2}$

n	I_n	R_n	n	I_n	R_n
2	.7471804289095		2	5.4901960784	
4	.7468553797910		4	2.4784313725	
8	.7468261205275	11.1	8	2.5725490196	-32.0
16	.7468242574357	15.7	16	2.6477345635	1.25
32	.7468241406070	15.9	32	2.6516272830	19.3
64	.7468241332997	16.0	64	2.6516352807	487
128	.7468241328429	16.0	128	2.6516353244	183
256	.7468241328143	16.0	256	2.6516353272	16.0

3.   The ratios R_n approached constants with increasing n. The values for $n=64$ are given in the following

α	.25	.5	.75	1.0
R_n	2.11	2.52	3.03	3.64

6. $I_h(f) = \frac{3}{4}h[f(0) + 3f(2h)]$

12. $\displaystyle\int_0^1 f(x)\ln\left(\frac{1}{x}\right) dx \approx w_1 f(x_1) + w_2 f(x_2)$

$$x_1 = \frac{15 - \sqrt{106}}{42} \doteq .1120088062 \qquad x_2 = \frac{15 + \sqrt{106}}{42} \doteq .6022769081$$

$$w_1 = \frac{21}{\sqrt{106}}\left[x_2 - \frac{1}{4}\right] \doteq .718539319 \qquad w_2 = 1 - w_1 = .281460681$$

13.

n	I_n	Error
2	1.26316	1.388
4	2.04729	.6044
6	2.41169	.2399
8	2.56008	.09156
10	2.61725	.03438

15.   For Gaussian quadrature, $|I - I_n| \leq \dfrac{.080}{n^4}\|f^{(4)}\|_\infty$. For Simpson's rule on $[-1,1]$, $|I - I_n| \leq .18\|f^{(4)}\|_\infty\, n^4$.

19. $\displaystyle\sum_1^\infty \frac{1}{n^{5/4}} = 4.59511254 + E,\ 0 < -E < 1.34 \times 10^{-6}$

21. Assuming that $I - I_n \doteq \frac{c}{n^p}$, and using I_{16}, I_{32}, and I_{64}, we obtain $p \doteq 3.44$, $c \doteq 0.0154$. For the Aitken extrapolate, $\tilde{I}_{16} = .28571428586$; and $I - I_{64} \doteq \tilde{I}_{16} - I_{64} = 9.43 \times 10^{-9}$. To have $|I - I_n| \leq 10^{-11}$ will require $n \geq 469$. The logical choice to use would be $n = 512$.

22. $\tilde{I}_n = \frac{1}{3}[I_n^{(T)} + 2I_n^{(M)}]$, which will be Simpson's rule.

27. (a) Split integral, $I = \left(\int_0^1 + \int_1^{4\pi} \right) \cos(x) \ln(x) dx$. Using a Taylor series,

$$\int_0^1 \cos(x) \ln(x) dx \approx -\left(1 - \frac{1}{2!9} + \frac{1}{4!25} - \frac{1}{6!49} + \frac{1}{8!81} \right)$$

$$\doteq -.9460830727$$

which is in error by less than 2.28×10^{-9}. For the remaining integral, a standard method can be used to compute it. For example, Romberg integration with $n = 129$ nodes leads to the value $-.5460781580$, which is in error by less than 2.8×10^{-9}. Thus the integral is $I = 1.4921612307$, in error by less than 5.1×10^{-9}.

(b) Since $|x^2 \sin(1/x)| \leq x^2$, pick $\epsilon > 0$ so that

$$\left| \int_0^\epsilon x^2 \sin\left(\frac{1}{x} \right) dx \right| \leq \int_0^\epsilon x^2 dx \leq .0005$$

A convenient choice is $\epsilon = .1$. Then compute $\int_{.1}^{2/\pi} (x^2) \sin(1/x) dx$ with an accuracy of at least .0005.

(c) Change the variable of integration to get $I = 3\int_0^1 \cos(x^3) dx$. Gauss quadrature with $n = 10$ nodes leads to more than 10 digits of accuracy, $I \doteq 2.7951133218$.

Chapter 6

2. (a) 2   (b) 1   (c) 0   (d) 20

3. True solution $Y(x) = 2e^x + \sin(x) - \cos(x)$. The Picard iterates are

$$Y_0(x) \equiv 1, \quad Y_1(x) = 1 + x + 2\sin(x)$$

$$Y_2(x) = 3 + x + \frac{x^2}{2} + 2\sin(x) - 2\cos(x)$$

$$Y_3(x) = 3 + 3x + \frac{1}{2}x^2 + \frac{1}{6}x^3 - 2\cos(x)$$

Compare them for $x \approx 0$ by using Taylor series expansions of them and $Y(x)$.

4. (b) $\begin{aligned} y_1' &= y_2 \\ y_2' &= 0.1(1 - y_1^2)y_2 - y_1 \end{aligned}$ $\begin{aligned} y_1(0) &= 1 \\ y_2(0) &= 0 \end{aligned}$

6. Assume the initial error $e_0 = 0$. For the bound from (6.30),

$$\underset{0 \le x \le b}{\text{Max}} |Y(x_n) - y_n| \le \frac{h}{2}(e^{2b} - 1)$$

The asymptotic error is given by

$$Y(x) - y_h(x) = h\frac{\ln(1+x)}{(1+x)^2} + O(h^2) \qquad x \ge 0$$

This shows the error decreases with increasing x, whereas the above bound predicts an increasing error as the interval $[0, b]$ increases.

8. **(b)** $y_{1,n+1} = y_{1,n} + hy_{2,n}$
$\qquad y_{2,n+1} = y_{2,n} + h[0.1(1 - y_{1,n}^2)y_{2,n} - y_{1,n}] \qquad n \ge 0$

10. $T_{n+1}(Y) = -\frac{5}{8}h^3 Y'''(x_n) + O(h^4)$

16. $a_0 = -9 - 3b_2 \qquad a_1 = 9 \qquad a_2 = 1 + 3b_2$
$\qquad b_0 = 6 + b_2 \qquad b_1 = 6 + 4b_2$

20. $u_n = c_1 + c_2(\frac{1}{2})^n + c_3(-\frac{1}{2})^n$,

with c_1, c_2, and c_3 arbitrary. The solution satisfying the given initial conditions is

$$u_n = 1 + \frac{1}{2^{n-1}} + \frac{(-1)^n}{2^n} \approx 1 \text{ for any large } n$$

24. **(a)** $0 \le a_0 < 2$ \qquad **(b)** $a_0 = 0$
\qquad **(c)** At $a_0 = 0$, there is no region of absolute stability. As a_0 goes from 0 toward 2, the region of absolute stability increases to $-2 < h\lambda < 0$ in the limit. This is only for the real values in the absolute stability region.

25. The characteristic equation for $h = 0$ is

$$\rho(r) \equiv r^3 - a_0 r^2 - 9r + a_0 + 8 = 0$$

Factoring out $r - 1$, the remaining two roots satisfy

$$r^2 + (1 - a_0)r - (a_0 + 8) = 0 \qquad\qquad (\#)$$

The discriminant is

$$(1 - a_0)^2 + 4(a_0 + 8) = (1 + a_0)^2 + 32 \ge 32 > 0$$

for all real values of a_0. Thus both of the roots r_1 and r_2 of equation $(\#)$ must be real, for any value of a_0. A direct examination of these roots will show that they cannot both satisfy $-1 < r < 1$ simultaneously for the same value of a_0.

28. $-\dfrac{2}{3} \leqq h\lambda < \infty$

29. $y_{n+1} = y_{n-1} + 2hf(x_{n-1}, y_{n-1})$ is first order and relatively stable, but it doesn't satisfy the strong root condition.

33.

$$\gamma_1 + \gamma_2 + \gamma_3 = 1$$

$$\gamma_2\alpha_2 + \gamma_3\alpha_3 = \frac{1}{2}$$

$$\gamma_2\beta_{21} + \gamma_3(\beta_{31} + \beta_{32}) = \frac{1}{2}$$

$$\gamma_2\alpha_2^2 + \gamma_3\alpha_3^2 = \frac{1}{3}$$

$$\gamma_2\alpha_2\beta_{21} + \gamma_3\alpha_3(\beta_{31} + \beta_{32}) = \frac{1}{3}$$

$$\gamma_2\beta_{21}^2 + \gamma_3(\beta_{31} + \beta_{32})^2 = \frac{1}{3}$$

$$\gamma_3\alpha_2\beta_{32} = \frac{1}{6}$$

$$\gamma_3\beta_{32}\beta_{21} = \frac{1}{6}$$

These equations are dependent and can be reduced to six independent equations. One particular solution is

$$Y_{n+1} = Y_n + \frac{h}{6}(V_1 + 4V_2 + V_3)$$

$$V_1 = f(x_n, y_n), \quad V_2 = f\left(x_n + \tfrac{1}{2}h, y_n + \tfrac{1}{2}hV_1\right)$$

$$V_3 = f\left[x_n + h, y_n + h(2V_2 - V_1)\right]$$

This is Simpson's rule if $f(x, y)$ does not depend on y.

Chapter 7

1. (a) dependent   **(b)** independent

4. (a) $(Ax, x) = x^T(Ax)$. But also,

$$(Ax, x) = (x, Ax) = (Ax)^T x = x^T A^T x = -x^T A x$$

Since $x^T A x = -x^T A x$, we must have $x^T A x = 0$, the desired result.

7. $u^{(3)} = (-7, 4, 1)$. To normalize, divide the respective vectors by their lengths,

$$\|u^{(1)}\|_2 = \sqrt{6} \qquad \|u^{(2)}\|_2 = \sqrt{11} \qquad \|u^{(3)}\|_2 = \sqrt{66}$$

8. (a)

$$A = \frac{1}{9} \begin{bmatrix} 7 & -4 & -4 \\ -4 & 1 & -8 \\ -4 & -8 & 1 \end{bmatrix}$$

9. (a) $\lambda_1 = -1, \qquad x = [2, -1]^T$
   $\lambda_2 = \quad 3, \qquad x = [2, 1]^T$

10. (a) $\|Ux\|_2^2 = (Ux, Ux) = (x, U^*Ux) = (x, Ix) = (x, x) = \|x\|_2^2$, showing the desired result: $\|Ux\|_2 = \|x\|_2$ for all x. For the distance between Ux and Uy,

$$\|Ux - Uy\|_2 = \|U(x - y)\|_2 = \|x - y\|_2$$

   from the earlier result.

11. Since A is Hermitian, let $x^{(1)}, \ldots, x^{(n)}$ be an orthonormal basis for C^n corresponding to the real eigenvalues $\lambda_1, \ldots, \lambda_n$. For any x in C^n, we can write

$$x = \sum_1^n \alpha_i x^{(i)} \qquad \text{with } \alpha_i = (x, x^{(i)}) \text{ for } i = 1, \ldots, n$$

   From the orthonormality, $\|x\|_2^2 = \sum_1^n |\alpha_i|^2$. Using this form for x, show

$$(Ax, x) = \sum_1^n \lambda_i |\alpha_i|^2$$

   Since $\alpha_1, \ldots, \alpha_n$ can be varied arbitrarily to obtain various elements x of C^n, this formula can be used to prove that A is positive definite if and only if all λ_i are positive.

18. First prove

$$\|x\|_\infty \leq \|x\|_p \leq n^{1/p} \|x\|_\infty \qquad \text{all } x \in C^n$$

   It then follows easily that $\|x\|_p \to \|x\|_\infty$ as $p \to \infty$.

21. (a) Let λ and x be an eigenvalue-eigenvector pair for the matrix A. Then the associated eigen pairs for various modifications of A are: (i) λ^m and x for A^m, (ii) $1/\lambda$ and x for A^{-1}, and (iii) $\lambda + c$ and x for $A + cI$.

22. Write $A = [A_{*1}, \ldots, A_{*n}]$ using the columns A_{*j} of A. Then

$$F(A) = \sqrt{\|A_{*1}\|_2^2 + \ldots + \|A_{*n}\|_2^2}$$

(a) $F(UA) = \sqrt{\|UA_{*1}\|_2^2 + \ldots + \|UA_{*n}\|_2^2}$
From Problem 10(a), $\|UA_{*j}\|_2 = \|A_{*j}\|_2$ for all j. Thus $F(UA) = F(A)$.

(b) Let $U^*AU = D = \mathrm{diag}[\lambda_1, \ldots, \lambda_n]$. Apply part (a) to obtain

$$F(A) = F(U^*AU) = F(D) = \sqrt{\lambda_1^2 + \ldots + \lambda_n^2}$$

Chapter 8

1. **(a)**

$$L = \begin{bmatrix} 1 & 0 & 0 \\ 1 & 1 & 0 \\ -2 & 3 & 1 \end{bmatrix} \quad U = \begin{bmatrix} 1 & 1 & -1 \\ 0 & 1 & -1 \\ 0 & 0 & 2 \end{bmatrix} \quad x = \begin{bmatrix} 2 \\ 2 \\ 3 \end{bmatrix}$$

2. **(a)** Without pivoting, the augmented matrix $[A|b]$ is reduced to

$$\begin{bmatrix} 6.000 & 2.000 & 2.000 & -2.000 \\ 0 & .0001000 & -.3333 & 1.667 \\ 0 & 0 & 5555 & -27790 \end{bmatrix} \begin{matrix} m_{21} = .3333 \\ m_{31} = .1667 \\ m_{32} = 16670 \end{matrix}$$

The solution by back substitution is

$$x_1 = 1.335 \qquad x_2 = 0 \qquad x_3 = -5.003$$

All arithmetic operations were carried out using four decimal digit floating-point arithmetic with rounding; and the same is true in the following part (b).

(b) $[A|b]$ is reduced, with pivoting, to

$$\begin{bmatrix} 6.000 & 2.000 & 2.000 & -2.000 \\ 0 & 1.667 & -1.333 & .3334 \\ 0 & 0 & .3332 & 1.667 \end{bmatrix} \begin{matrix} m_{21} = .3333 \\ m_{31} = .1667 \\ m_{32} = .00005999 \end{matrix}$$

The solution by back substitution is

$$x_1 = 2.602 \qquad x_2 = -3.801 \qquad x_3 = -5.003$$

6.

$$L = \begin{bmatrix} 1.5 & 0 & 0 \\ -2.0 & 1.0 & 0 \\ 3.0 & -4.0 & 3.0 \end{bmatrix}$$

8. $(Ax,x)=(LL^Tx,x)=(L^Tx,L^Tx)=\|L^Tx\|_2^2>0$ for all $x\neq0$, since L^T is non-singular. Also $\det(A)=\det(L)^2$.

9.

$$L=\begin{bmatrix} 2 & 0 & 0 & 0 & 0 \\ -1 & \frac{3}{2} & 0 & 0 & 0 \\ 0 & -1 & \frac{4}{3} & 0 & 0 \\ 0 & 0 & -1 & \frac{5}{4} & 0 \\ 0 & 0 & 0 & -1 & \frac{6}{5} \end{bmatrix} \qquad U=\begin{bmatrix} 1 & -\frac{1}{2} & 0 & 0 & 0 \\ 0 & 1 & -\frac{2}{3} & 0 & 0 \\ 0 & 0 & 1 & -\frac{3}{4} & 0 \\ 0 & 0 & 0 & 1 & -\frac{4}{5} \\ 0 & 0 & 0 & 0 & 1 \end{bmatrix}$$

$Lz=b$ has solution $z=\left[\frac{1}{2},\frac{1}{3},\frac{1}{4},\frac{1}{5},\frac{1}{6}\right]$.

$Ux=z$ has solution $x=\left[\frac{5}{6},\frac{2}{3},\frac{1}{2},\frac{1}{3},\frac{1}{6}\right]$.

11. (a) In partitioned form,

$$\begin{bmatrix} A_1 & -A_2 \\ A_2 & A_1 \end{bmatrix}\begin{bmatrix} x_1 \\ x_2 \end{bmatrix}=\begin{bmatrix} b_1 \\ b_2 \end{bmatrix}$$

(b) Let system 1 denote the real system of part (a), and let system 2 denote the original complex system $Ax=b$.

For the matrix storage requirements, system 1 requires $4n^2$ locations, and system 2 requires $2n^2$ locations (each complex number requires two storage locations). To solve system 1 requires about $\frac{1}{3}(2n)^3=\frac{8}{3}n^3$ multiplications and divisions. System 2 requires $\frac{1}{3}n^3$ complex multiplications and divisions. Since each complex multiplication requires four real multiplications, the actual operation count is $\frac{4}{3}n^3$. Thus system 2 requires half the storage requirements and about half the operation time of system 1.

12. (a) $\operatorname{cond}(A)_1=\operatorname{cond}(A)_\infty=39601$, $\operatorname{cond}(A)_2=39206$

18. For the rates of convergence μ of (8.66) and η of (8.75) in this case, $\mu=\frac{3}{4}$ and $\eta=\frac{2}{3}$. The actual observed rates of decrease are .61 for the Gauss-Jacobi and .37 for the Gauss-Seidel methods.

20. (a) Converges if and only if $r_\sigma(A^{-1}B)<1$; and the rate of convergence is essentially $r_\sigma(A^{-1}B)$ or $\|A^{-1}B\|$, depending on the norm used.

(b) Again it converges if and only if $r_\sigma(A^{-1}B)<1$. But the rate of convergence is about $r_\sigma(A^{-1}B)^2$ or $\|A^{-1}B\|^2$, which is much faster than (a).

21. (a) $A^{-1}-C_m=A^{-1}R_m=A^{-1}R_0^{2^m}$, $m\geq0$. C_m converges to A^{-1} if and only if $r_\sigma(R_0)<1$.

(b) Writing $(I-R_0)^{-1}$ as an infinite product,

$$A^{-1}=C_0(I+R_0)(I+R_0^2)(I+R_0^4)(I+R_0^8)\cdots$$
$$C_m=C_0(I+R_0)(I+R_0^2)(I+R_0^4)\cdots(I+R_0^{2^{m-1}})$$

Chapter 9

1. Applying the Gerschgorin theorem to the rows of A, the circles are $|\lambda - 1| \leq 1$, $|\lambda - 5| \leq 2$, and $|\lambda - 9| \leq 3$. The second and third circles intersect, and thus they must contain two eigenvalues in their union. The first circle is distinct from the others. Since the characteristic polynomial of A has real coefficients, the eigenvalues occur in conjugate pairs if they are complex. Thus $|\lambda - 1| \leq 1$ contains one real eigenvalue, in the interval $[0, 2]$.

   Applying the theorem to the columns of A, the circles are $|\lambda - 1| \leq 3$, $|\lambda - 5| \leq 2$, and $|\lambda - 9| \leq 1$. By the same type of argument as before, and using the previous results, there are real eigenvalues in each of the intervals $[0, 2]$, $[3, 7]$, and $[8, 10]$. The true eigenvalues are

$$1.331927689 \qquad 4.856933692 \qquad 8.811138616$$

2. (c) The n circles are

$$|r| \leq |a_0| \qquad |r| \leq 1 + |a_j| \qquad j = 1, \ldots, n-2 \qquad \text{and } |r + a_{n-1}| \leq 1$$

   (d) (i) There are nine roots satisfying $|r - 1| \leq 1$ and one real root satisfying $7 \leq r \leq 9$.

4. $\displaystyle \min_{1 \leq i \leq 4} |\lambda - \lambda_i| \leq .0000356$

7. (a) $\lambda = 15$, $x = [1, 1, 1, 1]^T$. The error decreases by a factor of $\frac{1}{3}$ with each iterate. Thus $\lambda = 5$ is likely to be the second largest eigenvalue.

10. (b) Using $w = [1/\sqrt{6}, -1/\sqrt{6}, -2/\sqrt{6}]^T$ to construct $P = I - 2ww^T$, we obtain $Px = e_1$. Thus

$$B = PAP^T = \begin{bmatrix} 9 & 0 & 0 \\ 0 & 18 & 0 \\ 0 & 0 & -9 \end{bmatrix}$$

In this case the matrix B is diagonal, but ordinarily it wouldn't be so simple. Using $w = (1/\sqrt{30})[5, 1, 2]^T$ will lead to a P for which $Px = -e_1$. But $B = PAP^T$ will still have the same form for column and row 1. In particular,

$$B = \begin{bmatrix} 9 & 0 & 0 \\ 0 & \dfrac{18}{25} & \dfrac{-324}{25} \\ 0 & \dfrac{-324}{25} & \dfrac{207}{25} \end{bmatrix}$$

The first form of P was constructed with the opposite sign to that in (9.70), while the second choice of P was constructed according to (9.70). Ordinarily the form of P agreeing with (9.70) would be the preferred form, because of the accuracy considerations pointed out following (9.70).

11.

$$Q = \frac{1}{3}\begin{bmatrix} -1 & 2 & 2 \\ -2 & 1 & -2 \\ -2 & -2 & 1 \end{bmatrix} \qquad R = 3\begin{bmatrix} -1 & 1 & -1 \\ 0 & 1 & -1 \\ 0 & 0 & 1 \end{bmatrix}$$

12. (a) $\alpha = \dfrac{b_k}{\gamma}, \ \beta = \dfrac{b_l}{\gamma}$

(b)

$$R^{(1,4)} = \begin{bmatrix} 1/\sqrt{2} & 0 & 0 & 1/\sqrt{2} \\ 0 & 1 & 0 & 0 \\ 0 & 0 & 1 & 0 \\ -1/\sqrt{2} & 0 & 0 & 1/\sqrt{2} \end{bmatrix}$$

$$R^{(1,3)} = \begin{bmatrix} \sqrt{2}/\sqrt{3} & 0 & 1/\sqrt{3} & 0 \\ 0 & 1 & 0 & 0 \\ -1/\sqrt{3} & 0 & \sqrt{2}/\sqrt{3} & 0 \\ 0 & 0 & 0 & 1 \end{bmatrix}$$

$$R^{(1,2)} = \begin{bmatrix} \sqrt{3}/2 & \frac{1}{2} & 0 & 0 \\ -\frac{1}{2} & \sqrt{3}/2 & 0 & 0 \\ 0 & 0 & 1 & 0 \\ 0 & 0 & 0 & 1 \end{bmatrix}$$

$$\hat{b} = Ub = [2,0,0,0]^T$$

$$U = R^{(1,2)}R^{(1,3)}R^{(1,4)} = \begin{bmatrix} \frac{1}{2} & \frac{1}{2} & \frac{1}{2} & \frac{1}{2} \\ -\frac{1}{2\sqrt{3}} & \frac{\sqrt{3}}{2} & -\frac{1}{2\sqrt{3}} & -\frac{1}{2\sqrt{3}} \\ -\frac{1}{\sqrt{6}} & 0 & \frac{\sqrt{2}}{\sqrt{3}} & -\frac{1}{\sqrt{6}} \\ -\frac{1}{\sqrt{2}} & 0 & 0 & \frac{1}{\sqrt{2}} \end{bmatrix}$$

16. $f_0(\lambda) = 1, \ f_1(\lambda) = -\lambda, \ f_j(\lambda) = (1-\lambda)f_{j-1}(\lambda) - f_{j-2}(\lambda)$ for $j = 2,3,4,$
$f_5(\lambda) = (2-\lambda)f_4(\lambda) - f_3(\lambda)$. Let $\lambda_1 < \lambda_2 < \lambda_3 < \lambda_4 < \lambda_5$

By the Gerschgorin theorem, all λ_i lie in $[-1,3]$. Then the roots can be separated using Theorem 9.5, as indicated in the following table. The actual roots can then be found by another method. For example, the secant method leads rapidly to $\lambda_5 \doteq 2.90211303$.

λ	$f_5(\lambda)$	$s(\lambda)$	Remark
-1.0	2.0	5	$\lambda_1 > -1$
3.0	-2.0	0	$\lambda_5 < 3$
1.0	0.0	2	$\lambda_3 = 1.0, \lambda_4 > 1$
0.0	1.0	3	$\lambda_2 < 0.0$
-0.5	-1.78	4	$-1 < \lambda_1 < .5 < \lambda_2 < 0$
2.0	-1.0	2	$\lambda_4 > 2$
2.5	1.78	1	$2 < \lambda_4 < 2.5 < \lambda_5 < 3$

17. (b) Using Householder transformations as in (9.81) to (9.83), we obtain

$$T = Q^T A Q$$

$$T = \begin{bmatrix} 5.0 & -4.2426406871 & 0.0 & 0.0 \\ -4.2426406871 & 6.0 & 1.4142135624 & 0.0 \\ 0.0 & 1.4142135624 & 5.0 & 0.0 \\ 0.0 & 0.0 & 0.0 & 2.0 \end{bmatrix}$$

$$Q = \begin{bmatrix} 1.0 & 0.0 & 0.0 & 0.0 \\ 0.0 & -.9428090416 & .3333333333 & 0.0 \\ 0.0 & -.2357022604 & -.6666666667 & -.7071067812 \\ 0.0 & -.2357022604 & -.6666666667 & .7071067812 \end{bmatrix}$$

Usually it would be wasteful of time to actually produce Q explicitly, as done here.

19. The eigenvalues are

$$\lambda_1 = 2 - \sqrt{3} \qquad \lambda_2 = 2 \qquad \lambda_3 = 2 + \sqrt{3}$$

(a) For the QR method without shift, the off diagonal elements converge to zero linearly. The elements in the $(1, 2)$ position decrease by a factor of .536 per iterate, and those in the $(2, 3)$ position decrease by a factor of $-.134$ per iterate.

(b) For the QR method with shift, using the choice of

$$c_m = a_{3,3}^{(m)}$$

we have rapid convergence. After five iterates, the element in position $(2, 3)$ is less than 5×10^{-11} in size. Denoting the original A by A_0, we have the result

$$A_5 = \begin{bmatrix} 3.7316925974 & .0249060210 & 0.0 \\ .0249060210 & 2.0003582102 & \epsilon \\ 0.0 & \epsilon & .2679491924 \end{bmatrix}$$

with $|\epsilon| < 5 \times 10^{-11}$. When the QR method with shift is applied to the reduced matrix obtained by deleting row and column 3 of A_5, it too converges very rapidly.

21. For $\lambda = 7.9329$, the eigenvector is

$$x = [.7211774817, .2724739430, 1.0, .2515508351]$$

For $\lambda = 5.6689$,

$$x = [.5736191640, .5489544105, -.8148078321, 1.0]$$

23. The values obtained are given in the following table.

n	$\lambda_1^{(n)}$	$\lambda_2^{(n)}$
2	.9	.1
4	.884566	.103957
8	.880971	.105360
16	.880085	.105715
32	.879865	.105805

By examining ratios of successive differences, it can be seen empirically that

$$\lambda_i - \lambda_i^{(n)} = O\left(\frac{1}{n^2}\right)$$

This rate of convergence can be justified theoretically. Richardson extrapolation can then be used to produce the error estimates

$$\lambda_1 - \lambda_1^{(n)} \doteq 7.3 \times 10^{-5} \qquad \lambda_2 - \lambda_2^{(n)} \doteq 3.0 \times 10^{-5}$$

INDEX

Note: (1) An asterisk (*) following a subentry name means that name is also listed separately with additional subentries of its own. (2) A page number followed by a number in parentheses, prefixed by 'P', refers to a problem on the given page. For example, 386(P19) refers to problem 19 on page 386.